Interdisciplinary Applied Mathematics

Problems in engineering, computational science, and the physical and biological sciences are using increasingly sophisticated mathematical techniques. Thus, the bridge between the mathematical sciences and other disciplines is heavily traveled. The correspondingly increased dialog between the disciplines has led to the establishment of the series: *Interdisciplinary Applied Mathematics*.

The purpose of this series is to meet the current and future needs for the interaction between various science and technology areas on the one hand and mathematics on the other. This is done, firstly, by encouraging the ways that mathematics may be applied in traditional areas, as well as point towards new and innovative areas of applications; and, secondly, by encouraging other scientific disciplines to engage in a dialog with mathematicians outlining their problems to both access new methods and suggest innovative developments within mathematics itself.

The series will consist of monographs and high-level texts from researchers working on the interplay between mathematics and other fields of science and technology.

Interdisciplinary Applied Mathematics

For other titles published in this series, go to
www.springer.com/series/1390

J. Michael McCarthy • Gim Song Soh

Geometric Design of Linkages

Second Edition

 Springer

J. Michael McCarthy
Department of Mechanical and
Aerospace Engineering
University of California, Irvine
Irvine, CA 92697
USA
jmmccart@uci.edu

Gim Song Soh
Singapore Institute of Manufacturing
Technology
71 Nanyang Drive
Singapore 638075
gsoh@uci.edu

Series Editors

S.S. Antman
Department of Mathematics
and
Institute for Physical Science
    and Technology
University of Maryland
College Park, MD 20742, USA
ssa@math.umd.edu

J.E. Marsden
Control and Dynamical Systems
Mail Code 107-81
California Institute of Technology
Pasadena, CA 91125, USA
marsden@cds.caltech.edu

L. Sirovich
Division of Applied Mathematics
Brown University
Providence, RI 02912, USA
chico@camelot.mssm.edu

ISSN 0939-6047
ISBN 978-1-4614-2767-4   ISBN 978-1-4419-7892-9 (eBook)
DOI 10.1007/978-1-4419-7892-9
Springer New York Dordrecht Heidelberg London

Mathematics Subject Classification (2010): 53A17, 13P15, 15A66

Springer is part of Springer Science+Business Media (www.springer.com)

For Kendall and Wyatt

Preface

2nd Edition

This second edition of *Geometric Design of Linkages* revises and updates our formulation of the kinematic theory of linkages. Four new chapters have been added that present the analysis and synthesis of multiloop planar and spherical linkages, and the synthesis theory for spatial serial chains.

An introduction to linkage graphs and linkage enumeration has been added to Chapter 1 to provide background for the new chapters on the synthesis of multiloop linkages, and the use of the Dixon determinant to analyze planar multiloop linkages has been added to Chapter two.

Chapters three, four, and five are the same as before with corrections of minor errors. Chapter six is new and presents a methodology for the synthesis of planar six-bar and eight-bar linkages by constraining three-degree-of-freedom 3R open chains or 6R closed chains, respectively. Examples are provided that demonstrate the technique.

Chapter seven is the same as Chapter six in the previous edition but now includes a section on the analysis of multiloop spherical linkages. Chapters eight and nine are the same as Chapters seven and eight in the previous edition, again with corrections of minor errors,. The new Chapter 10 parallels the new Chapter six and presents a way to design spherical six-bar and eight-bar linkages by constraining 3R open and 6R closed spherical chains.

Chapters 11, 12, and 13 are the same as Chapters 9, 10, and 11 in the previous edition. Chapters 14 and 15 on the synthesis of spatial serial chains are new and were the result of research with Haijun Su and Alba Perez-Gracia. Chapter 14 formulates and solves using numerical homotopy the synthesis equations for five-degree-of-freedom serial chains that can position a wrist center on specific algebraic surfaces, termed reachable surfaces. Chapter 15 introduces the Clifford algebra of dual quaternions and its use in formulating the synthesis equations for general spatial serial chains. Chapter 16 is the same as Chapter 12 in the previous edition.

I continue to benefit from the contributions of teachers, colleagues, and students toward a geometric synthesis theory for linkage systems, recently from Jeff Ge, Mohan Bodduluri, John Dooley, Pierre Larochelle, Andrew Murray, Fangli Hao, Curtis Collins, Alba Perez-Gracia, Haijun Su, Nina Robson, Duanling Li, and my coauthor Gim Song Soh. In addition, I am grateful for the continued inspiration of Qizheng Liao and Bernard Roth.

Finally, I gratefully acknowledge the support of the Engineering Design Program of the National Science Foundation that has made this book possible.

Irvine, CA,

J. Michael McCarthy
September 2010

First Edition

This book is an introduction to the mathematical theory of design for articulated devices that rely on simple mechanical constraints to provide a complex workspace for a workpiece or end-effector. Devices ranging from windshield wipers to robot manipulators and mechanical hands are examples of these systems each of which has a skeleton of links connected by joints called a linkage. The function or task for the device is defined as a set of positions to be reached by the end-effector. The goal is to determine the dimensions of all of the devices that can achieve a specific task. Formulated in this way the design problem is purely geometric in character.

This text blends two approaches to this design problem in order to develop the intuition needed to move from planar to spatial linkage design. One approach considers the geometric configurations of points and lines generated as a moving body is displaced through a finite set of positions. This is the foundation for graphical methods for planar linkage synthesis and can be generalized to spherical and spatial linkage design. A separate approach focuses directly on solving the nonlinear constraint equations that characterize a mechanical connection. This provides convenient equations for planar and spherical linkage design, and is crucial to addressing the geometric challenge of spatial linkage design.

This unified formulation requires a range of mathematical tools. The basic language is vector algebra and matrix theory, which should be familiar to junior and senior university students. However, something among the techniques ranging from graphical constructions, spherical trigonometry, complex vectors, and quaternions to line geometry and dual vector algebra is certain to be unfamiliar. For this reason, the presentation is designed to introduce these techniques, and additional background is provided in appendices.

The first chapter presents an overview of the articulated systems that we will be considering in this book. The generic mobility of a linkage is defined, and we separate them into the primary classes of planar, spherical, and spatial chains.

The second chapter presents the analysis of planar chains and details their movement and classification. Chapter three develops the graphical design theory for planar linkages and introduces many of the geometric principles that appear in the remainder of the book. In particular, geometric derivations of the pole triangle and the center-point theorem anticipate analytical results for the spherical and spatial cases.

Chapter four presents the theory of planar displacements, and Chapter five presents the algebraic design theory. The bilinear structure of the design equations provides a solution strategy that emphasizes the geometry underlying linear algebra. The five-position solution includes an elimination step that is probably new to most students, though it is understood and well received in the classroom.

Chapters six and seven introduce the properties of spherical linkages and detail the geometric theory of spatial rotations (now Chapters seven and eight, 2nd edition). Chapter eight presents the design theory for these linkages (now Chapter nine), which is analogous to the planar theory. This material exercises the student's use of vector methods to represent geometry in three dimensions. Perpendicular bisectors in the planar design theory become perpendicular bisecting planes that intersect to define axes. The analogue provides students with a geometric perspective of the linear equations that they are solving.

Chapter nine introduces the analysis of spatial linkages including open chains that are closely related to robot manipulators (now Chapter 11). The complexity of spatial linkages requires the introduction of new techniques. However, we maintain a point of view that emphasizes the similarity to the planar and spherical theories. For example, the constraint equations of planar and spherical linkages are shown to be special cases of those for spatial linkages.

Chapter 10 develops the geometry of spatial displacements (now Chapter 12). Here, we find that the screw triangle and the center-axis theorem must be formulated using lines rather than points. Dual vector algebra is introduced to provide vector operations for calculations with the Plücker coordinates of lines. The result is that geometric calculations with line coordinates are identical to the more familiar vector calculations with point coordinates.

Chapter 11 presents the design theory for spatial chains, and Chapter 12 introduces the geometry of linear combinations of lines that arise in the construction of spatial linkage systems (now Chapters 13 and 16). While the design techniques for planar linkages are well developed, there is room for much more work in the design and use of spatial linkages.

I am pleased to express my gratitude for the contribution of many teachers and colleagues whose work over the years has developed and clarified linkage design theory. This book includes results of the insight, commitment, and hard work by Jeff Ge, Mohan Bodduluri, John Dooley, Pierre Larochelle, Andrew Murray, Fangli Hao, Curtis Collins, Alan Ruth, Shawn Ahlers, and Alba Perez. I have also benefitted from the insight of Qizheng Liao and the inspiration of Bernard Roth. Finally, the support by the Division of Design, Manufacturing, and Industrial Innovation of the National Science Foundation that has made this book possible is gratefully acknowledged.

Irvine, CA,   *J. Michael McCarthy*

Contents

List of Figures

List of Tables

Chapter 1
Introduction

A mechanical system, or machine, generally consists of a power source and a mechanism for the controlled use of this power. The power may originate as the flow of water or the expansion of steam that drives a turbine and rotates an input shaft to the mechanism. It may be that instead the turbine rotates a generator and the resulting electricity is used to actuate a distant electric motor connected to the mechanism input. Another power source is the expansion of pressurized fluid or burning air–fuel mixture against a piston in order to drive its linear movement inside a cylinder. The purpose of the mechanism is to transform this input power into a useful application of forces combined with a desired movement. For this reason, machines are often defined abstractly as devices that transform energy from one form, such as heat or chemical energy, into another form, usually work.

In this book our focus is on devices that transform an input rotary or linear motion into more general movement. We assume that a power source is available that can provide the force and torque needed to drive the system. The primary concern is determining the mechanical constraints that provide a desired movement. This is known as the *kinematic synthesis*, or *geometric design*, of a mechanism.

A mechanism is often described as assembled from gears, cams, and linkages, though it usually contains other specialized components, such as springs, ratchets, brakes, and clutches, as well. Of these it is the linkage that provides versatile movement. Gears and cams generally rotate or translate, though with important torque and timing properties. And the other components are used to apply forces. Therefore, because our goal is to obtain a desired movement, we focus on the design of linkages.

1.1 Linkages

A *linkage* is a collection of interconnected components, individually called *links*. The physical connection between two links is called a *joint*. This definition is general enough to encompass gears and cams where the joint is formed by direct contact

J.M. McCarthy and G.S. Soh, *Geometric Design of Linkages*, Interdisciplinary Applied
Mathematics 11, DOI 10.1007/978-1-4419-7892-9_1,
© Springer Science+Business Media, LLC 2011

between two gear teeth or between a cam and follower. However, we limit our attention to joints that do not include the type of rolling and sliding contact that is found in gears and cams. In fact, the linkages that we study can be viewed as constructed from two elementary joints, the rotary hinge, called a *revolute* joint (denoted by R), and the linear slider, or *prismatic* joint (denoted by P). These joints allow *one-degree-of-freedom* movement between the two links that they connect. The configuration variable for a hinge is the angle measured around its axis between the two bodies, and for a slider it is the distance measured along the linear slide of the joint.

Other joints are available to form linkages, such as the universal joint, the ball-in-socket, and a circular cylinder on a rod. In these cases, it is possible to identify an equivalent assembly of hinges and sliders that provide the same geometric constraint. The universal joint, or T joint, can be viewed as constructed from two revolute joints that are at right angles to each other. This joint allows a two-degree-of-freedom movement between the two links that it connects. The ball-in-socket joint, or S joint, is formed from three revolute joints with concurrent axes. This joint allows three-degree-of-freedom rotational movement and is often found in robot wrists with each joint actuated. Finally, the cylindric joint, or C joint, is constructed by mounting a hinge on a slider so that the axis of the hinge is parallel to the direction of the slider. This joint allows two-degree-of-freedom movement.

These joints constrain the trajectories of points in one link to lie on simple geometric objects in the other link, assuming that the links do not bend or distort as they move. For example, the revolute joint constrains points in one body to follow circular trajectories relative to the other link, and a slider generates linear trajectories. The T and S joints both constrain these trajectories to lie on a sphere, while the cylindric joint forces them to lie on a cylinder. This feature leads to algebraic equations that characterize the mechanical constraint imposed by a linkage on the movement of a floating link, or workpiece. The design problem consists of solving these equations for a specified set of task positions for the workpiece.

1.2 Mobility

An important characteristic of an assembly of bodies forming a linkage is the *generic mobility* of the system. This is, generally speaking, the number of independent parameters such as joint angles and slide distances that are needed to specify the configuration of the linkage ignoring any deformation of the links. It is also known as the dimension of the *configuration space* of the system. The generic mobility of a linkage is the sum of the unconstrained degree of freedom for the links in the system less the constraints imposed by the joints.

In a linkage assembled from n bodies, one link is designated as the fixed frame, or ground, against which the movement of the remaining links is measured. This link has no freedom of movement. Thus, the unconstrained degree of freedom for a linkage constructed from n links is $(n-1)K$, where K is the number of parameters required to specify the position of a single link. In three-dimensional space $K = 6$

because three orientation parameters and three coordinates for a reference point are required to locate this body relative to the ground link. A body that can only rotate in space has $K = 3$; and one constrained to planar movement also has $K = 3$.

Revolute and prismatic joints, and joints constructed from them, reduce the dimension of the configuration space of a system by introducing constraint equations among the configuration parameters. These joints are said to impose *holonomic constraints* on the system. In contrast, nonholonomic constraints, such as the contact of a rolling wheel or knife edge on a surface, do not reduce the dimension of the configuration space of the system. They restrict the instantaneous movement from one configuration to the next. The number of constraint equations imposed by a joint is $u = K - f$, where f is known as the freedom allowed by the joint. Thus, the number of constraint equations imposed by j joints is $\sum_{i=1}^{j}(K - f_i)$, where f_i is the freedom of the ith joint.

The generic mobility of a linkage is the difference between the unconstrained freedom of the links and the number of constraints imposed by the joints, that is,

$$F = (n-1)K - \sum_{i=1}^{j}(K - f_i) = K(n - j - 1) + \sum_{i=1}^{j} f_i. \qquad (1.1)$$

This mobility formula places a lower bound on the degree of freedom of a linkage. The mobility may actually be greater due to special dimensions and internal symmetries in the linkage. An interesting example is the spatial 4R closed chain known as Bennett's linkage, which moves with one degree of freedom though this formula predicts that it is a structure with mobility $F = -2$. Such linkages are termed overconstrained (Waldron [141]).

If a linkage consists of a series of links separated by individual joints forming a *serial open chain*, then there is always one more link than the number of joints. This means that $n = j + 1$ and the mobility formula simplifies to

$$F = \sum_{i=1}^{j} f_i. \qquad (1.2)$$

Thus, the mobility of a serial open chain is simply the sum of the freedom at each joint.

We now consider separately the cases of planar, spherical, and spatial linkages.

1.2.1 Planar Linkages

A planar linkage has the property that all of its links move in parallel planes. Most linkages found in practice are, in fact, planar linkages. A body moving in the plane is located by the x and y components of a reference point and a rotation angle ϕ measured relative to ground. Thus, its unconstrained degree of freedom is $K = 3$.

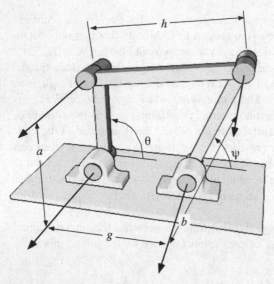

Fig. 1.1 The planar 4R linkage.

The only joints compatible with this movement are revolute joints with axes that are perpendicular to the plane, prismatic joints that move along lines parallel to it, and the direct contact joints of gears and cams that have lines of action parallel to the plane. Of the joints that we are considering, only the R and P joints can be used for planar linkages. Both of these joints have freedom $f = 1$. Thus, the mobility formula for planar linkages is

$$F = 3n - 2j - 3. \tag{1.3}$$

If the links are arranged to form a *single-loop closed chain*, then $j = n$ and the mobility formula becomes

$$F = n - 3. \tag{1.4}$$

It is easy to see that the four-bar linkage, which is a closed chain formed by four links and four joints, has mobility $F = 4 - 3 = 1$. Figure 1.1 is an example of a planar 4R closed chain. Another example is the RRRP slider-crank linkage.

It is interesting to consider how many joints j are needed to constrain a planar assembly of n links to a generic mobility of one. Set $F = 1$ in (1.3) and solve for j to obtain

$$j = \frac{3}{2}n - 2. \tag{1.5}$$

Clearly, n must be even, and we find, for example, that a one-degree-of-freedom six-bar linkage must have $j = 7$ revolute joints. Figure 1.2 shows the two classes of single-degree-of-freedom six-bar linkages.

(a)   (b)

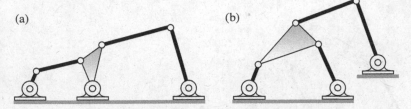

Fig. 1.2 Examples of the two classes of six bar linkages: (a) the Watt six-bar linkage; and (b) the Stephenson six-bar linkage.

1.2.2 Linkage Graphs

A linkage consists of links connected by joints, which can be represented as a graph the links as vertices and the joints as edges. A graph with n vertices and j edges is called an (n, j) graph, and following Tsai [138] we call the associated linkage an (n, j) linkage, usually termed an *n-bar linkage*.

Figure 1.3 shows a serial chain formed by four bodies including the ground frame which are connected by three revolute joints. Also shown is its linkage graph consisting of four vertices connected by three edges. Two 3R serial chains connected to the same workpiece form a 6R closed chain, Figure 1.4.

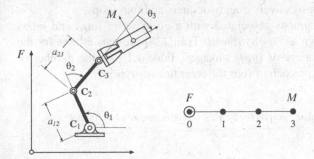

Fig. 1.3 A planar 3R serial chain formed from four links and three revolute joints, and its linkage graph.

The graph of a simple closed loop linkage has the same number of vertices as joints, and it divides the plane into inner and outer regions, or faces, F such that

$$F = j - n + 2. \tag{1.6}$$

This is Euler's equation, which relates the number of faces, edges, and vertices of a polyhedron. This equation applies to linkage graphs that are obtained by the projection of a polyhedron onto a plane.

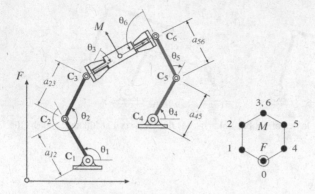

Fig. 1.4 A planar closed chain formed by two 3R serial chains connected to the same workpiece, and its linkage graph.

The faces of a linkage graph are the loops of the linkage. However, the outside face is usually not counted, so the graph is considered to consist of $L = F - 1$ loops. Thus, we obtain the relationship

$$L = j - n + 1. \tag{1.7}$$

We find that the four-, six-, eight- and ten-bar linkages given by $(4,4)$, $(6,7)$, $(8,10)$ and $(10,13)$ graphs have consecutively one, two, three, and four loops.

The enumeration of the graphs associated with a given set of links and joints is a useful design tool known as *type synthesis*. Tsai [138] presents atlases for the graphs associated with a variety of (n, j) linkages. Table 1.1 lists the number of distinct graphs associated with each of four different linkage types.

Table 1.1 Number of distinct linkage graphs for four-, six-, eight-, and ten-bar linkages.

(n, j)	Loops (L)	Graphs
$(4,4)$	1	1
$(6,7)$	2	2
$(8,10)$	3	16
$(10,13)$	4	230

A linkage graph can be viewed as a specialized application of constraint graphs used in geometric modeling, Bouma et al. [6]. A constraint graph maps geometric objects to vertices and geometric relationships to edges to provide a versatile tool for characterizing mechanical assemblies. For a four-bar linkage, the links and joints are represented as lines and points and form the vertices of the constraint graph. The distances between points, angles between lines, and the incidence of a point on a line form the edges of the constraint graph, Figure 1.5.

A subgraph of a constraint graph is called a *cluster* if the coordinates of all of its vertices can be computed once one pair of vertices is determined. For example, in Figure 1.5 the vertices $\mathbf{P}_O$, $\mathbf{L}_{OC}$, and $\mathbf{P}_C$ form the cluster C_{OC}. The reduced constraint graph that identifies clusters as vertices and shared geometric elements as edges becomes the linkage graph. This relationship between the constraint graph and the linkage graph provides a way to add dimensional information to the linkage graph, Li et al. [65].

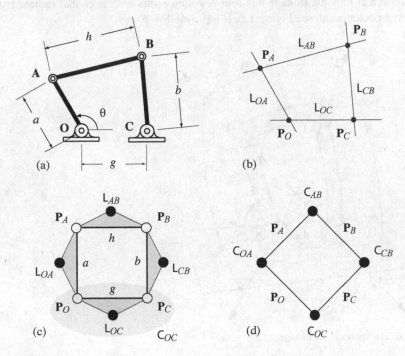

Fig. 1.5 A planar four-bar linkage (a) defined in terms of lines and points (b) is represented by the constraint graph (c). The constraint graph separates into clusters linked by points to become the linkage graph.

1.2.3 Spherical Linkages

Each of the links of a spherical linkage is constrained to rotate about the same fixed point in space, which means that the trajectories of points in each link lie on concentric spheres. The orientation of a link can be defined by three rotation angles often termed *roll*, *pitch*, and *yaw*. For convenience, we visualize yaw as a longitude angle on a globe, and pitch as latitude. Then the roll angle is the rotation about an axis through a given longitude and latitude. These three parameters are sufficient to define the orientation of a rigid body in space and, as in the plane, $K = 3$.

Only revolute joints assembled so their axes intersect the same fixed point are compatible with the geometric constraint of spatial rotation. Because this joint has freedom $f = 1$, the mobility formula for spherical linkages becomes

$$F = 3n - 2j - 3. \tag{1.8}$$

The spherical four-bar linkage has four revolute joints with axes that radiate from the fixed center point, see Figure 1.6. It has mobility $F = 1$.

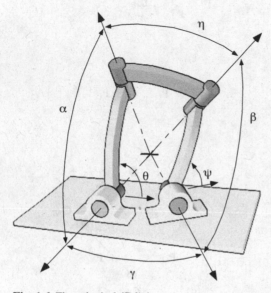

Fig. 1.6 The spherical 4R linkage.

1.2.4 Spatial Linkages

A spatial linkage is characterized by the feature that at least one link in the system moves between two general positions in space. Six parameters are needed to specify the location of this link, three for the coordinates of a reference point and three for the orientation of the body about this point. Therefore, $K = 6$ and we have

$$F = 6(n - j - 1) + \left(\sum_{i=1}^{j} f_i \right). \tag{1.9}$$

For the case of a single-loop closed chain, where $j = n$, the mobility formula becomes

$$F = \left(\sum_{i=1}^{j} f_i \right) - 6. \tag{1.10}$$

Thus, a spatial single-loop closed chain must have seven one-degree-of-freedom joints in order to have mobility $F = 1$. Notice that of these seven at most three can be prismatic joints, because these joints constrain only the translational freedom of the body. The spatial four-bar linkage constructed with four cylindric joints, denoted 4C, has mobility $F = 8 - 6 = 2$.

1.2.5 Platform Linkages

An important class of linkages consists of multiple serial chains that have their end-links connected to form a single floating link, or platform. Planar and spherical 4R linkages can each be viewed as platform linkages constructed from a pair of RR open chains. Similarly, the spatial 4C closed chain is obtained by rigidly connecting the end-links of two CC open chains. The 5TS platform linkage consists of a single floating link supported by five separate TS chains (Figure 1.7). The mobility of these systems is easily determined.

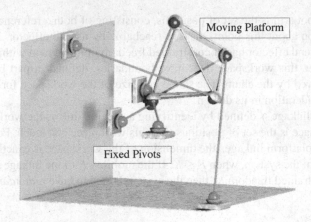

Moving Platform

Fixed Pivots

Fig. 1.7 The spatial 5TS platform linkage.

Consider a platform supported by c serial open chains connecting it to ground. Divide the platform among the c chains and let the mobility of the ith chain including its piece of the platform be F_i. Rejoining the end-links of these chains to form a single platform is the same as removing $c - 1$ bodies from the system. Thus, the mobility of a platform manipulator can be calculated to be

$$F = \sum_{i=1}^{c} F_i - K(c-1) = K - \sum_{i=1}^{c}(K - F_i). \tag{1.11}$$

A convenient interpretation of this equation is obtained by introducing the *degree of constraint* $U_i = K - F_i$ imposed by the ith open chain. If the mobility of a chain is greater than the unconstrained degree of freedom K, that is, $F_i > K$, then the chain does constrain the platform and $U_i = 0$. Let the degree of constraint of the platform itself be $U = K - F$. Then (1.11) can be written as

$$U = \sum_{i=1}^{c} U_i. \tag{1.12}$$

The degree of constraint imposed on the platform is the sum of the degrees of constraint of its supporting chains. For example, the TS open chain has five degrees of freedom, therefore it has one degree of constraint. Thus, the 5TS platform has five degrees of constraint, or one degree of freedom.

The formula (1.12) defines the maximum constraint that the supporting chains can impose. Special dimensions and symmetries can reduce this degree of constraint increasing the freedom of movement of the system.

1.3 Workspace

The *workspace* of a robot arm is the set of positions, consisting of both a reference point and the orientation about this point, that are reachable by its end effector. A robot is designed so its end effector has unconstrained freedom of movement within its workspace. However, this workspace does have boundaries, defined in part by the extreme reach allowed by the chain. The shape and size of the workspace for a robot is a primary consideration in its design.

The workspace of a linkage is defined by identifying a specific link as the workpiece. Then the workspace is the set of positions that this workpiece can reach. For serial open chains and platform linkages the dimension of the workspace is exactly the generic mobility F of the system, when $F \le K$. If the mobility F of the linkage is greater than the unconstrained freedom K, then the system is said to have *redundant* degrees of freedom.

1.4 Linkage Design

Linkage design is often divided into three categories of tasks, called motion generation, function generation, and point-path generation. Our approach, presented in later chapters, is based on techniques developed for motion generation. In this case, it is assumed that the designer has identified positions that represent the desired movement of a workpiece. This can be viewed as specifying positions that are to lie in the workspace of the linkage. Thus, a discrete representation of the workspace is known, but not the design parameters of the linkage. The constraint equations of the chain evaluated at each of the task positions provide design equations that are solved to determine the linkage.

This approach can also be applied to the design of linkages for function generation. In this case the goal is to coordinate an output crank rotation or slide with a specific input crank rotation or slide. Specialized design procedures have been developed for function generation for various linkage systems. However, we can transform this into a motion generation problem by holding the input link fixed and allowing the ground link to move. By doing this we obtain a set of task positions for the output link relative to the input link, which are defined by the desired coordinated set of input and output joint parameters. We use these positions to design another open chain that connects the input and output links. The result is a closed chain with the desired coordination between the input and output links.

The last category, point-path generation, is a classical problem in linkage design where the primary concern has been the generation of straight-line paths. It is possible to transform a path generation design problem into a motion generation problem by simply adding specifications for the orientation of a reference frame at each location of the moving point. However, this reduces the number of points that can be used to approximate a desired path. In particular, we will see that a planar 4R linkage can be designed to reach as many as five arbitrary task positions. In contrast, it is known that, by ignoring the orientation of the coupler, a four-bar linkage can be designed so the trajectory of point on the coupler passes through as many as nine specified points; see Roth and Freudenstein [108] and Wampler et al. [146]. Unfortunately, this is a difficult design problem even in the plane, and the theory for more general linkage systems has not been developed.

1.4.1 Approximate Straight-Line Mechanisms

One of the interesting problems in classical mechanism theory was the design of a linkage to generate straight-line motion. We take a moment here to present three important planar 4R linkages distinguished by the near straight-line movement of a point on the floating link. Let vertices of the quadrilateral formed by the linkage be labeled $OABC$ in a clockwise manner, such that O and C are the fixed pivots and A and B are the moving pivots; and let the link lengths be labeled $a = |OA|$, $b = |CB|$, $g = |OC|$, and $h = |AB|$.

1. *Watt's linkage*: Let the tracing point P be a distance x along the coupler AB of length h measured from A. If the link lengths OA and BC satisfy the ratio $x/(h - x) = a/b$ and the fixed link has length $g^2 = h^2 + (a+b)^2$, then the point P follows a near straight-line movement over part of its path. See Figure 1.8.
2. *Robert's linkage*: Let the link lengths satisfy the relations $a = b$ and $g = 2h$, and locate the tracing point P to form an isosceles triangle of side $x = a$ with the two

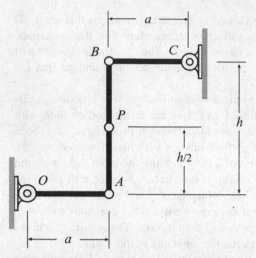

Fig. 1.8 Watt's straight-line linkage: $x = h/2, b = a$.

moving pivots. The result is that P moves on a nearly straight line along the part of its path between the two fixed pivots. The height d can be varied to modify this path. See Figure 1.9.

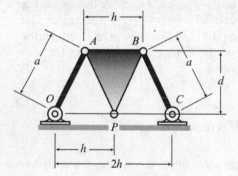

Fig. 1.9 Robert's straight-line linkage.

3. *Chebyshev's linkage*: Chebyshev's linkage has dimensions $g = 2h$ and $a = b = 2.5h$. In this case locate the point P midway between the two moving pivots to trace a near straight-line movement. See Figure 1.10.

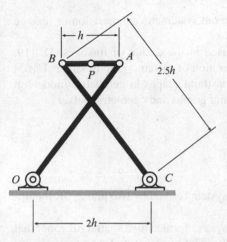

Fig. 1.10 Chebyshev's straight-line linkage: $a = 2.5h$, $b = 2.5h$, $g = 2h$.

1.5 Summary

This chapter has introduced the linkage systems that are the focus of the mathematical theory that follows. Linkages are classified by the type and number of joints that form the assembly. Joints that can be constructed from hinges and sliders to form planar, spherical, and spatial linkages provide a remarkably wide range of devices. The mobility formula defines the generic degree of freedom of a linkage, which is the dimension of its workspace. The size and shape of this workspace characterizes the performance of the device. The goal of the design theory presented in the following chapters is to formulate and solve the constraint equations that ensure that the workspace for a given open chain includes a discrete task space prescribed by the designer.

1.6 References

Dimarogonas [24] and Hartenberg and Denavit [46] present historical accounts of the origins of the theory of machines and mechanisms. The concept of a kinematic chain was introduced by Reuleaux [101] who also identified the central role played by the joints in classifying devices. A systematic notation to define kinematic chains constructed from elementary joints was introduced by Denavit and Hartenberg [22]. Freudenstein and Dobryankyj [36] and Crossley [17] applied graph theory to the problem of identifying all the linkage assemblies available for a certain number of links and type of joint. Crossley [18] described spatial linkages and the larger variety of joints available for these systems. Harrisberger [44] formulated a classification theory for spatial linkages and with Soni [45] enumerated the spatial four-bar linkages. See Kota [57] for more information about systematic enumeration of linkage systems.

Tsai[138] shows how graph theory is used in the design of linkages. Dai[19, 20] uses linkage graphs to describe the operation of metamorphic linkages. Li[65] shows that linkage graphs are related to constraint graphs in geometric modeling. See Hoffman[6] for a description of constraint graphs and geometric solvers.

Exercises

1. Determine the degree of freedom of the system formed by two planar 3R manipulators that hold the same workpiece.
2. Determine the degree of freedom of the system formed by a spatial 6R robot that (i) rotates a hinged lever; (ii) rubs a block against a planar surface.
3. Consider a linkage with c independent closed loops, show that $c = j + 1 - n$, and the mobility formula can be written as $F = \sum f_i - 6c$.
4. Show that while the planar and spherical 4R linkages have mobility $F = 1$, the spatial 4R Bennett linkage should not move at all.
5. The mobility formula assumes that a joint is a connection between two links. However, often three or more links are shown connected at one joint. Show that if p links appear connected at one joint, that the mobility formula requires the connection be counted as $p - 1$ joints.
6. Show that if a truss is a linkage with mobility $F = 0$, then its joint forces can be determined by elementary static analysis.
7. Let the links of a mechanism be the vertices of a graph, and its joints the edges. Show that there are only two nonisomorphic graphs of planar six-bar linkages with $F = 1$, the Watt and Stephenson chains.
8. Determine the number of revolute joints that a planar eight-bar linkage must so $F = 1$. Show that there are 16 nonisomorphic graphs for these linkages.
9. Enumerate the $F = 1$ spatial four-bar linkages that can be assembled using R, P, T, S, and C joints.

Chapter 2
Analysis of Planar Linkages

In this chapter we consider assemblies of links that move in parallel planes. Any one of these planes can be used to examine the movement since the trajectories of points in any link can be projected onto this plane without changing their properties. Our focus is on linkages constructed from revolute joints with axes perpendicular to this plane and prismatic joints that move along lines parallel to it. We examine the RR, PR, and RP open chains and the closed chains constructed from them, as well as the 3R and RPR planar robots. We determine the configuration of the linkage as a function of the independent joint parameters and the physical dimensions of the links.

2.1 Coordinate Planar Displacements

A revolute joint in a planar linkage allows rotation about a point, and a prismatic joint allows translation along a line. These movements are represented by transformations of point coordinates in the plane.

Consider the rotation of a link about the revolute joint $\mathbf{O}$ located at the origin of the fixed coordinate frame F. Let $\mathbf{x} = (x,y)^T$ be the coordinates of a point measured in the frame M of the link. If the moving frame has its origin also located at $\mathbf{O}$, and the angle between the x-axes of these two frames is θ, then the coordinates $\mathbf{X} = (X,Y)^T$ of this point in F are given by the matrix equation

$$\begin{Bmatrix} X \\ Y \\ 1 \end{Bmatrix} = \begin{bmatrix} \cos\theta & -\sin\theta & 0 \\ \sin\theta & \cos\theta & 0 \\ 0 & 0 & 1 \end{bmatrix} \begin{Bmatrix} x \\ y \\ 1 \end{Bmatrix}. \tag{2.1}$$

We introduce the extra column in this matrix to accommodate translations typical of prismatic joints as part of the matrix operation.

In particular, consider a prismatic joint that has the x-axis of F as its line of action. Let the distance between the origins of M and F along this line be s. Then

J.M. McCarthy and G.S. Soh, *Geometric Design of Linkages*, Interdisciplinary Applied Mathematics 11, DOI 10.1007/978-1-4419-7892-9_2, © Springer Science+Business Media, LLC 2011

we have

$$\begin{Bmatrix} X \\ Y \\ 1 \end{Bmatrix} = \begin{bmatrix} 1 & 0 & s \\ 0 & 1 & 0 \\ 0 & 0 & 1 \end{bmatrix} \begin{Bmatrix} x \\ y \\ 1 \end{Bmatrix}. \tag{2.2}$$

A translation along a prismatic joint parallel to the y-axis is defined in the same way as

$$\begin{Bmatrix} X \\ Y \\ 1 \end{Bmatrix} = \begin{bmatrix} 1 & 0 & 0 \\ 0 & 1 & s \\ 0 & 0 & 1 \end{bmatrix} \begin{Bmatrix} x \\ y \\ 1 \end{Bmatrix}. \tag{2.3}$$

The matrices in these equations define the three *coordinate displacements* of planar movement. Planar displacements are constructed from these three basic transformations.

We now introduce the notation $[Z(\theta)]$, $[X(s)]$, and $[Y(s)]$ for these coordinate displacements, so we have

$$\mathbf{X} = [Z(\theta)]\mathbf{x}, \quad \mathbf{X} = [X(s)]\mathbf{x}, \quad \text{and} \quad \mathbf{X} = [Y(s)]\mathbf{x}, \tag{2.4}$$

respectively. Notice that we do not distinguish symbolically between the coordinates $\mathbf{X}$ that are two-dimensional and those that have 1 as a third component. Some authors to refer to the former as vectors and the latter as affine points. We do not need this general distinction, and therefore will take the time to make the difference clear when needed in the context of our calculations.

2.1.1 The PR Open Chain

The benefit of this matrix formulation can be seen in considering the movement of a PR open chain. This chain consists of a link that slides along the linear guide of a P-joint relative to the ground. An end-link is attached to the slider by a revolute joint, Figure 2.1. We now determine the movement of a coordinate frame M attached to the end-link relative to a fixed frame F.

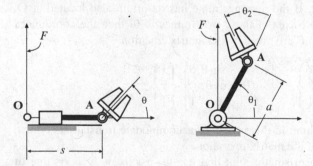

Fig. 2.1 The PR and RR open chain robots.

First, locate F so that its x-axis is parallel to the slide of the P-joint and denote its origin by $\mathbf{O}$. Locate M in the end-link so that its origin is centered on the revolute joint, which we denote by $\mathbf{A}$, and with its x-axis aligned initially with the x-axis of F.

The configuration of the PR chain is defined by the slide s from $\mathbf{O}$ to $\mathbf{A}$, and the rotation angle θ about $\mathbf{O}$ measured from the x-axis of F to the x-axis of M. The transformation of coordinates from M to F is given by the matrix product

$$\mathbf{X} = [X(s)][Z(\theta)]\mathbf{x}, \tag{2.5}$$

or

$$\begin{Bmatrix} X \\ Y \\ 1 \end{Bmatrix} = \begin{bmatrix} \cos\theta & -\sin\theta & s \\ \sin\theta & \cos\theta & 0 \\ 0 & 0 & 1 \end{bmatrix} \begin{Bmatrix} x \\ y \\ 1 \end{Bmatrix}. \tag{2.6}$$

The set of planar displacements $[D]$, given by

$$[D] = [X(s)][Z(\theta)], \tag{2.7}$$

is the workspace of the PR open chain. This matrix equation defines the *kinematics equations* for the chain.

An important question in the analysis of an open chain is what parameter values s and θ are needed to reach a given displacement $[D]$ in the workspace of the chain. Assuming the elements of the matrix

$$[D] = \begin{bmatrix} a_{11} & a_{21} & p_x \\ a_{21} & a_{22} & p_y \\ 0 & 0 & 1 \end{bmatrix} \tag{2.8}$$

are known, equation (2.7) can be solved to determine these parameters. Notice that $p_y = 0$ is required for the displacement $[D]$ to be in the workspace reachable by the PR chain. It is now easy to see that s and θ can be determined from the elements of $[D]$ by the formulas

$$s = p_x \quad \text{and} \quad \theta = \arctan\frac{a_{21}}{a_{11}}. \tag{2.9}$$

Its important to note here that the arctan function must keep track of the signs of both a_{21} and a_{11} so the correct value for θ is obtained in the range 0 to 2π.

The arctan function in calculators often incorporates the assumption that the denominator of the fraction a_{21}/a_{11} is positive. If this denominator is actually negative, as occurs when θ is in the second and third quadrants, then π must be added to the angle returned by the calculator in order to obtain the correct result.

2.1.2 The RR Open Chain

A planar RR open chain has a fixed revolute joint **O** that connects a rotating link, or *crank*, to the ground link. A second revolute joint **A** connects the crank to the end-link, or *floating link*, Figure 2.1.

Position the fixed frame F so that its origin is the fixed pivot **O** and its x-axis is directed toward **A** when the crank **OA** is in the zero position. Introduce the moving frame M in the end-link, so that its origin is located at **A** and its x-axis is also directed, initially, along the segment **OA**.

Let θ_1 be the angle measured from F to **OA** as the linkage moves, and let θ_2 be the angle measured from **OA** to M. Then the position of M relative to F is defined by the composition of coordinate displacements

$$\mathbf{X} = [Z(\theta_1)][X(a)][Z(\theta_2)]\mathbf{x}, \tag{2.10}$$

where $a = |\mathbf{A} - \mathbf{O}|$ is the length of the crank. Expanding this equation we obtain

$$\begin{Bmatrix} X \\ Y \\ 1 \end{Bmatrix} = \begin{bmatrix} \cos(\theta_1 + \theta_2) & -\sin(\theta_1 + \theta_2) & a\cos\theta_1 \\ \sin(\theta_1 + \theta_2) & \cos(\theta_1 + \theta_2) & a\sin\theta_1 \\ 0 & 0 & 1 \end{bmatrix} \begin{Bmatrix} x \\ y \\ 1 \end{Bmatrix}. \tag{2.11}$$

Notice that the position of the floating link of an RR chain is equivalent to a translation by the vector $\mathbf{d} = (a\sin\theta_1, a\sin\theta_1)^T$ followed by a rotation by the angle $\sigma = \theta_1 + \theta_2$.

The workspace of the RR chain is given by the set of displacements

$$[D] = [Z(\theta_1)][X(a)][Z(\theta_2)]. \tag{2.12}$$

This defines the kinematics equations of the RR chain. For a given position $[D]$ the parameter values θ_1 and θ_2 that reach it are obtained by equating (2.8) to the matrix in (2.11). The result is that the angles θ_1 and $\sigma = \theta_1 + \theta_2$ are given by

$$\theta_1 = \arctan\frac{p_y}{p_x} \quad \text{and} \quad \sigma = \arctan\frac{a_{21}}{a_{11}}. \tag{2.13}$$

Notice that the elements p_x and p_y must satisfy the relation

$$a = \sqrt{p_x^2 + p_y^2}. \tag{2.14}$$

in order for $[D]$ to be in the workspace of this chain.

2.1.3 The RPR and 3R Chains

If the distance a between the joints of an RR chain is allowed to vary, then we obtain the structure of a three-degree-of-freedom planar manipulator. This variation in length can be introduced either by a prismatic joint, forming an RPR open chain, or by a revolute joint to form a 3R open chain, Figure 2.2. The formulas for the RR chain can be used to analyze the RPR and 3R chains with minor modifications.

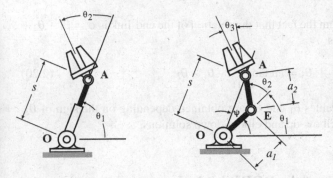

Fig. 2.2 The RPR and RRR open robots.

For the RPR, the link length a can be identified with the slide parameter s of the prismatic joint. The result is that (2.12) with $a = s$ defines the workspace of the RPR chain. Equations (2.14) and (2.13) define the values for s, θ_1, and θ_2 needed to reach a given goal displacement.

For the 3R case, we have an elbow joint **E** inserted between **O** and **A**. Let the lengths of the two links be $a_1 = |\mathbf{E} - \mathbf{O}|$ and $a_2 = |\mathbf{A} - \mathbf{E}|$. Denote the rotation angle about the elbow joint by θ_2 which is measured from **OE** counter-clockwise to **EA**. The rotation of the end-link around **A** is now denoted by θ_3. The kinematics equations of this chain become

$$[D] = [Z(\theta_1)][X(a_1)][Z(\theta_2)][X(a_2)][Z(\theta_3)]. \tag{2.15}$$

The variable length $s = |\mathbf{A} - \mathbf{O}|$ is given by the cosine law of the triangle $\triangle\mathbf{OEA}$,

$$s^2 = a_1^2 + a_2^2 + 2a_1 a_2 \cos\theta_2. \tag{2.16}$$

The positive sign for the cosine term in this equation arises because θ_2 is an exterior angle of the triangle $\triangle\mathbf{OEA}$. Notice that s must lie between the values $|a_2 - a_1|$ and $a_1 + a_2$.

For a given position $[D]$ of the end-link, we can determine the length s as we did for the RPR chain using (2.14). This allows us to compute the elbow joint angle as

$$\theta_2 = \arccos\frac{p_x^2 + p_y^2 - a_1^2 - a_2^2}{2a_1 a_2}. \tag{2.17}$$

The arccosine function yields two values for this angle $\pm\theta_2$. We can compute the joint angle θ_1 using (2.13), however, we must account for the presence of the angle $\psi = \angle\textbf{EOA}$, which is given by

$$\psi = \arctan\frac{a_2\sin\theta_2}{a_1 + a_2\cos\theta_2}. \tag{2.18}$$

The result is

$$\theta_1 = \arctan\frac{p_y}{p_x} - \psi. \tag{2.19}$$

Finally, θ_3 is obtained from the fact that the rotation of the end-link is $\sigma = \theta_1 + \theta_2 + \theta_3$ in (2.13), which yields

$$\theta_3 = \arctan\frac{a_{21}}{a_{11}} - \theta_1 - \theta_2. \tag{2.20}$$

Notice that two sets of values θ_1 and θ_3 are obtained depending on the sign of θ_2. These are known as the elbow-up and elbow-down solutions.

2.2 Position Analysis of the RRRP Linkage

The RRRP linkage is called a *slider-crank* and consists of a rotating crank linked to a translating slider by a connecting rod, or *coupler*. It is a fundamental machine element found in everything from automotive engines to door-closing mechanisms. We can also view this device as a platform linkage, in which case the coupler is a workpiece supported by an RR and a PR chain.

Denote the fixed and moving pivots of the input crank by $\textbf{O}$ and $\textbf{A}$, respectively, and let $\textbf{B}$ be the revolute joint attached to the slider. Position the fixed frame F so that its origin is $\textbf{O}$ and orient it so that its x-axis is perpendicular to the direction of slide, Figure 2.3. The input crank angle θ is measured from the x-axis of F around $\textbf{O}$ to $\textbf{OA}$, and the travel s of the slider is measured along the y-axis to $\textbf{B}$.

The length of the driving crank is $r = |\textbf{A} - \textbf{O}|$, and the length of the coupler is $L = |\textbf{B} - \textbf{A}|$. The distance e to the linear path of the pivot $\textbf{B}$ is called the *offset*. Notice that the dimensions r, L, and e are always positive.

2.2.1 The Output Slide

To analyze this linkage, we determine the output slide s as a function of the input crank angle θ. The linkage moves so the pivots $\textbf{A}$ and $\textbf{B}$ remain the constant distance L apart. The coordinates of these pivots in F are given by

$$\textbf{A} = \left\{\begin{matrix} r\cos\theta \\ r\sin\theta \end{matrix}\right\} \quad \text{and} \quad \textbf{B} = \left\{\begin{matrix} e \\ s \end{matrix}\right\}. \tag{2.21}$$

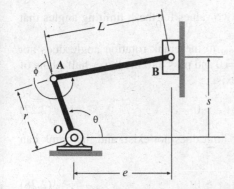

Fig. 2.3 The dimensions characterizing a slider-crank, or RRRP, linkage.

Thus, the length $L = |\mathbf{B} - \mathbf{A}|$ of the coupler provides the constraint equation

$$(\mathbf{B} - \mathbf{A}) \cdot (\mathbf{B} - \mathbf{A}) = L^2. \tag{2.22}$$

Substitute (2.21) into this expression and collect the coefficients of s to obtain the quadratic equation

$$s^2 - (2r\sin\theta)s + (r^2 + e^2 - 2er\cos\theta - L^2) = 0. \tag{2.23}$$

The quadratic formula yields the roots

$$s = r\sin\theta \pm \sqrt{L^2 - e^2 + 2er\cos\theta - r^2\cos^2\theta}. \tag{2.24}$$

Thus, for a given input crank angle θ there are two possible values of the slide s. They are, geometrically, the intersection of a circle of radius L centered on $\mathbf{A}$ with the line through $\mathbf{B}$ parallel to the y-axis of F. These two solutions define the two *assemblies* of the RRRP linkage. The positive solution generally has the slider moving above the crank, while the negative solution has it below.

2.2.2 The Range of Crank Rotation

We now consider the values of the crank angle θ for which a solution for the slider position s exists. The condition that the solution be a real number is

$$L^2 - e^2 + 2er\cos\theta - r^2\cos^2\theta \geq 0. \tag{2.25}$$

Set this to zero to obtain a quadratic equation in $\cos\theta$ that defines the minimum and maximum angular values for the crank angle θ, and obtain the roots

$$\theta_{\min} = \arccos\frac{e+L}{r} \quad \text{and} \quad \theta_{\max} = \arccos\frac{e-L}{r}. \tag{2.26}$$

Notice that the arccosine function returns two values for these limiting angles that are reflections through the x-axis of F.

If $\cos\theta_{min} > 1$, then the lower limit θ_{min} to the crank rotation angle does not exist. In which case the crank can reach $\theta = 0$ and pass into the lower half-plane of F. Thus, the condition that no lower limit exist is

$$S_1 = L - r + e > 0. \tag{2.27}$$

Similarly, if $\cos\theta_{max} < -1$ then the upper limit does not exist, and the crank can reach $\theta = \pi$. This yields the condition

$$S_2 = L - r - e > 0. \tag{2.28}$$

The signs of the parameters S_1 and S_2 identify four types of slider-crank linkage depending on the input rotation of the crank:

1. **A rotatable crank:** $S_1 > 0$ and $S_2 > 0$, in which case neither limit θ_{min} nor θ_{max} exists, and the input crank can fully rotate.
2. **A 0-rocker:** $S_1 > 0$ and $S_2 < 0$, for which θ_{max} exists but not θ_{min}, and the input crank rocks through $\theta = 0$ between the values $\pm\theta_{max}$.
3. **A π-rocker:** $S_1 < 0$ and $S_2 < 0$, which means that θ_{min} exists but not θ_{max}, and the input crank rocks through $\theta = \pi$ between the values $\pm\theta_{min}$.
4. **A rocker:** $S_1 < 0$ and $S_2 < 0$, in which case both upper and lower limit angles exist, and the crank cannot pass through either 0 or π. Instead, it rocks in one of two separate ranges: (i) $\theta_{min} \leq \theta \leq \theta_{max}$, or (ii) $-\theta_{max} \leq \theta \leq -\theta_{min}$.

The conditions $S_1 > 0$ and $S_2 > 0$ for a fully rotatable crank can be combined to define the formula

$$S_1 S_2 = (L - r + e)(L - r - e) = (L - r)^2 - e^2 > 0. \tag{2.29}$$

Notice that because e is always positive, $L - r$ must be positive for S_2 to be positive. This allows us to conclude that

$$L - r > e \tag{2.30}$$

is the condition that ensures that the crank of the RRRP linkage can fully rotate.

The parameters S_1 or S_2 can take on zero values as well. In these cases, the pivots **O**, **A**, and **B** line up along the x-axis of F, and the slider-crank linkage is said to *fold*.

2.2.3 The Coupler Angle

Let ϕ denote the angle around the moving pivot **A** measured counterclockwise from the line extending along the crank **OA** to the segment **AB** defining the coupler. Then the coordinates of the pivot $\mathbf{B} = (e, s)^T$ are also given by the vector

$$\mathbf{B} = \begin{Bmatrix} r\cos\theta + L\cos(\theta+\phi) \\ r\sin\theta + L\sin(\theta+\phi) \end{Bmatrix}. \tag{2.31}$$

We equate the two vectors defining $\mathbf{B}$ to obtain

$$r\cos\theta + L\cos(\theta+\phi) = e,$$
$$r\sin\theta + L\sin(\theta+\phi) = s. \tag{2.32}$$

These equations are called the *loop equations* of the slider-crank because they capture the fact that the linkage forms a closed loop. Solve these equations for $L\sin(\theta+\phi)$ and $L\cos(\theta+\phi)$ and use the arctan function to obtain

$$\theta + \phi = \arctan\frac{s - r\sin\theta}{e - r\cos\theta}. \tag{2.33}$$

This equation provides the value for ϕ associated with each solution for the slide s defined in (2.24).

2.2.4 The Extreme Slider Positions

The maximum translation of the slider, s_{max}, is reached when the coupler angle ϕ is equal to zero. In this instance the pivots $\mathbf{O}$, $\mathbf{A}$, and $\mathbf{B}$ fall on a line, so that $r+L$ forms the hypotenuse of a right triangle. This yields

$$s_{max} = \sqrt{(r+L)^2 - e^2}. \tag{2.34}$$

The crank angle θ_1 associated with s_{max} is obtained from the loop equations (2.32) as

$$\theta_1 = \arctan\frac{s_{max}}{e}. \tag{2.35}$$

Notice that the parameter s_{max} can be positive or negative, because the linkage can be assembled with the slider above over below the x-axis.

The minimum translation of the slider, s_{min}, occurs with the coupler angle ϕ is equal to π. In this configuration the pivots $\mathbf{A}$ and $\mathbf{B}$ are on opposite sides of $\mathbf{O}$ and $L - r$ is the hypotenuse of the triangle, so s_{min} is given by

$$s_{min} = \sqrt{(L-r)^2 - e^2}. \tag{2.36}$$

While s_{max} always exists, s_{min} exists only if this square root is real. There are two cases $L - r > 0$ and $L - r < 0$. In the first case the crank is fully rotatable and the associated crank angle is

$$\theta_2 = \pi + \arctan\frac{s_{min}}{e}. \tag{2.37}$$

The minimum slide results when the pivot **A** rotates to the position such that **O** lies
between it and **B**. If $L - r < 0$ then **A** and **B** are on the same side of **O** and the crank
angle is

$$\theta_2 = \arctan \frac{s_{\min}}{e}. \tag{2.38}$$

Notice that these extreme configurations can be reflected through the x-axis.

If the crank of the slider-crank is fully rotatable, then the angular travel of the
crank as the slider moves from $s_{\max}$ to $s_{\min}$ is $|\theta_2 - \theta_1|$. The angular travel of the
return from $s_{\min}$ to $s_{\max}$ is $|2\pi - (\theta_2 - \theta_1)|$. The ratio of these two ranges of travel
is known as the *time ratio*

$$r_t = \frac{|\theta_2 - \theta_1|}{|2\pi - (\theta_2 - \theta_1)|}. \tag{2.39}$$

Notice that if the offset e is nonzero then the time ratio is less than 1. This means that
the crank rotates a smaller angular distance as it pulls the slider to $s_{\min}$, than it does
when it pushes it out again to $s_{\max}$. This operation is known as *quick return* because
for a constant angular velocity the slider moves slowly toward $s_{\max}$ and quickly as
it returns to $s_{\min}$.

2.2.5 The RRPR Linkage

A slider-crank linkage is often used in an inverted configuration in which the P-joint
is connected to the floating link, Figure 2.4. In this form, the prismatic joint may be
the piston of a linear actuator that drives the rotation of the crank **OA**. This system
is analyzed as follows.

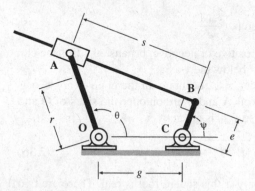

Fig. 2.4 The inverted slider-crank, or RRPR, linkage.

Let the driving RR crank be **OA** with length $r = |\mathbf{A} - \mathbf{O}|$, as before. Position the
frame F with its origin at **O** and its x-axis directed toward **C**, which denotes the fixed
pivot of the RP chain. The length of the ground link **OC** is $g = |\mathbf{C} - \mathbf{O}|$. Consider

the line through **C** perpendicular to the direction of the slider and the line through **A** parallel this direction. Let the intersection of these two lines be the point **B**. The length $e = |\mathbf{B} - \mathbf{C}|$ is the joint offset, and $s = |\mathbf{A} - \mathbf{B}|$ is the slide distance of the prismatic joint. Denote the input crank angle by θ and let ψ be the angle measured about **C** to the segment **CB**.

These conventions allow us to introduce the intermediate parameters b and β given by

$$b = \sqrt{s^2 + e^2} \quad \text{and} \quad \tan\beta = \frac{s}{e}. \tag{2.40}$$

The cosine law for the triangle $\triangle\mathbf{COA}$ yields the relation

$$b^2 = g^2 + r^2 - 2rg\cos\theta. \tag{2.41}$$

Substitute $s^2 + e^2$ to obtain

$$s = \sqrt{g^2 + r^2 - e^2 - 2rg\cos\theta}. \tag{2.42}$$

This defines the joint slide s for a given crank angle θ. Notice that this equation can also be solved to determine θ for a given slide:

$$\cos\theta = \frac{g^2 + r^2 - e^2 - s^2}{2rg}. \tag{2.43}$$

This latter situation arises when the slider is the piston in a linear actuator driving the RR crank.

The rotation angle ψ of the RP crank is determined using the fact that the coordinates of the pivot **A** can be written in two ways

$$\mathbf{A} = \left\{ \begin{array}{c} r\cos\theta \\ r\sin\theta \end{array} \right\} = \left\{ \begin{array}{c} g + b\cos(\psi+\beta) \\ b\sin(\psi+\beta) \end{array} \right\}. \tag{2.44}$$

These equations yield the formula

$$\psi + \beta = \arctan\frac{r\sin\theta}{r\cos\theta - g}. \tag{2.45}$$

Notice that β is determined from s by (2.40).

The range of movement of the cranks and the sliding joint for this linkage can be analyzed in the same way as shown above for the RRRP linkage.

2.3 Position Analysis of the 4R Linkage

Given a planar 4R closed chain, we can identify an input RR crank and an output RR crank, Figure 2.5. Let the fixed and moving pivots of the input crank be **O** and **A**, respectively, and that the fixed and moving pivots of the output crank be **C** and

B. The distances between these points characterize the linkage:

$$a = |\mathbf{A} - \mathbf{O}|, b = |\mathbf{B} - \mathbf{C}|, g = |\mathbf{C} - \mathbf{O}|, h = |\mathbf{B} - \mathbf{A}|. \qquad (2.46)$$

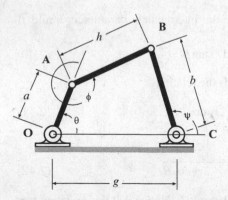

Fig. 2.5 The link lengths that define a 4R linkage.

To analyze the linkage, we locate the origin of the fixed frame F at $\mathbf{O}$, and orient it so that the x-axis passes through the other fixed pivot $\mathbf{C}$. Let θ be the input angle measured around $\mathbf{O}$ from the x-axis of F to $\mathbf{OA}$. Similarly, let ψ be the angular position of the output crank $\mathbf{CB}$.

2.3.1 Output Angle

The relationship between the input angle θ of the driving crank and the angle ψ of the driven crank is obtained from the condition that $\mathbf{A}$ and $\mathbf{B}$ remain a fixed distance apart throughout the motion of the linkage. Since $h = |\mathbf{B} - \mathbf{A}|$ is constant, we have the constraint equation

$$(\mathbf{B} - \mathbf{A}) \cdot (\mathbf{B} - \mathbf{A}) - h^2 = 0. \qquad (2.47)$$

The coordinates of $\mathbf{A}$ and $\mathbf{B}$ in F are given by

$$\mathbf{A} = \begin{Bmatrix} a\cos\theta \\ a\sin\theta \end{Bmatrix} \quad \text{and} \quad \mathbf{B} = \begin{Bmatrix} g + b\cos\psi \\ b\sin\psi \end{Bmatrix}. \qquad (2.48)$$

Substitute these coordinates into (2.47) to obtain

$$b^2 + g^2 + 2gb\cos\psi + a^2 - 2(a\cos\theta(g + b\cos\psi) + ab\sin\theta\sin\psi) - h^2 = 0. \qquad (2.49)$$

Gathering the coefficients of $\cos\psi$ and $\sin\psi$, we obtain the constraint equation for the 4R chain as

$$A(\theta)\cos\psi + B(\theta)\sin\psi = C(\theta), \qquad (2.50)$$

where

$$A(\theta) = 2ab\cos\theta - 2gb,$$
$$B(\theta) = 2ab\sin\theta,$$
$$C(\theta) = g^2 + b^2 + a^2 - h^2 - 2ag\cos\theta. \qquad (2.51)$$

The solution to this equation is

$$\psi(\theta) = \arctan\left(\frac{B}{A}\right) \pm \arccos\left(\frac{C}{\sqrt{A^2 + B^2}}\right). \qquad (2.52)$$

Equations of the form (2.50) arise many times in the analysis of linkages, so we present its solution in Appendix A for easy reference, see (A.1).

Notice that there are two angles ψ for each angle θ. This arises because the moving pivot **B** of the output crank can be assembled above or below the diagonal joining the moving pivot **A** of the input crank to the fixed pivot **C** of the output crank. The angle $\delta = \arctan(B/A)$ defines the location of this diagonal, and $\varepsilon = \arccos(C/\sqrt{A^2 + B^2})$ is the angle above and below this diagonal that locates the output crank.

The argument of the arccosine function must be in the range -1 to $+1$, which places a solvability constraint on the coefficients A, B, and C. Specifically, for a solution to exist we must have

$$A^2 + B^2 - C^2 \geq 0. \qquad (2.53)$$

If this constraint is not satisfied, then the linkage cannot be assembled for the specified input crank angle θ.

2.3.2 Coupler Angle

Let ϕ denote the angle of the coupler measured about **A** relative to the segment **OA**, so $\theta + \phi$ measures the angle to **AB** from the x-axis of F. The coordinates of **B** can also be defined in terms of ϕ as

$$\mathbf{B} = \left\{ \begin{array}{c} a\cos\theta + h\cos(\theta + \phi) \\ a\sin\theta + h\sin(\theta + \phi) \end{array} \right\}. \qquad (2.54)$$

Equating the two forms for **B**, we obtain the loop equations of the four-bar linkage

$$a\cos\theta + h\cos(\theta + \phi) = g + b\cos\psi,$$
$$a\sin\theta + h\sin(\theta + \phi) = b\sin\psi. \qquad (2.55)$$

For a given value of the drive crank θ, determine ψ using (2.52) then $\cos(\theta + \phi)$ and $\sin(\theta + \phi)$ are given by

$$\cos(\theta + \phi) = \frac{g + b\cos\psi - a\cos\theta}{h} \quad \text{and} \quad \sin(\theta + \phi) = \frac{b\sin\psi - a\sin\theta}{h}. \quad (2.56)$$

Thus, the value of the coupler angle is obtained as

$$\phi = \arctan\left(\frac{b\sin\psi - a\sin\theta}{g + b\cos\psi - a\cos\theta}\right) - \theta. \quad (2.57)$$

Notice that a unique value for ϕ is associated with each of the two solutions for the output angle ψ.

2.3.2.1 An Alternative Derivation

It is useful here to present a direct calculation of the coupler angle ϕ associated with a given crank angle θ. The derivation is identical to that above for the output angle. However, our standard frame is now F', positioned with its origin at $\mathbf{A}$ and its x-axis along the vector $\mathbf{O} - \mathbf{A}$. In this coordinate frame, the pivots $\mathbf{B}$ and $\mathbf{C}$ have the coordinates

$$F'\mathbf{B} = \begin{Bmatrix} h\cos(\phi - \pi) \\ h\sin(\phi - \pi) \end{Bmatrix} \quad \text{and} \quad F'\mathbf{C} = \begin{Bmatrix} a + g\cos(\pi - \theta) \\ g\sin(\pi - \theta) \end{Bmatrix}. \quad (2.58)$$

The constraint $(\mathbf{B} - \mathbf{C}) \cdot (\mathbf{B} - \mathbf{C}) = b^2$ yields the equation

$$A(\theta)\cos\phi + B(\theta)\sin\phi = C(\theta), \quad (2.59)$$

where

$$A(\theta) = 2ah - 2gh\cos\theta,$$
$$B(\theta) = 2gh\sin\theta,$$
$$C(\theta) = b^2 - a^2 - g^2 - h^2 + 2ag\cos\theta. \quad (2.60)$$

This equation is solved in exactly the same way as before (A.1). It results in two values for ϕ for each crank angle θ. The output angle ψ associated with each of these coupler angles can be determined from the loop equations of the linkage written for $\mathbf{C}$ in F'.

   This equation for the coupler angle is used in solutions for four and five position synthesis of a planar 4R linkage.

2.3.3 Transmission Angle

The angle ζ between the coupler and the driven crank at $\mathbf{B}$ is called the *transmission angle* of the linkage. If the only external loads on the linkage are torques on the input and output cranks, then the forces $\mathbf{F_A}$ and $\mathbf{F_B}$ acting on the coupler at the moving pivots must oppose each other along the line $\mathbf{AB}$, Figure 2.6. Thus, the force $\mathbf{F_B}$ is directed at the angle ζ relative to the driven crank, and $\sin \zeta$ measures the component of $\mathbf{F_B}$ that is transmitted as useful output torque. The $\cos \zeta$ component is absorbed as a reaction force at the fixed pivot of the driven crank.

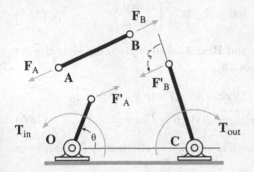

Fig. 2.6 The coupler is a two-force member connecting the input and output cranks.

To determine ζ in terms of θ, equate the cosine laws for the diagonal $d = |\mathbf{A} - \mathbf{C}|$ for the triangles $\triangle \mathbf{COA}$ and $\triangle \mathbf{ABC}$. Since ζ is the exterior angle at $\mathbf{B}$, we have

$$d^2 = g^2 + a^2 - 2ag\cos\theta = h^2 + b^2 + 2bh\cos\zeta. \qquad (2.61)$$

The result is

$$\cos\zeta = \frac{g^2 + a^2 - h^2 - b^2 - 2ag\cos\theta}{2bh}. \qquad (2.62)$$

2.3.4 Coupler Curves

As a linkage moves, points in the coupler trace curves in the fixed frame. The parameterized equation of this curve is obtained from the kinematics equations of the driving RR chain. Let $\mathbf{x} = (x, y)^T$ be the coordinates of a coupler point in the frame M located at $\mathbf{A}$ with its x-axis along $\mathbf{AB}$. The coordinates $\mathbf{X} = (X, Y)^T$ in F are given by the matrix equation

$$\begin{Bmatrix} X(\theta) \\ Y(\theta) \\ 1 \end{Bmatrix} = \begin{bmatrix} \cos(\theta + \phi) & -\sin(\theta + \phi) & a\cos\theta \\ \sin(\theta + \phi) & \cos(\theta + \phi) & a\sin\theta \\ 0 & 0 & 1 \end{bmatrix} \begin{Bmatrix} x \\ y \\ 1 \end{Bmatrix}. \qquad (2.63)$$

The coupler angle ϕ is a function of θ, thus the coupler curve is parametrized by the crank angle θ.

The algebraic equation for this curve, eliminating θ, is obtained by defining the coordinates of $\mathbf{X}$ from two points of view. Let the coupler triangle $\triangle \mathbf{XAB}$ (Figure

2.7) have lengths r and s given by

$$r = |\mathbf{X} - \mathbf{A}| = \sqrt{x^2 + y^2} \quad \text{and} \quad s = |\mathbf{X} - \mathbf{B}| = \sqrt{(x-h)^2 + y^2}. \qquad (2.64)$$

If λ is the angle to $\mathbf{AX}$ in F, and μ is the angle to $\mathbf{BX}$, then we have

$$\mathbf{X} - \mathbf{A} = \begin{Bmatrix} r\cos\lambda \\ r\sin\lambda \end{Bmatrix} \quad \text{and} \quad \mathbf{X} - \mathbf{B} = \begin{Bmatrix} s\cos\mu \\ s\sin\mu \end{Bmatrix}. \qquad (2.65)$$

Rearrange these equations to isolate $\mathbf{A}$ and $\mathbf{B}$, and substitute into the identities $\mathbf{A} \cdot \mathbf{A} = a^2$ and $(\mathbf{B} - \mathbf{C}) \cdot (\mathbf{B} - \mathbf{C}) = b^2$ to obtain

$$X^2 + Y^2 - 2Xr\cos\lambda - 2Yr\sin\lambda + r^2 = a^2,$$
$$X^2 + Y^2 - 2Xs\cos\mu - 2Ys\sin\mu + s^2 - 2gs\cos\mu - 2Xg + g^2 = b^2. \qquad (2.66)$$

The algebraic equation of the coupler curve is obtained by eliminating λ and μ from these two equations.

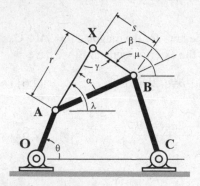

Fig. 2.7 The trajectory of a point in the floating link is known as a coupler curve of the 4R chain.

First, note that if α is the interior angle of the coupler triangle $\triangle\mathbf{XAB}$ at $\mathbf{A}$, then $\lambda = \alpha + \theta + \phi$. Similarly, if β is the exterior angle of this triangle at $\mathbf{B}$, then $\mu = \beta + \theta + \phi$, or equivalently

$$\mu - \lambda = \beta - \alpha. \qquad (2.67)$$

The angle $\gamma = \beta - \alpha$ is the interior angle of the coupler triangle at $\mathbf{X}$ given by the cosine law as

$$\cos\gamma = \frac{r^2 + s^2 - h^2}{2rs}. \qquad (2.68)$$

Substitute $\mu = \lambda + \gamma$ into (2.66) and rearrange these equations to obtain

$$A_1 \cos \lambda + B_1 \sin \lambda = C_1,$$
$$A_2 \cos \lambda + B_2 \sin \lambda = C_2, \tag{2.69}$$

where

$$A_1 = 2rX, \qquad\qquad A_2 = 2s(\cos \gamma(X-g) + Y \sin \gamma),$$
$$B_1 = 2rY, \qquad\qquad B_2 = 2s(-\sin \gamma(X-g) + Y \cos \gamma),$$
$$C_1 = X^2 + Y^2 + r^2 - a^2, \qquad C_2 = (X-g)^2 + Y^2 - b^2 + s^2. \tag{2.70}$$

Eliminate λ in these equations by solving linearly for $x = \cos \lambda$ and $y = \sin \lambda$. Then impose the condition $x^2 + y^2 = 1$. The result is

$$(C_1 B_2 - C_2 B_1)^2 + (A_2 C_1 - A_1 C_2)^2 - (A_1 B_2 - A_2 B_1)^2 = 0. \tag{2.71}$$

Notice that A_i and B_i are linear in the coordinates X and Y, and C_i are quadratic. Therefore, this equation defines a curve of degree six. See Hunt [50] for a detailed study of this curve, known as a tricircular sextic, and a description of its properties.

2.4 Range of Movement

2.4.1 Limits on the Input Crank Angle

The formula that defines the output angle ψ for a given input angle θ has a solution only when $A^2 + B^2 - C^2 \geq 0$. When this condition is violated, the crank is rotated to a positioned in which the mechanism cannot be assembled. The maximum and minimum values for θ are obtained by setting this condition to zero, which yields the quadratic equation in $\cos \theta$

$$4a^2 g^2 \cos^2 \theta - 4ag(g^2 + a^2 - h^2 - b^2) \cos \theta$$
$$+ \left((g^2 + a^2) - (h+b)^2\right)\left((g^2 + a^2) - (h-b)^2\right) = 0. \tag{2.72}$$

The roots of this equation are the upper and lower limiting angles θ_{max} and θ_{min} that define the range of movement of the input crank,

$$\cos \theta_{min} = \frac{(g^2 + a^2) - (h-b)^2}{2ag}, \quad \cos \theta_{max} = \frac{(g^2 + a^2) - (h+b)^2}{2ag}. \tag{2.73}$$

These equations are the cosine laws for the two ways that the triangle $\triangle AOC$ can be formed with the coupler **AB** aligned with the output crank **CB**, Figure 2.8. This alignment is what limits rotation of the input crank. The cosine function does not distinguish between $\pm\theta$ so there are actually two limits for each case $\pm\theta_{min}$ and $\pm\theta_{max}$ above and below **OC**.

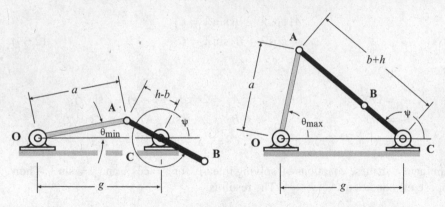

Fig. 2.8 The angles $\theta_{\min}$ and $\theta_{\max}$ are the limits to the range of movement of the input link.

The arccosine function yields a real angle only if its argument is between -1 and 1. This provides conditions that determine whether these crank limits exist.

2.4.1.1 The Lower Limit: $\theta_{\min}$

If $\theta_{\min}$ does not exist, then the crank has no lower limit to its movement and it rotated through $\theta = 0$ to reach negative values below the segment **OC**. Thus, $\cos\theta_{\min} > 1$ is the condition that there is no lower limit to the input crank rotation, that is,

$$\frac{(g^2 + a^2) - (h - b)^2}{2ag} > 1. \tag{2.74}$$

This simplifies to yield

$$(g - a)^2 - (h - b)^2 > 0. \tag{2.75}$$

Factor the difference of two squares to obtain

$$(g - a + h - b)(g - a - h + b) > 0,$$
$$T_1 T_2 > 0, \tag{2.76}$$

where

$$T_1 = g - a + h - b \quad \text{and} \quad T_2 = g - a - h + b. \tag{2.77}$$

Thus, T_1 and T_2 must both be either positive or negative for there to be no lower limit to the rotation of the input crank.

2.4.1.2 The Upper Limit: θ_{max}

If θ_{max} does not exist, then the crank has no upper limit to its movement and it will be able to rotate through $\theta = \pi$. Thus, $\cos\theta_{max} < -1$, or

$$\frac{(g^2 + a^2) - (h+b)^2}{2ag} < -1, \tag{2.78}$$

is the condition that this limit does not exist. This inequality simplifies to

$$(h+b)^2 - (g+a)^2 > 0, \tag{2.79}$$

which factors to become

$$(h+b-g-a)(h+b+g+a) > 0,$$
$$T_3 T_4 > 0, \tag{2.80}$$

where

$$T_3 = h+b-g-a, \quad \text{and} \quad T_4 = h+b+g+a. \tag{2.81}$$

The sum of the link lengths T_4 is always positive. Therefore, the condition that there is no upper limit to the rotation of the input crank is $T_3 > 0$.

2.4.1.3 Input Crank Types

We can now identify four types of movement available to the input crank of a 4R linkage:

1. **A crank:** $T_1 T_2 > 0$ and $T_3 > 0$, in which case neither θ_{min} nor θ_{max} exists, and the input crank can fully rotate.
2. **A 0-rocker:** $T_1 T_2 > 0$ and $T_3 < 0$, for which θ_{max} exists but not θ_{min}, and the input crank rocks through $\theta = 0$ between the values $\pm\theta_{max}$.
3. **A π-rocker:** $T_1 T_2 < 0$ and $T_3 > 0$, which means that θ_{min} exists but not θ_{max}, and the input crank rocks through $\theta = \pi$ between the values $\pm\theta_{min}$.
4. **A rocker:** $T_1 T_2 < 0$ and $T_3 < 0$, which means that both upper and lower limiting angles exist, and the crank cannot pass through either 0 or π. Instead, it rocks in one of two separate ranges: (i) $\theta_{min} \leq \theta \leq \theta_{max}$, or (ii) $-\theta_{max} \leq \theta \leq -\theta_{min}$.

2.4.2 Limits on the Output Crank Angle

The range of movement of the output crank can be analyzed in the same way. The limiting positions occur when the input crank **OA** and coupler **AB** become aligned, see Figure 2.9. The limits ψ_{min} and ψ_{max} are defined by the equations

$$\cos \psi_{min} = \frac{(h+a)^2 - (g^2 + b^2)}{2bg}, \cos \psi_{max} = \frac{(h-a)^2 - (g^2 + b^2)}{2bg}. \quad (2.82)$$

Note that in this case ψ is the exterior angle, which changes the sign of the cosine term in the cosine law formula.

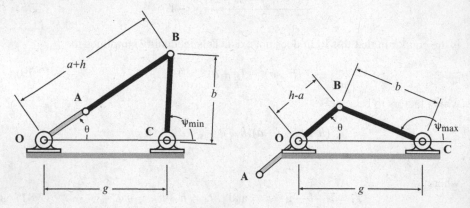

Fig. 2.9 The angles ψ_{min} and ψ_{max} are the limits to the range of motion of the output link.

Examining the existence of solutions to arccosine in (2.82), we find that the condition for no lower limit ψ_{min} is

$$(h+a-g-b)(h+a+g+b) > 0,$$
$$(-T_2)(T_4) > 0, \quad (2.83)$$

where T_2 and T_4 are the same parameters used above for the input crank. Because T_4 is always greater than zero, the condition that there be no lower limit to the range of movement of the output crank is $T_2 < 0$.

Similarly, in order for there to be no upper limit ψ_{max}, we have

$$(g-b-h-a)(g-b+h-a) > 0,$$
$$(-T_3)(T_1) > 0. \quad (2.84)$$

Again, the parameters T_3 and T_1 are the same as were defined above and there is no upper limit to the movement of the output crank when $T_1 T_3 < 0$.

2.4.2.1 Output Crank Types

We can now identify four types of movement available to the output crank of a four-bar linkage:

1. **A rocker:** $T_1 T_3 > 0$ and $T_2 > 0$. In this case both limits ψ_{min} and ψ_{max} exist, and the crank cannot not pass through either 0 or π. Instead, it rocks in one of two separate ranges: (i) $\psi_{min} \leq \psi \leq \psi_{max}$, or (ii) $-\psi_{max} \leq \psi \leq -\psi_{min}$.
2. **A 0-rocker:** $T_1 T_3 < 0$ and $T_2 > 0$, for which ψ_{max} exists but not ψ_{min}, and the output crank rocks through $\psi = 0$ between the values $\pm \psi_{max}$.
3. **A π-rocker:** $T_1 T_3 > 0$ and $T_2 < 0$, which means that ψ_{min} exists but not ψ_{max}, and the output crank rocks through $\psi = \pi$ between the values $\pm \psi_{min}$.
4. **A crank:** $T_1 T_3 < 0$ and $T_2 < 0$. Then neither limit ψ_{min} nor ψ_{max} exists, and the output crank can fully rotate.

2.4.3 The Classification of Planar 4R Linkages

A planar 4R linkage is classified by the movement of its input and output cranks. For example, a crank-rocker has a fully rotatable input link, and an output link that rocks between two limits. On the other hand a rocker-crank has an input link that rocks and an output link that fully rotates. The combinations of positive and negative signs for the parameters T_1, T_2, T_3 identify eight basic linkage types. These parameters can take zero values as well, in which case the linkage folds.

2.4.3.1 The Eight Basic Types

The link lengths a, b, g, and h for a 4R chain define the three parameters T_1, T_2, and T_3. Our classification scheme requires only the signs of these parameters, therefore we assemble the array $(\operatorname{sgn} T_1, \operatorname{sgn} T_2, \operatorname{sgn} T_3)$. The eight possible arrays identify the eight basic types of 4R linkages.

We separate the linkage types into two general classes depending upon the sign of the product $T_1 T_2 T_3$. If $T_1 T_2 T_3 > 0$ then the linkage is called *Grashof*; otherwise, it is called *nonGrashof*. There are four Grashof and four non-Grashof linkage types.

We consider the Grashof cases first:

1. $(+,+,+)$: Because $T_1 T_2 > 0$ and $T_3 > 0$ the input link can fully rotate. Similarly, because $T_1 T_3 > 0$ and $T_2 > 0$ the output link is a rocker with two output ranges. This linkage is a *crank-rocker*.
2. $(+,-,-)$: With $T_1 T_2 < 0$ and $T_3 < 0$ the input is a rocker, and with $T_1 T_3 < 0$ and $T_2 < 0$ the output is a crank. This defines the *rocker-crank* linkage.
3. $(-,-,+)$: In this case, $T_1 T_2 > 0$ and $T_3 > 0$, so the input link is a crank, and $T_1 T_3 < 0$ and $T_2 < 0$, which means that the output link is also a crank. This defines the *double-crank* linkage.
4. $(-,+,-)$: $T_1 T_2 < 0$ and $T_3 < 0$ define the input as a rocker, and with $T_1 T_3 > 0$ and $T_2 > 0$ the output is also a rocker. This defines the *Grashof double-rocker* linkage type.

Now consider the nonGrashof cases:

5. $(-,-,-)$: Here we have $T_1 T_2 > 0$ and $T_3 < 0$, and the input link rocks through the value $\theta = 0$. With $T_1 T_3 > 0$ and $T_2 < 0$, the output link rocks through the value $\psi = 0$. This type of linkage is termed a 00 *double-rocker*.

6. $(+,+,-)$: In this case, the input rocks through $\theta = 0$. However, with $T_1 T_3 < 0$ and $T_2 > 0$ the output rocks through $\psi = \pi$. This linkage is called a 0π *double-rocker*.

7. $(+,-,+)$: With $T_1 T_2 > 0$ and $T_3 > 0$ the input link rocks through π, and because $T_1 T_3 < 0$ and $T_2 < 0$ the output link rocks through 0. This is the $\pi 0$ *double-rocker*.

8. $(-,+,+)$: Finally, the input again rocks through π, as does the output, defining the $\pi\pi$ *double-rocker*.

The parameters associated with these linkages are listed in Table 2.1.

Table 2.1 Basic Planar 4R Linkage types

	Linkage type	T_1	T_2	T_3
1	Crank-rocker	+	+	+
2	Rocker-crank	+	−	−
3	Double-crank	−	−	+
4	Grashof double-rocker	−	+	−
5	00 double-rocker	−	−	−
6	0π double-rocker	+	+	−
7	$\pi 0$ double-rocker	+	−	+
8	$\pi\pi$ double-rocker	−	+	+

2.4.4 Grashof Linkages

If a linkage is to be used in a continuous operation, the input crank should be able to fully rotate so that it can be driven by a rotating power source. A study of the configurations of a 4R linkage lead Grashof to conclude that, for a shortest link of length s and longest link of length l, the shortest link will fully rotate if

$$s + l < p + q, \qquad (2.85)$$

where p and q are the lengths of the other two links. This is known as Grashof's criterion and linkages that have a rotatable crank are called *Grashof linkages*.

There are four linkage types that satisfy Grashof's criterion. If the input or output link is the shortest, then we have the crank-rocker or the rocker-crank, respectively. If the ground link is the shortest, then both the input and output links will fully rotate relative to the ground; this is the double-crank linkage. Finally, if the floating link is the shortest link, then the input and output links are rockers; this is the Grashof double-rocker. By examining Table 2.1, it is easy to see that these four linkage types satisfy the condition

$$T_1 T_2 T_3 > 0, \tag{2.86}$$

which can be shown to be equivalent to Grashof's criterion.

The rockers of each of the Grashof linkage types are distinguished by the fact that both upper and lower limits exist. The means that they have two distinct angular ranges of movement, one in the upper half plane and one in the lower relative to the fixed link. If the linkage is assembled so that the rocker is in one angular range, then it cannot reach the other range without disassembly. Thus, Grashof linkages have two distinct sets of configurations called *assemblies*. The linkage can move between the configurations in only one of these assemblies and cannot reach the others.

2.4.5 Folding Linkages

If any one of the parameters T_1, T_2, or T_3 has the value zero, then the linkage can take a configuration in which all four joints lie on a line. The linkage is said to fold.

If we consider the positive, negative, and zero values for the array (T_1, T_2, T_3), then we find that there are 27 types of planar 4R linkages, 19 of which fold. Furthermore, the number of parameters T_i that are zero defines the number of folding configurations of the linkage. It is often useful to have a linkage fold. However, while it is easy to drive the linkage into a folded configuration, it may be difficult to get it out of the this configuration.

Consider, for example, the *parallelogram linkage* defined by $a = b$ and $g = h$. This linkage has $T_2 = 0$ and $T_3 = 0$. Thus, it has two folding configurations, which occur for the input crank angles of $\theta = 0, \pi$. Another doubly folding example is the *kite linkage* with $g = a$ and $h = b$, which yields $T_1 = 0$ and $T_2 = 0$. This linkage folds when $\theta = 0$, at which point the output link can freely rotate because the joints **A** and **C** coincide; the second folding position occurs when $\psi = \pi$.

There is one triply folding case, the *rhombus linkage*, for which $a = b = g = h$. This linkage is a combination of the parallelogram and kite linkages. It folds at the two configurations $\theta = 0, \pi$ like the parallelogram. When $\theta = 0$ the output link is free to rotate because the joints **A** and **C** coincide as with the kite linkage. The third folding configuration occurs when $\psi = \pi$. The linkage can also reach this configuration with $\theta = \pi$, in which case it is the input crank that can freely rotate because **O** and **B** coincide.

Linkages that have small values for any of the parameters T_i are termed *near folding*. These linkages have configurations in which the joints can lie close to a line. In nearly folded configurations the transmission angle of the linkage is near 0 or π, and the output crank is difficult to move using the input crank.

2.5 Velocity Analysis

The velocity analysis of a linkage determines the angular rates of the various joint parameters as a function of the configuration of the linkage and the input joint rate. This analysis can be used in combination with the principle of virtual work to provide an important technique for determining the force and torque transmission properties of these systems.

2.5.1 Velocity of a Point in a Moving Link

Points $\mathbf{x}$ fixed in a moving link M trace *trajectories* $\mathbf{X}(t) = [T(t)]\mathbf{x}$ in the fixed frame F. The velocity of a point along its trajectory is the time derivative of its coordinate vector, that is, $\mathbf{V} = \dot{\mathbf{X}}$. Of importance to us is the relationship between this velocity and the movement of the linkage as a whole.

The usual convention in velocity calculations is to focus on the trajectories in F rather than coordinates in M of the moving point. For this reason, the trajectory of the of a segment $\mathbf{AB}$ fixed in M is defined by coordinates $\mathbf{A}$ and $\mathbf{B}$ measured in F. A general trajectory $\mathbf{X}$ of M has the property that $r = |\mathbf{X} - \mathbf{A}|$ and $\alpha = \angle\mathbf{BAX}$ are constants, because the three points $\mathbf{A}$, $\mathbf{B}$, and $\mathbf{X}$ are part of the same link.

Let the orientation of M be defined by the angle θ of $\mathbf{AB}$ measured relative to the x-axis of F. Then we can determine the relative position vector $\mathbf{X} - \mathbf{A}$ as

$$\mathbf{X} - \mathbf{A} = \begin{Bmatrix} r\cos(\theta + \alpha) \\ r\sin(\theta + \alpha) \end{Bmatrix}. \tag{2.87}$$

The time derivative of this vector yields

$$\mathbf{V} = \dot{\mathbf{X}} = \dot{\mathbf{A}} + \dot{\theta}[J](\mathbf{X} - \mathbf{A}), \quad \text{where} \quad [J] = \begin{bmatrix} 0 & -1 \\ 1 & 0 \end{bmatrix}. \tag{2.88}$$

This defines the velocity of a general point in terms of the velocity of a reference point and the rate of rotation of the body. We now show that $\dot{\theta}[J]$ is directly related to the angular velocity of this link.

Each body in a planar linkage rotates about an axis that is perpendicular to the plane of movement. Denote this direction by $\vec{k} = (0,0,1)^T$. Then the usual vector cross product yields $\vec{k} \times \vec{\imath} = \vec{\jmath}$ and $\vec{k} \times \vec{\jmath} = -\vec{\imath}$, where $\vec{\imath}$ and $\vec{\jmath}$ are unit vectors along the x- and y-axes of the fixed frame F. We now define the *angular velocity* of the link $\mathbf{AB}$ to be the vector

$$\mathbf{w}_{AB} = \dot{\theta}\vec{k}. \tag{2.89}$$

where θ defines the orientation of $\mathbf{AB}$ in F. Notice that for any vector $\mathbf{y}$, the angular velocity vector satisfies the identity

$$\mathbf{w}_{AB} \times \mathbf{y} = \dot{\theta}[J]\mathbf{y}. \tag{2.90}$$

This allows us to write equation (2.88) for the velocity of a point in the form

$$\mathbf{V} = \dot{\mathbf{A}} + \mathbf{w}_{AB} \times (\mathbf{X} - \mathbf{A}). \tag{2.91}$$

The angular velocity vector can be viewed as an operator that computes the component of velocity that arises from the rotation of the link.

Notice that if the link $\mathbf{AB}$ simply rotates about $\mathbf{A}$, then $\dot{\mathbf{A}} = 0$, and we have

$$\mathbf{V} = \mathbf{w}_{AB} \times (\mathbf{X} - \mathbf{A}). \tag{2.92}$$

In this case, the velocity of a point $\mathbf{X}$ is directed $90°$ to the line joining it to $\mathbf{A}$.

2.5.2 Instant Center

It interesting to note that there is a point in every moving link that has zero velocity. This point $\mathbf{I}$, known as the *instant center*, is found by setting (2.91) to zero, that is,

$$\dot{\mathbf{A}} + \mathbf{w}_{AB} \times (\mathbf{I} - \mathbf{A}) = 0. \tag{2.93}$$

Take the cross product by $\mathbf{w}_{AB}$ and solve for $\mathbf{I}$ to obtain

$$\mathbf{I} - \mathbf{A} = \frac{\mathbf{w}_{AB} \times \dot{\mathbf{A}}}{\mathbf{w}_{AB} \cdot \mathbf{w}_{AB}}. \tag{2.94}$$

This calculation uses the vector identity $\mathbf{a} \times (\mathbf{b} \times \mathbf{c}) = \mathbf{b}(\mathbf{a} \cdot \mathbf{c}) - \mathbf{c}(\mathbf{a} \cdot \mathbf{b})$.

The geometric meaning of $\mathbf{I}$ is found by substituting $\dot{\mathbf{A}}$ from (2.93) into (2.91) to obtain

$$\mathbf{V} = \mathbf{w}_{AB} \times (\mathbf{X} - \mathbf{I}). \tag{2.95}$$

Compare this to (2.92) to see that the distribution of velocities in this link, at this instant, is the same as is generated by a rotation about the instant center $\mathbf{I}$.

2.6 Velocity Analysis of an RR Chain

The kinematics equations of an open chain define the set of positions it can reach as a function of its joint parameters. If each of these parameters is given as a function of time, then we obtain a curve in its workspace that defines the trajectory of the end-link. The time derivative of the kinematics equations defines the velocity along this trajectory.

The 3×3 transform $[D] = [A, \mathbf{P}]$ for planar open chains separate into a 2×2 rotation matrix $[A]$ and a 2×1 translation vector $\mathbf{P}$. The translation vector $\mathbf{P}$ is defined by the position of reference point in the end-link. The orientation ϕ of this link is the sum of the relative rotation angles at each joint. Thus, the velocity of any trajectory

$\mathbf{X}(t)$ of any point in the end-link is given by the equation

$$\mathbf{V} = \dot{\mathbf{P}} + \mathbf{w}_M \times (\mathbf{X} - \mathbf{P}),\tag{2.96}$$

where $\mathbf{w}_M$ is, the angular velocity vector of the end-link, is the sum of the angular velocities at each joint.

2.6.1 The Jacobian

For an RR chain let $\theta_1(t)$ and $\theta_2(t)$ be the rotation angles at each joint. We can compute

$$\mathbf{w} = (\dot{\theta}_1 + \dot{\theta}_2)\vec{k} \quad \text{and} \quad \dot{\mathbf{P}} = \left\{ \begin{array}{c} -a\dot{\theta}_1 \sin\theta_1 \\ a\dot{\theta}_1 \cos\theta_1 \end{array} \right\}.\tag{2.97}$$

These two equations are considered to define the velocity of the end-link as a whole, as opposed the velocity of trajectories traced by its points. The $\dot{\mathbf{P}}$ and $\dot{\phi}$ are assembled into a vector and (2.97) is written in the matrix form

$$\left\{ \begin{array}{c} \dot{\mathbf{P}} \\ \dot{\phi} \end{array} \right\} = \left[\begin{array}{cc} -a\sin\theta_1 & 0 \\ a\cos\theta_1 & 0 \\ 1 & 1 \end{array} \right] \left\{ \begin{array}{c} \dot{\theta}_1 \\ \dot{\theta}_2 \end{array} \right\}\tag{2.98}$$

In robotics literature this 3×2 matrix is called the *Jacobian* of the RR chain. Given a desired velocity for the end effector, we can solve these equations to obtain the required joint rates $\dot{\theta}_1$ and $\dot{\theta}_2$.

Another form of the Jacobian is obtained by considering the trajectory of a general point $\mathbf{x}$ in M given by

$$\mathbf{X}(t) = [D(t)]\mathbf{x} = [Z(\theta_1)][X(a)][Z(\theta_2)]\mathbf{x}.\tag{2.99}$$

Compute the velocity $\mathbf{V} = \dot{\mathbf{X}}$ then eliminate the M-frame coordinates using $\mathbf{x} = [D^{-1}]\mathbf{X}$. The result is

$$\mathbf{V} = \dot{\mathbf{X}} = [\dot{D}][D^{-1}]\mathbf{X}.\tag{2.100}$$

The matrix $[S] = [\dot{D}][D^{-1}]$ can be viewed as operating on a trajectory $\mathbf{X}(t)$ to compute its velocity $\mathbf{V}$.

For the RR chain, we use (2.11) and compute

$$[S] = \left[\begin{array}{ccc} 0 & -\dot{\theta}_1 - \dot{\theta}_2 & a\dot{\theta}_2 \sin\theta_1 \\ \dot{\theta}_1 + \dot{\theta}_2 & 0 & -a\dot{\theta}_2 \cos\theta_1 \\ 0 & 0 & 0 \end{array} \right].\tag{2.101}$$

The upper left 2×2 matrix is $(\dot{\theta}_1 + \dot{\theta}_2)[J]$, which is the matrix that we have associated with the angular velocity of the end-link. The third column is the velocity of the trajectory $\mathbf{Y}(t)$ that passes through the origin of F. Assemble this into the matrix equation

$$\left\{ \begin{matrix} \mathbf{v} \\ \phi \end{matrix} \right\} = \begin{bmatrix} 0 & a\sin\theta_1 \\ 0 & -a\cos\theta_1 \\ 1 & 1 \end{bmatrix} \left\{ \begin{matrix} \dot\theta_1 \\ \dot\theta_2 \end{matrix} \right\}. \tag{2.102}$$

This alternative form for the Jacobian is the focus of our study in the last chapter of this text.

2.6.2 The Centrode

We now compute the instant center for the instantaneous movement of the end-link of the RR chain. From (2.94) we have

$$\mathbf{I} = \mathbf{P} + \frac{\mathbf{w}_M \times \dot{\mathbf{P}}}{\mathbf{w}_M \cdot \mathbf{w}_M}. \tag{2.103}$$

Simplify this equation and introduce the vector $\vec{e} = (\cos\theta_1, \sin\theta_1)^T$ to obtain

$$\mathbf{I} = a\left(\frac{\dot\theta_2}{\dot\theta_1 + \dot\theta_2} \right) \vec{e}. \tag{2.104}$$

This shows that the instant center lies on the line through the two revolute joints of the RR chain.

Equation (2.103) defines an instant center for every configuration of the chain. If the joint angles θ_1 and θ_2 are related by a function $f(\theta_1, \theta_2) = 0$, then the set of instant centers forms a curve in F known as the *centrode*.

Other planar open chains can be analyzed in the same way to relate the velocity of the end-link to the rate of change of the configuration parameters.

2.7 Velocity Analysis of a Slider-Crank

If the input crank to an RRRP linkage is driven at the rate $\dot\theta$, then we can determine the rotation rate $\dot\phi$ of the coupler link, and the linear velocity $\dot{s}$ of the slider using the *velocity loop equations*. These equations are obtained by computing the time derivative of the loop equations (2.32)

$$\dot\theta \left\{ \begin{matrix} -r\sin\theta \\ r\cos\theta \end{matrix} \right\} + (\dot\theta + \dot\phi) \left\{ \begin{matrix} -L\sin(\theta + \phi) \\ L\cos(\theta + \phi) \end{matrix} \right\} = \dot{s} \left\{ \begin{matrix} 0 \\ 1 \end{matrix} \right\}. \tag{2.105}$$

Rearrange the terms so this equation takes the form

$$\begin{bmatrix} 0 & L\sin(\theta + \phi) \\ 1 & -L\cos(\theta + \phi) \end{bmatrix} \left\{ \begin{matrix} \dot{s} \\ \dot\phi \end{matrix} \right\} = \dot\theta \left\{ \begin{matrix} -r\sin\theta - L\sin(\theta + \phi) \\ r\cos\theta + L\cos(\theta + \phi) \end{matrix} \right\}. \tag{2.106}$$

Notice that to solve these equations we must have previously determined the parameters ϕ and s. Then Cramer's rule yields

$$\frac{\dot{s}}{\dot{\theta}} = \frac{r\sin\phi}{\sin(\theta+\phi)} \quad \text{and} \quad \frac{\dot{\phi}}{\dot{\theta}} = \frac{r\sin\theta+L\sin(\theta+\phi)}{L\sin(\theta+\phi)}. \qquad (2.107)$$

It is useful to note that we can obtain the slider velocity directly from the constraint (2.23) and avoid the need to determine ϕ or $\dot{\phi}$. To do this, simply compute the time derivative of this constraint equation to obtain

$$\dot{s}(s - r\sin\theta) - \dot{\theta}r(s\cos\theta - e\sin\theta) = 0. \qquad (2.108)$$

The result is

$$\frac{\dot{s}}{\dot{\theta}} = \frac{r(s\cos\theta - e\sin\theta)}{s - r\sin\theta}. \qquad (2.109)$$

This equation is used to determine the mechanical advantage of this linkage.

2.7.1 Mechanical Advantage

The ratio of the static force generated at the slider to the input torque applied at the crank is known as *mechanical advantage*. We compute this using the *principle of virtual work* which states that the work done by input forces and torques must equal the work done by output forces and torques during a virtual displacement. For the RRRP linkage, we assume the weight of each link and the friction in each joint are negligible compared to the applied forces and torques. In which case, the principle of virtual work requires that the work done by the torque applied to the input crank must equal the work done by the slider on an external load during a virtual displacement.

A *virtual displacement* is a small movement of the system over which the applied forces and torques are considered to be constant. This small movement is easily defined in terms of the velocities of each link. The angular velocity of the input crank $\dot{\theta}$ acting over a small increment of time δt generates the virtual crank displacement $\delta\theta = \dot{\theta}\delta t$. Similarly, a virtual displacement of the slider is $\delta s = \dot{s}\delta t$.

Let the input torque to the crank be $\mathbf{T} = F_{in}p\vec{k}$, where F_{in} is a force applied perpendicular to the link at a distance p along it. Then the virtual work of this torque is $F_{in}p\delta\theta$. The virtual work done by the slider as it applies a force $\mathbf{F} = F_{out}\vec{j}$ along its direction of movement is $F\delta s$. Thus, we have

$$F_{out}\dot{s}\delta t = F_{in}p\dot{\theta}\delta t. \qquad (2.110)$$

Because the virtual time increment δt is not zero, we can equate coefficients to obtain the relationship

$$\frac{F_{out}}{F_{in}} = \frac{\dot{\theta}}{\dot{s}} = \frac{p(s - r\sin\theta)}{r(s\cos\theta - e\sin\theta)}. \tag{2.111}$$

This ratio defines the mechanical advantage of the slider-crank. Notice that it depends on the configuration of the linkage, as well as the ratio p/r, which defines the point of application of the input force F_{in}.

This formula has an interesting geometric interpretation. Let $\mathbf{I}$ be the intersection of the line through the crank $\mathbf{OA}$ and the line $y = s$ that locates the slider. We now determine the distances $|\mathbf{IA}|$ and $|\mathbf{IB}|$ from the geometry of the linkage and obtain

$$|\mathbf{IA}| = r - \frac{s}{\sin\theta} \quad \text{and} \quad |\mathbf{IB}| = e - \left(\frac{s}{\sin\theta}\right)\cos\theta. \tag{2.112}$$

Thus, we find that the mechanical advantage for the slider-crank can be written as

$$\frac{F_{out}}{F_{in}} = \frac{p|\mathbf{IA}|}{r|\mathbf{IB}|}. \tag{2.113}$$

Decreasing the distance $|\mathbf{IB}|$ increases the mechanical advantage. In fact, as $\tan\theta$ approaches s/e, the extreme position of the slider, the distance $|\mathbf{IB}|$ approaches zero, and the mechanical advantage becomes very large.

2.8 Velocity Analysis of a 4R Chain

The velocity loop equations of the 4R chain are obtained by computing the time derivative of the loop equations (2.55) to obtain

$$\dot{\theta}\begin{Bmatrix} -a\sin\theta \\ a\cos\theta \end{Bmatrix} + (\dot{\theta} + \dot{\phi})\begin{Bmatrix} -h\sin(\theta + \phi) \\ h\cos(\theta + \phi) \end{Bmatrix} = \dot{\psi}\begin{Bmatrix} -b\sin\psi \\ b\cos\psi \end{Bmatrix}. \tag{2.114}$$

For a given input angular velocity $\dot{\theta}$, these equations are linear in the angular velocities $\dot{\phi}$ and $\dot{\psi}$ of the coupler and output link. Notice that we must have already determined the angles ϕ and ψ. Assemble these equations into the matrix equation

$$\begin{bmatrix} -b\sin\psi & h\sin(\theta + \phi) \\ b\cos\psi & -h\cos(\theta + \phi) \end{bmatrix}\begin{Bmatrix} \dot{\psi} \\ \dot{\phi} \end{Bmatrix} = \dot{\theta}\begin{Bmatrix} -a\sin\theta - h\sin(\theta + \phi) \\ a\cos\theta + h\cos(\theta + \phi) \end{Bmatrix}. \tag{2.115}$$

Solve this equation to determine the velocity ratios

$$\frac{\dot{\psi}}{\dot{\theta}} = \frac{a\sin\phi}{b\sin(\theta + \phi - \psi)} \quad \text{and} \quad \frac{\dot{\phi}}{\dot{\theta}} = \frac{a\sin(\psi - \theta) - h\sin(\theta + \phi - \psi)}{h\sin(\theta + \phi - \psi)}. \tag{2.116}$$

2.8.1 Output Velocity Ratio

We now examine the velocity properties of the 4R chain in terms of the angular velocity vectors $\mathbf{w_O} = \dot{\theta}\vec{k}$ and $\mathbf{w_C} = \dot{\psi}\vec{k}$, where $\vec{\imath}$, $\vec{\jmath}$, and $\vec{k}$ are the unit vectors along the coordinate axes of a three dimensional frame. The time derivative of the constraint equation $(\mathbf{B} - \mathbf{A}) \cdot (\mathbf{B} - \mathbf{A}) = b^2$ yields

$$(\dot{\mathbf{B}} - \dot{\mathbf{A}}) \cdot (\mathbf{B} - \mathbf{A}) = 0. \tag{2.117}$$

Since $\dot{\mathbf{B}} = \mathbf{w_C} \times (\mathbf{B} - \mathbf{C})$ and $\dot{\mathbf{A}} = \mathbf{w_O} \times \mathbf{A}$, this can be written as

$$\left(\dot{\psi}\vec{k} \times (\mathbf{B} - \mathbf{C}) - \dot{\theta}\vec{k} \times \mathbf{A}\right) \cdot (\mathbf{B} - \mathbf{A}) = 0. \tag{2.118}$$

Interchange the dot and cross operations and expand this equation to obtain

$$\dot{\psi}\vec{k} \cdot \mathbf{B} \times (\mathbf{B} - \mathbf{A}) - \dot{\theta}\vec{k} \cdot \mathbf{A} \times (\mathbf{B} - \mathbf{A}) - \dot{\psi}\vec{k} \cdot \mathbf{C} \times (\mathbf{B} - \mathbf{A}) = 0. \tag{2.119}$$

Notice that the cross products $\mathbf{A} \times (\mathbf{B} - \mathbf{A})$ and $\mathbf{B} \times (\mathbf{B} - \mathbf{A})$ are equal, and, in fact, any point on the line $L_{AB}: \mathbf{Y}(t) = \mathbf{A} + t(\mathbf{B} - \mathbf{A})$ yields the same result. In particular, both $\mathbf{A}$ and $\mathbf{B}$ can be replaced by the point $\mathbf{I} = r\vec{\imath}$, which is the intersection of L_{AB} with the x-axis. Since $\mathbf{C} = g\vec{\imath}$, this equation takes the form

$$\vec{k} \cdot \left(\dot{\psi}(r - g) - \dot{\theta}r\right)\vec{\imath} \times (\mathbf{B} - \mathbf{A}) = 0. \tag{2.120}$$

It is now easy to see that the output velocity ratio is given by

$$\frac{\dot{\psi}}{\dot{\theta}} = \frac{-r}{g - r}. \tag{2.121}$$

The distance r to the point $\mathbf{I}$ along the x-axis can be computed by finding the parameter t that satisfies the relation $\vec{\jmath} \cdot (\mathbf{A} + t(\mathbf{B} - \mathbf{A})) = 0$. Substitute this into $r = \vec{\imath} \cdot (\mathbf{A} + t(\mathbf{B} - \mathbf{A}))$ to obtain

$$r = \frac{ab\sin(\theta - \phi)}{b\sin\phi - a\sin\theta}. \tag{2.122}$$

Notice that the velocity ratio between the output and input links can be viewed as instantaneously equivalent to the speed ratio between two gears in contact at the instant center $\mathbf{I}$ that have the radii $g - r$ and r respectively, see Figure 2.10.

2.8.2 Coupler Velocity Ratio

A similar relationship for the coupler velocity ratio is obtained by computing the velocity of $\mathbf{B}$ in the fixed frame using vector operations. Combining this with the fact that $\dot{\mathbf{B}} \cdot (\mathbf{B} - \mathbf{C}) = 0$, we obtain a geometric representation of the velocity ratio.

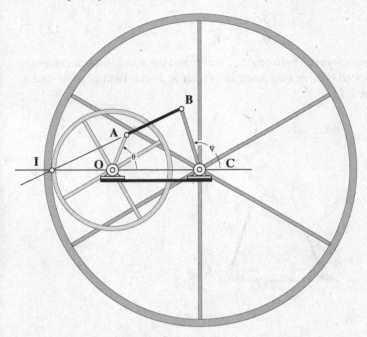

Fig. 2.10 The angular velocities of the input and output links are instantaneously equivalent to gears in contact at the instant center **I**.

The coupler has the angular velocity $\mathbf{w}_A = (\dot{\phi} + \dot{\theta})\vec{k}$, so the velocity of **B** is given by

$$\dot{\mathbf{B}} = \dot{\mathbf{A}} + \mathbf{w}_A \times (\mathbf{B} - \mathbf{A}). \tag{2.123}$$

Since $\dot{\mathbf{A}} = \dot{\theta}\vec{k} \times \mathbf{A}$, this equation becomes

$$\dot{\mathbf{B}} = \dot{\theta}\vec{k} \times \mathbf{A} + (\dot{\phi} + \dot{\theta})\vec{k} \times (\mathbf{B} - \mathbf{A}) = \dot{\theta}\vec{k} \times \mathbf{B} + \dot{\phi}\vec{k} \times (\mathbf{B} - \mathbf{A}). \tag{2.124}$$

Substitute this into the condition $\dot{\mathbf{B}} \cdot (\mathbf{B} - \mathbf{C}) = 0$ to obtain

$$\left(\dot{\theta}\vec{k} \times \mathbf{B} + \dot{\phi}\vec{k} \times (\mathbf{B} - \mathbf{A})\right) \cdot (\mathbf{B} - \mathbf{C}) = 0. \tag{2.125}$$

Notice that **B** can be replaced by any point on the line L_{CB}: $\mathbf{Y}(t) = \mathbf{B} + t(\mathbf{B} - \mathbf{C})$ because $\vec{k} \times t(\mathbf{B} - \mathbf{C}) \cdot (\mathbf{B} - \mathbf{C}) = 0$. In particular, consider the point **J** that is the intersection of L_{CB} with the line L_{OA} that joins **O** and **A**. Let $\vec{e}$ be the unit vector in the direction **A**, so $\mathbf{A} = a\vec{e}$ and $\mathbf{J} = r\vec{e}$. Substitute this into (2.125) and obtain

$$\left(\dot{\theta}r + \dot{\phi}(r - a)\right)\vec{k} \times \vec{e} \cdot (\mathbf{B} - \mathbf{C}) = 0. \tag{2.126}$$

For this equation to be zero, the coupler velocity ratio must satisfy the relation

$$\frac{\dot{\phi}}{\dot{\theta}} = \frac{r}{a - r}. \tag{2.127}$$

Thus, the angular velocity ratio between the coupler and input link is instantaneously equivalent to the speed ratio of two gears in contact at $\mathbf{J}$ with radii of $a - r$ and a, respectively, Figure 2.11.

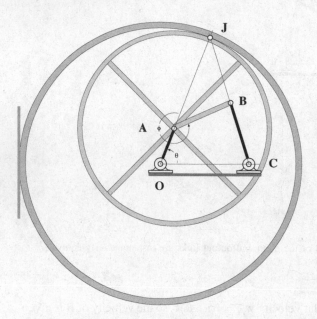

Fig. 2.11 The angular velocities of the input and coupler links are instantaneously equivalent to gears in contact at the instant center $\mathbf{J}$.

The value of r defining $\mathbf{J}$ along the line L_{OA} is obtained by solving for t such that $\vec{e}^{\perp} \cdot (\mathbf{B} + t(\mathbf{B} - \mathbf{C})) = 0$. Note that $\vec{e}^{\perp} = (-\sin\theta, \cos\theta)^T$ is the unit vector perpendicular to $\vec{e}$. Then, substitute the result into the relation $r = \vec{e} \cdot (\mathbf{B} + t(\mathbf{B} - \mathbf{A}))$ to compute r.

2.8.3 Kennedy's Theorem

We have seen that the output velocity ratio of a 4R linkage can be viewed as generated instantaneously by a pair of gears connecting the input and output links. The points in contact along the pitch circles of the two gears, $\mathbf{Q}$ on the input link and $\mathbf{P}$ on the output link, must have the same velocity, that is, $\dot{\mathbf{Q}} = \dot{\mathbf{P}}$. The point in F that coincides with these two points is the instant center $\mathbf{I}$. We now show that an instant center with this property exists for any two links moving relative to a ground frame F.

Consider the movement of two independent links and let their instant centers in F be $\mathbf{O}$ and $\mathbf{C}$. We now ask whether there are points, $\mathbf{Q}$ on one and $\mathbf{P}$ on the other, that have both the same coordinates $\mathbf{I} = (X,Y)^T$ in F and the same velocity.

Let $g = |\mathbf{C} - \mathbf{O}|$ be the distance between the instant centers, and let $\mathbf{Q} - \mathbf{O} = (r\cos\theta, r\sin\theta)^T$ and $\mathbf{P} - \mathbf{C} = (g + k\cos\psi, k\sin\psi)^T$ be the relative vectors locating $\mathbf{Q}$ and $\mathbf{P}$. The velocities of these points are

$$\dot{\mathbf{Q}} = \dot{\theta}\left\{\begin{matrix} -r\sin\theta \\ r\cos\theta \end{matrix}\right\}, \quad \dot{\mathbf{P}} = \dot{\psi}\left\{\begin{matrix} -k\sin\psi \\ k\cos\psi \end{matrix}\right\}. \tag{2.128}$$

If $\mathbf{Q} = \mathbf{P} = (X,Y)^T$, then $Y = r\sin\theta = k\sin\psi$, and to have the same velocity

$$\dot{\mathbf{P}} - \dot{\mathbf{Q}} = \left\{\begin{matrix} -\dot{\psi}Y + \dot{\theta}Y \\ \dot{\psi}(r\cos\theta - g) - \dot{\theta}r\cos\theta \end{matrix}\right\} = \left\{\begin{matrix} 0 \\ 0 \end{matrix}\right\}. \tag{2.129}$$

The first component of this equation shows that $Y = 0$. Set $\theta = \psi = 0$ and let r and k take positive and negative values so that $X = r = k + g$. From the second component of (2.129) we find that r must satisfy the equation

$$\dot{\psi}(r - g) - \dot{\theta}r = 0,$$

that is,

$$\frac{\dot{\psi}}{\dot{\theta}} = \frac{r}{r - g}. \tag{2.130}$$

Thus, $\mathbf{I} = (r,0)^T$ is the desired instant center. The fact that the point $\mathbf{I}$ must lie on the line joining the $\mathbf{O}$ and $\mathbf{C}$ is known as *Kennedy's theorem*.

2.8.4 Mechanical Advantage in a 4R Linkage

The relationship between an applied input torque and the torque generated at the output crank of a 4R linkage is easily determined by considering the equivalent set of gears and the principle of virtual work. From the velocity ratios determined above, we have the virtual displacement of the output crank defined as

$$\delta\psi = \dot{\psi}\delta t = \dot{\theta}\frac{r}{r - g}\delta t, \tag{2.131}$$

where r is the distance to the instant center $\mathbf{I}$ from the fixed pivot $\mathbf{O}$ of the drive crank.

The virtual work of the input torque $\mathbf{T_O} = T_O\vec{k}$ is $T_O\delta\theta$. Similarly, the virtual work of the output torque, $\mathbf{T_C} = T_C\vec{k}$, is $T_C\delta\psi$. From the principle of virtual work we obtain

$$T_O\delta\theta = T_C\delta\psi, \quad \text{or} \quad T_O\dot{\theta}\delta t = T_C\frac{r}{r - g}\dot{\theta}\delta t. \tag{2.132}$$

The virtual time increment δt is nonzero, so we equate coefficients to obtain the relationship

$$\frac{T_C}{T_O} = \frac{r - g}{r}. \tag{2.133}$$

Note that the distance r has a sign associated with its direction along the x-axis from **O**. The conclusion is that the torque ratio of a linkage is the inverse of its velocity ratio, which is exactly the torque ratio of the equivalent gear train. Notice that the value of this torque ratio changes with the configuration of the linkage.

Let the input torque be generated by a *couple*, which is a pair of forces in opposite directions but of equal magnitude F_O separated by the perpendicular distance a, so $M_O = aF_O$. Similarly, let the output torque result in a couple with magnitude $M_C = bF_C$. Then, the ratio of output force F_C to input force F_O is obtained from (2.133) as

$$\frac{F_C}{F_O} = \frac{a}{b}\left(\frac{r - g}{r}\right). \tag{2.134}$$

This ratio is called the *mechanical advantage* of the linkage. For a given set of dimensions a and b the mechanical advantage is directly proportional to the velocity ratio of the input and output links.

2.9 Analysis of Multiloop Planar Linkages

The study of mulitiloop planar linkages is covered in detail in a later chapter. Here we summarize how to use Dixon's determinant to solve for the configuration angles that define the assemblies of the linkage (Wampler [148]).

2.9.1 Complex Loop Equations

Given an input angle θ_0, the vector equations for each independent loop L of a planar linkage can be written as

$$\mathscr{F}_k : \alpha_k + \sum_{j=1}^{2L} \beta_{kj}\cos\theta_j + \sum_{j=1}^{2L} \gamma_{kj}\sin\theta_j = 0, \quad k = 1,\ldots,2L. \tag{2.135}$$

Here, α_k, β_{kj}, and γ_{kj} are real quantities that depend on the dimensions of the links, and θ_j denotes the rotation angle of link j. Because the input angle θ_0 is known, we absorb it into the coefficients of α_k, and these $2L$ equations are solved for $2L$ configuration angles θ_j.

Combine the loop equations (2.135) into a single complex equation by using complex vectors $\mathbf{x} = x + iy$, where $i^2 = -1$, rather than vectors $\mathbf{x} = (x, y)$. Introduce the complex vector $\Theta_j = e^{i\theta_j}$, $j = 1,\ldots,2L$, so the $2L$ loop equations become

$$\mathcal{C}_k: \quad c_{k0} + \sum_{j=1}^{2L} c_{k,j}\Theta_j = 0, \quad k = 1,\ldots,L. \tag{2.136}$$

We obtain a second set of loop equations by computing the complex conjugates

$$\mathcal{C}_k^\star: \quad c_{k0}^\star + \sum_{j=1}^{2L} c_{k,j}^\star \Theta_j^{-1} = 0, \quad k = 1,\ldots,L. \tag{2.137}$$

These equations combine to provide a set of $2L$ complex equations for $2L$ complex configuration angles Θ_j that are solved using the Dixon determinant.

2.9.2 The Dixon Determinant

Suppress the joint angle Θ_{2L}, so we have $2L$ complex equations in $2L-1$ variables Θ_j, labeled $\mathcal{C}_k$ and $\mathcal{C}_k^\star$. These equations form the first row of the Dixon determinant. The second row consists of the same functions but with the variable θ_1 replaced by α_1. Similarly, row three has Θ_1 and Θ_2 replaced by α_1 and α_2. This continues for the remaining rows in the determinant, so that we obtain

$$\Delta(\Theta_1, \Theta_2, \ldots, \Theta_{2L-1})$$
$$= \begin{vmatrix} \mathcal{C}_1(\Theta_1, \Theta_2, \ldots, \Theta_{2L-1}) & \cdots & \mathcal{C}_L^\star(\Theta_1, \Theta_2, \ldots, \Theta_{2L-1}) \\ \mathcal{C}_1(\alpha_1, \Theta_2, \ldots, \Theta_{2L-1}) & \cdots & \mathcal{C}_L^\star(\alpha_1, \Theta_2, \ldots, \Theta_{2L-1}) \\ \vdots & & \\ \mathcal{C}_1(\alpha_1, \alpha_2, \ldots, \alpha_{2L-1}) & \cdots & \mathcal{C}_L^\star(\alpha_1, \alpha_2, \ldots, \alpha_{2L-1}) \end{vmatrix}. \tag{2.138}$$

his determinant is zero when $\Theta_1, \Theta_2, \ldots, \Theta_{2L-1}$ satisfy the loop equations, because the elements in the first row become zero.

Insight into the structure of the determinant Δ is obtained by noting that the complex equations for each loop k have the form

$$\mathcal{C}_k: c_{k0} + c_{k,2L}x + \sum_{j=1}^{2L-1} c_{k,j}\Theta_j \quad \text{and} \quad \mathcal{C}_k^\star: c_{k0}^\star + c_{k,2L}^\star x^{-1} + \sum_{j=1}^{2L-1} c_{k,j}^\star \Theta_j^{-1}, \tag{2.139}$$

where x denotes the suppressed variable Θ_{2L}. These equations maintain this form when α_j replaces Θ_j. Thus, we can row reduce Δ by subtracting the second row from the first row, then the third from the second, the fourth from the third, and so on to obtain

$$\begin{vmatrix} c_{1,1}(\Theta_1 - \alpha_1) & c_{1,1}^*(\Theta_1^{-1} - \alpha_1^{-1}) & \cdots & c_{L,1}^*(\Theta_1^{-1} - \alpha_1^{-1}) \\ c_{1,2}(\Theta_2 - \alpha_2) & c_{1,2}^*(\Theta_2^{-1} - \alpha_2^{-1}) & \cdots & c_{L,2}^*(\Theta_2^{-1} - \alpha_2^{-1}) \\ \vdots & \vdots & & \vdots \\ c_{1,2L-1}(\Theta_{2L-1} - \alpha_{2L-1}) & c_{1,2L-1}^*(\Theta_{2L-1}^{-1} - \alpha_{2L-1}^{-1}) & \cdots & c_{L,2L-1}^*(\Theta_{2L-1}^{-1} - \alpha_{2L-1}^{-1}) \\ \mathscr{C}_1(\alpha_1,\alpha_2,\ldots,\alpha_{2L-1}) & \mathscr{C}_1^*(\alpha_1,\alpha_2,\ldots,\alpha_{2L-1}) & \cdots & \mathscr{C}_L^*(\alpha_1,\alpha_2,\ldots,\alpha_{2L-1}) \end{vmatrix} . \qquad (2.140)$$

Because $\Theta_j - \alpha_j = -\Theta_j \alpha_j (\Theta_j^{-1} - \alpha_j^{-1})$, we can divide out these extraneous roots $(\Theta_j^{-1} - \alpha_j^{-1})$ to define the determinant

$$\delta = \frac{\Delta(\Theta_1, \Theta_2, \ldots, \Theta_{2L-1})}{\prod_{k=1}^{2L-1}(\Theta_k^{-1} - \alpha_k^{-1})}, \qquad (2.141)$$

that is,

$$\delta = \begin{vmatrix} -c_{1,1}\Theta_1\alpha_1 & c_{1,1}^* & \cdots & c_{L,1}^* \\ -c_{1,2}\Theta_2\alpha_2 & c_{1,2}^* & \cdots & c_{L,2}^* \\ \vdots & & & \\ -c_{1,2L-1}\Theta_{2L-1}\alpha_{2L-1} & c_{1,2L-1}^* & \cdots & c_{L,2L-1}^* \\ \mathscr{C}_1(\alpha_1,\alpha_2,\ldots,\alpha_{2L-1}) & \mathscr{C}_1^*(\alpha_1,\alpha_2,\ldots,\alpha_{2L-1}) & \cdots & \mathscr{C}_L^*(\alpha_1,\alpha_2,\ldots,\alpha_{2L-1}) \end{vmatrix} . \qquad (2.142)$$

This determinant is a polynomial of the form

$$\delta = \mathbf{a}^T[W]\mathbf{t} = 0, \qquad (2.143)$$

where $\mathbf{a} = (\mathbf{a}_1, \mathbf{a}_2)^T$ contains the monomials formed from α_j, $\mathbf{t} = (\mathbf{t}_1, \mathbf{t}_2)^T$ are formed from monomials of Θ_j, and $[W]$ is the $2L \times 2L$ matrix is given by

$$[W] = \begin{bmatrix} D_1 x + D_2 & A^T \\ A & s(D_1^* x^{-1} + D_2^*) \end{bmatrix}, \qquad (2.144)$$

where D_1 and D_2 are diagonal matrices and the elements of A obey the relations $a_{ij} = sa_{ji}^\star$ and $s = (-1)^{L-1}$.

The vectors of monomials $\mathbf{a} = (\mathbf{a}_1, \mathbf{a}_2)^T$ and $\mathbf{t} = (\mathbf{t}_1, \mathbf{t}_2)^T$ in (2.143) are generated as follows. Starting with $\mathbf{a}$, find all combinations $\binom{2L-1}{L-1}$ of distinct variables of degree $L-1$ from the set $(\alpha_1, \alpha_2, \ldots, \alpha_{2L-1})^T$. Assemble these into the vector $\mathbf{a}_1$, and then form $\mathbf{a}_2$ using the complement of degree L corresponding to each monomial in $\mathbf{a}_1$. The vector $\mathbf{t}$ is obtained in the same way.

Values Θ_j that satisfy the loop equations (2.136) and (2.137) also yield $\delta = 0$ for arbitrary values of the auxiliary variables α_j. Thus, solutions for these loop equations must also satisfy the matrix equation

$$[W]\mathbf{t} = 0. \qquad (2.145)$$

This equation has nonzero solutions only if $\det[W] = 0$. Expand this determinant to obtain a polynomial in $x = \Theta_{2L}$.

The structure of $[W]$ yields

$$[W]\mathbf{t} = \left[\begin{pmatrix} D_1 & 0 \\ A & -D_2^* \end{pmatrix} x - \begin{pmatrix} -D_2 & -A^T \\ 0 & D_1^* \end{pmatrix} \right] \mathbf{t} = [Mx - N]\mathbf{t} = 0. \qquad (2.146)$$

Notice that the values of x that satisfy $\det[W] = 0$ are the roots of the characteristic polynomial $p(x) = \det(Mx - N)$ of the generalized eigenvalue problem

$$N\mathbf{t} = xM\mathbf{t}. \qquad (2.147)$$

Each value of $x = \Theta_{2L}$ has an associated eigenvector $\mathbf{t}$, which yields the values of the remaining joint angles Θ_j, $j = 1, 2, \ldots, 2L - 1$.

2.9.3 Tangent Sorting

For each value of the input angle Θ_0, the solution of the Dixon determinant yields multiple roots for the configuration angles $\vec{\Theta} = (\Theta_1, \ldots, \Theta_{2L})$. Each root defines one way that the linkage can be assembled. For example, a six-bar linkage can have as many as six values for Θ_j, or six assemblies, for each input angle. An eight-bar linkage can have as many as 20 assemblies. Presented here is a way to sort these roots to define these assemblies.

Assume that for the kth value of the input angle Θ_0 we calculate the configuration angles $\vec{\Theta}_i^k, i = 1, \ldots, m$ that define m assemblies of the linkage. When we increment Θ_0 and solve the loop equations, we to obtain $\vec{\Theta}_i^{k+1}$, and our goal is to sort these roots so to match those of $\vec{\Theta}_i^k$.

Use Newton's method to approximate the loop equations by computing the Jacobian $[\nabla \mathscr{C}]$, in order to obtain

$$[\nabla \mathscr{C}(\vec{\Theta}_i^k)](\vec{\Psi} - \vec{\Theta}_i^k) = 0, \qquad (2.148)$$

where $\vec{\Psi}$ is an approximation to $\vec{\Theta}_i^{k+1}$ in the assembly that we seek. Solve these equations to obtain $\vec{\Psi}$, and select from the available roots $\vec{\Theta}_i^{k+1}$ the one that is closest to $\vec{\Psi}$, in order to match the assemblies.

The configurations traced by one assembly of a multiloop linkage is called a *circuit*. The solution of (2.148) identifies a value Ψ on the tangent to the ith circuit through $\vec{\Theta}_i^k$. This allows rapid and exact calculation of the configuration angles for each assembly of a multiloop planar linkage for a range of values of the input angle.

2.10 Summary

This chapter presented the position and velocity analysis of planar open chains and the closed chain slider-crank and four-bar linkages. Conditions on the existence of solutions to the input-output equations for the closed chains provide a classification scheme for these devices based on the range of movement of their cranks. The velocity analysis of these systems lead to the introduction of instant centers and Kennedy's theorem, which can be used to compute the mechanical advantage of the linkage. The position analysis of multiloop planar linkages using complex number coordinates and the Dixon determinant elimination method were also discussed.

2.11 References

The position and velocity analysis of planar open chains follows the approach used in robotics as found in Craig [15] and Paul [92]. The analysis of planar linkages including the study of accelerations and dynamic forces can be found in many textbooks. See, for example, Waldron and Kinzel [144], Erdman and Sandor [30], Mabie and Reinholtz [70], Mallik et al. [71], and Shigley and Uicker [114]. For further study of the dynamics of these systems see Krishnaprasad and Yang [58] and Sreenath et al. [125]. The strategy used to classify planar slider-crank and 4R linkages follows Murray and Larochelle [87]. The closed form kinematic analysis of planar multiloop mechanism was presented by Wampler [148]. Also see Nielsen and Roth [90], and Wampler [147].

Exercises

1. Consider the PRRP *elliptic trammel* formed from two PR chains connected so that the directions of the two sliders are at right angles in the ground link. Derive the coupler angle ϕ as a function of the input slider translation s and show that a general coupler curve is an ellipse.
2. *Oldham's coupling* is an RPPR linkage with the directions of the two sliders oriented at a right angle to form the coupler link. Analyze this linkage to determine the output crank angle ψ as a function of the input angle θ.
3. The *Scotch yoke* mechanism is an RRPP linkage with the ways of the sliders at right angles. Analyze this linkage to determine the output slide s as a function of the input θ.
4. Derive the algebraic equation of the coupler curve of an RRRP linkage and show that it is a quartic curve.
5. Analyze (i) Watt's linkage, (ii) Robert's linkage, (iii) Chebyshev's linkage, and determine the coupler angle ϕ as a function of the input crank angle θ. Generate the coupler curve of the point that traces an approximately straight line.
6. Derive the algebraic form of the 4R coupler curve and show that its highest degree terms are $(x^2 + y^2)^3$, and that those of fifth and fourth degree contain the factors $(x^2 + y^2)^2$ and $x^2 + y^2$, respectively. These features identify this curve to as a *tricircular* sextic.

7. Select a coupler point **X** on a 4R linkage **OABC**. Construct the triangle $\triangle$**OCY** that is similar to the coupler triangle $\triangle$**ABX**. Show that the coupler curve traced by **X** has a double point at its intersections with the circle circumscribing $\triangle$**OCY**.

8. Show that the centrode for an RR chain becomes a circle when $\dot{\phi} = \mu\dot{\theta}$ and μ is constant. Because this curve lies in the fixed frame F it is called the *fixed centrode*.

9. Transform the coordinates of the centrode of an RR chain to the moving frame M by $\mathbf{m} = [T^{-1}]\mathbf{I}$. This defines a curve known as the *moving centrode*. Show that for $\dot{\phi} = \mu\dot{\theta}$ and constant μ, the moving centrode is a circle.

Chapter 3
Graphical Synthesis in the Plane

The geometric principles that are fundamental to linkage design can be found in simple and efficient graphical constructions for RR and PR chains. As the floating link of one of these chains reaches various task positions, points in it define sets of corresponding points in the fixed frame. The design problem is to find a circle for the RR chain and a straight line for the PR chain that passes through these corresponding points. This usually results in multiple RR and PR chains that can be combined to form slider-crank or four-bar linkages.

3.1 Displacement of a Planar Body

The positions of moving body M can be specified by simply drawing the body in various locations in the background plane F. The features needed to define a position M in F are a point D and a directed line segment $\vec{e}$, which we denote by $M : (D, \vec{e})$.

Consider two positions of M, that is, $M_1 : (\vec{e}_1, D^1)$ and $M_2 : (\vec{e}_2, D^2)$. The *displacement* of M from the first position to the second position consists of the *translation* of D from D^1 to D^2 along the segment $D^1 D^2 = \vec{d}_{12}$ combined with a *rotation* that carries $\vec{e}_1$ to $\vec{e}_2$ defined by the angle ϕ_{12} measured in a counterclockwise sense. See Figures 3.1 and 3.2.

3.1.1 The Pole of a Displacement

The displacement between two positions M_1 and M_2 can be achieved by a pure rotation about a special point P_{12} called the *pole* of the displacement. The pole has the property that it is located in exactly the same place in the ground frame F, whether the moving body is in position M_1 or M_2.

To find P_{12}, we first consider the point of intersection C of the line through D^1 along $\vec{e}_1$ and the line through D^2 along $\vec{e}_2$. Notice that the angle between these

J.M. McCarthy and G.S. Soh, *Geometric Design of Linkages*, Interdisciplinary Applied Mathematics 11, DOI 10.1007/978-1-4419-7892-9_3,
© Springer Science+Business Media, LLC 2011

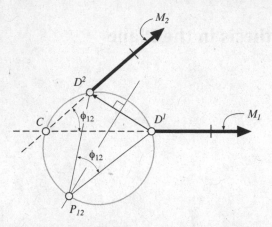

Fig. 3.1 The location of the displacement pole P_{12} when CD^1 and CD^2 have the same signs relative to $\vec{e}_1$ and $\vec{e}_2$.

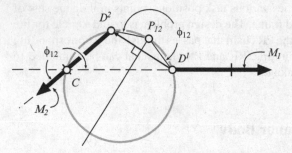

Fig. 3.2 The location of the pole P_{12} when CD^1 and CD^2 have opposite signs relative to $\vec{e}_1$ and $\vec{e}_2$.

lines about C is the rotation angle ϕ_{12} of the displacement. Now construct the circle through the vertices of the triangle $\triangle CD^1D^2$; see Appendix B for this construction. Let the angle $\angle D^1CD^2$ at C measured counterclockwise be denoted by κ. C is said to view the segment D^1D^2 in the angle κ. Notice that every point on the circular arc that contains C views D^1D^2 in the same angle. The points on the opposite circular arc view D^1D^2 in the angle $\kappa + \pi$.

If the segments CD^1 and CD^2 are directed along $\vec{e}_1$ and $\vec{e}_2$, respectively, then $\kappa = \phi_{12}$, see Figure 3.1. This remains true if both segments CD^1 and CD^2 are directed opposite to $\vec{e}_1$ and $\vec{e}_2$. On the other hand, if CD^1 is directed along $\vec{e}_1$ and CD^2 is directed along $-\vec{e}_2$, as shown in Figure 3.2, or equivalently $CD^1 = -\vec{e}_1$ and $CD^2 = \vec{e}_2$, then $\phi_{12} = \kappa + \pi$.

Now consider the perpendicular bisector $V = (D^1D^2)^\perp$ which intersects the two arcs of this circle in points P and P'. Let P be the intersection with the arc that contains C. If $\kappa = \phi_{12}$, then P is the pole P_{12} of the displacement. If $\phi_{12} = \kappa + \pi$, then P' is the pole P_{12}. In either case, a rotation about P_{12} by ϕ_{12} carries the segment $P_{12}D^1$ into $P_{12}D^2$, and the line $\vec{e}_1$ into $\vec{e}_2$.

3.1.2 Determining the Position of a Point

For any point Q in a moving body let Q^1 and Q^2 denote its corresponding points in positions M_1 and M_2. We can use Q^1 and the pole P_{12} to construct the point Q^2, using the fact that the displacement from M_1 to M_2 is a rotation about P_{12}. Join Q^1 to P_{12} to define the line L^1, and then duplicate the angle $\angle D^1 P_{12} D^2$ from this line around P_{12} to define L^2; see Appendix B. Simply measure the distance $P_{12}Q^1$ along the line L^2 in order to define Q^2, Figure 3.3. This procedure is reversed to determine Q^1 from Q^2, in which case we construct the angle $-\phi_{12}$ about P_{12} from L^2 to define L^1.

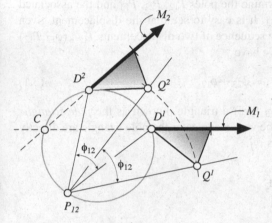

Fig. 3.3 A general point Q^1 is displaced by a pure rotation about P_{12} by the angle ϕ_{12} to the point Q^2.

An alternative technique for rotating Q^1 about P_{12} to Q^2 uses the fact that a rotation by the angle ϕ_{12} can be achieved by a pair of reflections through lines separated by the angle $\phi_{12}/2$. Consider the line L joining D^1 to the pole P_{12} and the perpendicular bisector $V = (D^1 D^2)^\perp$. These lines lie at the angle $\phi_{12}/2$ around P_{12}. Two reflections, one through L and a second through V, will move Q^1 to its displaced position Q^2.

3.1.3 An Alternative Construction for the Pole

We now use the fact that every point in the displaced body must move in a circle about the pole P_{12} in order to define an alternative construction for this point. The two points D^1 and D^2 and the perpendicular bisector $V = (D^1 D^2)^\perp$ are already known. (See Appendix B for the construction of a perpendicular bisector.) Now choose another point A in the moving body and identify its corresponding points A^1

and A^2 in F. The perpendicular bisector of the segment A^1A^2 must pass through the pole, which means that its intersection with V is the desired point P_{12}.

Because this must be true for any point Q in the moving body, we find that P_{12} is the intersection of the perpendicular bisectors of all segments Q^1Q^2 generated by two positions of M.

3.1.4 The Pole Triangle

If we have three positions M_i, $i = 1,2,3$, for a moving body M, we can consider the displacements in pairs and determine the poles P_{12}, P_{23}, P_{13} and the associated relative rotation angles ϕ_{12}, ϕ_{23}, ϕ_{13}. It is easy to see that the displacement given by $T_{13} : (\phi_{13}, P_{13})$ is obtained by the sequence of two displacements $T_{12} : (\phi_{12}, P_{12})$ followed by $T_{23} : (\phi_{23}, P_{23})$. Thus, we have

$$\phi_{13} = \phi_{23} + \phi_{12}. \tag{3.1}$$

The three poles P_{12}, P_{23}, and P_{13} form a triangle, known as the *pole triangle*. See Figure 3.4. We now show that the vertex angles of the pole triangle are directly related to the relative rotation angles ϕ_{ij}.

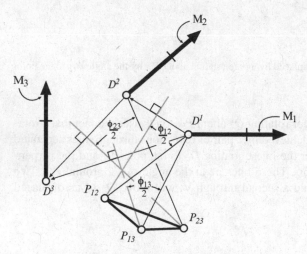

Fig. 3.4 The pole triangle associated with the three positions M_1, M_2, and M_3.

Consider the point in the moving body M that coincides with the pole P_{23} when the body is in position M_2, Figure 3.5. Notice that in position M_3 this point is fixed in place by definition of the pole of the displacement T_{23}. Now consider its location in position M_1 denoted by P_{23}^1 and called the *image pole*. P_{23}^1 moves to the location P_{23} after a rotation by ϕ_{12} about P_{12} and after a rotation by ϕ_{13} about P_{13}. Thus,

P_{23}^1 must be the reflection of P_{23} through the side $\mathsf{N}_1 = P_{12}P_{13}$ of the pole triangle. Furthermore, N_1 bisects the rotation angle ϕ_{12} as well as the rotation angle ϕ_{13}.

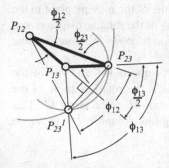

Fig. 3.5 The pole P_{23} and its image P_{23}^1 are reflections through the side $P_{12}P_{13}$ of the pole triangle.

We can distinguish two cases: (i) the pole P_{13} is to the left of the directed line segment $P_{23}P_{12}$, or (ii) P_{13} is to the right of $P_{23}P_{12}$. In the first case the sum of the angles ϕ_{12} and ϕ_{23} is less than 2π, and in the second case their sum is greater than 2π.

1. If $\phi_{12} + \phi_{23} < 2\pi$, then the interior angles of the pole triangle at P_{12} and P_{23} are the angles $\phi_{12}/2$ and $\phi_{23}/2$, and the exterior angle at P_{13} is $\phi_{13}/2 = \phi_{12}/2 + \phi_{23}/2$.
2. If $\phi_{12} + \phi_{23} > 2\pi$, then $\phi_{12}/2$ and $\phi_{23}/2$ are the exterior angles of the pole triangle at P_{12} measured from N_1 to the segment $P_{12}P_{23}$, and $\kappa = \phi_{12}/2 + \phi_{23}/2 - \pi$ is the interior angle at P_{13}. The angle $\phi_{13}/2 = \kappa + \pi$ measured counterclockwise around P_{13}.

The pole triangle provides a geometric way to determine the rotation angle and pole of a displacement T_{13}, given the rotation angles and poles of two relative displacements T_{12} and T_{23}.

3.2 The Geometry of an RR Chain

The displacement of the end-link of an RR chain is the result of a rotation first about the moving pivot W^1 followed by a rotation about the fixed pivot G. This is equivalent to the composition of rotations about the relative poles P_{12} and P_{23}, and we find that the *dyad triangle* $\triangle W^1GP_{12}$ has the same properties as the pole triangle.

3.2.1 The Dyad Triangle

Consider the displacement of the floating link of an RR chain from position M_1 to position M_2. Let W^1 and W^2 be the corresponding points of the moving pivot in the two positions, and let G be the fixed pivot. The pole P_{12} of the relative displacement of the M forms a triangle with W^1 and G, called the *dyad triangle*. We now examine the geometry of this triangle.

Let L^1 and L^2 be the lines joining G to the two positions W^1 and W^2 of the moving pivot. The angle β_{12} between these two lines is the rotation angle of the crank. Choose the direction $\vec{e}_1$ defining the orientation of M_1 to be along L^1. This allows us to identify the relative rotation of the floating link around W^2 as the angle α_{12} between L^2 and $\vec{e}_2$, see Figure 3.6.

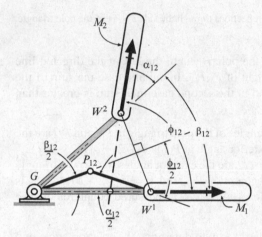

Fig. 3.6 The fixed and moving pivots G and W^1 of an RR chain and the pole P_{12} form the dyad triangle $\triangle W^1 G P_{12}$.

The relative rotation ϕ_{12} of the end-link between positions M_1 and M_2 is the sum of the relative rotations about the fixed and moving pivots, that is,

$$\phi_{12} = \beta_{12} + \alpha_{12}. \tag{3.2}$$

Now identify the point R_{12} on the perpendicular bisector $V_{12} = (W^1 W^2)^\perp$ that corresponds to the location of the moving pivot when the crank angle is $\beta_{12}/2$, that is, $\angle W^1 G R_{12} = \beta_{12}/2$. Notice that R_{12} lies on the arc of the circle through W^1 and W^2 around G. The angle measured counterclockwise from $P_{12} W^1$ to the perpendicular bisector V_{12} must be one-half the rotation of the end-link, $\phi_{12}/2$. There are two forms for the triangle $\triangle W^1 G P_{12}$ depending on the location of P_{12} relative to G along V_{12}. It lies either on the same side as the point R_{12}, or on the opposite side. We have the two cases:

1. If P_{12} is on the same side of G as R_{12}, then $\beta_{12}/2 < \phi_{12}/2$ and $\alpha_{12} + \beta_{12}$ is less than 2π. In this case the interior angles of the dyad triangle at W^1 and G are $\angle P_{12}W^1G = \alpha_{12}/2$ and $\angle W^1GP_{12} = \beta_{12}/2$, respectively, and the exterior angle at P_{12} between the segment $P_{12}W^1$ and V_{12} is $\phi_{12}/2 = \alpha_{12}/2 + \beta_{12}/2$.

2. If P_{12} is on the opposite side of G as R_{12}, then the sum $\alpha_{12} + \beta_{12}$ is greater than 2π, and $\phi_{12}/2 > \pi$. The angles $\alpha_{12}/2$ and $\beta_{12}/2$ are the exterior angles of the dyad triangle at W^1 and G, and $\kappa = \alpha_{12}/2 + \beta_{12}/2 - \pi$ is the interior angle at P_{12}. Thus, we have $\phi_{12}/2 = \kappa + \pi$ measured counterclockwise around P_{12}.

3.2.2 The Center-Point Theorem

We now examine the case of three positions M_1, M_2, and M_3 of the floating link of an RR chain. A result fundamental to Burmester's techniques for the design of these chains is that the angle $\angle P_{ij}GP_{jk}$ is directly related to the crank rotation angle β_{ik}.

We know from the geometry of the dyad triangle that $\angle W^1GP_{12}$ is either $\beta_{12}/2$ or $\beta_{12}/2 + \pi$ depending on the location of P_{12} relative to G. Similarly, for the dyad triangle $\triangle W^2GP_{23}$ we have that $\angle W^2GP_{23}$ is either $\beta_{23}/2$ or $\beta_{23}/2 + \pi$. Now, notice that $\angle W^1GP_{12} = \angle P_{12}GW^2$. Considering each of the possible cases for the angle $\angle P_{12}GP_{23} = \angle P_{12}GW^2 + \angle W^2GP_{23}$, we see that this angle must be either $\beta_{13}/2$ or $\beta_{13}/2 + \pi$. Thus, G views the segment $P_{12}P_{23}$ in either the angle $\beta_{13}/2$ or $\beta_{13}/2 + \pi$, see Figure 3.7.

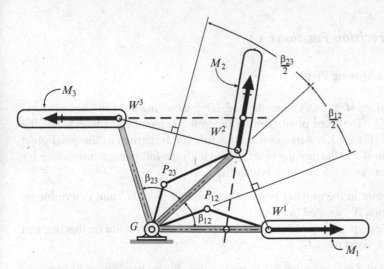

Fig. 3.7 The fixed pivot G of an RR chain views the poles P_{12} and P_{23} in the angle $\beta_{13}/2$, where β_{13} is the crank rotation from position M_1 to M_3.

This generalizes to the following result central to Burmester's theory of linkage synthesis:

Theorem 1 (The Center-Point Theorem). *The center point G of an RR chain that reaches three positions M_i, M_j, and M_k views the relative poles P_{ij} and P_{jk} in the angle $\beta_{ik}/2$ or $\beta_{ik}/2 + \pi$, where β_{ik} is the crank rotation angle from position M_i to M_k.*

Another way of saying this is that β_{ik} is the central angle of the circle circumscribing the triangle $\triangle P_{ij}GP_{jk}$ measured counterclockwise from P_{ij} to P_{jk}. For a given set of task positions the relative poles P_{ij} are known, and this theorem provides a condition on the possible locations of center points.

3.3 Finite-Position Synthesis of RR Chains

We now consider the design of RR chains that reach a specific set of task positions M_i. The positions are specified by drawing each reference point D^i and direction vector $\vec{e}^i$, $i = 1, \ldots, n$, on the background plane F.

The fixed pivot G of the RR chain is located in F and attached by a link to the moving pivot W in the moving body M. The moving pivot defines the corresponding points W^i, $i = 1, \ldots, n$, in each of the task positions. The points W^i must lie on a circle about G, because the crank connecting the G and W has a constant length. Thus, the goal of the design process is to find points in the moving body that have n corresponding positions on a circle.

3.3.1 Two Precision Positions

3.3.1.1 Select a Moving Pivot

Given two positions $M_1 : (\vec{e}_1, D^1)$ and $M_2 : (\vec{e}_2, D^2)$, a moving pivot W^1 has a second position W^2 in F. The fixed pivot G must lie on the perpendicular bisector of the segment W^1W^2, Figure 3.8. Any point on this line can be chosen as the fixed pivot of the chain with W^1 as the moving pivot. This yields the following construction for an RR chain that can reach two task positions:

1. Select any point in the moving body as the moving pivot W^1 and determine its second position W^2 located in F.
2. Construct the perpendicular bisector $V_{12} = (W^1W^2)^\perp$. Any point on this line can be used as the fixed pivot G.

For each choice of a moving pivot W^1 there is a one-dimensional set of fixed pivots G. Thus, there is a three-dimensional set of RR chains compatible with the two positions.

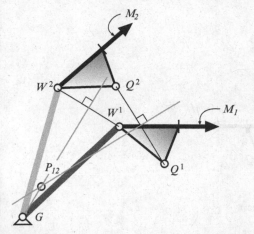

Fig. 3.8 The fixed pivot G of an RR chain lies on the perpendicular bisector of the segment W^1W^2.

3.3.1.2 Select a Fixed Pivot

Rather than select the moving pivot, it is may be preferable to select the fixed pivot G and construct the associated moving pivot. To do this, we locate the pole P_{12} and determine the relative rotation ϕ_{12} of the displacement from $M_1 : (\vec{e}_1, D^1)$ to $M_2 : (\vec{e}_2, D^2)$. For a specific choice of the fixed pivot G the line $\mathsf{V} = GP_{12}$ must be the perpendicular bisector to the segment W^1W^2 for all possible moving pivots W^1, Figure 3.9. This leads to the following construction for the moving pivot:

1. Construct the pole P_{12} and the relative rotation angle ϕ_{12} from the two given positions M_1 and M_2.
2. Select any point in F as the fixed pivot G and draw the line $\mathsf{V} = GP_{12}$.
3. Duplicate the angle $\phi_{12}/2$ on either side of V around P_{12} to determine the lines L^1 and L^2.
4. Choose any point on L^1 as the moving pivot W^1. The circle about P_{12} with radius $P_{12}W^1$ intersects L^2 in the corresponding point W^2.

The two-dimensional set fixed pivots combines with the one-dimensional set moving pivots on the line L^1 to yield a three-dimensional set of RR chains that reach a two position task.

3.3.2 Three Precision Positions

3.3.2.1 Select a Moving Pivot

Given three positions $M_i : (\vec{e}_i, D^i)$, $i = 1, 2, 3$, the moving pivot W^1 for M_1 moves to the points W^2 and W^3 in the other two positions. The desired fixed pivot G is the

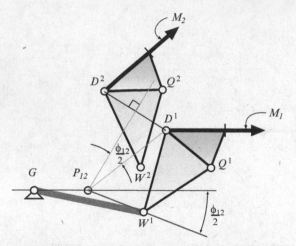

Fig. 3.9 Given the fixed pivot G the moving pivot W^1 lies on a line through P_{12} at the angle $\phi_{12}/2$ to the GP_{12}.

center of the circle through $\triangle W^1 W^2 W^2$, Figure 3.10. The construction for the fixed pivot given a moving pivot is

1. Select an arbitrary point W^1 in F to be the moving pivot and determine the corresponding point W^2 and W^3 in positions M_2 and M_3.
2. Construct the perpendicular bisectors $V_{12} = (W^1 W^2)^\perp$ and $V_{23} = (W^2 W^3)^\perp$.
3. The intersection of these lines is the fixed pivot G.

Notice that for every choice of the moving pivot there is a unique fixed pivot. Thus, there is a two-dimensional set of RR chains compatible with three task positions.

3.3.2.2 Select a Fixed Pivot

In order to determine a moving pivot W^1 given the fixed pivot for a three position task we construct poles P_{12} and P_{13} and rotation angles ϕ_{12} and ϕ_{13}, Figure 3.11. Join G to these poles to define lines V_{12} and V_{13} that are the perpendicular bisectors of segments $W^1 W^2$ and $W^1 W^2$, respectively. The moving pivot W^1 is constructed as follows:

1. Construct the poles P_{12} and P_{13} and the rotation angles ϕ_{12} and ϕ_{13}.
2. Select a fixed pivot G and join it to the poles P_{12} and P_{13} by the lines V_{12} and V_{13}.
3. Duplicate the angle $\phi_{12}/2$ on either side of V_{12} to define the lines L^1 and L^2, and the angle $\phi_{13}/2$ on either side of V_{13} to define the lines M^1 and M^2.
4. The intersection of the lines L^1 and M^1 is the moving pivot W^1.

Thus, three-position synthesis yields a unique moving pivot for each fixed pivot.

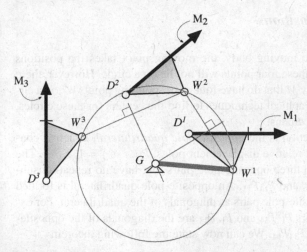

Fig. 3.10 For three specified positions the fixed pivot G is the center of the circle through W^1, W^2, and W^3.

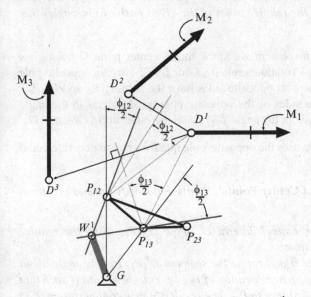

Fig. 3.11 For three specified positions the moving pivot W^1 can be constructed using the selected fixed pivot G and the pole triangle.

3.3.3 Four Precision Positions

Given four positions for the moving body, the moving pivot takes the positions $W^i, i = 1,2,3,4$. In general, these four points will not lie on a circle. However, there are points in the moving body M that do have four corresponding points W^i on a circle. Burmester introduced graphical techniques to find the centers for these circles, which define the desired RR chains.

The tool for this construction is the *opposite-pole quadrilateral*, which is constructed from four of the six relative displacement poles, P_{ij}, $i < j = 1,2,3,4$. The six poles are assembled into three opposite-pole pairs that have no repeating subscript, that is, $P_{12}P_{34}$, $P_{13}P_{24}$, and $P_{14}P_{23}$. An opposite-pole quadrilateral is formed by identifying any two opposite-pole pairs as diagonals of the quadrilateral. For example, the opposite-pole pairs $P_{12}P_{34}$ and $P_{14}P_{23}$ are the diagonals of the opposite-pole quadrilateral $\mathscr{Q} : P_{12}P_{23}P_{34}P_{14}$. We can now state the following theorem:

Theorem 2 (Burmester's Theorem). *The center point of an RR chain that can reach four specified positions in the plane views opposite sides of an opposite-pole quadrilateral obtained from the relative poles of the given positions in angles that are equal, or differ by π.*

Proof. From the center-point theorem we know that a center point G must view the segment $P_{ij}P_{ik}$ in the crank rotation angle $\beta_{jk}/2$ or $\beta_{jk}/2 + \pi$. An opposite-pole quadrilateral is constructed so that opposite sides have the form $P_{ij}P_{ik}$ and $P_{mj}P_{mk}$. Thus, G must view opposite sides of the opposite-pole quadrilateral in the angle $\beta_{jk}/2$ or $\beta_{jk}/2 + \pi$, or one side in the angle $\beta_{jk}/2$ and the other in $\beta_{jk}/2 + \pi$. $\square$

The following construction uses the opposite-pole quadrilateral to generate a center point, Figure 3.12.

Theorem 3 (Construction of Center Points). *Points that satisfy Burmester's theorem are obtained as follows:*

1. *Construct the opposite-pole quadrilateral $\mathscr{Q} : P_{12}P_{23}P_{34}P_{14}$ using the relative poles of the four task positions.*
2. *Choose an arbitrary angle θ and rotate the segment $P_{12}P_{23}$ by this angle about P_{12} to obtain P'_{23}. Construct a new location of P_{34} on a circle about P_{14} such that P'_{34} maintains its original distance to P'_{23}. The result is a new configuration $\mathscr{Q}'$ of the quadrilateral $\mathscr{Q}$.*
3. *The pole G of the displacement of the segment $P'_{23}P'_{34}$ relative to its original location $P_{23}P_{34}$ satisfies Burmester's theorem and is a center point.*

Proof. Let G be the intersection of the perpendicular bisectors $V_1 = (P_{23}P'_{23})^{\perp}$ and $V_2 = (P_{34}P'_{34})^{\perp}$ that are used to define the pole of the displacement of the segment $P_{23}P_{34}$. The input RR chain formed by $P_{12}P_{23}$ has the dyad triangle $\triangle P_{23}P_{12}G$. Let κ be the rotation of $P_{23}P_{34}$ to $P'_{23}P'_{34}$ around G, then G must view the segment $P_{12}P_{23}$ in the angle $\kappa/2$ or $\kappa/2 + \pi$, depending on form of the dyad triangle. Similarly, the geometry of the dyad triangle $\triangle P_{34}P_{14}G$ requires that G view the segment $P_{14}P_{34}$ in either $\kappa/2$ or $\kappa/2 + \pi$. Thus, the pole G views the opposite sides $P_{12}P_{23}$ and $P_{14}P_{34}$ in angles that are equal, or differ by π. The same argument shows that G views the other two sides $P_{23}P_{34}$ and $P_{12}P_{14}$ in angles that are equal, or differ by π. Thus, G satisfies Burmester's theorem. □

Once a center point is obtained by this construction, any three of the four positions can be used to construct the associated moving pivot. The result is an RR chain that reaches the four specified positions.

Notice that for a given increment of rotation of the crank $P_{12}P_{23}$ of the opposite-pole quadrilateral there are actually two center points, one for each assembly of the quadrilateral as a linkage. The relative poles obtained from all of the configurations of the opposite-pole quadrilateral form a cubic curve known as the *center-point curve*.

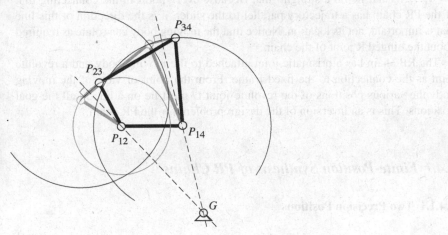

Fig. 3.12 A fixed pivot G for four specified positions is constructed using the quadrilateral formed by the poles $P_{12}P_{23}P_{34}P_{14}$.

3.3.4 Five Precision Positions

Burmester's theorem can be used to identify center points for RR chains that reach five specified task positions for a moving body. This is equivalent to finding points W^i, $i = 1, \ldots, 5$, that lie on a circle. Given $M_i : (\vec{e}_i, D^i)$, $i = 1, \ldots, 5$, construct two opposite-pole quadrilaterals $\mathscr{Q}_{14} : P_{12}P_{23}P_{34}P_{14}$ and $\mathscr{Q}_{15} : P_{12}P_{23}P_{35}P_{15}$. Associated with each opposite-pole quadrilateral is a center-point curve. The intersection of these two curves is a finite number of points that are the desired center points, known as *Burmester points*. We will see later that there are at most four Burmester points, therefore there can be at most four RR chains that reach the five task positions.

While the constructions for two- and three-position synthesis are easy and convenient, four- and five-position synthesis problems are more efficiently solved using the algebraic techniques presented in a later chapter.

3.4 The Geometry of the PR and RP Chains

A PR chain consists of a prismatic joint connected by a link to a revolute joint, which in turn is attached to a moving body. The prismatic joint guides the moving pivot of the PR chain along a line in the fixed frame.

The design of a PR chain that reaches a specified set of goal positions M_i, $i = 1, \ldots, n$, requires finding the moving pivot W^1 that has corresponding positions W^i, $i = 1, \ldots, n$, that lie on a straight line. Because every point in the connecting link of the PR chain has a trajectory parallel to the slider, it is the direction of this line that is important, not its location. Notice that the moving body can rotate as required about the hinged R joint of the chain.

The RP chain has a prismatic joint attached to the moving body and a revolute joint as the connection to the fixed frame. From the point of view of the moving body the various positions of the revolute joint G must lie on a line in all the goal positions. This is an inversion of the design problem for the PR chain.

3.4.1 Finite-Position Synthesis of PR Chains

3.4.1.1 Two Precision Positions

Given two positions of the moving body M_1 and M_2, we select the moving pivot W^1 and find the corresponding position W^2. These two points define the direction of the prismatic joint parallel to $W^1 W^2$. For any choice of a moving pivot there is a unique PR chain. Thus, there is a two-dimensional set of PR chains that reach two specified positions.

It is possible to choose the direction of the prismatic joint and determine the associated moving pivot. Let the pole of the relative displacement from M_1 to M_2 be P_{12} and determine the relative rotation angle ϕ_{12}.

1. Draw a line $\vec{s}$ in the direction of the slider and drop a perpendicular line V from the pole P_{12} to $\vec{s}$.
2. Now duplicate the angle $\phi_{12}/2$ on either side of V to define the lines L^1 and L^2.
3. The two positions of the moving pivot W^1 and W^2 are the intersections of L^1 and L^2 with $\vec{s}$.

3.4.1.2 Three Precision Positions

For three task positions M_i, $i = 1, 2, 3$, we seek a moving pivot W^1 that has the property that the three corresponding points W^1, W^2, and W^3 lie on a straight line in F. We now show that to have this property W^1 views the side $P_{13}P_{12}$ of the pole triangle in $\phi_{23}/2$.

Theorem 4 (The Slider-Point Theorem). *Given three positions M_1, M_2, and M_3, a point that has three corresponding positions W^1, W^2, and W^3 on a line $\vec{s}$ has the property that the angle measured counterclockwise from P_{ik} to P_{ij} around W^i is $\angle P_{ik}W^iP_{ij} = \phi_{jk}/2$, where ϕ_{jk} is the relative rotation for each displacement.*

Proof. If the points W^1, W^2, and W^3 lie on a line, then the perpendicular bisectors V_{12}, V_{13} are parallel to each other and perpendicular to the line along $\vec{s}$. Recall from the properties of a pole that the angle measured counterclockwise from $P_{ij}W^i$ to the perpendicular bisector V_{ij} is $\phi_{ij}/2$. Let V_{ij} be the midpoint of the segment W^iW^j, and consider the two right triangles $\triangle P_{12}V_{12}W^1$ and $\triangle P_{13}V_{13}W^1$ that share the vertex W^1. Assume first that $\phi_{12} + \phi_{23} < 2\pi$, which means that $\phi_{13}/2 > \phi_{12}/2$. Then considering the various configurations available for these triangles we find $\angle P_{13}W^1P_{12} = \phi_{23}/2$. On the other hand, if $\phi_{12} + \phi_{23} > 2\pi$, then $\phi_{13}/2 > \pi$ and the vertex angle of the triangle $\triangle P_{13}V_{13}W^1$ at P_{13} is $\phi_{13} - \pi$. This case yields the same result $\angle P_{13}W^1P_{12} = \phi_{23}/2$. □

The points that view the segment $P_{13}P_{12}$ in the angle $\phi_{23}/2$ form a circle $\mathscr{C}_1$. This circle can be obtained as the reflection through of $N_1 = P_{12}P_{13}$ of the $\mathscr{C}^*$ that circumscribes the pole triangle $\triangle P_{12}P_{23}P_{13}$. We can define similar circles $\mathscr{C}_2$ and $\mathscr{C}_3$ as reflections of $\mathscr{C}^*$ through the line $N_2 = P_{12}P_{23}$ and $N_3 = P_{13}P_{23}$, respectively.

The three circles $\mathscr{C}_i$, $i = 1, 2, 3$, intersect each other at the point H called the *orthocenter* of the pole triangle. This point is also the intersection of the altitudes dropped from each vertex of the pole triangle to the opposite side. Each line $\vec{s}$ through H intersects the three circles $\mathscr{C}_i$ in the points W^i, $i = 1, 2, 3$. These three points are the three locations of the moving pivot of an PR chain as the slider moves along the line $\vec{s}$.

The following construction yields the moving pivot W^1 that has three positions on a line $\vec{s}$:

1. Given three positions, determine the pole triangle $\triangle P_{12}P_{23}P_{13}$ and construct its circumcircle $\mathscr{C}^*$.
2. Reflect this circle through the side $N_1 = P_{12}P_{13}$ to obtain $\mathscr{C}_1$, which is the set of points available to be the moving pivot W^1.
3. Construct the orthocenter H of the pole triangle as the intersection of the altitudes of the pole triangle.
4. Either choose a line $\vec{s}$ through H and find its intersection with $\mathscr{C}_1$ to determine W^1, or choose W^1 on $\mathscr{C}_1$ and construct $\vec{s}$ as its join with H.
5. The remaining positions W^2 and W^3 are obtained as the intersection of $\vec{s}$ with the circles $\mathscr{C}_2$ and $\mathscr{C}_3$.

The result is that for three positions of the moving body there is a one-dimensional set of PR chains.

3.4.1.3 Four Precision Positions

Given four positions of the moving body, we have six poles that can be assembled into six pole triangles. Choose two of these triangles, for example $\triangle P_{12}P_{23}P_{13}$ and $\triangle P_{12}P_{24}P_{14}$, and construct their two orthocenters H_{123} and H_{124}. The line $\vec{s} = H_{123}H_{124}$ joining these orthocenters contains the four corresponding points $W^i, i = 1,\ldots,4$. To determine W^1, intersect $\vec{s}$ with the circle $\mathscr{C}_1$ described in the previous section. The remaining points W^2, W^3, and W^4 are obtained from the intersections with the appropriate circles $\mathscr{C}_i$, $i = 2,3,4$.

Thus, there is a single PR chain that can reach four arbitrary positions of a moving body.

3.4.2 Finite-Position Synthesis of RP Chains

To design an RP chain, we invert the design process for a PR chain. Our goal is to construct the point G that has the inverted positions g^i, $i = 1,\ldots,n$, lying on a straight line as seen from the moving body. Each inverted location g^i is constructed using the pole and rotation angle of the relative inverse displacement obtained from the task positions.

3.4.2.1 Two Precision Positions

Given two positions of the moving body M_1 and M_2, we can compute the relative pole P_{12}. Choose any point in the plane to be g^1, which coincides with G in position M_1. Compute the second position g^2 by a rotation of $-\phi_{12}$ about P_{12}. The line through these two points determines the direction of the prismatic joint in M. There is a unique RP chain for each choice of the fixed pivot.

Rather than select the point G, we can choose the direction of the line $\vec{s}$ parallel to the prismatic joint. The construction is essentially identical to that for a PR chain:

1. Draw a line $\vec{s}$ in the direction of the slider and drop a perpendicular line V from the pole P_{12} to $\vec{s}$.
2. Now duplicate the relative inverse rotation angle $-\phi_{12}/2$ on either side of V to define the lines L^1 and L^2.
3. The inverted positions of the fixed pivot g^1 and g^2 are the intersections of L^1 and L^2 with $\vec{s}$.

3.4.2.2 Three Precision Positions

For three task positions M_i, $i = 1, 2, 3$, we now seek fixed points that have the inverted locations g^1, g^2, and g^3 that lie on a straight line in M. The pole triangle associated with the relative inverse displacements is $\triangle P_{12} P_{23}^1 P_{13}$. It is the reflection through the side $\mathsf{N}_1 = P_{12} P_{13}$ of the pole triangle $\triangle P_{12} P_{23} P_{13}$ obtained from the task positions.

The slider-point theorem shows g^1 must view the side $P_{12} P_{13}$ of the pole triangle in the relative inverse rotation angle $-\phi_{23}/2$. Similarly, g^2 must view $P_{12} P_{23}^1$ in the angle $-\phi_{13}/2$ and g^3 must view $P_{13} P_{23}^1$ in the angle $-\phi_{12}/2$.

Introduce the circumcircle $\mathcal{K}^*$ of the image pole triangle $\triangle P_{12} P_{23}^1 P_{13}$. This circle is reflected through the side $\mathsf{N}_1 = P_{12} P_{13}$ to define $\mathcal{K}_1$. This circle is the reflection of the circle $\mathcal{C}_1$ obtained in the previous section, and happens to be the circumcircle $\mathcal{C}^*$ of the original pole triangle. In fact, it is the circle $\mathcal{C}_1$ that forms the circumcircle of the inverted pole triangle, now denoted by $\mathcal{K}^*$. The point g^2 lie on the circle $\mathcal{K}_2$ that is the reflection of $\mathcal{K}^*$ through the line $\mathsf{N}_2 = P_{12} P_{23}^1$. And finally, g^3 lies on circle $\mathcal{K}_3$ that is the reflection of $\mathcal{K}^*$ through the line $\mathsf{N}_3 = P_{13} P_{23}^1$.

The three circles $\mathcal{K}_i$, $i = 1, 2, 3$, intersect each other at the orthocenter h of the image pole triangle. Each line $\vec{s}$ through h intersects the three circles $\mathcal{K}_i$ in the points g^i. These are the inverted locations of the fixed pivot G that lie on the line $\vec{s}$.

Thus, we have the following construction for the inverted point g^1 that has with three corresponding positions on a line $\vec{s}$ as seen from M:

1. Given three positions, determine the inverted pole triangle $\triangle P_{12} P_{23}^1 P_{13}$ and construct its circumcircle $\mathcal{K}^*$.
2. Reflect this circle through the side $\mathsf{N}_1 = P_{12} P_{13}$ to obtain $\mathcal{K}_1$ that is the set of available points for g^1.
3. Construct the orthocenter h of the pole triangle as the intersection of the altitudes to $\triangle P_{12} P_{23}^1 P_{13}$.
4. Either choose a line $\vec{s}$ through h and find its intersection with $\mathcal{K}_1$ to determine g^1, or choose g^1 and construct $\vec{s}$ as its join with h.
5. In F the fixed pivot G is g^1 the direction of $\vec{s}$ in each position is obtained by constructing the location of h in each of the positions M_2 and M_3.

The result is a one-dimensional set of PR chains given three task positions for the moving body.

3.4.2.3 Four Precision Positions

The construction for the RP chain through four specified positions is the same as for the PR chain presented above. The only difference is the use of the image pole triangles $\triangle P_{12} P_{23}^1 P_{13}$ and $\triangle P_{12} P_{24}^1 P_{14}$. Construct the two orthocenters h_{123} and h_{124}. Then the line $\vec{s}$ joining these orthocenters contains the four corresponding points g^i, $i = 1, \ldots, 4$. To determine g^1 intersect $\vec{s}$ with the circle $\mathcal{K}_1$ as shown in the previous section. The result is a unique RP chain.

3.5 The Design of Four-Bar Linkages

Planar four-bar linkages are formed by connecting the end-links of any two of the RR, PR, and RP chains. If both chains are designed to reach a given set of task positions, then the four-bar chain can be assembled in each of these positions.

The 4R linkage is constructed by rigidly connecting the floating links of two RR chains, Figure 3.13. Because as many as four RR chains may exist that can reach five positions, at most $\binom{4}{2}$, or six, 4R linkages exist that also reach these positions. For fewer than five task positions, the dimensionality of the space of 4R linkages is twice the dimensionality of solutions for the RR chains.

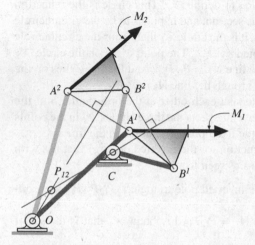

Fig. 3.13 Two-position synthesis of a 4R chain is obtained by constructing two different RR open chains and connecting their end-links.

The RRRP linkage is formed by connecting the floating links of an RR and PR chain. For four task positions there is a one dimensional set of RRRP linkages. They are obtained from the single PR chain and the one dimensional set of RR chains obtained from the center-point curve. For the cases of two and three task positions there are, respectively, five- and three-dimensional sets of RRRP chains.

A two-dimensional set of double-slider PRRP linkages can be constructed for three task positions. Two task positions have an associated four-dimensional set of these linkages.

Four-bar linkages have one degree of freedom, and therefore require only one actuator. However, the interaction of the two chains forming the linkage can generate singular configurations that introduce limits to the movement of the system.

3.6 Summary

In this chapter we have presented graphical techniques for the design of planar RR and PR chains. The geometry of the pole triangle and the center-point theorem that arise in this study are central to the geometric theory of linkage design. In the following chapters we develop an algebraic formulation to broaden the application of these results, and to lay the foundation for their generalization to the design of spherical and spatial linkages.

3.7 References

Our approach to the design of linkages is called the design of a guiding linkage by Hall [42], rigid body guidance by Suh and Radcliffe [134], motion generation by Sandor and Erdman [112], and finite-position synthesis by Roth [106]. It is inspired by ideas introduced by Ludwig Burmester [7] for planar movement and Arthur Schoenflies [113] for spatial movement. Also see Beyer [4]. The texts by Hartenberg and Denavit [46], Hall [42], and Kimbrell [56] provide detailed development of graphical linkage synthesis. The use of the opposite-pole quadrilateral to construct the center-point curve can be found in Luck and Modler [69].

Exercises

1. Figure 3.14 shows two goal positions for the cover of a box. (i) Construct the pole of the displacement; then (ii) design a 4R linkage to move the cover from one position to the other. Place the fixed pivots inside the box and attach the moving pivots outside the boundary of the cover.

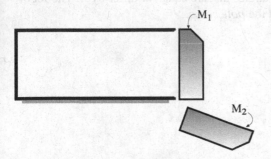

Fig. 3.14 Design a 4R linkage to move the cover.

2. Use three-position synthesis to design a 4R linkage to move the scoop in Figure 3.15 from the first position to the third. Choose the intermediate position M_2 so that it clears the obstacle.

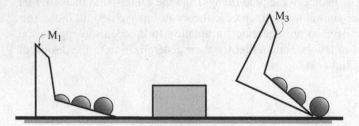

Fig. 3.15 Design a 4R linkage to move the scoop so it clears the obstacle.

3. Reflect a point A^* through each side of the pole triangle $\triangle P_{12}P_{23}P_{13}$ to obtain $A^i, i = 1,2,3$. Show that these points are correspond to the same point A in M for each of the positions M_i, $i = 1,2,3$. The point A^* is called the *cardinal point* associated with a point in M.
4. Show that the circle that circumscribes a pole triangle is the set of cardinal points for those that have three positions on a straight line.
5. Show that the orthocenter H^* of the pole triangle is the cardinal point for a point that has its three positions H^i, $i = 1,2,3$, on the circumscribing circle of the pole triangle.
6. Assume that five points are given as five of the six relative position poles associated with four task positions, $M_i, i = 1,\ldots,4$. Suppose the missing pole is P_{24}. Construct this pole as the common vertex of the two pole triangles $\triangle P_{24}P_{41}P_{12}$ and $\triangle P_{24}P_{34}P23$.
7. Let the points P_{12}, P_{23}, P_{34}, and P_{14} be the relative poles for four positions M_i, $i = 1,\ldots,4$. For a fifth point to be the pole P_{13} it must form the two pole triangles $\triangle P_{12}P_{23}P_{13}$ and $P_{14}P_{34}P_{23}$. Show that P_{13} must view the opposite sides of the quadrilateral $\mathscr{Q} : P_{12}P_{23}P_{34}P_{14}$ in angles that are equal, or differ by π. The set of points with this property is called the *pole curve*.

Chapter 4
Planar Kinematics

In this chapter we study the geometry of planar displacements. The position of a moving body is defined by a coordinate transformation. Associated with each of these transformations is an invariant point called the pole of the displacement. We examine the relationship between relative positions of points in the moving body and the location of this pole. We also consider the triangle formed by the poles of two displacements and the pole of their composite displacement. The geometry of this triangle describes the relationship between the three displacements.

4.1 Isometry

We are concerned with the movement of bodies in the plane such that the distances between points in the body are unchanged. A transformation with this property is called an *isometry*. The measurement of distance is given by the usual *distance formula*, which is also known as the *Euclidean metric* in the plane.

Consider the two points $\mathbf{P} = (P_x, P_y)^T$ and $\mathbf{Q} = (Q_x, Q_y)^T$ in a fixed frame F. The distance between these points is the magnitude of the vector $\mathbf{Q} - \mathbf{P}$, given by

$$|\mathbf{Q} - \mathbf{P}| = \sqrt{(Q_x - P_x)^2 + (Q_y - P_y)^2}. \tag{4.1}$$

Square both sides of this formula to obtain the Pythagorean theorem

$$|\mathbf{Q} - \mathbf{P}|^2 = (Q_x - P_x)^2 + (Q_y - P_y)^2. \tag{4.2}$$

Using vector notation, this equation can be written

$$|\mathbf{Q} - \mathbf{P}|^2 = (\mathbf{Q} - \mathbf{P}) \cdot (\mathbf{Q} - \mathbf{P}), \tag{4.3}$$

where the dot denotes the scalar product between two vectors. It is useful to consider the vector $\mathbf{Q} - \mathbf{P}$ as a column matrix, so that the scalar product can also be written

J.M. McCarthy and G.S. Soh, *Geometric Design of Linkages*, Interdisciplinary Applied
Mathematics 11, DOI 10.1007/978-1-4419-7892-9_4,
© Springer Science+Business Media, LLC 2011

as the matrix product

$$|\mathbf{Q}-\mathbf{P}|^2 = (\mathbf{Q}-\mathbf{P})^T(\mathbf{Q}-\mathbf{P}). \tag{4.4}$$

In order to study the movement of a body we attach a coordinate frame M to it and determine the position of this frame relative to a fixed coordinate frame F. Assume that initially M coincides with F. Then a *displacement* of the body moves M so it takes the position M' relative to F. This displacement is defined by a transformation of coordinates from M' to F. In addition, because the transformation is an isometry, the distances between points in M' are the same as when measured in F. Planar displacements are composed of planar translations and rotations.

4.1.1 Planar Translations

Let the coordinates of a point in the moving body M be denoted by the vector $\mathbf{x} = (x,y)^T$, and let the coordinates of the same point in F be $\mathbf{X} = (X,Y)^T$. If the moving frame M coincides initially with F, then for every point in M we have $\mathbf{X} = \mathbf{x}$. Now add a constant vector $\mathbf{d} = (d_x, d_y)^T$ to the coordinates of all the points in the body in order to translate it to a new position M' relative to F, that is,

$$\mathbf{X} = \mathbf{x} + \mathbf{d}. \tag{4.5}$$

The vector $\mathbf{x}$ defines a point in the initial position and $\mathbf{X}$ is the coordinate vector of this point after M is translated to the new position M'.

To see that a *translation* is an isometry, we compute the distance between two points before and after the translation. If $\mathbf{p}$ and $\mathbf{q}$ are the coordinates of two points in M, then in the position M' we have $\mathbf{P} = \mathbf{p} + \mathbf{d}$ and $\mathbf{Q} = \mathbf{q} + \mathbf{d}$. Compute the distance $|\mathbf{Q} - \mathbf{P}|$ to obtain

$$|\mathbf{Q}-\mathbf{P}| = |(\mathbf{q}+\mathbf{d})-(\mathbf{p}+\mathbf{d})| = |\mathbf{q}-\mathbf{p}|. \tag{4.6}$$

Thus, the distance between these points is the same before and after the translation.

4.1.2 Planar Rotations

The rotation of a body relative to the fixed frame F can occur around any point. However, for now we consider only rotations about the origin $\mathbf{O}$ of F. Let M be aligned initially with F, then a rotation about $\mathbf{O}$ introduces an angle ϕ between the x-axis of F and the x-axis of the rotated frame M'. Thus, a rotation changes the direction of the unit vectors $\vec{\imath} = (1,0)^T$ and $\vec{\jmath} = (0,1)^T$ along the x and y axes of M so they are directed along the coordinate axes of M'.

Let $\mathbf{e}_x$ and $\mathbf{e}_y$ be the unit vectors along the x- and y-axes of M', then

$$\mathbf{e}_x = \cos\phi\vec{\imath} + \sin\phi\vec{\jmath} \quad \text{and} \quad \mathbf{e}_y = -\sin\phi\vec{\imath} + \cos\phi\vec{\jmath}. \tag{4.7}$$

The rotation from M to M' transforms the vector $\mathbf{x} = x\vec{\imath} + y\vec{\jmath}$ into

$$\mathbf{X} = X\vec{\imath} + Y\vec{\jmath} = x\mathbf{e}_x + y\mathbf{e}_y. \tag{4.8}$$

This can be written in matrix form as

$$\begin{Bmatrix} X \\ Y \end{Bmatrix} = \begin{bmatrix} \cos\phi & -\sin\phi \\ \sin\phi & \cos\phi \end{bmatrix} \begin{Bmatrix} x \\ y \end{Bmatrix}, \tag{4.9}$$

or

$$\mathbf{X} = [A(\phi)]\mathbf{x}. \tag{4.10}$$

Notice that the first column of the 2×2 matrix $[A(\phi)]$ is simply $\mathbf{e}_x$ and the second column is $\mathbf{e}_y$. All planar rotations about the origin of F can be written in this way, and the matrix $[A(\phi)]$ is called a *rotation matrix* and ϕ the *rotation angle*.

We now show that rotations preserve the distances between points. Let $\mathbf{p}$ and $\mathbf{q}$ be the coordinates of two points in M, and let $\mathbf{P}$ and $\mathbf{Q}$ be their coordinates in F when M is rotated to the position M'. We can compute

$$|\mathbf{Q} - \mathbf{P}|^2 = (\mathbf{Q} - \mathbf{P})^T(\mathbf{Q} - \mathbf{P}) = ([A]\mathbf{q} - [A]\mathbf{p})^T([A]\mathbf{q} - [A]\mathbf{q})$$
$$= (\mathbf{q} - \mathbf{p})^T[A^T][A](\mathbf{q} - \mathbf{p}). \tag{4.11}$$

The last step uses the linearity of matrix multiplication, $[A]\mathbf{q} - [A]\mathbf{p} = [A](\mathbf{q} - \mathbf{p})$.

Notice that $|\mathbf{Q} - \mathbf{P}| = |\mathbf{q} - \mathbf{p}|$ only if the matrix $[A]$ has the property

$$[A^T][A] = [I], \tag{4.12}$$

where $[I]$ denotes the 2×2 identity matrix. This is equivalent to saying that the transpose of $[A]$ is also its inverse, that is, $[A^{-1}] = [A^T]$. Matrices with this property are called *orthogonal*, because their columns must be orthogonal unit vectors. The columns of $[A(\phi)]$ in (4.9) are the orthogonal unit vectors $\mathbf{e}_x$ and $\mathbf{e}_y$, so a planar rotation is an isometry.

The product of two planar rotations $[A(\phi_1)]$ and $[A(\phi_2)]$ is the rotation of angle $\phi_1 + \phi_2$, that is,

$$[A(\phi_1)][A(\phi_2)] = [A(\phi_1 + \phi_2)]. \tag{4.13}$$

From this relation, we see that the inverse of $[A(\phi)]$ is the rotation by the angle $-\phi$, and $[A(\phi)^T] = [A(-\phi)]$.

4.1.3 Planar Displacements

A general planar displacement consists of rotation of the coordinate frame M to M' followed by a translation of M' to a position M''. Rather than distinguish between its initial and final positions, M and M'', we assume that the moving frame is always aligned initially with F, and denote the final position by M, Figure 4.1. Thus, the displacement that defines the position of the moving frame M relative to the fixed frame F is the coordinate transformation

$$\left\{ \begin{matrix} X \\ Y \end{matrix} \right\} = \begin{bmatrix} \cos\phi & -\sin\phi \\ \sin\phi & \cos\phi \end{bmatrix} \left\{ \begin{matrix} x \\ y \end{matrix} \right\} + \left\{ \begin{matrix} d_x \\ d_y \end{matrix} \right\}, \tag{4.14}$$

or

$$\mathbf{X} = [A(\phi)]\mathbf{x} + \mathbf{d}. \tag{4.15}$$

It is convenient to assemble the rotation matrix and translation vector into the single 3×3 matrix $[T]$, so equation (4.14) takes the form

$$\left\{ \begin{matrix} X \\ Y \\ 1 \end{matrix} \right\} = \begin{bmatrix} \cos\phi & -\sin\phi & 0 \\ \sin\phi & \cos\phi & 0 \\ 0 & 0 & 1 \end{bmatrix} \left\{ \begin{matrix} x \\ y \\ 1 \end{matrix} \right\}, \tag{4.16}$$

or

$$\mathbf{X} = [T]\mathbf{x}. \tag{4.17}$$

The context of our calculations will make it clear whether we consider the coordinate vectors $\mathbf{X}$ and $\mathbf{x}$ have a third component of 1. We use the matrix-vector pair $[T]$ $= [A(\phi), \mathbf{d}]$ to define the *position* of the moving frame M relative to F.

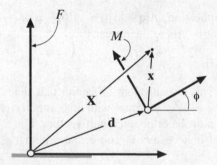

Fig. 4.1 The fixed and moving frames defining a planar displacement.

4.1.3.1 Complex Vectors

Complex numbers provide a convenient way to manipulate the coordinates of points in the plane. Let the coordinate vector $\mathbf{X}$ be the complex number $\mathbf{X} = X + iY$, where i is the imaginary unit ($i^2 = -1$). A translation by the vector $\mathbf{d}$ is defined by complex addition, so $\mathbf{X} = \mathbf{x} + \mathbf{d}$. Rotation by the angle ϕ is achieved by multiplication by the complex exponential $e^{i\phi}$, $\mathbf{X} = e^{i\phi}\mathbf{x}$. Thus, a general planar displacement is defined by the equation

$$\mathbf{X} = e^{i\phi}\mathbf{x} + \mathbf{d}. \tag{4.18}$$

The advantage of using complex numbers lies in the ease of manipulating the complex exponential $e^{i\phi}$, as opposed to the rotation matrix $[A(\phi)]$. For example, the product of two rotations $e^{i\phi_1}$ and $e^{i\phi_2}$ is seen to be $e^{i(\phi_1+\phi_2)}$, and the inverse of $e^{i\phi}$ is $e^{-i\phi}$.

4.1.4 The Composition of Two Displacements

Two planar displacements can combine to define a third displacement called the *composition* of the two displacements. Consider $[T_1]$ to define the displacement from F to M_1, so we have

$$\mathbf{X} = [A_1]\mathbf{y} + \mathbf{d}_1 = [T_1]\mathbf{y}. \tag{4.19}$$

Notice that $\mathbf{y}$ is in M_1 and $\mathbf{X}$ is in F. If the displacement $[T_2]$ defines the position of M_2 relative to M_1, then we have

$$\mathbf{y} = [A_2]\mathbf{x} + \mathbf{d}_2 = [T_2]\mathbf{x}, \tag{4.20}$$

where $\mathbf{x}$ is in M_2. The composite displacement from F to M_2 is obtained by direct substitution

$$\mathbf{X} = [A_1 A_2]\mathbf{x} + \mathbf{d}_1 + A_1 \mathbf{d}_2 = [T_1][T_2]\mathbf{x}. \tag{4.21}$$

This computation defines a product operation $[T_3] = [T_1][T_2]$ for displacements given by

$$[T_1][T_2] = [A_1, \mathbf{d}_1][A_2, \mathbf{d}_1] = [A_1 A_2, \mathbf{d}_1 + A_1 \mathbf{d}_2]. \tag{4.22}$$

The *inverse displacement* is obtained by solving for $\mathbf{x}$ in M in terms of the point $\mathbf{X}$ in F, that is,

$$\mathbf{x} = [A(\phi)^T]\mathbf{X} - [A(\phi)^T]\mathbf{d} = [T^{-1}]\mathbf{X}. \tag{4.23}$$

Thus, the inverse of a displacement $[T] = [A(\phi), \mathbf{d}]$ is

$$[T^{-1}] = [A(\phi)^T, -A(\phi)^T \mathbf{d}]. \tag{4.24}$$

The composition of a displacement with its inverse yields the identity displacement $[I] = [I, 0]$.

4.1.4.1 Changing Coordinates of a Displacement

Consider the planar displacement $\mathbf{X} = [T]\mathbf{x}$ that defines the position of M relative to F. We now consider the transformation $[T']$ between the frames M' and F' that are displaced by the same amount from both M and F. In particular, let $[R] = [B, \mathbf{c}]$ be the displacement that transforms the coordinates between the primed and unprimed frames, that is, $\mathbf{Y} = [R]\mathbf{X}$ and $\mathbf{y} = [R]\mathbf{x}$ are the coordinates in F' and M', respectively.

Then, from $\mathbf{X} = [T]\mathbf{x}$ we can compute

$$\mathbf{Y} = [R][T][R^{-1}]\mathbf{y}. \tag{4.25}$$

Thus, the original displacement $[T]$ is transformed by the change of coordinates into $[T'] = [R][T][R^{-1}]$.

4.1.5 Relative Displacements

Consider two positions M_1 and M_2 of a rigid body defined by the displacements $[T_1]$ and $[T_2]$ relative to F. Let $\mathbf{X} = [T_1]\mathbf{x}$ and $\mathbf{Y} = [T_2]\mathbf{x}$, where both M_1 and M_2 initially coincide with F. The transformation $[T_{12}]$ that carries the coordinates $\mathbf{X}$ of M_1 in F into the coordinates $\mathbf{Y}$ of M_2 in F is defined by

$$\mathbf{Y} = [T_{12}]\mathbf{X},$$

or

$$[T_2]\mathbf{x} = [T_{12}][T_1]\mathbf{x}. \tag{4.26}$$

Thus, $[T_{12}]$ is given by

$$[T_{12}] = [T_2][T_1^{-1}]. \tag{4.27}$$

This defines the *relative displacement* from M_1 to M_2 measured in F.

The relative rotation and translation components of $[T_{12}] = [A_{12}, \mathbf{d}_{12}]$ are determined by the rotation and translation components of $[T_1] = [A_1, \mathbf{d}_1]$ and $[T_2] = [A_2, \mathbf{d}_2]$. Expand the composition of these displacements to obtain

$$[T_{12}] = [A_{12}, \mathbf{d}_{12}] = [A_2 A_1^T, \mathbf{d}_2 - A_2 A_1^T \mathbf{d}_1]. \tag{4.28}$$

If the rotation angles of these displacements are ϕ_1 and ϕ_2, respectively, then we see that the rotation angle of $[A_{12}]$ is $\phi_{12} = \phi_2 - \phi_1$.

For a set of positions $[T_i] = [A(\phi_i), \mathbf{d}_i], i = 1, \ldots, n$, of a moving body M measured in F, we have the relative displacement $[T_{jk}] = [T_k][T_j^{-1}]$ given by

$$[T_{jk}] = [A_{jk}, \mathbf{d}_{jk}] = [A_k A_j^T, \mathbf{d}_k - A_k A_j^T \mathbf{d}_j], \tag{4.29}$$

where

$$\phi_{jk} = \phi_k - \phi_j, \tag{4.30}$$

is the relative rotation angle.

The complex exponential simplifies the calculation of the rotation terms in these transformation equations. In particular, the relative displacement $[T_{12}] = [e^{i\phi_{12}}, \mathbf{d}_{12}]$ is given by

$$[T_{12}] = [T_2][T_1^{-1}] = [e^{i\phi_2}, \mathbf{d}_2][e^{-i\phi_1}, -e^{-i\phi_1}\mathbf{d}_1]$$
$$= [e^{i\phi_{12}}, \mathbf{d}_2 - e^{i\phi_{12}}\mathbf{d}_1]. \tag{4.31}$$

4.1.6 Relative Inverse Displacements

The inverse $[T^{-1}]$ of a displacement $[T]$ defines the position of F measured relative to the moving frame M. Given two positions of a moving body M_1 and M_2 defined by the displacements $[T_1]$ and $[T_2]$, then $[T_1^{-1}]$ and $[T_2^{-1}]$ define displacements that locate the fixed frame in two positions F_1 and F_2 relative to M. We now consider the relative inverse displacement from F_1 to F_2 in M.

Let $\mathbf{X}$ be a point in F that corresponds to $\mathbf{x}$ in M_1 and $\mathbf{y}$ in M_2. Then, we have $\mathbf{x} = [T_1^{-1}]\mathbf{X}$ and $\mathbf{y} = [T_2^{-1}]\mathbf{X}$. The relative displacement $[T_{12}^\dagger]$ transforms the coordinates in F_1 to coordinates in F_2 measured in M, that is,

$$\mathbf{y} = [T_{12}^\dagger]\mathbf{x},$$

or

$$[T_2^{-1}]\mathbf{x} = [T_{12}][T_1^{-1}]\mathbf{x}. \tag{4.32}$$

Thus, $[T_{12}^\dagger]$ is given by

$$[T_{12}^\dagger] = [T_2^{-1}][T_1]. \tag{4.33}$$

This is called the *relative inverse displacement*. The formula for the rotation and translation terms of $[T_{12}^\dagger]$ is obtained as

$$[T_{12}^\dagger] = [A_{12}^\dagger, \mathbf{d}_{12}^\dagger] = [A_2^T A_1, -A_2^T(\mathbf{d}_2 - \mathbf{d}_1)]. \tag{4.34}$$

Notice that this is not the inverse of the relative displacement $[T_{12}]$, which would be $[T_1][T_2^{-1}]$.

Associated with the set of positions M_i, $i = 1, \ldots, n$, is the set of relative inverse displacements $[T_{ik}^\dagger]$. Each of these transformation is defined from the point of view of the moving frame M. We can choose a specific position M_j in F and transform the coordinates of the inverse displacement $[T_{ik}^\dagger]$ to obtain

$$[T_{ik}^j] = [T_j][T_{ik}^\dagger][T_j^{-1}]. \tag{4.35}$$

This is known as the *image* of the relative inverse transformation for position M_j in F.

If M_j is one of the frames of the relative inverse displacement, that is $j = i$, then we have

$$[T_{ik}^i] = [T_i]([T_k^{-1}][T_i])[T_i^{-1}] = [T_i][T_k^{-1}] = [T_{ik}^{-1}]. \tag{4.36}$$

This is also true for $j = k$. Thus, for $j = i$ or $j = k$, the image of the relative inverse displacement $[T_{ik}^j]$ is the inverse of the relative displacement.

4.2 The Geometry of Displacement Poles

An important feature of any transformation is the set of points that it leaves invariant. For a general planar displacement there is a single point that has the same coordinates in M and in F. This point is called the *pole* of the displacement. A planar displacement is equivalent to a pure rotation about this pole.

4.2.1 The Pole of a Displacement

Let $\mathbf{P}$ be the coordinates of a point that are unchanged by the planar displacement $[T] = [A(\phi), \mathbf{d}]$, that is,

$$\mathbf{P} = [A(\phi)]\mathbf{P} + \mathbf{d}. \tag{4.37}$$

We can solve this equation to determine the coordinates of $\mathbf{P}$ as

$$\mathbf{P} = [I - A(\phi)]^{-1}\mathbf{d}. \tag{4.38}$$

Notice that if the displacement is a pure translation, then $A = I$ and $[I - A]$ is not invertible. In this instance the pole is said to be at infinity in the direction orthogonal to $\mathbf{d}$.

For the case of a nontrivial 2×2 rotation matrix $[A]$, the matrix $[I - A]$ is always invertible. This is not the case when $[A]$ is a 3×3 rotation matrix, as we will see for spatial rotations.

The pole $\mathbf{p}$ of the inverse displacement $[T^{-1}] = [A^T, -A^T\mathbf{d}]$ can be determined by substituting the inverted rotation and translation into (4.38). The calculation

$$\mathbf{p} = -[I - A^T(\phi)]^{-1}[A^T]\mathbf{d} = [I - A(\phi)]^{-1}\mathbf{d} = \mathbf{P} \tag{4.39}$$

shows that $[T]$ and its inverse $[T^{-1}]$ have the same pole $\mathbf{P}$.

The formula (4.38) can be used to define the translation component of the displacement in terms of the coordinates of the pole, that is,

$$\mathbf{d} = [I - A(\phi)]\mathbf{P}. \tag{4.40}$$

Thus, a planar displacement $[T] = [A(\phi), \mathbf{d}]$ can be defined directly in terms of a rotation angle ϕ and pole $\mathbf{P}$ such that

$$[T(\phi, \mathbf{P})] = [A(\phi), [I - A(\phi)]\mathbf{P}]. \tag{4.41}$$

Let the translation vector **d** be defined in terms of the coordinates of the pole, so the coordinate transformation (4.14) becomes

$$\mathbf{X} = [A(\phi)]\mathbf{x} + [I - A(\phi)]\mathbf{P}. \tag{4.42}$$

Rewrite this equation in terms of vectors $\mathbf{X} - \mathbf{P}$ and $\mathbf{x} - \mathbf{P}$ measured relative to the pole to obtain

$$\mathbf{X} - \mathbf{P} = [A(\phi)](\mathbf{x} - \mathbf{P}). \tag{4.43}$$

This shows that the displaced position **X** of any point **x** is obtained by rotating the vector $\mathbf{x} - \mathbf{P}$ by the angle ϕ. Thus, a general planar displacement is a pure rotation about its pole.

4.2.2 Perpendicular Bisectors and the Pole

Because the relative vectors $\mathbf{x} - \mathbf{P}$ and $\mathbf{X} - \mathbf{P}$ are related by pure rotation (4.43) their magnitudes $|\mathbf{X} - \mathbf{P}|$ and $|\mathbf{x} - \mathbf{P}|$ must be equal. This is equivalent to the statement

$$(\mathbf{X} - \mathbf{P})^2 - (\mathbf{x} - \mathbf{P})^2 = 0. \tag{4.44}$$

Consider this to be the difference of two squares using the scalar product and factor to obtain

$$(\mathbf{X} - \mathbf{x}) \cdot \left(\frac{\mathbf{X} + \mathbf{x}}{2} - \mathbf{P} \right) = 0. \tag{4.45}$$

Notice that $\mathbf{V} = (\mathbf{X} + \mathbf{x})/2$ is the midpoint of the segment $\mathbf{X} - \mathbf{x}$, and $\mathbf{V} - \mathbf{P}$ is the vector from the pole to **V**. This allows us to interpret (4.45) geometrically as stating that $\mathbf{V} - \mathbf{P}$ is the perpendicular bisector of $\mathbf{X} - \mathbf{x}$. Thus, because this equation is true for any point **x**, we find that the pole **P** lies on the perpendicular bisector of every segment joining the initial and final positions of points in a displaced body.

This last result provides another way to compute the pole of a planar displacement. Let **r**, **R** and **s**, **S** be two sets of initial and final positions of points associated with a displacement $[T]$. Then these points satisfy (4.45) and we have

$$(\mathbf{R} - \mathbf{r}) \cdot \left(\frac{\mathbf{R} + \mathbf{r}}{2} - \mathbf{P} \right) = 0,$$

$$(\mathbf{S} - \mathbf{s}) \cdot \left(\frac{\mathbf{S} + \mathbf{s}}{2} - \mathbf{P} \right) = 0. \tag{4.46}$$

These two equations expand to define two linear equations in the coordinates of **P**, which are easily solved.

For a general point **x** in M the triangle $\triangle \mathbf{x}\mathbf{P}\mathbf{X}$ is isosceles with the rotation angle ϕ at the vertex **P**. The altitude of this triangle is the perpendicular bisector $\mathbf{V} - \mathbf{P}$, therefore

$$\tan\frac{\phi}{2} = \frac{|\mathbf{X}-\mathbf{V}|}{|\mathbf{V}-\mathbf{P}|}. \tag{4.47}$$

Introduce the operator $\vec{k}\times$, which performs a rotation by $90°$ in the plane. Then we have

$$\mathbf{X}-\mathbf{V} = \tan\frac{\phi}{2}\vec{k}\times(\mathbf{V}-\mathbf{P}). \tag{4.48}$$

Replace the midpoint $\mathbf{V}$ by its definition in terms of the vectors $\mathbf{x}$ and $\mathbf{X}$ in order to obtain *Rodrigues's equation* in the plane,

$$\mathbf{X}-\mathbf{x} = \tan\frac{\phi}{2}\vec{k}\times(\mathbf{X}+\mathbf{x}-2\mathbf{P}). \tag{4.49}$$

4.2.3 Pole of a Relative Displacement

The pole of the relative displacement from M_1 to M_2 is an important tool in linkage design. This pole $\mathbf{P}_{12}$ is found by applying (4.38) to the transformation $[T_{12}] = [A(\phi_{12}), \mathbf{d}_{12}]$. This can be written in terms of the components of the two displacements $[T_1]$ and $[T_2]$ as

$$\mathbf{P}_{12} = [I - A(\phi_{12})]^{-1}(\mathbf{d}_2 - A(\phi_{12})\mathbf{d}_1). \tag{4.50}$$

Recall that ϕ_{12} is measured from the x-axis of M_1 to the x-axis of M_2.

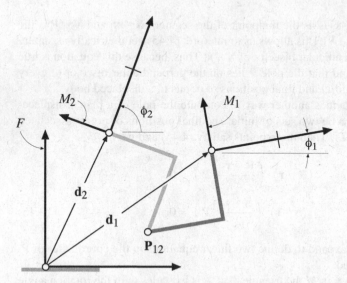

Fig. 4.2 The relative pole $\mathbf{P}_{12}$ of the two positions M_1 and M_2.

The relative translation vector $\mathbf{d}_{12}$ is given in terms of the pole $\mathbf{P}_{12}$ by the relation

$$\mathbf{d}_{12} = [I - A(\phi_{12})]\mathbf{P}_{12}. \qquad (4.51)$$

This can be substituted into the transformation equation (4.29) and simplified to yield

$$\mathbf{X}^2 - \mathbf{P}_{12} = [A(\phi_{12})](\mathbf{X}^1 - \mathbf{P}_{12}). \qquad (4.52)$$

Thus, the relative displacement is a pure rotation about the relative pole, Figure 4.2.
From the fact that the magnitudes of $\mathbf{X}^1 - \mathbf{P}_{12}$ and $\mathbf{X}^1 - \mathbf{P}_{12}$ are equal we have

$$(\mathbf{X}^2 - \mathbf{P}_{12})^2 - (\mathbf{X}^1 - \mathbf{P}_{12})^2 = 0, \qquad (4.53)$$

which can be manipulated into the form

$$(\mathbf{X}^2 - \mathbf{X}^1) \cdot \left(\frac{\mathbf{X}^2 + \mathbf{X}^1}{2} - \mathbf{P}_{12} \right) = 0. \qquad (4.54)$$

Thus, the relative pole $\mathbf{P}_{12}$ lies on the perpendicular bisector of the segment $\mathbf{X}^2 - \mathbf{X}^1$.
For a general pair of positions M_j and M_k defined by transformations $[T_j]$ and $[T_k]$ we have

$$\mathbf{P}_{jk} = [I - A_{jk}]^{-1}(\mathbf{d}_k - A_{jk}\mathbf{d}_j). \qquad (4.55)$$

The relative displacement $[T_{jk}]$ is a rotation about the pole $\mathbf{P}_{jk}$.

4.2.4 The Pole of a Relative Inverse Displacement

The pole $\mathbf{p}_{ik}$ of the relative inverse displacement $[T_{ik}^{\dagger}]$ is located in M and given by (4.50) as

$$\mathbf{p}_{ik} = [I - A_{ik}^{\dagger}]^{-1}\mathbf{d}_{ik}^{\dagger} = -[A_k - A_i]^{-1}(\mathbf{d}_k - \mathbf{d}_i). \qquad (4.56)$$

The relative angle of rotation about $\mathbf{p}_{ik}$ is $-\phi_{ik}$, where $\phi_{ik} = \phi_k - \phi_i$ is the angle of M_k relative to M_i.
We now compute the point in F that corresponds to $\mathbf{p}_{ik}$ when the body is in position M_j. This point $\mathbf{P}_{ik}^j$ is the image of the pole $\mathbf{p}_{ik}$, and is called the *image pole* in the jth position. A formula for $\mathbf{P}_{ik}^j$ is obtained by using the transformation $[T_j] = [A_j, \mathbf{d}_j]$ to compute

$$\mathbf{P}_{ik}^j = [T_j]\mathbf{p}_{ik} = [A_{ij} - A_{jk}]^{-1}(\mathbf{d}_k - \mathbf{d}_i) + \mathbf{d}_j. \qquad (4.57)$$

When $j = i$, the image pole $\mathbf{P}_{ik}^i$ is given by

$$\mathbf{P}_{ik}^i = [I - A_{ik}]^{-1}(\mathbf{d}_k - \mathbf{d}_i) + \mathbf{d}_i = [I - A_{ik}]^{-1}(\mathbf{d}_k - A_{ik}\mathbf{d}_i) = \mathbf{P}_{ik}. \qquad (4.58)$$

Thus, the image pole $\mathbf{P}_{ik}^i$ is the pole $\mathbf{P}_{ik}$ for all k. A similar calculation shows that $\mathbf{P}_{ik}^k = \mathbf{P}_{ik}$ for all positions i.

4.3 The Pole Triangle

4.3.1 The Pole of a Composite Displacement

The poles of two displacements $[T_a]$ and $[T_b]$ form a triangle with the pole of their composition $[T_c] = [T_b][T_a]$. Let α and β be the rotation angles of $[T_a]$ and $[T_b]$, respectively, and let γ be the rotation angle for $[T_c]$. If we write these transformations in terms of the coordinates of the poles, we have

$$[A(\gamma), [I - A(\gamma)]\mathbf{C}] = [A(\beta), [I - A(\beta)]\mathbf{B}][A(\alpha), [I - A(\alpha)]\mathbf{A}]. \tag{4.59}$$

Expand this expression and equate the rotation and translation terms to obtain

$$[A(\gamma)] = [A(\beta)][A(\alpha)] = [A(\beta + \alpha)],$$
$$[I - A(\gamma)]\mathbf{C} = [I - A(\beta)]\mathbf{B} + [A(\beta)][I - A(\alpha)]\mathbf{A}. \tag{4.60}$$

The first equation states that $\gamma = \beta + \alpha$. The second provides a formula for the pole $\mathbf{C}$ in terms of the coordinates of the poles $\mathbf{B}$ and $\mathbf{A}$. We will see that this second equation is the formula for a planar triangle with $\alpha/2$ and $\beta/2$ as the angles at the vertices $\mathbf{A}$ and $\mathbf{B}$.

The computation of the product $[T_c] = [T_b][T_a]$ is simplified by using complex vectors, that is,

$$[e^{i\gamma}, (1 - e^{i\gamma})\mathbf{C}] = [e^{i\beta}, (1 - e^{i\beta})\mathbf{B}][e^{i\alpha}, (1 - e^{i\alpha})\mathbf{A}], \tag{4.61}$$

which expands to yield

$$e^{i\gamma} = e^{i\beta} e^{i\alpha},$$
$$(1 - e^{i\gamma})\mathbf{C} = (1 - e^{i\beta})\mathbf{B} + e^{i\beta}(1 - e^{i\alpha})\mathbf{A}. \tag{4.62}$$

We have already seen that the rotation terms yield $\gamma = \alpha + \beta$. The second equation is the complex number form of the equation of a triangle formed by the poles $\mathbf{A}$, $\mathbf{B}$, and $\mathbf{C}$, Figure 4.3.

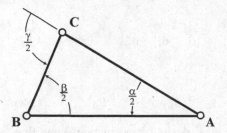

Fig. 4.3 The triangle formed by three points $\mathbf{A}$, $\mathbf{B}$, and $\mathbf{C}$.

4.3.1.1 The Composite Pole Theorem

A fundamental result in the geometry of planar displacements is the relationship between the vertex angles of the triangle formed the poles of a composite displacement and its two factors and the rotation angles of these displacements.

Theorem 5 (The Composite Pole Theorem). *The pole* **C** *of a composite displacement* $[T(\gamma, \mathbf{C})] = [T(\beta, \mathbf{B})][T(\alpha, \mathbf{A})]$ *forms a triangle with the poles* **B** *and* **A**. *If* $\alpha + \beta < 2\pi$, *then* $\alpha/2$ *and* $\beta/2$ *are the interior angles at the vertices* **A** *and* **B**, *respectively, and* $\gamma/2$ *is the exterior angle at* **C**. *If* $\alpha + \beta > 2\pi$, *then* $\alpha/2$ *and* $\beta/2$ *are the exterior angles at* **A** *and* **B**. *Denote by* κ *the interior angle at* **C**, *then* $\gamma/2 = \kappa + \pi$.

Proof. The two cases are distinguished by the location of **C** relative to the segment **BA**. If the sum of α and β is less than 2π, then **C** lies to the left of the directed segment **BA**, in which case the angle $\angle \mathbf{ABC}$ is less than π. If the sum of α and β is greater than 2π, then the angle $\angle \mathbf{ABC}$ is greater than π and **C** lies to the right of **BA**. We consider these cases separately:

Case 1. $\alpha + \beta < 2\pi$

In this case the angles $\alpha/2$ and $\beta/2$ are the interior angles of $\triangle \mathbf{ABC}$ at the vertices **B** and **A**. The exterior angle at **C** is $\gamma/2 = \alpha/2 + \beta/2$. The vector $\mathbf{C} - \mathbf{B}$ defining one side of this triangle can be obtained by rotating the vector $\mathbf{A} - \mathbf{B}$ by the angle $\beta/2$ and rescaling it using the law of sines. The result is

$$\mathbf{C} - \mathbf{B} = \left(\frac{\sin \frac{\alpha}{2}}{\sin(\pi - \frac{\gamma}{2})} \right) e^{i\beta/2} (\mathbf{A} - \mathbf{B}). \tag{4.63}$$

Use the identities

$$\sin \alpha/2 = \frac{-1}{2i}(1 - e^{i\alpha})e^{-i\alpha/2} \quad \text{and} \quad \sin(\frac{\pi - \gamma}{2}) = \sin \gamma/2 \tag{4.64}$$

to obtain the formula

$$(1 - e^{i\gamma})\mathbf{C} = (1 - e^{i\beta})\mathbf{B} + e^{i\beta}(1 - e^{i\alpha})\mathbf{A}. \tag{4.65}$$

This equation is identical to (4.62) and defines the vertex **C** in terms of the vertices **A** and **B** and their interior angles.

Case 2. $\alpha + \beta > 2\pi$

In this case the angles $\alpha/2$ and $\beta/2$ are the exterior angles at the vertices **B** and **A**. The interior angle at **C** is $\kappa = \alpha/2 + \beta/2 - \pi$. The side $\mathbf{C} - \mathbf{B}$ of this triangle is obtained by rotating $\mathbf{A} - \mathbf{B}$ by the angle $\beta/2$ and then rescaling its length by using the sine law. The result is

$$\mathbf{C} - \mathbf{B} = -\left(\frac{\sin(\pi - \frac{\alpha}{2})}{\sin \kappa}\right) e^{i\beta/2}(\mathbf{A} - \mathbf{B}). \tag{4.66}$$

Now, notice that $\sin(\pi - \alpha/2) = \sin \alpha/2$, therefore this equation becomes identical to (4.63) for $\gamma/2 = \kappa + \pi$, that is, $\sin(\gamma/2 - \pi) = -\sin \gamma/2$.

The result is that in both cases we obtain (4.65), which is exactly the equation defining the composite translation (4.62). $\square$

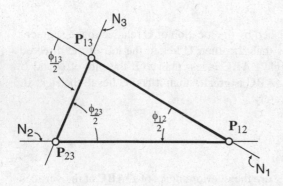

Fig. 4.4 The pole triangle $\triangle \mathbf{P}_{12}\mathbf{P}_{23}\mathbf{P}_{13}$ with sides N_1, N_2, and N_3.

4.3.2 The Triangle of Relative Displacement Poles

Given three positions of a body M_i, $i = 1,2,3$, we have the three relative displacements $[T_{12}]$, $[T_{23}]$, and $[T_{13}]$ between pairs of these positions. The product $[T_{23}][T_{12}]$ is the relative displacement $[T_{13}]$ as can be seen from

$$[T_{23}][T_{12}] = ([T_3][T_2^{-1}])([T_2][T_1^{-1}]) = [T_3][T_1^{-1}] = [T_{13}]. \tag{4.67}$$

Thus, the three relative displacement poles are related by the composite pole theorem, and we have

$$(1 - e^{i\phi_{13}})\mathbf{P}_{13} = (1 - e^{i\phi_{23}})\mathbf{P}_{23} + e^{i\phi_{23}}(1 - e^{i\phi_{12}})\mathbf{P}_{12}. \tag{4.68}$$

This is the equation of the *pole triangle*.

For a general set of three positions M_i, M_j, and M_k we have the relative transformations $[T_{ik}] = [T_{jk}][T_{ij}]$ and the pole triangle

$$(1 - e^{i\phi_{ik}})\mathbf{P}_{ik} = (1 - e^{i\phi_{jk}})\mathbf{P}_{jk} + e^{i\phi_{jk}}(1 - e^{i\phi_{ij}})\mathbf{P}_{ij}. \tag{4.69}$$

The composite pole theorem shows that each of the poles $\mathbf{P}_{ij}$, $\mathbf{P}_{jk}$, and $\mathbf{P}_{ik}$ views the opposite side of the triangle of relative poles in angles directly related to one-half of the associated relative rotation angle, Figure 4.4.

4.3.3 The Image Pole Triangle

Consider relative inverse displacements associated with three positions M_i, M_j, and M_k. Let M be in position M_i and transform coordinates to obtain the image of the relative inverse displacements, $[T_{ij}^i]$, $[T_{jk}^i]$, and $[T_{ik}^i]$. Notice that

$$[T_{ik}^i] = [T_{jk}^i][T_{ij}^i]. \tag{4.70}$$

The composite pole theorem yields the equation of the image pole triangle

$$(1 - e^{-i\phi_{ik}})\mathbf{P}_{ik}^i = (1 - e^{-i\phi_{jk}})\mathbf{P}_{jk}^i + e^{-i\phi_{jk}}(1 - e^{-i\phi_{ij}})\mathbf{P}_{ij}^i. \tag{4.71}$$

Compare this to the pole triangle for three relative displacements (4.69). Recall that $\mathbf{P}_{ik}^i = \mathbf{P}_{ik}$ and $\mathbf{P}_{ij}^i = \mathbf{P}_{ij}$ are the poles of the original relative displacements. Thus, the image pole $\mathbf{P}_{jk}^i$ is the reflection of $\mathbf{P}_{jk}$ through the line joining $\mathbf{P}_{ij}$ and $\mathbf{P}_{ik}$, Figure 4.5.

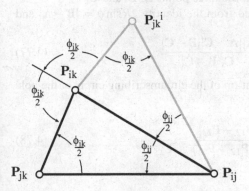

Fig. 4.5 A comparison of the image pole triangle $\triangle\mathbf{P}_{ij}^i\mathbf{P}_{jk}^i\mathbf{P}_{ik}^i$ in position M_i with the pole triangle $\triangle\mathbf{P}_{ij}\mathbf{P}_{jk}\mathbf{P}_{ik}$.

4.3.4 The Circumscribing Circle

The equation of the circle that circumscribes the triangle $\triangle\mathbf{ABC}$ can be obtained as follows. The basic principle we use is that a general point $\mathbf{X}$ on the circle forms the angle $\angle\mathbf{AXB}$ that is equal to $\phi = \angle\mathbf{ACB}$ or $\phi + \pi$.

The cosine of the angle ϕ at $\mathbf{C}$ can be computed from the vectors $\mathbf{A} - \mathbf{C}$ and $\mathbf{B} - \mathbf{C}$ using the formula

$$\cos\phi = \frac{(\mathbf{A}-\mathbf{C})\cdot(\mathbf{B}-\mathbf{C})}{|\mathbf{A}-\mathbf{C}||\mathbf{B}-\mathbf{C}|}. \tag{4.72}$$

The sine of this angle is obtained by using the determinant of the matrix $[\mathbf{A}-\mathbf{C},\mathbf{B}-\mathbf{C}]$, that is,

$$\sin\phi = \frac{|\mathbf{A}-\mathbf{C},\,\mathbf{B}-\mathbf{A}|}{|\mathbf{A}-\mathbf{C}||\mathbf{B}-\mathbf{C}|}. \tag{4.73}$$

Notice that we are using vertical bars to denote the determinant of a matrix. Thus, we have

$$\tan\phi = \frac{|\mathbf{A}-\mathbf{C},\,\mathbf{B}-\mathbf{C}|}{(\mathbf{A}-\mathbf{C})\cdot(\mathbf{B}-\mathbf{C})} = \frac{L_{AB}}{C_{AB}}. \tag{4.74}$$

The same formula defines the angle $\angle\mathbf{AXB}$, and because $\tan\phi = \tan(\phi + \pi)$, we have

$$\mathscr{C}: \frac{|\mathbf{A}-\mathbf{X},\,\mathbf{B}-\mathbf{X}|}{(\mathbf{A}-\mathbf{X})\cdot(\mathbf{B}-\mathbf{X})} = \frac{L_{AB}}{C_{AB}}, \tag{4.75}$$

or

$$\mathscr{C}: |\mathbf{A}-\mathbf{X},\,\mathbf{B}-\mathbf{X}|C_{AB} - (\mathbf{A}-\mathbf{X})\cdot(\mathbf{B}-\mathbf{X})L_{AB} = 0. \tag{4.76}$$

This is the equation of the circle through the three points $\mathbf{A}$, $\mathbf{B}$, and $\mathbf{C}$.

The radius R of this circle is obtained from the identity $2R\sin\phi = |\mathbf{B}-\mathbf{A}|$ and (4.73),

$$R = \frac{1}{2}\frac{|\mathbf{B}-\mathbf{A}||\mathbf{A}-\mathbf{C}||\mathbf{B}-\mathbf{C}|}{|\mathbf{A}-\mathbf{C},\,\mathbf{B}-\mathbf{C}|}. \tag{4.77}$$

As an example, we compute the equation of the circumscribing circle of the pole triangle $\triangle\mathbf{P}_{12}\mathbf{P}_{23}\mathbf{P}_{13}$, given by

$$\frac{|\mathbf{P}_{13}-\mathbf{P}_{23},\,\mathbf{P}_{12}-\mathbf{P}_{23}|}{(\mathbf{P}_{13}-\mathbf{P}_{23})\cdot(\mathbf{P}_{12}-\mathbf{P}_{23})} = \frac{L_{23}}{C_{23}}. \tag{4.78}$$

Thus, we obtain

$$\mathscr{C}: |\mathbf{P}_{13}-\mathbf{X},\,\mathbf{P}_{12}-\mathbf{X}|C_{23} - (\mathbf{P}_{13}-\mathbf{X})\cdot(\mathbf{P}_{12}-\mathbf{X})L_{23} = 0 \tag{4.79}$$

as the equation of this circle.

4.4 Summary

This chapter has presented the algebraic form the geometric concepts introduced in the previous chapter. Of particular importance are the definition of a relative displacement and the properties of the pole triangle. It is interesting that these concepts are relatively easy to understand graphically, while they are difficult to define algebraically. However, we need this latter approach in order to obtain similar results for spatial rotations and general spatial displacements.

4.5 References

The geometric theory presented here can be found in Hartenberg and Denavit [46] as well as in Bottema and Roth [5]. The results on the equation of a triangle are drawn from [78], while the complex vector formulation follows Erdman and Sandor [30].

Exercises

1. Determine the 3×3 homogeneous transform $[T_{12}]$ that defines the planar displacement by constructing the matrix equation using homogeneous coordinates $[\mathbf{A}^2, \mathbf{B}^2, \mathbf{C}^2] = [T_{12}][\mathbf{A}^1, \mathbf{B}^1, \mathbf{C}^1]$. Solve this equation for $[T_{12}]$, using the coordinates in Table 4.1 (Suh and Radcliffe [134]).

Table 4.1 Point coordinates defining two planar positions.

Point	M_1	M_2
$\mathbf{A}^i$	$(2,4,1)^T$	$(5,1,1)^T$
$\mathbf{B}^i$	$(2,6,1)^T$	$(7,1,1)^T$
$\mathbf{C}^i$	$(1,5,1)^T$	$(6,2,1)^T$

2. Show that the coordinates of the pole $\mathbf{P} = (p_x, p_y)^T$ of a displacement $[T] = [A, \mathbf{d}]$ are given by

$$p_x = \frac{\frac{d_x}{2} \sin \frac{\phi}{2} - \frac{d_y}{2} \cos \frac{\phi}{2}}{\sin \frac{\phi}{2}} \quad \text{and} \quad p_y = \frac{\frac{d_x}{2} \cos \frac{\phi}{2} + \frac{d_y}{2} \sin \frac{\phi}{2}}{\sin \frac{\phi}{2}}. \tag{4.80}$$

3. Given two displacements $[T_1] = [A_1, \mathbf{d}_1]$ and $[T_2] = [A_2, \mathbf{d}_2]$, derive a formula for the coordinates of the pole $\mathbf{P}_{12}$ of the relative displacement $[T_{12}]$.

4. Let two planar positions be $M_1 = (0°, 1, 1)$ and $M_2 = (60°, 3, 2)$, and determine the relative position pole $\mathbf{P}_{12}$ for these two displacements (Suh and Radcliffe [134]).

5. Complete the derivation of the equation of the planar triangle using complex numbers to obtain (4.65).

6. Given the coordinates $\mathbf{A} = (3,3)^T$ and $\mathbf{B} = (1,1)^T$ and interior angles $\alpha = 30°$ and $\beta = 60°$, compute the coordinates of the point $\mathbf{C}$.

7. Let $\mathscr{C}^i$ be the reflection of the circumscribing circle of the pole triangle through the side $N_i = \mathbf{P}_{ij}\mathbf{P}_{ik}$. Show that the three circles $\mathscr{C}^i$, $i = 1, 2, 3$, intersect in the *orthocenter* of the pole triangle.

8. Given three positions and the associated pole triangle $\triangle \mathbf{P}_{12}\mathbf{P}_{23}\mathbf{P}_{13}$ determine the image pole triangle $\triangle \mathbf{P}_{12}^1\mathbf{P}_{23}^1\mathbf{P}_{13}^1$ for M in position M_1. Show that this triangle has $\mathbf{P}_{12}^1 = \mathbf{P}_{12}$, $\mathbf{P}_{13}^1 = \mathbf{P}_{13}$ and that $\mathbf{P}_{23}^1$ is the reflection of $\mathbf{P}_{23}$ through the side $N_1 = \mathbf{P}_{12}\mathbf{P}_{13}$.

Chapter 5
Algebraic Synthesis of Planar Chains

In this chapter we examine the design of RR, PR and RP planar open chains that reach a specified set of task positions. A constraint equation is defined for each chain that characterizes the set of positions that it can reach. This relationship is inverted by considering the positions as known and the fixed and moving pivots of the chain as unknowns. The result is a set of design equations that are solved to design the chain.

Two of these chains can be connected to the moving body to form a one-degree-of-freedom planar four-bar linkage. This closed chain can be used as a function generator which provides coordinated movement of the input and output links. Such a connection, however, limits the movement of the two chains and can interfere with the smooth travel of the workpiece through the task positions. Techniques used to avoid this problem are known as solution rectification.

5.1 A Single Revolute Joint

A revolute, or hinged, joint provides pure rotation about a point. Given two positions of a rigid body, M_1 and M_2, we can locate a revolute joint such that it moves the body between the two positions. This is easily done by locating the hinged joint at the pole of the relative displacement.

Let the two positions be specified by the transformations $[T_1] = [A(\phi_1), \mathbf{d}_1]$ and $[T_2] = [A(\phi_2), \mathbf{d}_2]$. Then, locate the revolute joint $\mathbf{G}$ at the relative pole

$$\mathbf{G} = [I - A(\phi_{12})]^{-1}(\mathbf{d}_2 - A(\phi_{12})\mathbf{d}_1), \tag{5.1}$$

where $\phi_{12} = \phi_2 - \phi_1$. Notice that this joint does not exist if the relative displacement is a pure translation.

The point $\mathbf{g}$ in the moving body M that is to be connected to the hinge $\mathbf{G}$ is the pole of the relative inverse displacement $[T_{12}^{\dagger}]$, which is obtained from (4.39) to define

J.M. McCarthy and G.S. Soh, *Geometric Design of Linkages*, Interdisciplinary Applied Mathematics 11, DOI 10.1007/978-1-4419-7892-9_5,
© Springer Science+Business Media, LLC 2011

$$\mathbf{g} = -[A_2 - A_1]^{-1}(\mathbf{d}_2 - \mathbf{d}_1). \tag{5.2}$$

The locations of the point $\mathbf{G}$ in F and $\mathbf{g}$ in M are uniquely defined by the two task positions. In the following sections we show that we can design an RR chain that reaches as many as five task positions.

5.2 The Geometry of RR Chains

An RR chain consists of a fixed revolute joint located at a point $\mathbf{G} = (x, y)^T$ in F connected by a link to a moving revolute joint located at $\mathbf{w}$ in M. Let $[T] = [A, \mathbf{d}]$ be a displacement that locates M. Then the point $\mathbf{W}$ in F that coincides with $\mathbf{w}$ is given by

$$\mathbf{W} = [A]\mathbf{w} + \mathbf{d}. \tag{5.3}$$

Clearly, $\mathbf{W} = (\lambda, \mu)^T$ must lie on a circle about the fixed pivot $\mathbf{G}$, that is,

$$(\mathbf{W} - \mathbf{G}) \cdot (\mathbf{W} - \mathbf{G}) = (\lambda - x)^2 + (\mu - y)^2 = R^2, \tag{5.4}$$

where R is the length of the link. This geometric constraint characterizes the RR chain.

5.2.1 Perpendicular Bisectors

Let n positions of the end-link of an RR chain be defined by the transformations $[T_i]$, $i = 1, \ldots, n$. The coordinates $\mathbf{W}^i$ of the moving pivot must satisfy (5.4) for each position M_i, and we have the n equations

$$(\mathbf{W}^i - \mathbf{G}) \cdot (\mathbf{W}^i - \mathbf{G}) = |\mathbf{W}^i|^2 - 2\mathbf{W}^i \cdot \mathbf{G} + |\mathbf{G}|^2 = R^2, i = 1, \ldots, n. \tag{5.5}$$

Subtract the first equation from the others to cancel the terms $|\mathbf{G}|^2$ and R^2. The result is

$$(\mathbf{W}^i - \mathbf{W}^1) \cdot \mathbf{G} - \frac{1}{2}(|\mathbf{W}^i|^2 - |\mathbf{W}^1|^2) = 0, i = 2, \ldots, n. \tag{5.6}$$

We now show that (5.6) defines the perpendicular bisector of the segment joining $\mathbf{W}^1$ to $\mathbf{W}^i$. Rewrite the second term in this equation as

$$\mathbf{W}^i \cdot \mathbf{W}^i - \mathbf{W}^1 \cdot \mathbf{W}^1 = (\mathbf{W}^i + \mathbf{W}^1) \cdot (\mathbf{W}^i - \mathbf{W}^1). \tag{5.7}$$

Introduce the midpoint $\mathbf{V}_{1i} = (\mathbf{W}^i + \mathbf{W}^1)/2$ and substitute this into (5.6) to obtain

$$(\mathbf{W}^i - \mathbf{W}^1) \cdot (\mathbf{G} - \mathbf{V}_{1i}) = 0. \tag{5.8}$$

Thus, $\mathbf{G} - \mathbf{V}_{1i}$ is perpendicular to the vector $\mathbf{W}^i - \mathbf{W}^1$ and passes through its mid-point, Figure 5.1. This is an algebraic expression of the fact that the perpendicular bisectors of all chords of a circle must pass through its center.

The pole $\mathbf{P}_{1i}$ of the relative displacement $[T_{1i}]$ of the end-link of the RR chain also lies on the perpendicular bisector of $\mathbf{W}^i - \mathbf{W}^1$. This means that we can replace the vector $\mathbf{G} - \mathbf{V}_{1i}$ in (5.8) by $\mathbf{G} - \mathbf{P}_{1i}$ to obtain

$$(\mathbf{W}^i - \mathbf{W}^1) \cdot (\mathbf{G} - \mathbf{P}_{1i}) = 0, i = 2, \ldots, n. \qquad (5.9)$$

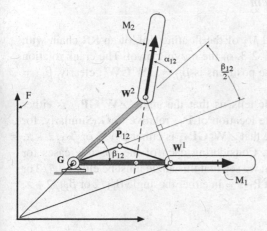

Fig. 5.1 The fixed pivot $\mathbf{G}$ lies on the perpendicular bisector of the segment $\mathbf{W}^1\mathbf{W}^2$ formed by two positions of the moving pivot.

5.2.2 The Dyad Triangle

The displacement of the end-link from M_1 to M_i is the composite of a rotation of angle α_{1i} about $\mathbf{W}^1$ followed by a rotation β_{1i} about $\mathbf{G}$. The result is a rotation by ϕ_{1i} about the pole $\mathbf{P}_{1i}$. Thus, $[T_{1i}]$ is given by

$$[T(\phi_{1i}, \mathbf{P}_{1i})] = [T(\beta_{1i}, \mathbf{G})][T(\alpha_{1i}, \mathbf{W}^1)]. \qquad (5.10)$$

This equation can be obtained from the kinematics equations of the RR chain, see equation (E.4). The composite pole theorem connects the geometry of the triangle $\triangle \mathbf{W}^1\mathbf{G}\mathbf{P}_{1i}$, called the *dyad triangle*, and the rotation angles α_{1i}, β_{1i}, and ϕ_{1i}.

Expand (5.10) using the notation of complex vectors to obtain

$$e^{i\phi_{1i}} = e^{i(\beta_{1i}+\alpha_{1i})},$$
$$(1 - e^{i\phi_{1i}})\mathbf{P}_{1i} = (1 - e^{i\beta_{1i}})\mathbf{G} + e^{i\beta_{1i}}(1 - e^{i\alpha_{1i}})\mathbf{W}^1. \qquad (5.11)$$

The first equation shows that $\phi_{1i} = \beta_{1i} + \alpha_{1i}$. The second equation is the equation of the dyad triangle $\triangle \mathbf{W}^1 \mathbf{G} \mathbf{P}_{1i}$. Thus, from the composite pole theorem we have:

1. If $\alpha_{1i} + \beta_{1i} < 2\pi$, then the interior angles at $\mathbf{W}^1$ and $\mathbf{G}$ are $\alpha_{1i}/2$ and $\beta_{1i}/2$, respectively; and the exterior angle at $\mathbf{P}_{1i}$ is $\phi_{1i}/2$.
2. If $\alpha_{1i} + \beta_{1i} > 2\pi$, then $\alpha_{1i}/2$ and $\beta_{1i}/2$ are the exterior angles at $\mathbf{W}^1$ and $\mathbf{G}$, respectively. Denote by κ the interior angle at $\mathbf{P}_{1i}$, then $\phi_{1i}/2 = \kappa + \pi$.

5.2.3 The Center-Point Theorem

Consider three positions M_1, M_2, and M_3 of the floating link of an RR chain with the corresponding positions $\mathbf{W}^i$, $i = 1, 2, 3$, of the moving pivot. The crank rotation angle around $\mathbf{G}$ between each of these positions is $\beta_{ij} = \angle \mathbf{W}^i \mathbf{G} \mathbf{W}^j$; clearly, $\beta_{13} = \beta_{23} + \beta_{12}$.

The geometry of the dyad triangle tells us that the angle $\angle \mathbf{W}^1 \mathbf{G} \mathbf{P}_{12}$ is either $\beta_{12}/2$ or $\beta_{12}/2 + \pi$, depending on the location of $\mathbf{P}_{12}$ relative to $\mathbf{G}$. Similarly, for the dyad triangle $\triangle \mathbf{W}^2 \mathbf{G} \mathbf{P}_{23}$ we have that $\angle \mathbf{W}^2 \mathbf{G} \mathbf{P}_{23}$ is either $\beta_{23}/2$ or $\beta_{23}/2 + \pi$. Notice that $\angle \mathbf{W}^1 \mathbf{G} \mathbf{P}_{12} = \angle \mathbf{P}_{12} \mathbf{G} \mathbf{W}^2$. Considering each of the possible cases for $\angle \mathbf{P}_{12} \mathbf{G} \mathbf{P}_{23} = \angle \mathbf{P}_{12} \mathbf{G} \mathbf{W}^2 + \angle \mathbf{W}^2 \mathbf{G} \mathbf{P}_{23}$, we see that this angle must be either $\beta_{13}/2$ or $\beta_{13}/2 + \pi$. Thus, $\mathbf{G}$ views the segment $\mathbf{P}_{12} \mathbf{P}_{23}$ in either the angle $\beta_{13}/2$ or $\beta_{13}/2 + \pi$, see Figure 5.2.

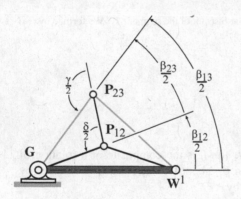

Fig. 5.2 The two dyad triangles $\triangle \mathbf{W}^1 \mathbf{G} \mathbf{P}_{12}$ and $\triangle \mathbf{W}^1 \mathbf{G} \mathbf{P}_{13}$ define a third triangle $\mathbf{P}_{12} \mathbf{G} \mathbf{P}_{13}$.

This generalizes to the theorem already presented in chapter 3, which is the foundation for Burmester's approach to RR chain design:

Theorem 6 (The Center-Point Theorem). *The center point of an RR chain that reaches positions M_i, M_j, and M_k views the relative displacement poles $\mathbf{P}_{ij}$ and $\mathbf{P}_{jk}$ in the angle $\beta_{ik}/2$ or $\beta_{ik}/2 + \pi$, where β_{ik} is the crank rotation angle from position M_i to M_k.*

We now determine the equation of the triangle formed by $\mathbf{G}$ and the two poles $\mathbf{P}_{12}$ and $\mathbf{P}_{23}$. Let the vector $\mathbf{P}_{12} - \mathbf{G}$ be the base of the triangle with interior angles at vertices $\mathbf{P}_{12}$ and $\mathbf{G}$ given by $\delta/2$ and $\beta_{13}/2$. Then the exterior angle $\gamma/2$ at $\mathbf{P}_{23}$ is given by $\gamma/2 = \delta/2 + \beta_{13}/2$, and the equation of the triangle (4.62) yields

$$(1 - e^{i\gamma})\mathbf{P}_{23} = (1 - e^{i\beta_{13}})\mathbf{G} + e^{i\beta_{13}}(1 - e^{i\delta})\mathbf{P}_{12}. \tag{5.12}$$

Substitute for δ and solve for $\mathbf{G}$ to obtain

$$\mathscr{C}: \mathbf{G} = \frac{\mathbf{P}_{23} - e^{i\beta_{13}}\mathbf{P}_{12}}{1 - e^{i\beta_{13}}} + \left(\frac{\mathbf{P}_{23} - \mathbf{P}_{12}}{1 - e^{i\beta_{13}}}\right)e^{i\gamma}. \tag{5.13}$$

If we fix β_{13} and let γ vary in this equation, then we obtain a circle with center $\mathbf{C}$ given by

$$\mathbf{C} = \frac{\mathbf{P}_{12} - e^{i\beta_{13}}\mathbf{P}_{23}}{1 - e^{i\beta_{13}}}. \tag{5.14}$$

This circle has the property that the central angle measured from $\mathbf{P}_{12}$ to $\mathbf{P}_{23}$ is β_{13}. Thus, any point $\mathbf{G}$ on this circle views the segment $\mathbf{P}_{12}\mathbf{P}_{23}$ in either $\beta_{13}/2$ or $\beta_{13}/2 + \pi$.

5.3 Finite Position Synthesis of RR Chains

In order to design an RR chain we identify a set of task positions M_i, $i = 1, \ldots, n$, for the end-link of the chain. This means that the displacements $[T_i], i = 1, \ldots, n$, are known, and the angles ϕ_{1i} and the relative poles $\mathbf{P}_{1i}$ can be determined at the outset. The unknowns are the two coordinates of the fixed pivot $\mathbf{G} = (x, y)^T$ and the two coordinates of the moving pivot $\mathbf{W}^1 = (\lambda, \mu)^T$, four in all.

5.3.1 The Algebraic Design Equations

The equations (5.9) can be formulated in a way that yields a convenient set of algebraic design equations for an RR chain. Starting with the relation

$$\mathbf{W}^i - \mathbf{P}_{1i} = [A(\phi_{1i})](\mathbf{W}^1 - \mathbf{P}_{1i}), \tag{5.15}$$

we subtract $\mathbf{W}^1 - \mathbf{P}_{1i}$ from both side to obtain

$$\mathbf{W}^i - \mathbf{W}^1 = [A(\phi_{1i}) - I](\mathbf{W}^1 - \mathbf{P}_{1i}). \tag{5.16}$$

Substitute this into (5.9) to obtain the equations

$$\mathscr{D}_{1i}: (\mathbf{G} - \mathbf{P}_{1i}) \cdot [A(\phi_{1i}) - I](\mathbf{W}^1 - \mathbf{P}_{1i}) = 0, i = 2, \ldots, n, \tag{5.17}$$

which we call the design equations for the RR chain.

Notice that when $n = 5$, we have four design equations in four unknowns. Thus, an RR chain can be designed to reach five arbitrarily specified precision positions.

5.3.1.1 The Bilinear Structure

The design equations (5.17) are quadratic in the four unknowns $\mathbf{G} = (x,y)^T$ and $\mathbf{W}^1 = (\lambda,\mu)^T$. However, they have the important property that they are linear when considered separately in the unknowns x, y and λ, μ. This structure provides a convenient strategy for the solution the five position problem.

However, before considering five-position synthesis, we examine the subproblems of design for two-, three- and four-precision positions. In these cases, the bilinear structure provides alternative solutions that we describe as "select the fixed pivot" or "select the moving pivot." These solution strategies correspond to the two ways the design equations can be used to design RR chains.

Let the coordinates of the relative pole be $\mathbf{P}_{1i} = (p_i,q_i)^T$. Then we can expand (5.17) to obtain

$$\begin{Bmatrix} x - p_i \\ y - q_i \end{Bmatrix}^T \begin{bmatrix} \cos\phi_{1i} - 1 & -\sin\phi_{1i} \\ \sin\phi_{1i} & \cos\phi_{1i} - 1 \end{bmatrix} \begin{Bmatrix} \lambda - p_i \\ \mu - q_i \end{Bmatrix} = 0, i = 2,\ldots,n. \tag{5.18}$$

If we select the fixed pivot $\mathbf{G}$ then the coordinates x, y are known, and we can collect the coefficients of λ and μ to obtain the design equations

$$A_i(x,y)\lambda + B_i(x,y)\mu = C_i(x,y), i = 2,\ldots,n, \tag{5.19}$$

where

$$A_i(x,y) = (\cos\phi_{1i} - 1)(x - p_i) + \sin\phi_{1i}(y - q_i),$$
$$B_i(x,y) = -\sin\phi_{1i}(x - p_i) + (\cos\phi_{1i} - 1)(y - q_i),$$
$$C_i(x,y) = (\cos\phi_{1i} - 1)(p_i(x - p_i) + q_i(y - q_i)) + \sin\phi_{1i}(p_i y - q_i x).$$

The coordinates of the moving pivot (λ,μ) are obtained by solving this set of linear equations.

On the other hand, if we select the moving pivot $\mathbf{W}^1$, then λ, μ are known, and we can collect the coefficients of x and y to obtain

$$A_i'(\lambda,\mu)x + B_i'(\lambda,\mu)y = C_i'(\lambda,\mu), i = 2,\ldots,n, \tag{5.20}$$

where

$$A_i'(\lambda,\mu) = (\cos\phi_{1i} - 1)(\lambda - p_i) - \sin\phi_{1i}(\mu - q_i),$$
$$B_i'(\lambda,\mu) = \sin\phi_{1i}(\lambda - p_i) + (\cos\phi_{1i} - 1)(\mu - q_i),$$
$$C_i'(\lambda,\mu) = (\cos\phi_{1i} - 1)(p_i(\lambda - p_i) + q_i(\mu - q_i)) - \sin\phi_{1i}(p_i\mu - q_i\lambda).$$

Thus, the fixed pivot coordinates x, y are obtained by solving a set of linear equations, as well.

5.3.2 Parameterized Form of the Design Equations

The equations of the dyad triangle provide a set of design equations that include the crank rotation angles β_{1i}. These equations provide a way to select a crank angle in the design process.

For a set of task positions M_i, $i = 1, \ldots, n$, we have the $n-1$ dyad triangle equations

$$(1 - e^{i\phi_{1i}})\mathbf{P}_{1i} = (1 - e^{i\beta_{1i}})\mathbf{G} + e^{i\beta_{1i}}(1 - e^{i\alpha_{1i}})\mathbf{W}^1, \; i = 2, \ldots, n, \tag{5.21}$$

which are linear in the in the unknown complex vectors $\mathbf{G} = x + iy$ and $\mathbf{W}^1 = \lambda + i\mu$. If the crank angles β_{1i} are specified, then the rotation angles $\alpha_{1i} = \phi_{1i} - \beta_{1i}$ are known as well.

5.3.3 Two Precision Positions

If two positions M_1 and M_2 of the end-link are specified, then the displacements $[T_1] = [A(\phi_1), \mathbf{d}_1]$ and $[T_2] = [A(\phi_2), \mathbf{d}_2]$ are given. The relative rotation angle ϕ_{12} and the pole $\mathbf{P}_{12}$ can be determined. Because $n = 2$, there is a single design equation

$$(\mathbf{G} - \mathbf{P}_{12}) \cdot [A(\phi_{12}) - I](\mathbf{W}^1 - \mathbf{P}_{12}) = 0. \tag{5.22}$$

To design the RR chain we can select either the fixed or moving pivot, and still have a free parameter.

5.3.3.1 Select the Fixed Pivot

Choose values for the coordinates of the fixed pivot $\mathbf{G} = (x, y)^T$, then (5.19) yields the equation

$$A_2(x,y)\lambda + B_2(x,y)\mu = C_2(x,y), \tag{5.23}$$

for the coordinates $\mathbf{W}^1 = (\lambda, \mu)^T$ of the moving pivot. This is a single equation relating λ and μ. Simply choose one and compute the other.

5.3.3.2 Select the Moving Pivot

The bilinearity of the design equations allows us to select values for the coordinates $\mathbf{W}^1 = (\lambda, \mu)^T$ of the moving pivot, and use (5.20) to define

$$A'(\lambda,\mu)x + B'(\lambda,\mu)y = C'(\lambda,\mu). \tag{5.24}$$

This is the equation of the perpendicular bisector of the segment $\mathbf{W}^1\mathbf{W}^2$. Any point on this line can be used as the center point $\mathbf{G}$.

5.3.3.3 Select the Crank Angle

The equation of the dyad triangle can be used to specify the crank angle β_{12} between the two positions M_1 and M_2. In this case (5.22) yields a linear equation in the complex vectors $\mathbf{G} = x + iy$ and $\mathbf{W}^1 = \lambda + i\mu$:

$$(1 - e^{i\phi_{12}})\mathbf{P}_{12} = (1 - e^{i\beta_{12}})\mathbf{G} + e^{i\beta_{12}}(1 - e^{i\alpha_{12}})\mathbf{W}^1. \tag{5.25}$$

Recall that $\alpha_{12} = \phi_{12} - \beta_{12}$. We can choose either of the vectors $\mathbf{G}$ or $\mathbf{W}^1$ and solve this equation for the other one. Notice that the free parameter that existed in the previous solutions has been used here to define the crank angle β_{12}.

5.3.4 Three Precision Positions

For the case of three specified positions of the floating link, we have the three displacements $[T_i] = [A(\phi_i), \mathbf{d}_i], i = 1, 2, 3$. Compute the relative angles ϕ_{12}, ϕ_{13} and the poles $\mathbf{P}_{12}, \mathbf{P}_{13}$ in order to obtain the pair of design equations

$$(\mathbf{G} - \mathbf{P}_{1i}) \cdot [A(\phi_{1i}) - I](\mathbf{W}^1 - \mathbf{P}_{1i}) = 0, i = 2, 3. \tag{5.26}$$

These equations yield a unique solution for either the fixed pivot $\mathbf{G}$ or the moving pivot $\mathbf{W}^1$ for an arbitrary choice of the other.

5.3.4.1 Select the Fixed Pivot

Choose values for the coordinates of $\mathbf{G}$ and assemble the two design equations (5.19) into the matrix equation

$$\begin{bmatrix} A_1(x,y) & B_1(x,y) \\ A_2(x,y) & B_2(x,y) \end{bmatrix} \begin{Bmatrix} \lambda \\ \mu \end{Bmatrix} = \begin{Bmatrix} C_1(x,y) \\ C_2(x,y) \end{Bmatrix}. \tag{5.27}$$

Solve these equations to obtain a unique moving pivot $\mathbf{W}^1$.

5.3.4.2 Select the Moving Pivot

We may specify the coordinates of $\mathbf{W}^1 = (\lambda, \mu)^T$ and write the design equations (5.20) in matrix form to obtain

$$\begin{bmatrix} A_1'(\lambda, \mu) & B_1'(\lambda, \mu) \\ A_2'(\lambda, \mu) & B_2'(\lambda, \mu) \end{bmatrix} \begin{Bmatrix} x \\ y \end{Bmatrix} = \begin{Bmatrix} C_1'(\lambda, \mu) \\ C_2'(\lambda, \mu) \end{Bmatrix}. \tag{5.28}$$

These equations define the two perpendicular bisectors $\mathscr{D}_{12}$ and $\mathscr{D}_{13}$ that intersect at the point $\mathbf{G}$, see Figure 5.3.

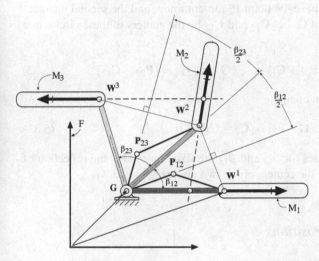

Fig. 5.3 The fixed pivot $\mathbf{G}$ is the intersection of the two bisectors $\mathbf{V}_{12}$ and $\mathbf{V}_{23}$.

5.3.4.3 Select the Crank Angles

The relative displacements from M_1 to M_2 and from to M_3 yield the two dyad triangle equations

$$\begin{bmatrix} (1 - e^{i\beta_{12}}) & e^{i\beta_{12}}(1 - e^{i\alpha_{12}}) \\ (1 - e^{i\beta_{13}}) & e^{i\beta_{13}}(1 - e^{i\alpha_{13}}) \end{bmatrix} \begin{Bmatrix} \mathbf{G} \\ \mathbf{W}^1 \end{Bmatrix} = \begin{Bmatrix} (1 - e^{i\phi_{12}})\mathbf{P}_{12} \\ (1 - e^{i\phi_{13}})\mathbf{P}_{13} \end{Bmatrix}. \tag{5.29}$$

Choose values for the crank angles β_{12} and β_{13} and compute $\alpha_{1i} = \phi_{1i} - \beta_{1i}$. Then Cramer's rule yields unique coordinate vectors $\mathbf{G}$ and $\mathbf{W}^1$.

Another approach to this problem uses the center-point theorem to determine the fixed pivot $\mathbf{G}$ that has selected values for the crank angles β_{12} and β_{13}. To be the desired fixed pivot $\mathbf{G}$ must view the sides $\mathbf{P}_{13}\mathbf{P}_{23}$ and $\mathbf{P}_{12}\mathbf{P}_{23}$ of the pole triangle in the angles $\beta_{12}/2$ and $\beta_{13}/2$, respectively.

This is achieved by determining the circle that has the segment $\mathbf{P}_{12}\mathbf{P}_{23}$ as a chord with arc length β_{13}. From (5.13) we have

$$\mathscr{C}_{13} : \mathbf{G} = \frac{\mathbf{P}_{23} - e^{i\beta_{13}}\mathbf{P}_{12}}{1 - e^{i\beta_{13}}} + \left(\frac{\mathbf{P}_{23} - \mathbf{P}_{12}}{1 - e^{i\beta_{13}}}\right)e^{i\gamma}. \tag{5.30}$$

In the same way, the circle that has the $\mathbf{P}_{13}\mathbf{P}_{23}$ as a chord with arc length β_{12} is given by

$$\mathscr{C}_{12} : \mathbf{G} = \frac{\mathbf{P}_{23} - e^{i\beta_{12}}\mathbf{P}_{13}}{1 - e^{i\beta_{12}}} + \left(\frac{\mathbf{P}_{23} - \mathbf{P}_{13}}{1 - e^{i\beta_{12}}}\right)e^{i\gamma}. \tag{5.31}$$

The circles $\mathscr{C}_{13}$ and $\mathscr{C}_{12}$ have the point $\mathbf{P}_{23}$ in common, and the second intersection is the desired fixed pivot $\mathbf{G}$. Let $\mathbf{C}_{13}$ and $\mathbf{C}_{12}$ be the centers of these circles and note that

$$(\mathbf{G} - \mathbf{C}_{13})^2 = (\mathbf{P}_{23} - \mathbf{C}_{13})^2, \quad (\mathbf{G} - \mathbf{C}_{12})^2 = (\mathbf{P}_{23} - \mathbf{C}_{12})^2. \tag{5.32}$$

From these relations, we can compute

$$(\mathbf{G} - \mathbf{P}_{23}) \cdot (\mathbf{C}_2 - \mathbf{C}_1) = 0. \tag{5.33}$$

Thus, for a given set of values for β_{12} and β_{13} the fixed pivot $\mathbf{G}$ is the reflection of $\mathbf{P}_{23}$ through the line joining the centers of the two circles, $\mathscr{C}_{13}$ and $\mathscr{C}_{12}$.

5.3.5 Four Precision Positions

In order to find RR cranks that reach four design positions we must find points $\mathbf{W}^i$, $i = 1, 2, 3, 4$, that lie on a circle. Clearly, an arbitrary point will not satisfy this condition. However, this does not mean that no such points exist. In fact, there is a cubic curve of moving pivots called the *circle-point* curve that have four positions on a circle. The centers of all of these circles form the *center-point curve*.

5.3.5.1 The Center-Point Curve

Given four specified positions M_i, $i = 1, 2, 3, 4$, we can determine the relative displacements $[T_{1i}] = [A(\phi_{1i}, \mathbf{d}_{1i})]$ and define the matrix form of the design equations

$$\begin{bmatrix} A_2(x,y) & B_2(x,y) \\ A_3(x,y) & B_3(x,y) \\ A_4(x,y) & B_4(x,y) \end{bmatrix} \begin{Bmatrix} \lambda \\ \mu \end{Bmatrix} = \begin{Bmatrix} C_2(x,y) \\ C_3(x,y) \\ C_4(x,y) \end{Bmatrix}. \tag{5.34}$$

There is a solution for the moving pivot $\mathbf{W}^1$ only if these three equations are linearly dependent.

For these equations to have a solution, the fixed pivot $\mathbf{G}$ must be selected so the 3×3 augmented coefficient matrix $[M] = [A_i, B_i, C_i]$ is of rank two. This means that the determinant $|M|$ equals zero, which yields a cubic polynomial

$$\mathcal{R}(x,y): \quad |M| = a_{30}y^3 + (a_{21}x + a_{20})y^2 + (a_{12}x^2 + a_{11}x + a_{10})y$$
$$+ a_{03}x^3 + a_{02}x^2 + a_{01}x + a_{00} = 0. \tag{5.35}$$

This polynomial defines a cubic curve in the fixed frame, and any point on this curve may be chosen as the center point $\mathbf{G}$ for the RR chain. This is the *center-point curve*.

Formulas for the coefficients in (5.35) are obtained by noting that each of the elements of $[M]$ are linear in the components of $\mathbf{G} = (x,y)^T$. Introducing the column vectors $\mathbf{a}_i$, $\mathbf{b}_i$, and $\mathbf{c}_i$, we have

$$\det[M] = \left| \mathbf{a}_1 x + \mathbf{b}_1 y + \mathbf{c}_1, \ \mathbf{a}_2 x + \mathbf{b}_2 y + \mathbf{c}_2, \ \mathbf{a}_3 x + \mathbf{b}_3 y + \mathbf{c}_3 \right| = 0. \tag{5.36}$$

The linearity of the determinant allows us to expand this expression to define the coefficients of the center-point curve as

$$a_{30} = |\mathbf{b}_1 \mathbf{b}_2 \mathbf{b}_3|,$$
$$a_{21} = |\mathbf{a}_1 \mathbf{b}_2 \mathbf{b}_3| + |\mathbf{b}_1 \mathbf{a}_2 \mathbf{b}_3| + |\mathbf{b}_1 \mathbf{b}_2 \mathbf{a}_3|,$$
$$a_{20} = |\mathbf{b}_1 \mathbf{b}_2 \mathbf{c}_3| + |\mathbf{b}_1 \mathbf{c}_2 \mathbf{b}_3| + |\mathbf{c}_1 \mathbf{b}_2 \mathbf{b}_3|,$$
$$a_{12} = |\mathbf{a}_1 \mathbf{a}_2 \mathbf{b}_3| + |\mathbf{a}_1 \mathbf{b}_2 \mathbf{a}_3| + |\mathbf{b}_1 \mathbf{a}_2 \mathbf{a}_3|,$$
$$a_{11} = |\mathbf{a}_1 \mathbf{b}_2 \mathbf{c}_3| + |\mathbf{a}_1 \mathbf{c}_2 \mathbf{b}_3| + |\mathbf{b}_1 \mathbf{a}_2 \mathbf{c}_3| + |\mathbf{b}_1 \mathbf{c}_2 \mathbf{a}_3| + |\mathbf{c}_1 \mathbf{a}_2 \mathbf{b}_3| + |\mathbf{c}_1 \mathbf{b}_2 \mathbf{a}_3|,$$
$$a_{10} = |\mathbf{b}_1 \mathbf{c}_2 \mathbf{c}_3| + |\mathbf{c}_1 \mathbf{b}_2 \mathbf{c}_3| + |\mathbf{c}_1 \mathbf{c}_2 \mathbf{b}_3|,$$
$$a_{03} = |\mathbf{a}_1 \mathbf{a}_2 \mathbf{a}_3|,$$
$$a_{02} = |\mathbf{a}_1 \mathbf{a}_2 \mathbf{c}_3| + |\mathbf{a}_1 \mathbf{c}_2 \mathbf{a}_3| + |\mathbf{c}_1 \mathbf{a}_2 \mathbf{a}_3|,$$
$$a_{01} = |\mathbf{a}_1 \mathbf{c}_2 \mathbf{c}_3| + |\mathbf{c}_1 \mathbf{a}_2 \mathbf{c}_3| + |\mathbf{c}_1 \mathbf{c}_2 \mathbf{a}_3|,$$
$$a_{00} = |\mathbf{c}_1 \mathbf{c}_2 \mathbf{c}_3|. \tag{5.37}$$

5.3.5.2 Burmester's Theorem

The center-point theorem provides a geometric condition that characterizes center points for four precision positions. Given four positions, there are six relative displacement poles $\mathbf{P}_{ij}$, $i < j = 1, 2, 3, 4$, and the center-point theorem requires that a fixed pivot $\mathbf{G}$ view the pole pairs $\mathbf{P}_{ij}\mathbf{P}_{ik}$ and $\mathbf{P}_{mj}\mathbf{P}_{mk}$ in the angle $\beta_{jk}/2$ or $\beta_{jk}/2 + \pi$.

Burmester [7] assembled the six relative poles into the three complementary pairs $\mathbf{P}_{12}\mathbf{P}_{34}$, $\mathbf{P}_{13}\mathbf{P}_{24}$, and $\mathbf{P}_{14}\mathbf{P}_{23}$ such that each pair has the numbers 1 through 4 in its indices. He then introduced the *opposite-pole quadrilateral* that has any two of these complementary pairs as its diagonals, Figure 5.4. This construction ensures that the opposite sides of the opposite-pole quadrilateral have the form needed to apply the center-point theorem. The result is Burmester's theorem presented in the chapter 3, which we repeat here:

Theorem 7 (Burmester's Theorem). *The center point* **G** *of an RR chain that can reach four specified positions in the plane views opposite sides of an opposite-pole quadrilateral obtained from the relative poles of the given positions in angles that are equal, or differ by* π.

Proof. Burmester's definition of the opposite-pole quadrilateral ensures that opposite sides have the form $\mathbf{P}_{ij}\mathbf{P}_{ik}$ and $\mathbf{P}_{mj}\mathbf{P}_{mk}$. The center-point theorem states that **G** must view $\mathbf{P}_{ij}\mathbf{P}_{ik}$ in the angle $\beta_{jk}/2$ or $\beta_{jk}/2 + \pi$, where β_{jk} is the angle from position M_j to M_k. Similarly, it must view the $\mathbf{P}_{mj}\mathbf{P}_{mk}$ in either $\beta_{jk}/2$ or $\beta_{jk}/2 + \pi$. Consider the various combinations to see that **G** views these sides in angles that are equal, or differ by π. □

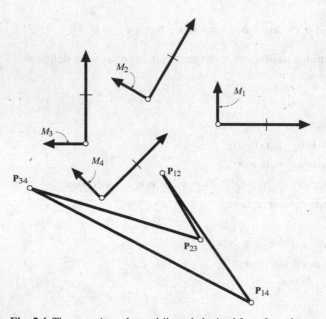

Fig. 5.4 The opposite-pole quadrilateral obtained from four planar positions.

Burmester's theorem provides a way to derive the center-point curve in terms of the coordinates of the relative displacement poles. Let the opposite-pole quadrilateral be formed with vertices $\mathscr{Q}: \mathbf{P}_{12}\mathbf{P}_{23}\mathbf{P}_{34}\mathbf{P}_{14}$, and assume the fixed pivot **G** views $\mathbf{P}_{23}\mathbf{P}_{12}$ in the angle κ. Then **G** must view $\mathbf{P}_{34}\mathbf{P}_{14}$ in either κ or $\kappa + \pi$.

We determine the angle $\angle \mathbf{P}_{12}\mathbf{G}\mathbf{P}_{23} = \kappa$ by separately determining $\sin \kappa$ and $\cos \kappa$. Introduce, for the moment, a third coordinate direction $\vec{k}$ perpendicular to the plane, and consider our coordinate vectors to be three-dimensional with zeros as the third component. This allows us to compute $\sin \kappa$ using the vector cross product

$$\vec{k} \cdot (\mathbf{P}_{12} - \mathbf{G}) \times (\mathbf{P}_{23} - \mathbf{G}) = \sin \kappa |\mathbf{P}_{12} - \mathbf{G}||\mathbf{P}_{23} - \mathbf{G}|. \qquad (5.38)$$

This quantity is the determinant of the 2×2 matrix $[\mathbf{P}_{12} - \mathbf{G}, \mathbf{P}_{23} - \mathbf{G}]$, so we have

$$\sin \kappa |\mathbf{P}_{12} - \mathbf{G}||\mathbf{P}_{23} - \mathbf{G}| = |\mathbf{P}_{12} - \mathbf{G}, \mathbf{P}_{23} - \mathbf{G}|. \tag{5.39}$$

The cosine of κ is obtained using the dot product

$$\cos \kappa |\mathbf{P}_{12} - \mathbf{G}||\mathbf{P}_{23} - \mathbf{G}| = (\mathbf{P}_{12} - \mathbf{G}) \cdot (\mathbf{P}_{23} - \mathbf{G}). \tag{5.40}$$

Divide these two equations, and substitute $\mathbf{P}_{12} = (p_2, q_2)^T$, $\mathbf{P}_{23} = (a_1, b_1)^T$ to obtain

$$\tan \kappa = \frac{(b_1 - q_2)x + (a_1 - p_2)y + p_2 b_1 - q_2 a_1}{x^2 + y^2 - (p_2 + a_1)x - (q_2 + b_1)y + p_2 a_1 + q_2 b_1} = \frac{L_{12}}{C_{12}}. \tag{5.41}$$

The numerator in this equation is a linear function of the coordinates x, y, and the denominator is the equation of a circle.

A similar calculation yields the angle $\angle \mathbf{P}_{14}\mathbf{G}\mathbf{P}_{34}$, which must be either κ or $\kappa + \pi$. However, since $\tan \kappa = \tan(\kappa + \pi)$, we have

$$\tan \kappa = \frac{(b_2 - q_4)x + (a_2 - p_4)y + p_4 b_2 - q_4 a_2}{x^2 + y^2 - (p_4 + a_2)x - (q_4 + b_2)y + p_4 a_2 + q_4 b_2} = \frac{L_{34}}{C_{34}}, \tag{5.42}$$

where the coordinates of the relative poles are $\mathbf{P}_{14} = (p_4, q_4)^T$ and $\mathbf{P}_{34} = (a_2, b_2)^T$.

Equate (5.41) and (5.42) to obtain a formula for the center-point curve, given by

$$\mathscr{R}(x, y): \ L_{12}C_{34} - L_{34}C_{12} = 0. \tag{5.43}$$

It is easy to see that the cubic terms of $\mathscr{R}(x, y)$ in (5.35) are

$$a_{30}y^3 + a_{21}y^2 x + a_{12}yx^2 + a_{03}x^3$$
$$= ((b_1 - q_2 - b_2 + q_4)x + (a_1 - p_2 - a_2 + p_4)y)(x^2 + y^2). \tag{5.44}$$

The factor $x^2 + y^2$ in this term identifies this cubic polynomial as a *circular cubic*. This also means that the ten coefficients a_{ij} are not independent. In fact, it is easy to see that

$$a_{30} = a_{12} = a_1 - p_2 - a_2 + p_4, \quad a_{03} = a_{21} = b_1 - q_2 - b_2 + q_4. \tag{5.45}$$

5.3.5.3 The Parameterized Center-Point Curve

Burmester's theorem is also the basis for a derivation of a parameterized version of the center-point curve. The construction presented in chapter 3 uses the opposite-pole quadrilateral $\mathscr{Q}: \mathbf{P}_{12}\mathbf{P}_{23}\mathbf{P}_{34}\mathbf{P}_{14}$ to generate points that satisfy Burmester's theorem. We formulate this construction analytically in terms of the dyad triangle for the RR chain $\mathbf{P}_{12}\mathbf{P}_{23}$ and compute the center points $\mathbf{G}$ as the relative displacement poles of the segment $\mathbf{P}_{23}\mathbf{P}_{34}$, Figure 5.5.

Identify the vertices of the opposite-pole quadrilateral $\mathscr{Q}$ with the pivots of a 4R linkage so $\mathbf{O} = \mathbf{P}_{12}$, $\mathbf{A} = \mathbf{P}_{23}$, $\mathbf{B} = \mathbf{P}_{34}$, and $\mathbf{C} = \mathbf{P}_{14}$. The formulas in chapter 2 are

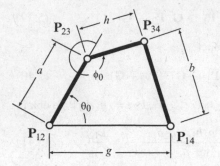

Fig. 5.5 The dimensions of an opposite-pole quadrilateral considered as a four-bar linkage.

used to determine the angles at each vertex. Let θ be the interior angle at $\mathbf{P}_{12}$, then we can determine the coupler angle $\phi(\theta)$ at $\mathbf{P}_{23}$ using equation (2.57).

Rotate the segment $\mathbf{P}_{12}\mathbf{P}_{23}$ an angle $\Delta\theta$ from the initial configuration of $\mathcal{Q}$. This requires a corresponding rotation of angle $\Delta\phi$ of the segment $\mathbf{P}_{23}\mathbf{P}_{34}$ about $\mathbf{P}_{23}$. The composition of these two displacement is a relative rotation of angle $\kappa = \Delta\theta + \Delta\phi$ about the pole $\mathbf{G}$ (Figure 5.6), given by

$$[T(\kappa, \mathbf{G})] = [T(\Delta\theta, \mathbf{P}_{12})][T(\Delta\phi, \mathbf{P}_{23})]. \tag{5.46}$$

This composition yields the dyad triangle equation

$$(1 - e^{i(\Delta\theta + \Delta\phi)})\mathbf{G} = (1 - e^{i\Delta\theta})\mathbf{P}_{12} + e^{i\Delta\theta}(1 - e^{i\Delta\phi})\mathbf{P}_{23}. \tag{5.47}$$

Let θ_0 and ϕ_0 be the initial values for these angles in $\mathcal{Q}$, then we have

$$\Delta\theta = \theta - \theta_0, \quad \Delta\phi = \phi(\theta) - \phi_0. \tag{5.48}$$

The result is that this equation defines a point $\mathbf{G}$ on the center-point curve for every value of the parameter θ.

To complete this formulation, we need the initial configuration angles θ_0 and ϕ_0, which are computed by using formulas (5.39) and (5.40) to obtain

$$\theta_0 = \arctan\left(\frac{|\mathbf{P}_{14} - \mathbf{P}_{12}, \mathbf{P}_{23} - \mathbf{P}_{12}|}{(\mathbf{P}_{14} - \mathbf{P}_{12}) \cdot (\mathbf{P}_{23} - \mathbf{P}_{12})}\right) \tag{5.49}$$

and

$$\phi_0 = \arctan\left(\frac{|\mathbf{P}_{23} - \mathbf{P}_{12}, \mathbf{P}_{34} - \mathbf{P}_{23}|}{(\mathbf{P}_{23} - \mathbf{P}_{12}) \cdot (\mathbf{P}_{34} - \mathbf{P}_{23})}\right). \tag{5.50}$$

A benefit of this parameterization is that a center-point curve can be classified by the linkage type of the opposite-pole quadrilateral $\mathcal{Q}$ that generates it. In particular, center-point curves generated by nonGrashof opposite-pole quadrilaterals have a single circuit, while those generated by Grashof opposite-pole quadrilater-

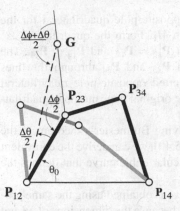

Fig. 5.6 A center point **G** is the pole of the displacement of the coupler $\mathbf{P}_{23}\mathbf{P}_{34}$ of the opposite-pole quadrilateral when driven as a four-bar linkage.

als have two circuits. Furthermore, all three opposite-pole quadrilaterals that can be constructed from the six relative poles generate the same curve.

5.3.5.4 The Circle-Point Curve

For each point of the center-point curve we have a solution to the design equation (5.19), which yields a moving pivot $\mathbf{W}^1$. These points form the *circle-point curve*. We can obtain an equation for this curve directly by using (5.20) to define the matrix equation

$$\begin{bmatrix} A_2'(\lambda,\mu) & B_2'(\lambda,\mu) \\ A_3'(\lambda,\mu) & B_3'(\lambda,\mu) \\ A_4'(\lambda,\mu) & B_4'(\lambda,\mu) \end{bmatrix} \begin{Bmatrix} x \\ y \end{Bmatrix} = \begin{Bmatrix} C_2'(\lambda,\mu) \\ C_3'(\lambda,\mu) \\ C_4'(\lambda,\mu) \end{Bmatrix}. \tag{5.51}$$

These equations have a solution for the fixed pivot $\mathbf{G} = (x,y)^T$ only if the augmented coefficient matrix $[M']$ has rank two. Here, as above, the elements of $[M']$ are linear functions of coordinates of the moving pivot $\mathbf{W}^1 = (\lambda,\mu)$ and can be assembled into column vectors, so we have

$$\det[M'] = \left| \mathbf{a}_1'\lambda + \mathbf{b}_1'\mu + \mathbf{c}_1', \ \mathbf{a}_2'\lambda + \mathbf{b}_2'\mu + \mathbf{c}_2', \ \mathbf{a}_3'\lambda + \mathbf{b}_3'\mu + \mathbf{c}_3' \right| = 0. \tag{5.52}$$

The expansion of this determinant yields a polynomial $\mathscr{R}(\lambda,\mu)$ that has the same form as $\mathscr{R}(x,y)$ in (5.35). The coefficients of $\mathscr{R}(\lambda,\mu)$ are given by (5.37) using $\mathbf{a}'$, $\mathbf{b}'$ and $\mathbf{c}'$.

If the four displacements of M relative to F are inverted, then we can compute the circling-point curve simply as the center-point curve of the inverted displacements. In this case, the curve is defined in the moving reference frame M. We can then transform these coordinates to the fixed frame in the first position to obtain the curve of moving pivots $\mathbf{W}^1$.

This result is achieved by determining the opposite-pole quadrilateral for the relative inverse displacements with M in position M_1. Form the quadrilateral $\mathscr{Q}^\dagger$ from the image poles $\mathbf{P}_{12}^1\mathbf{P}_{23}^1\mathbf{P}_{34}^1\mathbf{P}_{14}^1$. Recall that $\mathbf{P}_{12}^1 = \mathbf{P}_{12}$ and $\mathbf{P}_{14}^1 = \mathbf{P}_{14}$. The image poles $\mathbf{P}_{23}^1$ and $\mathbf{P}_{34}^1$ are the reflections of $\mathbf{P}_{23}$ and $\mathbf{P}_{34}$ through the lines $\mathbf{P}_{12}\mathbf{P}_{13}$ and $\mathbf{P}_{13}\mathbf{P}_{14}$, respectively. Thus, the inverted opposite-pole quadrilateral $\mathscr{Q}^\dagger : \mathbf{P}_{12}\mathbf{P}_{23}^1\mathbf{P}_{34}^1\mathbf{P}_{14}$ has the same dimensions as the original opposite-pole quadrilateral $\mathscr{Q}$.

The circle-point curve is constructed by applying Burmester's theorem to the quadrilateral $\mathscr{Q}^\dagger$. Using the equations (5.39) and (5.40), we can derive the equivalent to equations (5.41) and (5.42). The result is a circular cubic curve that defines the moving pivots $\mathbf{W}^1$.

A parameterized version of the circle-point curve is obtained using the same procedure as above for the center-point curve. In fact, because the dimensions of $\mathscr{Q}$ and $\mathscr{Q}^\dagger$ are the same, the only difference is the initial configuration of the quadrilateral. Compute new the values for θ_0 and ϕ_0 using $\mathbf{P}_{23}^1$ and $\mathbf{P}_{34}^1$ in equations (5.49) and (5.50), then (5.47) yields the circle-point curve.

5.3.6 Five Precision Positions

Given five task positions for the moving body, we obtain four design equations (5.17) that are quadratic in four unknowns $\mathbf{G} = (x,y)^T$ and $\mathbf{W}^1 = (\lambda,\mu)^T$. We use a two-step procedure to eliminate the variable in these equations. The goal is a single polynomial in one unknown. The solutions of this polynomial are then used to determine the remaining unknowns. We also solve this problem by finding the intersections of two center-point curves.

5.3.6.1 Algebraic Elimination

Let the coordinates of the relative poles be $\mathbf{P}_{1i} = (p_i, q_i)^T$, and expand the design equations (5.17) for the case $n = 5$. Assemble these equations into four linear equations in the two unknowns (λ,μ)

$$\begin{bmatrix} A_2(x,y) & B_2(x,y) \\ A_3(x,y) & B_3(x,y) \\ A_4(x,y) & B_4(x,y) \\ A_5(x,y) & B_5(x,y) \end{bmatrix} \begin{Bmatrix} \lambda \\ \mu \end{Bmatrix} = \begin{Bmatrix} C_2(x,y) \\ C_3(x,y) \\ C_4(x,y) \\ C_5(x,y) \end{Bmatrix}. \tag{5.53}$$

In order for this system of equations to have a solution the rank of the 4×3 augmented coefficient matrix $[M] = [A_i, B_i, C_i]$ must be two. For this to occur, each the four 3×3 minors of this matrix must equal zero. Let $\mathscr{R}_j$ be the determinant of the 3×3 matrix formed by removing row $5 - j$; so $\mathscr{R}_1$ is the computed from the first three rows, $\mathscr{R}_2$ from the first two and last row, and so on. The result is four cubic polynomials in x and y

$$\mathscr{R}_j(x,y): a_{30,j}y^3 + (a_{21,j}x + a_{20,j})y^2 + (a_{12,j}x^2 + a_{11,j}x + a_{10,j})y$$
$$+ a_{03,j}x^3 + a_{02,j}x^2 + a_{01,j}x + a_{00,j} = 0, \ j = 1,2,3,4. \qquad (5.54)$$

Thus, our four equations in four unknowns are transformed into four equations in two unknowns. The next step eliminates y to obtain a single polynomial in x.

At this point we assume that determinants $\mathscr{R}_j$ have the structure of a general cubic polynomial in two variables. Collect the coefficients of y in each $\mathscr{R}_j$ to define

$$\mathscr{R}_j: d_{j0}y^3 + d_{j1}y^2 + d_{j2}y + d_{j3} = 0, \ j = 1,2,3,4. \qquad (5.55)$$

The coefficient d_{jk} is a polynomial in x of degree k.

Assemble these polynomials into a matrix equation with the vector of unknowns $(y^3, y^2, y)^T$, given by

$$\begin{bmatrix} d_{10} & d_{11} & d_{12} \\ \vdots & \vdots & \vdots \\ d_{40} & d_{41} & d_{42} \end{bmatrix} \begin{Bmatrix} y^3 \\ y^2 \\ y \end{Bmatrix} = - \begin{Bmatrix} d_{13} \\ \vdots \\ d_{43} \end{Bmatrix}. \qquad (5.56)$$

This equation has a solution only if the rank of the 4×4 augmented coefficient matrix $[D] = [\mathbf{d}_0, \mathbf{d}_1, \mathbf{d}_2, \mathbf{d}_3]$, where $\mathbf{d}_j = (d_{1j}, d_{2j}, d_{3j}, d_{4j})^T$, is three. This means that the determinant $|D|$ of this matrix must be zero.

The determinant $|D|$ is a polynomial in the single variable x. The degree of this polynomial is the sum of the degrees of each of the columns of $[D]$, that is, $0 + 1 + 2 + 3 = 6$. Thus, we obtain a single sixth-degree polynomial in x

$$\mathscr{P}(x): |D| = \sum_{i=0}^{6} a_i x^i = 0. \qquad (5.57)$$

It happens that $a_5 = a_6 = 0$ and $\mathscr{P}(x)$ is a quartic polynomial. This is due to the circular cubic structure (5.45) of the polynomials $\mathscr{R}_j$. This quartic polynomial has four roots of which zero, two, or four will be real. Thus, there can be as many as four RR chains that reach five positions.

To determine the RR chains that reach five task positions, first formulate the polynomial $\mathscr{P}(x)$ and determine its roots x_i, $i = 1,2,3,4$. For each real root x_i, solve (5.56) to determine the coordinate y_i. This defines as many as four fixed pivots $\mathbf{G}_i = (x_i, y_i)^T$. Determine the associated moving pivots $\mathbf{W}_i^1 = (\lambda, \mu)^T$ by solving two of the linear constraint equations (5.17).

5.3.6.2 Intersecting Two Center-Point Curves

Five task positions determine ten relative displacement poles $\mathbf{P}_{ij}$, $i < j = 1, \ldots, 5$. Consider the two opposite-pole quadrilaterals $\mathscr{Q}_{14}: \mathbf{P}_{12}\mathbf{P}_{23}\mathbf{P}_{34}\mathbf{P}_{14}$ and $\mathscr{Q}_{15}: \mathbf{P}_{12}\mathbf{P}_{23}\mathbf{P}_{35}\mathbf{P}_{15}$.

A fixed pivot compatible with five positions lies on the center-point curve defined by $\mathscr{Q}_{14}$ and on the center-point curve defined by $\mathscr{Q}_{15}$. This provides another way to determine the fixed pivot $\mathbf{G}$.

The opposite-pole quadrilaterals $\mathscr{Q}_{14}$ and $\mathscr{Q}_{15}$ share the side $\mathbf{P}_{12}\mathbf{P}_{23}$, Figure 5.7. Thus, the pivot $\mathbf{G}$ must satisfy the two equations

$$\mathbf{G} = \frac{\mathbf{P}_{12}(1 - e^{i\Delta\theta_1}) + \mathbf{P}_{23}(1 - e^{i\Delta\phi_1})e^{i\Delta\theta_1}}{1 - e^{i(\Delta\phi_1 + \Delta\theta_1)}} \tag{5.58}$$

and

$$\mathbf{G} = \frac{\mathbf{P}_{12}(1 - e^{i\Delta\theta_2}) + \mathbf{P}_{23}(1 - e^{i\Delta\phi_2})e^{i\Delta\theta_2}}{1 - e^{i(\Delta\phi_2 + \Delta\theta_2)}}. \tag{5.59}$$

The angles $\Delta\phi_1$ and $\Delta\phi_2$ are functions of $\Delta\theta_1$ and $\Delta\theta_2$ defined by the dimensions of the two opposite-pole quadrilaterals.

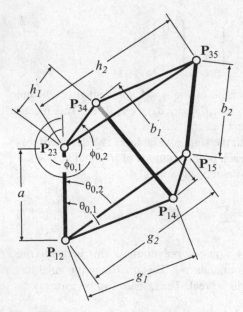

Fig. 5.7 The reference configuration for the planar compatibility platform.

Notice that equations (5.58) and (5.59) define the same point $\mathbf{G}$ when

$$\Delta\theta_1 = \Delta\theta_2 \quad \text{and} \quad \Delta\phi_1 = \Delta\phi_2. \tag{5.60}$$

The first condition is satisfied by using the same parameter θ to drive $\mathbf{P}_{12}\mathbf{P}_{23}$ for both curves. The second condition requires that the triangle $\triangle\mathbf{P}_{23}\mathbf{P}_{34}\mathbf{P}_{35}$ have the same shape in each solution configuration. Thus, the fixed pivots $\mathbf{G}$ are the poles of the displacement of $\triangle\mathbf{P}_{23}\mathbf{P}_{34}\mathbf{P}_{35}$ to each of the assemblies of the platform, Figure 5.8. We call this assembly of relative displacement poles the *planar compatibility platform* and obtain the following theorem:

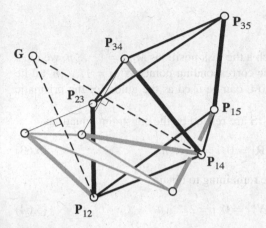

Fig. 5.8 The relative displacement poles of the assemblies of the planar compatibility platform define the fixed pivots **G**.

Theorem 8 (The Planar Compatibility Platform). *The fixed pivot of an RR chain compatible with five specified planar positions is a pole of the displacement of the planar compatibility platform from its original configuration another of its assemblies.*

The analysis of the two 4R linkages in this platform yields two equations of the form

$$A_i \cos\phi + B_i \sin\phi = C_i, i = 1, 2. \tag{5.61}$$

that are easily solved as shown in (A.11). This 3RR platform is known to have six assemblies. One is the initial configuration, therefore we obtain a relative pole to each of the remaining five assemblies. However, one of these is the pole P_{13}, thus the remaining four are the desired fixed pivots.

5.4 The Design of PR Chains

A PR chain consists of a prismatic joint in the fixed frame F connected by a slider to a revolute joint in the moving frame M. Let the trajectory of the moving pivot **W** lie on the line L: $Y(t) = R + tS$, which must be parallel to the guide of the prismatic joint. The condition that **W** lie on L is simply that the vector $W - R$ be aligned with the direction **S**. This is expressed by the equation

$$|S, W - R| = 0. \tag{5.62}$$

This constraint characterizes the PR chain.

5.4.1 The Design Equations

In order to design a PR chain to reach a the task positions M_i, $i = 1,\ldots,n$, we must find a moving pivot $\mathbf{W}^1$ such that the corresponding points $\mathbf{W}^i$, $i = 1,\ldots,n$, all lie on a line L in F. Any line parallel to L can be used as the guide for the prismatic joint.

The points $\mathbf{W}^i$ and L: $\mathbf{Y}(t) = \mathbf{R} + t\mathbf{S}$ are related by the constraint equations

$$|\mathbf{S}, \mathbf{W}^i - \mathbf{R}| = 0, i = 1,\ldots,n. \tag{5.63}$$

Subtract the first equation from those remaining to obtain

$$|\mathbf{S}, \mathbf{W}^i - \mathbf{W}^1| = 0, i = 2,\ldots,n. \tag{5.64}$$

Notice that $\mathbf{R}$ has cancelled in these equations. This is because the point $\mathbf{W}^1$ is sufficient to locate the line L.

Now recall from our derivation of the design equations for an RR chain that

$$\mathbf{W}^i - \mathbf{W}^1 = [A(\phi_{1i}) - I](\mathbf{W}^1 - \mathbf{P}_{1i}). \tag{5.65}$$

Substitute this into (5.64) to obtain the design equations

$$|\mathbf{S}, [A(\phi_{1i}) - I](\mathbf{W}^1 - \mathbf{P}_{1i})| = 0, i = 2,\ldots,n. \tag{5.66}$$

The unknowns in these equations are the direction $\mathbf{S}$ of the guide and the coordinates of the moving pivot $\mathbf{W}^1$.

The design equations are homogeneous in the components of $\mathbf{S}$, which means that if $\mathbf{S}$ is a solution then $k\mathbf{S}$ is also a solution. Therefore, only one component of $\mathbf{S}$ is independent, so we set $\mathbf{S} = (1, m)^T$, where m is the *slope* of L. The components of $\mathbf{W}^1 = (\lambda, \mu)^T$ are independent unknowns. Thus, the design equations have a total of three unknowns, m, λ, and μ. We will see in what follows that the three design equations obtained from four task positions define one PR chain.

5.4.1.1 The Bilinear Structure

The design equations are linear in m and separately linear in λ and μ. This bilinear structure allows us to follow a solution procedure similar to that used for RR chains. However, the different number of unknowns introduces an asymmetry into the analysis.

The coefficients of m can be isolated by introducing the vectors $\mathbf{C}_i = [A(\phi_{1i}) - I](\mathbf{W}^1 - \mathbf{P}_{1i})$. Then (5.66) can be written as

$$[m, -1] [\mathbf{C}_2, \ldots, \mathbf{C}_n] = 0, \tag{5.67}$$

where the components of $\mathbf{C}_i = (A_i, B_i)^T$ are given by

$$\begin{aligned} A_i &= (\cos\phi_{1i} - 1)\lambda - \sin\phi_{1i}\mu - ((\cos\phi_{1i} - 1)p_i - \sin\phi_{1i}q_i), \\ B_i &= \sin\phi_{1i}\lambda + (\cos\phi_{1i} - 1)\mu - (\sin\phi_{1i}p_i + (\cos\phi_{1i} - 1)q_i). \end{aligned} \tag{5.68}$$

Note that the components of the poles are given by $\mathbf{P}_{1i} = (p_i, q_i)^T$.

We can also collect coefficients of λ and μ, so (5.66) becomes

$$D_i\lambda + E_i\mu + F_i = 0, i = 2,\ldots,n, \tag{5.69}$$

where

$$\begin{aligned} D_i &= m(\cos\phi_{1i} - 1) - \sin\phi_{1i}, \\ E_i &= -m\sin\phi_{1i} - (\cos\phi_{1i} - 1), \\ F_i &= -m((\cos\phi_{1i} - 1)p_i - \sin\phi_{1i}q_i) + (\sin\phi_{1i}p_i + (\cos\phi_{1i} - 1)q_i). \end{aligned} \tag{5.70}$$

We now determine the solutions to these equations for the cases $n = 2, 3, 4$.

5.4.2 Two Positions

For two specified positions for the moving body, we have the single design equation

$$[m, -1] \mathbf{C}_2 = mA_2 - B_2 = 0. \tag{5.71}$$

This equation is solved by selecting either a point $\mathbf{W}^1$ or a direction m.

In the first case, given values for components of $\mathbf{W}^1 = (\lambda, \mu)^T$ the components of the vector $\mathbf{C}_2 = (A_2, B_2)^T$ are specified, and equation (5.71) has the single unknown m. Thus, this slope is uniquely determined.

If instead we select a direction m, then (5.71) becomes a linear equation in λ and μ

$$D_2\lambda + E_2\mu + F_2 = 0. \tag{5.72}$$

Choose either λ or μ and solve for the other.

The result is a two-dimensional set of PR chains, which are obtained by selecting either the moving pivot $\mathbf{W}^1$ or a direction m and one component of $\mathbf{W}^1$.

5.4.3 Three Positions

When three positions of the moving body are specified, we have two design equations that form the matrix equation

$$[m, -1] [\mathbf{C}_2, \mathbf{C}_3] = 0. \tag{5.73}$$

This equation has a solution if the 2×2 coefficient matrix $[\mathbf{C}_2, \mathbf{C}_3]$ has rank 1, that is, when

$$\left| [A(\phi_{12}) - I](\mathbf{W}^1 - \mathbf{P}_{12}), \ [A(\phi_{13}) - I](\mathbf{W}^1 - \mathbf{P}_{13}) \right| = 0. \tag{5.74}$$

This provides a condition on the selection of the moving pivots $\mathbf{W}^1$.

To simplify (5.74), we introduce the identity

$$[A(\phi_{1i}) - I] = 2 \sin \frac{\phi_{1i}}{2} [J][A(\frac{\phi_{1i}}{2})], \tag{5.75}$$

where $[J]$ is a rotation by $90°$. This combines with the linearity of the determinant to yield

$$\left| [A(\tfrac{\phi_{12}}{2})](\mathbf{P}_{12} - \mathbf{W}^1), \ [A(\tfrac{\phi_{13}}{2})](\mathbf{P}_{13} - \mathbf{W}^1) \right| = 0. \tag{5.76}$$

Multiply both columns by $[A(\phi_{12}/2)]^T$, so (5.74) finally takes the form

$$\left| \mathbf{P}_{12} - \mathbf{W}^1; \ [A(\tfrac{\phi_{23}}{2})](\mathbf{P}_{13} - \mathbf{W}^1) \right| = 0. \tag{5.77}$$

This determinant is zero when the two columns are colinear vectors. Therefore, the rotation by $\phi_{23}/2$ around $\mathbf{W}^1$ must bring $\mathbf{P}_{13} - \mathbf{W}^1$ into alignment with $\mathbf{P}_{12} - \mathbf{W}^1$. Thus, $\mathbf{W}^1$ must view the segment $\mathbf{P}_{12}\mathbf{P}_{13}$ in the angle $\phi_{23}/2$, and (5.77) is the equation of a circle. This is an algebraic proof of the slider-point theorem.

Expand equation (5.77) to obtain the circle $\mathscr{C}_1$ given by

$$\mathscr{C}_1 : \lambda^2 + \mu^2 + a_{11}\lambda + a_{12}\mu + a_{13} = 0. \tag{5.78}$$

Any point on this circle may be chosen as the pivot $\mathbf{W}^1$, in which case the first design equation (5.71) can be solved to determine m.

Rather than select $\mathbf{W}^1$, we can choose the direction m and obtain the pair of linear equations

$$D_2\lambda + E_2\mu + F_2 = 0,$$
$$D_3\lambda + E_3\mu + F_3 = 0. \tag{5.79}$$

The solution of these equations defines a unique point $\mathbf{W}^1$ associated with a given direction m.

The result is one PR chain for each direction m or point on the circle $\mathscr{C}_1$. This is a one-dimensional set of solutions.

5.4.4 Four Positions

If four task positions for the moving body are specified, then the design equations become

$$[m, -1]\,[\mathbf{C}_2, \mathbf{C}_3, \mathbf{C}_4] = 0. \tag{5.80}$$

In order for this equation to have a solution, the three minors of the matrix $[\mathbf{C}_2, \mathbf{C}_3, \mathbf{C}_4]$ must be simultaneously zero. This provides three conditions on the coordinates of the point $\mathbf{W}^1$, each of which has the same form as (5.77).

The condition $|\mathbf{C}_2, \mathbf{C}_3| = 0$ yields the circle described in the previous section. Setting the other two minors to zero, we obtain

$$|\mathbf{C}_2, \mathbf{C}_4| = \left| \mathbf{P}_{12} - \mathbf{W}^1,\ [A(\tfrac{\phi_{24}}{2})](\mathbf{P}_{14} - \mathbf{W}^1) \right| = 0,$$
$$|\mathbf{C}_3, \mathbf{C}_4| = \left| \mathbf{P}_{13} - \mathbf{W}^1,\ [A(\tfrac{\phi_{34}}{2})](\mathbf{P}_{14} - \mathbf{W}^1) \right| = 0. \tag{5.81}$$

This defines two additional circles,

$$\mathscr{C}_2 : \lambda^2 + \mu^2 + a_{21}\lambda + a_{22}\mu + a_{23} = 0,$$
$$\mathscr{C}_3 : \lambda^2 + \mu^2 + a_{31}\lambda + a_{32}\mu + a_{33} = 0, \tag{5.82}$$

which combine with $\mathscr{C}_1$ above to determine the components of $\mathbf{W}^1 = (\lambda, \mu)^T$.

Collect the coefficients of μ and write the three circle equations as the matrix equation

$$\begin{bmatrix} 1 & a_{12} & \lambda^2 + a_{11}\lambda + a_{13} \\ 1 & a_{22} & \lambda^2 + a_{21}\lambda + a_{23} \\ 1 & a_{32} & \lambda^2 + a_{31}\lambda + a_{33} \end{bmatrix} \begin{Bmatrix} \mu^2 \\ \mu \\ 1 \end{Bmatrix} = \begin{Bmatrix} 0 \\ 0 \\ 0 \end{Bmatrix}. \tag{5.83}$$

This system of equations has a solution if the determinant $|C|$ of the coefficient matrix is zero. This determinant can be simplified by subtracting the first row from the other two to obtain a linear equation in λ, given by

$$\begin{vmatrix} a_{22} - a_{12} & (a_{21} - a_{11})\lambda + a_{23} - a_{13} \\ a_{32} - a_{12} & (a_{31} - a_{11})\lambda + a_{33} - a_{13} \end{vmatrix} = 0. \tag{5.84}$$

Solve this equation for λ and then (5.83) for μ in order to determine the point $\mathbf{W}^1$. Using this point, the design equations (5.80) yield the direction m for the slider. The result is a unique PR chain that can reach four specified positions.

5.5 The Design of RP Chains

An RP chain consists of a revolute joint $\mathbf{G}$ in the fixed frame F connected to a prismatic joint in the moving frame M. The prismatic joint follows a fixed line in M. From the point of view of M, the point $\mathbf{G}$ must follow the line L: $\mathbf{y}(t) = \mathbf{r} + t\mathbf{s}$ parallel to the guide of the prismatic joint as the body moves through a set of positions M_i. Let $\mathbf{g}$ be the coordinates of $\mathbf{G}$ measured in M. The condition that $\mathbf{g}$ lie on L is simply that $\mathbf{g} - \mathbf{r}$ must be aligned with $\mathbf{s}$, which is given by

$$\left| \mathbf{s}, \mathbf{g} - \mathbf{r} \right| = 0. \tag{5.85}$$

This constraint characterizes an RP chain.

5.5.1 The Design Equations

In order for the end-link of an RP chain to reach the task positions M_i, $i = 1, \ldots, n$, we must find a point $\mathbf{G}$ in F that has its corresponding n points $\mathbf{g}^i$, $i = 1, \ldots, n$, in M on a line L. Any line parallel to L can be used as the guide for the prismatic joint.

Let the n positions of the moving body be defined by the transformations $[T_i]$, $i = 1, \ldots, n$, so we have the inverse transformations $\mathbf{g}^i = [T_i^{-1}]\mathbf{G}$. Then, the points $\mathbf{g}^i$ and the line L: $\mathbf{y}(t) = \mathbf{r} + t\mathbf{s}$ in M are related by the equations

$$\left| \mathbf{s}, \mathbf{g}^i - \mathbf{r} \right| = 0, i = 1, \ldots, n. \tag{5.86}$$

Subtract the first equation from those remaining to obtain

$$\left| \mathbf{s}, \mathbf{g}^i - \mathbf{g}^1 \right| = 0, i = 2, \ldots, n. \tag{5.87}$$

Now introduce the relative inverse transformation $[T_{1i}^\dagger] = [A_{1i}^\dagger, \mathbf{d}_{1i}^\dagger]$ so that this equation becomes

$$\left| \mathbf{s}, [A_{1i}^\dagger - I](\mathbf{g}^1 - \mathbf{p}_{1i}) \right| = 0, i = 2, \ldots, n, \tag{5.88}$$

where $\mathbf{p}_{1i}$ is the pole of the relative inverse displacement in M. This is simply the inverted version of the PR design equation (5.66).

Consider the moving frame to be in position M_1 and transform coordinates to the frame F. This requires multiplying (5.88) by $[A_1]$, so we obtain

$$\left| [A_1]\mathbf{s}, [A_1][A_{1i}^\dagger - I](\mathbf{g}^1 - \mathbf{p}_{1i}) \right| = 0, i = 2, \ldots, n. \tag{5.89}$$

Let $\mathbf{S} = [A_1]\mathbf{s}$ be the direction of the line L in F. Notice that $[A_1](\mathbf{g}^1 - \mathbf{p}_{1i}) = \mathbf{G} - \mathbf{P}_{1i}$, where $\mathbf{P}_{1i}$ is the relative pole in F. We obtain the RP design equations in the form

$$\left| \mathbf{S}, [A_{1i}^\dagger - I](\mathbf{G} - \mathbf{P}_{1i}) \right| = 0, i = 2, \ldots, n. \tag{5.90}$$

Note that $[A_1][A_{1i}^\dagger][A_1^T] = [A_{1i}^\dagger]$, because planar rotations commute.

This set of equations is identical in form to those for the PR chain, and their analysis is the same except for one fundamental difference. The relative inverse rotation $[A_{1i}^\dagger]$ is the inverse of the relative rotation $[A_{1i}]$, which means that the rotation angles $\phi_{1i}^\dagger$ are $-\phi_{1i}$. From this we conclude that the circles derived in the previous section for PR chains will be reflected through through the segments $\mathbf{P}_{1i}\mathbf{P}_{1k}$. Use these circles and the design procedure for PR chains to design RP chains.

5.6 Planar Four-Bar Linkages

The design equations developed in this chapter determine RR, PR and RP chains that reach a specified set of positions. These chains can be connected in pairs to construct various four-bar linkages, each of which has one degree of freedom. The connection between two chains also allows us to coordinate the movement of the input and output links in order to design a linkage known as a *function generator*.

The coupling between two open chains introduces limits on their relative movement that can interfere with the smooth travel of the floating link between the specified positions. For example, a Grashof linkage may reach one set of task positions in one assembly and the other positions in the other assembly. This is referred to as the *branching problem*.

In what follows, we consider the solution to the branching problem for 4R chains, and then present a strategy for the design of function generators.

5.6.1 Solution Rectification

Filemon [32] introduced a construction for the moving pivot of the input crank of a 4R linkage that ensures that the linkage moves smoothly through the task positions. This construction assumes that an output crank $\mathbf{CB} = \mathbf{G}_{out}\mathbf{W}_{out}^1$ has been selected. It is then possible to determine how this crank rotates to reach each of the design positions. Viewed from the coupler, this link sweeps out two wedge shaped regions centered on $\mathbf{W}_{out}^1$, Figure 5.9. Filemon showed that all that is necessary is that we choose the moving pivot $\mathbf{W}_{in}^1$ of the input crank outside of these regions. Recall that the limit positions of the input crank $\mathbf{OA} = \mathbf{G}_{in}\mathbf{W}_{in}^1$ of a 4R linkage occur when the coupler $\mathbf{AB} = \mathbf{W}_{in}^1\mathbf{W}_{out}^1$ lines up with the output crank.

5.6.1.1 Filemon's Construction

We will focus on three-position synthesis to develop rectification theory, though it can be applied more generally. For positions defined by the displacements $[T_1]$, $[T_2]$, and $[T_3]$ we have the design equations (5.28). For any choice of the output moving pivot $\mathbf{W}_{out}^1$, we can determine a unique fixed pivot $\mathbf{G}_{out}$.

The positions that $\mathbf{G}_{out}$ can take relative to the moving frame are computed using the relative inverse displacements $[T_{1i}^1]$ with M in position M_1, that is,

$$\mathbf{G}_{out}^i = [T_{1i}^1]\mathbf{G}_{out}. \tag{5.91}$$

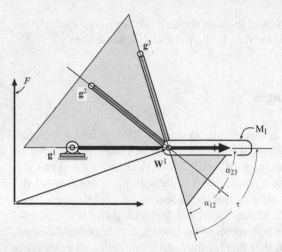

Fig. 5.9 Filemon's construction uses the three positions of the output crank to define a wedge-shaped region; the input moving pivot is selected from outside this region.

Recall that $[T_{1i}^1] = [T_{1i}^{-1}]$.

Now consider the angle α_{12} measured from $\mathbf{G}_{out}^1$ to $\mathbf{G}_{out}^2$ around $\mathbf{W}_{out}^1$. Similarly, we have $\alpha_{23} = \angle \mathbf{G}_{out}^2 \mathbf{W}_{out}^1 \mathbf{G}_{out}^3$ measured around $\mathbf{W}_{out}^1$. These angles combine to form the wedge swept by the driven crank relative to the moving frame M. Assume these angles are between π and $-\pi$, then the angle τ of this wedge is the sum $\alpha_{12} + \alpha_{23}$ if these angles have the same sign. If these angles have different signs then τ is the angle with the largest absolute value.

Choose the input moving pivot $\mathbf{W}_{in}^1$ from *outside* of this wedge-shaped region. The resulting 4R linkage will pass through the design positions before the coupler lines up with the output crank, which defines the limit to the movement of the input crank.

5.6.1.2 Waldron's Construction

Waldron [142, 143] shows that if the output pivot $\mathbf{W}_{out}^1$ rotates so that any of the angles α_{12}, α_{23}, or α_{13} is greater than or equal to π, then there is no solution to Filemon's construction. This lead him to consider the points that view each side of the image pole triangle in $\pi/2$ and define the *three-circle diagram*.

The poles of the relative inverse displacements $[T_{12}^1]$, $[T_{23}^1]$, and $[T_{13}^1]$ define the image pole triangle $\triangle \mathbf{P}_{12} \mathbf{P}_{23}^1 \mathbf{P}_{13}$. The center-point theorem applied to the image pole triangle yields the result the moving pivot $\mathbf{W}_{out}^1$ views the sides of this triangle the rotation angle $-\alpha_{ik}/2$ of the coupler relative to the RR chain. Thus, for a point $\mathbf{W}$ and side $\mathbf{P}_{ij}^1 \mathbf{P}_{jk}^1$ of the image pole triangle, we have the relation

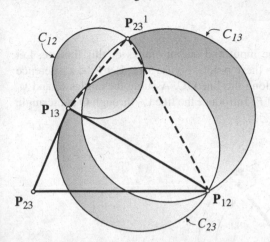

Fig. 5.10 Waldron's three-circle diagram identifies regions of driven moving pivots that ensure that Filemon's construction yields useful driving pivots.

$$\cos\frac{\alpha_{ik}}{2} = \frac{(\mathbf{P}^1_{ij} - \mathbf{W}) \cdot (\mathbf{P}^1_{jk} - \mathbf{W})}{|\mathbf{P}^1_{ij} - \mathbf{W}||\mathbf{P}^1_{jk} - \mathbf{W}|}. \tag{5.92}$$

The points that have $\alpha_{ik}/2 = \pi/2$ lie on the circles

$$\mathscr{C}_{ik} : (\mathbf{P}^1_{ij} - \mathbf{W}) \cdot (\mathbf{P}^1_{jk} - \mathbf{W}) = 0. \tag{5.93}$$

The diameters of these circles are the segments $\mathbf{P}^1_{ij}\mathbf{P}^1_{jk}$.

The three circles $\mathscr{C}_{12}$, $\mathscr{C}_{23}$, and $\mathscr{C}_{13}$ bound regions of points for which $\alpha_{ij} > \pi/2$. Points outside of these circles, as well as points in regions where they overlap can be used as moving pivots $\mathbf{W}^1_{out}$, Figure 5.10. For these points the output crank $\mathbf{G}_{out}\mathbf{W}^1_{out}$ has a solution for Filemon's construction. The result is a 4R linkage that moves through the three specified positions before it reaches a limit to the range of movement of the input crank.

5.6.2 Function Generation

A four-bar linkage is often designed to coordinate the movement of the input and output links. Suppose that we are given a table of coordinated angular values $\theta_i, \psi_i, i = 1, \ldots, n$, for the input and output cranks of a 4R linkage, or θ_i, s_i for the input and output of a slider-crank. We can apply the design theory developed in preceding sections to find a linkage that provides this functional relationship.

5.6.2.1 The 4R Chain

Select fixed pivots **O** and **C** for the input and output cranks of this linkage. Let $g = |\mathbf{C} - \mathbf{O}|$ be the distance between these points, Figure 5.11. Let F be a reference frame located at **O** with its x-axis along the line **OC**. Assume the angles θ_i and ψ_i are measured relative to the x-axis of F. Introduce the line $\mathsf{L_O}$ through **O** at the angle θ_1.

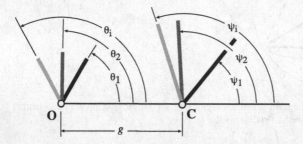

Fig. 5.11 A 4R function generator has a prescribed set of input and output angles θ_i, and ψ_i.

We now invert this problem by defining the exterior angles $\bar{\theta}_i = \pi - \theta_i$ at **O**, Figure 5.12. Introduce the coordinate frame F' attached to the input crank so that its x-axis is aligned with $\mathsf{L_O}$ such that $\bar{\theta}_1$ is the angle to **OC**. The angles $\bar{\theta}_i$ and ψ_i can now be viewed as the joint angles of the RR open chain formed by **OC** as it moves in F'. Using the kinematics equations of this chain, we compute the n positions

$$[D_i] = [Z(\bar{\theta}_i)][X(g)][Z(\psi_i)], i = 1,\ldots,n. \tag{5.94}$$

The positions $[D_i]$ can now be used to design an RR chain **AB** to close the 4R chain.

The result is a 4R chain that has the desired set of coordinated angles θ_i and ψ_i, $i = 1,\ldots,n$. Clearly, $n \leq 5$, because this is the maximum number of positions for which solutions exist to the finite position problem.

5.6.2.2 The RRRP Chain

The same strategy can be used to design a slider-crank that has the input angle θ coordinated with an output slide s. Let θ_i and s_i, $i = 1,\ldots,n$, be the desired table of values. Select a fixed pivot **O** and the direction **S**. Introduce a fixed frame F at **O** with its x-axis perpendicular to **S**, where e is the offset distance from **O** to the guide **S**. We assume the angles θ_i are measured from the x-axis of F, and s_i are measured along the y-axis. Let $\mathsf{L_O}$ be the line through **O** at the angle θ_1 to the x-axis of F.

We invert the problem and determine the exterior angles $\bar{\theta}_i = \pi - \theta_i$ and introduce the frame F' with its x-axis along $\mathsf{L_O}$ at the angle $\bar{\theta}_1$. In this frame the kinematic equations of the RP chain can be written as

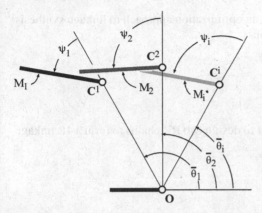

Fig. 5.12 Holding the input crank fixed the output crank takes positions $M_1, \ldots, M_i$ for each of the prescribed input and output angles.

$$[D_i] = [Z(\bar{\theta}_i)][X(e)][Y(s_i)], i = 1, \ldots, n. \tag{5.95}$$

The positions $[D_i]$ can be used to design the RR chain connecting the input and output cranks of the RRRP linkage. Again, at most five sets of coordinated input angles and output slides can be specified.

5.7 Summary

This chapter has presented algebraic techniques for the design of planar RR, PR, and RP chains. The results are a direct reflection of the graphical constructions presented earlier. In fact, many of the graphical results are simply reproduced here in algebraic form. This provides a transition to the formulation we need for the design of spatial linkages. Because the geometric principles are the same, it is useful to appeal to understanding gained in planar synthesis theory to provide insight to the design of spherical and spatial chains.

5.8 References

The algebraic formulation of linkage design was introduced by Freudenstein and Sandor [37] using a complex vector formulation and is developed in detail in Erdman and Sandor [30]. See also Waldron and Kinzel [144]. The polynomial elimination procedure used to solve the RR and PR design equations was inspired by Innocenti [52] and Liao [66]. Computer implementations of planar linkage synthesis originated with Kaufman's KINSYN [54], and include Waldron and Song's RECSYN [145], Erdman and Gustafson's LINCAGES [28], and Ruth's SphinxPC

[109]. Ravani and Roth [100] present an optimization approach to linkage synthesis that allows more that five task positions.

Exercises

1. Use the positions listed in Table 5.1 to design two RR chains to form a 4R linkage (Sandor and Erdman [112]).

Table 5.1 Three positions

M_i	θ_i	$d_{x,i}$	$d_{y,i}$
1	293°	1.55	−0.90
2	138°	1.75	−0.30
3	348°	0.80	1.60

2. Use the positions in Table 5.2 to determine the pole triangle $\triangle P_{12}P_{23}P_{13}$. Determine the equation of the circle that circumscribes this pole triangle (Sandor and Erdman [112]).

Table 5.2 Three more positions

M_i	θ_i	$d_{x,i}$	$d_{y,i}$
1	0°	0	0
2	22°	−6	11
3	68°	−17	13

3. Design a slider-crank linkage to guide the workpiece through the three positions in Table 5.2.
4. Design a 4R linkage that moves a workpiece through the three positions in Table 5.2.
5. Determine the equation of the curve of moving pivots $\mathbf{W}^1$ that lie on circles of the same radius R. Show that this is a tricircular sextic.
6. Show that if the opposite-pole quadrilateral has the shape of a kite, then the center-point curve degenerates into a circle and a line.
7. Determine the center-point and circle-point curves for the positions listed in Table 5.3 (Suh and Radcliffe [134]).
8. Determine the Burmester points for the five positions listed in Table 5.4 (Sandor and Erdman [112]).

Table 5.3 Four positions

M_i	θ_i	$d_{x,i}$	$d_{y,i}$
1	$0°$	1.0	1.0
2	$0°$	2.0	0.5
3	$45°$	3.0	1.5
4	$90°$	2.0	2.0

Table 5.4 Five positions

M_i	θ_i	$d_{x,i}$	$d_{y,i}$
1	$0°$	0	0
2	$10°$	1.5	.8
3	$20°$	1.6	1.5
4	$60°$	2.0	3.0
5	$90°$	2.3	3.5

Chapter 6
Multiloop Planar Linkages

In this chapter, we will formulate the systematic design of multiloop planar linkages in a way that combines with traditional robotics and four-bar linkage synthesis theory to obtain innovative articulated robotic systems. First, we will show how mechanical constraints can be introduced to a planar 3R serial chain to guide the movement of its end effector through a set of five specified task positions to obtain a six-bar linkage, as illustrated in Figure 6.1. Then, we will show how mechanical constraints can be introduced to a planar 6R loop to obtain an eight-bar linkage. An example on a Novel Steering Linkage and Depolyable Bed is presented as an example for each six- and eight-bar linkage respectively to illustrate the design process.

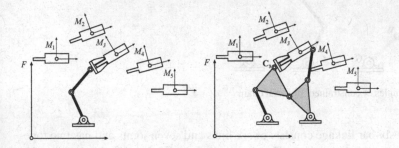

Fig. 6.1 A planar 3R serial chain is constrained by two RR chains to define a six-bar linkage.

6.1 Synthesis of Six-Bar Linkages

A planar 3R chain consists of four links and three revolute joints $\mathbf{C}_i$, $i = 1, 2, 3$ as shown in Figure 6.2. We assume that this chain has full mobility in the plane with

J.M. McCarthy and G.S. Soh, *Geometric Design of Linkages*, Interdisciplinary Applied
Mathematics 11, DOI 10.1007/978-1-4419-7892-9_6,
© Springer Science+Business Media, LLC 2011

its configuration defined joint angles θ_1, θ_2, and θ_3, and can reach a specified set of task positions.

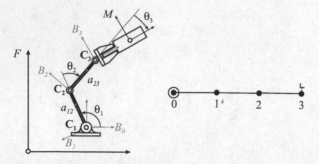

Fig. 6.2 A schematic of a planar 3R serial chain. The graph of this chain forms a straight line with a vertex for each link in the manipulator.

Our goal is to add constraints consistent with the task positions. While any of the constraints PR, RP, and RR can be used, our focus is on RR constraints, Figure 6.3.

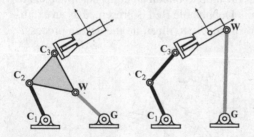

Fig. 6.3 Examples of constrained 3R serial chain.

A planar six-bar linkage consists of six links and seven joints and has two topologically distinct configurations. These are called the Watt and Stephenson six-bar chains, Figure 6.4. Inversions of these chains yield the Watt I, Watt II, Stephenson I, Stephenson II, and Stephenson III linkages.

6.1.1 Adding RR Constraints

We design a six-bar linkage by adding constraints to 3R serial chain to obtain a one-degree-of-freedom linkage. We will focus on the synthesis of RR constraints, but PR or RP chains can be used to constrain the 3R chain using the same methodology.

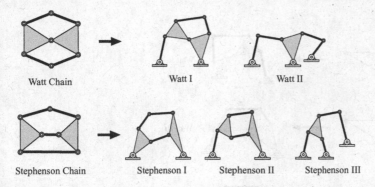

Fig. 6.4 The Watt and Stephenson six-bar chains, and the different forms obtained by selecting different links as the base.

In order to identify the ways to add RR constraints to a 3R serial chain, we denote links of the 3R chain, as B_i, $i = 0, 1, 2, 3$. An RR chain cannot constrain consecutive links, therefore the two links that can be connected by the first RR constraint are (i) B_0B_2, (ii) B_0B_3, or (iii) B_1B_3. Once the first constraint is attached, the second RR constraint can be connected either to one of the original links, or to the link created by the first RR constraint.

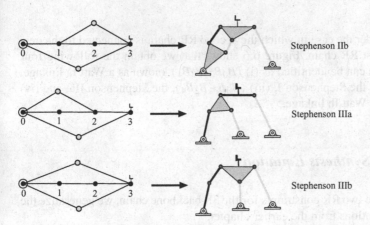

Fig. 6.5 The linkage graphs show the synthesis sequence for the three constrained 3R chains in which the two RR chains are attached independently.

Figure 6.5 shows the various systems that result from independent RR constraints applied to a 3R chain. Notice that while two RR chains can connect B_3 to ground, there are no other cases in which the two RR cranks are connected to the same bodies. The result is three six-bar linkages, the Stephenson IIb, Stephenson IIIa, and Stephenson IIIb. The notation "a" and "b" is used to distinguish the input crank, which we consider to be the first link of the 3R chain.

128   6 Multiloop Planar Linkages

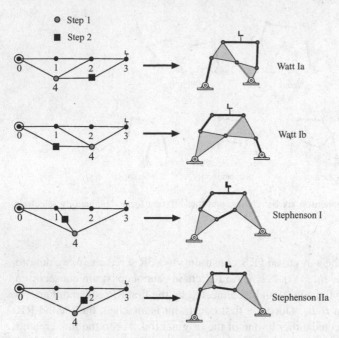

Fig. 6.6 The linkage graphs show the synthesis sequence for the four constrained 3R chains in which the second RR chain connects to the first RR chain.

We now consider the case in which the second RR chain is connected to the new link B_4 of the first RR chain. Figure 6.6 shows that we obtain the following four topologies, which can be identified as (i) (B_0B_2, B_3B_4) known as a Watt Ia linkage, (ii) (B_0B_3, B_1B_4), the Stephenson I, (iii) (B_0B_3, B_2B_4), the Stephenson IIa, and (iv) (B_1B_3, B_0B_4), the Watt Ib linkage.

6.1.2 The RR Synthesis Equation

In order to size the two RR constraints for the 3R backbone chain, we generalize the RR synthesis equations from the earlier chapter.

Let $[B_{l,j}]$, $j = 1, \ldots, 5$, be the five positions of the lth moving link, and $[B_{k,j}]$, $j = 1, \ldots, 5$, the five positions of the kth moving link measured in a world frame F. Let $\mathbf{g}$ be the coordinates of the R joint attached to the lth link measured in the link frame B_l. Similarly, let $\mathbf{w}$ be the coordinates of the other R joint measured in the link frame B_k. The five positions taken by these points in the moving frame as the two bodies move relative to each other are given by

$$\mathbf{G}^j = [B_{l,j}]\mathbf{g} \quad \text{and} \quad \mathbf{W}^j = [B_{k,j}]\mathbf{w}. \tag{6.1}$$

Now introduce the relative displacements

$$[R_{1j}] = [B_{l,j}][B_{l,1}]^{-1} \quad \text{and} \quad [S_{1j}] = [B_{k,j}][B_{k,1}]^{-1}, \tag{6.2}$$

so these equations become

$$\mathbf{G}^j = [R_{1j}]\mathbf{G}^1 \quad \text{and} \quad \mathbf{W}^j = [S_{1j}]\mathbf{W}^1, \tag{6.3}$$

where $[R_{11}] = [S_{11}] = [I]$ are the identity transformations.

The points $\mathbf{G}^j$ and $\mathbf{W}^j$ define the ends of a rigid link of length R; therefore we have the constraint equations

$$([S_{1j}]\mathbf{W}^1 - [R_{1j}]\mathbf{G}^1) \cdot ([S_{1j}]\mathbf{W}^1 - [R_{1j}]\mathbf{G}^1) = R^2. \tag{6.4}$$

These five equations can be solved to determine the five design parameters of the RR constraint, $\mathbf{G}^1 = (u, v, 1)^T$, $\mathbf{W}^1 = (x, y, 1)^T$, and R. We will refer to these equations as the *general synthesis equations* for the planar RR link.

To solve the general synthesis equations, it is convenient to introduce the displacements

$$[D_{1j}] = [R_{1j}]^{-1}[S_{1j}] = [B_{l,1}][B_{l,j}]^{-1}[B_{k,j}][B_{k,1}]^{-1}, \tag{6.5}$$

so we obtain

$$([D_{1j}]\mathbf{W}^1 - \mathbf{G}^1) \cdot ([D_{1j}]\mathbf{W}^1 - \mathbf{G}^1) = R^2, j = 1, \dots, 5. \tag{6.6}$$

Subtract the first of these equations from the remaining ones to cancel R^2 and the square terms in the variables u, v and x, y. The resulting four bilinear equations can be solved algebraically to obtain the desired pivots.

6.1.3 Algebraic Elimination

The general synthesis equations (6.6) can be solved using the algebraic elimination procedure presented previously. Recall that this consists of constructing four bilinear equations and extracting four 3×3 minors M_j to obtain four cubic polynomials in x and y, given by

$$\mathcal{R}_j : d_{j0}y^3 + d_{j1}y^2 + d_{j2}y + d_{j3} = 0, \quad j = 1, \dots 4, \tag{6.7}$$

where the coefficient d_{kj} is a polynomial in x of degree k. The four polynomials $\mathcal{R}_j$ are assembled into the matrix equation

$$\mathcal{R} = [D(x)]\mathbf{m} = \begin{bmatrix} d_{10}(x) & d_{11}(x) & d_{12}(x) & d_{13}(x) \\ \vdots & \vdots & \vdots & \vdots \\ d_{40}(x) & d_{41}(x) & d_{42}(x) & d_{43}(x) \end{bmatrix} \begin{Bmatrix} y^3 \\ y^2 \\ y \\ 1 \end{Bmatrix} = \begin{Bmatrix} 0 \\ 0 \\ 0 \\ 0 \end{Bmatrix}. \tag{6.8}$$

As we have seen, this equation can be solved for a non-zero $\mathbf{m} = (y^3, y^2, y, 1)$, only if the resultant matrix $[D(\mathbf{x})]$ had a determinant equal zero.

Here we present a method to compute the roots of $\det[D(x)] = 0$ using the eigenvalue technique described in [72]. For convenience, rename x as λ, and expand $[D(\lambda)]$ in matrix polynomial form as

$$[D(\lambda)] = [D_0] + [D_1]\lambda + [D_2]\lambda^2 + [D_3]\lambda^3, \qquad (6.9)$$

where D_0, D_1, D_2, and D_3 are 4×4 matrices.

The roots of $\det D(\lambda) = 0$ are the finite eigenvalues of the generalized system

$$[B]\mathbf{x} = \lambda[A]\mathbf{x}, \qquad (6.10)$$

where A, B, and $\mathbf{x}$ are given by

$$[A] = \begin{bmatrix} I_4 & 0 & 0 \\ 0 & I_4 & 0 \\ 0 & 0 & D_3 \end{bmatrix}, \quad [B] = \begin{bmatrix} 0 & I_4 & 0 \\ 0 & 0 & I_4 \\ -D_0 & -D_1 & -D_2 \end{bmatrix}, \quad \mathbf{x} = \left\{ \begin{matrix} \mathbf{m} \\ \lambda\mathbf{m} \\ \lambda^2\mathbf{m} \end{matrix} \right\}, \qquad (6.11)$$

where I_4 is the 4×4 identity matrix and 0 is a 4×4 matrix of zeros. The solution of this eigenvalue problem yields four finite solutions.

For each real value of λ we computes its eigenvector and obtain a value for y. Notice that each eigenvector $\mathbf{x}$ is defined up to a constant multiple μ, therefore, it is convenient to determine the coordinate of y by computing the ratio of elements of $\mathbf{x}$, such as

$$y = \frac{x_3}{x_4} = \frac{\mu y}{\mu}. \qquad (6.12)$$

The remaining variables u and v are obtained by solving two of the four bilinear synthesis equations formulated from Eq (6.6).

6.1.4 The Number of Six-Bar Linkage Designs

The synthesis of an RR constraint yields as many as four designs; therefore two constraints can yield 16 designs. However, our process always has one link of the 3R chain as one of the solutions, so there are at most 12 designs.

The Stephenson I structure has all 12 designs. The Stephenson II has two different sets of designs that differ in the workpiece link, and so yields 24 designs. The Stephenson III also has two ways to connect to the end effector, one of which yields 12 designs and the other six, a total of 18.

The Watt I has two different sets of designs that differ in the input crank, and both sets include a link on the 3R chain for both RR constraints, so there are 2×9 candidates. The result is as many as 72 different six-bar linkage designs obtained using this process.

6.2 Analysis of a Six-Bar Linkage

Given any six-bar linkage, we can perform closed-form kinematic analysis to simulate its movement using complex vectors and the Dixon determinant. Our focus in this section is on the Watt Ia six-bar, but the approach can be applied to all of the six-bar linkages obtained from our synthesis methodology. All that is needed are the loop equations for the linkage.

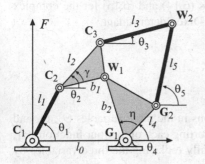

Fig. 6.7 The joint angle and link length parameters for the Watt Ia six-bar linkage.

6.2.1 Complex Loop Equations

Consider the general Watt Ia linkage shown in Figure 6.7. Introduce a coordinate frame F such that the base pivot of the 3R chain, $\mathbf{C_1}$, is the origin with its x-axis directed toward $\mathbf{G_1}$. Let the configuration angles θ_i, $i = 1, 2, \ldots, 5$ be as shown in Figure 6.7.

Using the notation in Figure 6.7, we formulate the vector equations of the loops formed by $\mathbf{C_1 C_2 W_1 G_1}$ and $\mathbf{C_1 C_2 C_3 W_2 G_2 G_1}$, that is,

$$
\begin{aligned}
\mathscr{F}_1: \quad & l_1 \cos\theta_1 + b_1 \cos(\theta_2 - \gamma) - b_2 \cos(\theta_4 + \eta) - l_0 = 0, \\
\mathscr{F}_2: \quad & l_1 \sin\theta_1 + b_1 \sin(\theta_2 - \gamma) - b_2 \sin(\theta_4 + \eta) = 0, \\
\mathscr{F}_3: \quad & l_1 \cos\theta_1 + l_2 \cos\theta_2 + l_3 \cos\theta_3 - l_4 \cos\theta_4 - l_5 \cos\theta_5 - l_0 = 0, \\
\mathscr{F}_4: \quad & l_1 \sin\theta_1 + l_2 \sin\theta_2 + l_3 \sin\theta_3 - l_4 \sin\theta_4 - l_5 \sin\theta_5 = 0. \qquad (6.13)
\end{aligned}
$$

Select the angle θ_1 as the input to the six-bar linkage, then these four equations $\mathscr{F}_i$ determine the joint angles θ_j, $j = 2, 3, 4, 5$.

Now introduce the complex numbers $\Theta_j = e^{i\theta_j}$, so the four loop equations (6.13) become two complex loop equations,

$$\mathscr{C}_1: \quad l_1\Theta_1 + b_1\Theta_2 e^{-i\gamma} - b_2\Theta_4 e^{i\eta} - l_0 = 0,$$
$$\mathscr{C}_2: \quad l_1\Theta_1 + l_2\Theta_2 + l_3\Theta_3 - l_4\Theta_4 - l_5\Theta_5 - l_0 = 0. \tag{6.14}$$

Take the complex conjugate of these two equations to define

$$\mathscr{C}_1^*: \quad l_1\Theta_1^{-1} + b_1\Theta_2^{-1} e^{i\gamma} - b_2\Theta_4^{-1} e^{-i\eta} - l_0 = 0,$$
$$\mathscr{C}_2^*: \quad l_1\Theta_1^{-1} + l_2\Theta_2^{-1} + l_3\Theta_3^{-1} - l_4\Theta_4^{-1} - l_5\Theta_5^{-1} - l_0 = 0. \tag{6.15}$$

We now solve the four complex loop equations (6.14) and (6.15) for the complex configuration angles Θ_j, $j = 2,3,4,5$ using the Dixon determinant.

6.2.2 The Dixon Determinant

Suppress Θ_3, so we have four complex equations in the three variables Θ_2, Θ_4 and Θ_5. We formulate the Dixon determinant by inserting each of the four functions $\mathscr{C}_1$, $\mathscr{C}_1^*$, $\mathscr{C}_2$, $\mathscr{C}_2^*$ as the first row, and then sequentially replacing the three variables by α_j in the remaining rows, to obtain

$$\Delta(\mathscr{C}_1, \mathscr{C}_1^*, \mathscr{C}_2, \mathscr{C}_2^*)$$
$$= \begin{vmatrix} \mathscr{C}_1(\Theta_2,\Theta_4,\Theta_5) & \mathscr{C}_1^*(\Theta_2,\Theta_4,\Theta_5) & \mathscr{C}_2(\Theta_2,\Theta_4,\Theta_5) & \mathscr{C}_2^*(\Theta_2,\Theta_4,\Theta_5) \\ \mathscr{C}_1(\alpha_2,\Theta_4,\Theta_5) & \mathscr{C}_1^*(\alpha_2,\Theta_4,\Theta_5) & \mathscr{C}_2(\alpha_2,\Theta_4,\Theta_5) & \mathscr{C}_2^*(\alpha_2,\Theta_4,\Theta_5) \\ \mathscr{C}_1(\alpha_2,\alpha_4,\Theta_5) & \mathscr{C}_1^*(\alpha_2,\alpha_4,\Theta_5) & \mathscr{C}_2(\alpha_2,\alpha_4,\Theta_5) & \mathscr{C}_2^*(\alpha_2,\alpha_4,\Theta_5) \\ \mathscr{C}_1(\alpha_2,\alpha_4,\alpha_5) & \mathscr{C}_1^*(\alpha_2,\alpha_4,\alpha_5) & \mathscr{C}_2(\alpha_2,\alpha_4,\alpha_5) & \mathscr{C}_2^*(\alpha_2,\alpha_4,\alpha_5) \end{vmatrix}. \tag{6.16}$$

This determinant is zero when Θ_j, $j = 2,4,5$, satisfy the loop equations, because the elements of the first row become zero.

Notice that each complex loop equation has the form

$$\mathscr{C}_k: \ c_{k0} + c_{k3}x + \sum_{j=2,4,5} c_{k,j}\Theta_j \quad \text{and} \quad \mathscr{C}_k^*: \ c_{k0}^* + c_{k3}^* x^{-1} + \sum_{j=2,4,5} c_{k,j}^* \Theta_j^{-1}, \tag{6.17}$$

where x denotes the suppressed variable Θ_3. Clearly, the equations maintain this form when α_j replaces Θ_j. Now row reduce Δ by subtracting the second row from the first row, then the third from the second, and the fourth from the third, to obtain

$$\begin{vmatrix} c_{12}(\Theta_2 - \alpha_2) & c_{12}^*(\Theta_2^{-1} - \alpha_2^{-1}) & c_{22}(\Theta_2 - \alpha_2) & c_{22}^*(\Theta_2^{-1} - \alpha_2^{-1}) \\ c_{14}(\Theta_4 - \alpha_4) & c_{14}^*(\Theta_4^{-1} - \alpha_4^{-1}) & c_{24}(\Theta_4 - \alpha_4) & c_{24}^*(\Theta_4^{-1} - \alpha_4^{-1}) \\ c_{15}(\Theta_5 - \alpha_5) & c_{15}^*(\Theta_5^{-1} - \alpha_5^{-1}) & c_{25}(\Theta_5 - \alpha_5) & c_{25}^*(\Theta_5^{-1} - \alpha_5^{-1}) \\ \mathscr{C}_1(\alpha_2,\alpha_4,\alpha_5) & \mathscr{C}_1^*(\alpha_2,\alpha_4,\alpha_5) & \mathscr{C}_2(\alpha_2,\alpha_4,\alpha_5) & \mathscr{C}_2^*(\alpha_2,\alpha_4,\alpha_5) \end{vmatrix}. \tag{6.18}$$

This determinant contains extraneous roots of the form $\Theta_j = \alpha_j$, which we can remove by dividing out the factor $(\Theta_j^{-1} - \alpha_j^{-1})$ using $\Theta_j - \alpha_j = -\Theta_j\alpha_j(\Theta_j^{-1} - $

α_j^{-1}), in order to define the determinant

$$\delta = \frac{\Delta(\mathscr{C}_1, \mathscr{C}_1^\star, \mathscr{C}_2, \mathscr{C}_2^\star)}{(\Theta_2^{-1} - \alpha_2^{-1})(\Theta_4^{-1} - \alpha_4^{-1})(\Theta_5^{-1} - \alpha_5^{-1})}, \qquad (6.19)$$

that is,

$$\delta = \begin{vmatrix} -c_{12}\Theta_2\alpha_2 & c_{12}^* & -c_{22}\Theta_2\alpha_2 & c_{22}^* \\ -c_{14}\Theta_4\alpha_4 & c_{14}^* & -c_{24}\Theta_4\alpha_4 & c_{24}^* \\ -c_{15}\Theta_5\alpha_5 & c_{15}^* & -c_{25}\Theta_5\alpha_5 & c_{25}^* \\ \mathscr{C}_1(\alpha_2,\alpha_4,\alpha_5) & \mathscr{C}_1^*(\alpha_2,\alpha_4,\alpha_5) & \mathscr{C}_2(\alpha_2,\alpha_4,\alpha_5) & \mathscr{C}_2^*(\alpha_2,\alpha_4,\alpha_5) \end{vmatrix}. \qquad (6.20)$$

This determinant expands to form a polynomial of the form

$$\delta = \mathbf{a}^T[W]\mathbf{t} = 0, \qquad (6.21)$$

where $\mathbf{a}$ and $\mathbf{t}$ are vectors of monomials,

$$\mathbf{a} = \begin{Bmatrix} \alpha_2 \\ \alpha_4 \\ \alpha_5 \\ \alpha_4\alpha_5 \\ \alpha_2\alpha_5 \\ \alpha_2\alpha_4 \end{Bmatrix} \quad \text{and} \quad \mathbf{t} = \begin{Bmatrix} \Theta_2 \\ \Theta_4 \\ \Theta_5 \\ \Theta_4\Theta_5 \\ \Theta_2\Theta_5 \\ \Theta_2\Theta_4 \end{Bmatrix}. \qquad (6.22)$$

The 6×6 matrix $[W]$ is given by

$$[W] = \begin{bmatrix} D_1 x + D_2 & A^T \\ A & -(D_1^* x^{-1} + D_2^*) \end{bmatrix}. \qquad (6.23)$$

The matrices $[D_1]$ and $[D_2]$ are 3×3 diagonal matrices, given by

$$[D_1] = \begin{bmatrix} b_1 b_2 l_3 l_5 e^{-i(\gamma+\eta)} & 0 & 0 \\ 0 & -b_1 b_2 l_3 l_5 e^{i(\gamma+\eta)} & 0 \\ 0 & 0 & 0 \end{bmatrix}, \quad [D_2] = \begin{bmatrix} d_1 & 0 & 0 \\ 0 & d_2 & 0 \\ 0 & 0 & d_3 \end{bmatrix}, \qquad (6.24)$$

where

$$d_1 = (l_1\Theta_1 - l_0)(b_1 b_2 l_5 e^{-i(\gamma+\eta)} - b_2 l_2 l_5 e^{-i\eta}),$$
$$d_2 = (l_0 - l_1\Theta_1)(b_1 b_2 l_5 e^{i(\gamma+\eta)} - b_1 l_4 l_5 e^{i\gamma}),$$
$$d_3 = (l_1\Theta_1 - l_0)(b_2 l_2 l_5 e^{-i\eta} - b_1 l_4 l_5 e^{i\gamma}).$$

And the 3×3 matrix $[A]$ is given by

$[A]$

$$
= \begin{bmatrix}
0 & b_1 b_2 l_5^2 e^{i(\gamma+\eta)} & -b_2^2 l_2 l_5 + b_1 b_2 l_4 l_5 e^{i(\gamma+\eta)} \\
-b_1 b_2 l_5^2 e^{-i(\gamma+\eta)} & 0 & b_1 b_2 l_2 l_5 e^{-i(\gamma+\eta)} - b_1^2 l_4 l_5 \\
b_2^2 l_2 l_5 - b_1 b_2 l_4 l_5 e^{-i(\gamma+\eta)} & -b_1 b_2 l_2 l_5 e^{i(\gamma+\eta)} + b_1^2 l_4 l_5 & 0
\end{bmatrix}.
$$

$$(6.25)$$

A set of values Θ_j that satisfy the loop equations (6.14) and (6.15) will also yield $\delta = 0$, which will be true for arbitrary values of the auxiliary variables α_j. Thus, solutions for these loop equations must also satisfy the matrix equation

$$[W]\mathbf{t} = 0. \tag{6.26}$$

The matrix $[W]$ has the structure

$$[W]\mathbf{t} = \left[\begin{pmatrix} D_1 & 0 \\ A & -D_2^* \end{pmatrix} x - \begin{pmatrix} -D_2 & -A^T \\ 0 & D_1^* \end{pmatrix} \right] \mathbf{t} = [Mx - N]\mathbf{t} = 0. \tag{6.27}$$

Because $[W]$ is a square matrix, this equation has non-zero solutions only if $\det[W] = 0$. Expanding this determinant, we obtain a polynomial in $x = \Theta_3$.

Notice that the values of $x = \Theta_3$ that satisfy $\det[W] = 0$ are also eigenvalues of the characteristic polynomial $p(x) = \det(Mx - N)$ of the generalized eigenvalue problem

$$[N]\mathbf{t} = x[M]\mathbf{t}. \tag{6.28}$$

Each value of $x = \Theta_3$ has an associated eigenvector $\mathbf{t}$, which yields the values of the remaining joint angles Θ_j, $j = 2, 4, 5$.

It is useful to notice that an eigenvector $\mathbf{t} = (t_1, t_2, t_3, t_4, t_5, t_6)^T$ is defined only up to a constant multiple, μ. Therefore, it is convenient to determine the values Θ_j by the computing the ratios,

$$\Theta_2 = \frac{t_5}{t_3} = \frac{\mu \Theta_2 \Theta_5}{\mu \Theta_5}, \quad \Theta_4 = \frac{t_6}{t_1} = \frac{\mu \Theta_2 \Theta_4}{\mu \Theta_2}, \quad \Theta_5 = \frac{t_4}{t_2} = \frac{\mu \Theta_4 \Theta_5}{\mu \Theta_4}. \tag{6.29}$$

For a given input angle, a six-bar linkage can have as many as six roots for the configuration variables Θ_j, $j = 2, 3, 4, 5$. Each of these roots defines an *assembly* of the six-bar linkage.

6.2.3 Sorting Assemblies

As we analyze a six-bar linkage for a sequence of input angles Θ_1^k, there are as many as six sets of configuration angles $\vec{\Theta} = (\Theta_2, \Theta_3, \Theta_4, \Theta_5)_i$, $i = 1, \ldots, 6$ that define the assemblies of the linkage associated with each input angle. In order to sort the roots among the assemblies, we use the Jacobian of the loop equations.

Compute the derivative of the loop equations (6.14) and (6.15) to obtain

$$\nabla \mathscr{C}_1: \quad l_1 \dot{\Theta}_1 + b_1 \dot{\Theta}_2 e^{-i\gamma} - b_2 \dot{\Theta}_4 e^{i\eta} = 0,$$

$$\nabla \mathscr{C}_2: \quad l_1 \dot{\Theta}_1 + l_2 \dot{\Theta}_2 + l_3 \dot{\Theta}_3 - l_4 \dot{\Theta}_4 - l_5 \dot{\Theta}_5 = 0,$$

$$\nabla \mathscr{C}_1^\star: \quad -l_1 \dot{\Theta}_1 \Theta_1^{-2} - b_1 \dot{\Theta}_2 e^{i\gamma} \Theta_2^{-2} + b_2 \dot{\Theta}_4 e^{-i\eta} \Theta_4^{-2} = 0,$$

$$\nabla \mathscr{C}_2^\star: \quad -l_1 \dot{\Theta}_1 \Theta_1^{-2} - l_2 \dot{\Theta}_2 \Theta_2^{-2} - l_3 \dot{\Theta}_3 \Theta_3^{-2} + l_4 \dot{\Theta}_4 \Theta_4^{-2} + l_5 \dot{\Theta}_5 \Theta_5^{-2} = 0. \quad (6.30)$$

Factor out the derivative vector $\dot{\vec{\Theta}} = (\dot{\Theta}_1, \dot{\Theta}_2, \dot{\Theta}_3, \dot{\Theta}_4, \dot{\Theta}_5)$, and obtain the Jacobian matrix

$$[\nabla \mathscr{C}(\vec{\Theta}_k)] = \begin{bmatrix} l_1 & b_1 e^{-i\gamma} & 0 & -b_2 e^{i\eta} & 0 \\ l_1 & l_2 & l_3 & -l_4 & -l_5 \\ -l_1 \Theta_1^{-2} & -b_1 e^{i\gamma} \Theta_2^{-2} & 0 & b_2 e^{-i\eta} \Theta_4^{-2} & 0 \\ -l_1 \Theta_1^{-2} & -l_2 \Theta_2^{-2} & -l_3 \Theta_3^{-2} & l_4 \Theta_4^{-2} & l_5 \Theta_5^{-2} \end{bmatrix}. \quad (6.31)$$

In order to sort the roots to among assemblies of the six-bar linkage, we approximate the complex loop equations using the Jacobian, and obtain

$$[\nabla \mathscr{C}(\vec{\Theta}_i^k)](\vec{\Psi} - \vec{\Theta}_i^k) = 0, \quad (6.32)$$

where Ψ approximates the value $\vec{\Theta}_i^{k+1}$ associated with the input angle Θ_1^{k+1} and is near the assembly defined by $\vec{\Theta}_i^k$. It is then a matter of identifying which of the root $\vec{\Theta}_i^{k+1}$ is closest to Ψ on the ith circuit, in order to match the assemblies. This provides a rapid and exact method to determine a sequence of configuration angles for each assembly in order to animate the six-bar linkage.

6.3 Example: A Steering Linkage

As an example of our six-bar linkage design methodology, we consider a steering linkage. Generally, a steering linkage controls the direction of the two wheels around king-pin axes attached to the frame of a vehicle. Our goal is a design that does not use a king-pin, and, instead, allows the wheels to move laterally relative to the vehicle frame as they change direction. This is intended for a gravity racer in which extra stability in turning reduces the need for braking.

For this design, we use a Watt Ia linkage, see Figure 6.8. We will add the RR constraints $\mathbf{G}_1 \mathbf{W}_1$, and $\mathbf{G}_2 \mathbf{W}_2$ to the 3R backbone chain $\mathbf{C}_1 \mathbf{C}_2 \mathbf{C}_3$ to obtain the Watt Ia linkage.

Step 1.

The task positions that define the positions of the left wheel, $[T_i], i = 1, \ldots, 5$ are given in Table 6.1. These positions provide coordinated tracking of the front wheels with turning radius shown in the table.

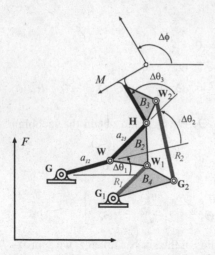

Fig. 6.8 A 3R chain, $C_1C_2C_3$, constrained by two RR chains, G_1W_1 and G_2W_2, to form a Watt Ia linkage.

Table 6.1 Five task positions for the end effector of the 3R chain (dimensions in mm)

Task	Position (ϕ, x, y)	Turning radius
1	$(43.9°, -325, 1511)$	-1895
2	$(20.0°, -432, 1562)$	-4724
3	$(0°, -559, 1588)$	no turn
4	$(-15.9°, -689, 1542)$	4724
5	$(-29.1°, -751, 1473)$	1895

The dimensions of the 3R chain are chosen to be $a_{12} = a_{23} = 274$mm. The location of the base and moving pivots of the 3R chain, $[G]$ and $[H]$, are specified by the translation vectors $(0°, -153, 1241)$ and $(0°, -133, 0)$, respectively. Use this data to formulate the inverse kinematics equations of the 3R chain and solve for the configuration angles $\mathbf{q}_j = (\theta_1, \theta_2, \theta_3)_j$, $j = 1, \ldots, 5$, that reach the specified task positions $[T_j]$, $j = 1, \ldots, 5$.

Notice that the inverse kinematics equations yield two sets of configuration angles corresponding to a 3R chain with its elbow up and elbow down. In this example, we chose the elbow configuration defined by the coordinates $C_1 = (-153, 1241)$, $C_2 = (6.6, 1463.7)$, and $C_3 = (-229.2, 1603.2)$

The five configurations of the 3R chain provide the coordinate transformations for each link relative to the ground frame. Compute $[B_{1j}] = [R(\theta_{1j}), C_{1j}]$, which defines the five positions of the first link in F. The positions of the second and third links are given by

$$[B_{2j}] = [R(\theta_{1j} + \theta_{2j}), C_{2j}],$$
$$\text{and } [B_{3j}] = [R(\theta_{1j} + \theta_{2j} + \theta_{3j}), C_{3j}], \quad j = 1, \ldots, 5. \quad (6.33)$$

Table 6.2 Step 1. Select the joint coordinates for the 3R chain; Step 2. Solve for the first RR chain; Step 3. Solve for the second RR chain; The selected values are highlighted in bold (dimensions in mm)

Step 1	C_1	C_2	C_3
	$(-153, 1241)$	$(6.6, 1463.7)$	$(-229.2, 1603.2)$

Step 2	G_1	W_1
1	$(-153, 1241)$	$(6.6, 1463.7)$
2	$(-93.6, 936.0)$	$(-64.9, 918.0)$
3	$(\mathbf{-84.3, 1294.9})$	$(\mathbf{150.7, 1570.3})$
4	$(74.5, 861.6)$	$(172.6, 805.0)$

Step 3	G_2	W_2
1	$(\mathbf{-148.9, 1442.1})$	$(\mathbf{-198.2, 1735.8})$
2	$(424.6, 962.0)$	$(-244.2, 1668.3)$
3	$(-221.7, 1794.7)$	$(-197.7, 1669.5)$
4	$(150.7, 1570.3)$	$(-229.2, 1603.2)$

The 15 coordinate transformations $[B_{ij}]$ form the tasks requirements that can be used to synthesize the planar RR constraints.

Step 2.

Using the five positions $B_{2j}, j = 1, \ldots, 5$ of the link B_2 relative to the ground frame, assemble the design equations for an RR chain, denoted $G_1 W_1$ in Figure 6.8. In this case, we obtain (6.6) with $[D_{1i}] = [B_{2i}][B_{21}]^{-1}, i = 1, \ldots, 5$. The solution yields the values listed in Table 6.2.

Notice that the RR chain formed by $C_1 C_2$ is among the computed RR constraints. This means we must choose of the remaining three sets to be the first constraint $G_1 W_1$. Because real solutions to the design equations occur in pairs, the presence of the RR chain $C_1 C_2$, guarantees at least one real solution to this set of design equations.

Step 3.

The RR chain $G_1 W_1$ introduces a new link B_4, which takes the positions $[B_{4j}], j = 1, \ldots, 5$, when the end effector is in each of the specified task positions. This defines the coordinate transformations $[B_{4j}] = [R(\theta_{4j}), G_{1j}]$

Using the positions of the end effector of the 3R chain and the positions of the link B_4, we assemble the design equations for an RR constraint $G_2 W_2$. The solution to these equations are shown in Table 6.2. In this case, the chain $C_2 C_3$ appears

Fig. 6.9 The example solution for a Watt Ia linkage reaching each of the five task positions.

among the solutions, which guarantees a second real solution. Thus, this design process results in at least one real Watt 1a six-bar linkage.

Analysis

In order to animate the movement of this system, solve the loop equations of the six-bar linkage for a sequence of input angles interpolated between those defined by the task positions. Let these input angles θ_1^k, $k = 1 \ldots, N$, then, for each value, θ_1^k, solve the loop equations (6.14) and (6.15) to determine the remaining configuration angles $\Theta^k = (\theta_2^k, \theta_3^k, \theta_4^k, \theta_5^k)^T$.

Recall that there can be as many as six roots for each Θ^k, only one of which corresponds to the assembly of the six-bar linkcage specified in each of the task positions. Identify the root Θ^k for the correct assembly by formulating and solving the approximation to the loop equations (6.32). This calculation can be checked against the known configurations at each of the task positions. The results for our design are shown in Table 6.3.

Table 6.3 Analysis solutions for the steering mechanism

	θ_1	θ_2	θ_3	θ_4	θ_5
1	16.26°	111.28°	−43.06°	66.37°	−63.49°
2	31.19°	118.18°	−36.16°	70.35°	−55.85°
3	53.76°	126.41°	−27.93°	83.67°	−39.56°
4	81.50°	136.38°	−17.96°	106.03°	−16.23°
5	101.28°	144.20°	−10.14°	125.43°	2.08°

Figure 6.9 shows the six-bar linkage that results when the pivots highlighted in Table 6.2 are selected for the design. In order to complete the steering linkage

provide a mirror-image of this six-bar linkage, and connect the input cranks to the steering wheel. The result is a single degree-of-freedom 14-bar linkage, Figure 6.10. This linkage achieves five Ackermann steering positions and extends the outside wheel to increase the reaction to the roll-over moment during a turn.

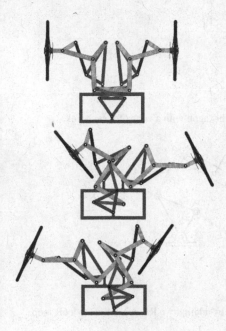

Fig. 6.10 The 14-bar steering system formed by the six-bar linkage and its mirror image at the front wheels which are connected by a driving crank and two coupler links.

6.4 Synthesis of a Planar Eight-Bar Linkage

Two 3R serial chains that share an end effector form a planar 6R loop with three degrees of freedom, Figure 6.11. The graph of this linkage is a hexagon. We will an design eight-bar linkage by adding two RR constraints to a 6R loop, Figure 6.12. It is known that there are sixteen topologically distinct eight-bar linkages, as compared to two for six-bar linkages. Inversions of these topologies yields a large number different design options for an eight-bar linkage.

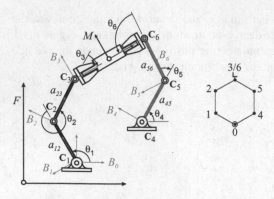

Fig. 6.11 The graph of the planar 6R loop forms a hexagon with a vertex for each link.

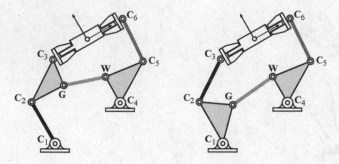

Fig. 6.12 Examples of eight-bar linkages obtained by adding two RR constraints to a 6R loop.

6.4.1 Adding RR Constraints

In order to design an eight-bar chain, we choose a 6R loop that moves through the task positions, and then attach two RR constraints. Notice that RP and PR chains can be used instead of the RR chain to constrain the 6R loop, in which case the design equations for these constraints are used in what follows. The selection of the 6R loop provides the designer the freedom to choose the two connections to ground and the two connections to the moving workpiece. We choose not to add a constraint to ground in order to avoid a third ground pivot.

The various ways that two RR constraints can be added to a 6R loop are labeled by listing the links that are connected. Denote the ground link as B_0, the three links of the left 3R chain as B_i, $i = 1,2,3$, and the links of the right 3R chain as B_i, $i = 4,5,6$, which means that $B_3 = B_6$. Because we cannot constrain two neighboring links and prefer not to constrain a link to ground, the first RR constraint is one of the six cases (i) B_1B_3, (ii) B_1B_4, (iii) B_1B_5, (iv) B_2B_4, (v) B_2B_5, and (vi) B_3B_4, Figure 6.13.

As we saw in the design of a six-bar linkage, the second RR constraint can be attached independently to existing links of the 6R loop, as well as the case where the

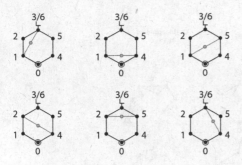

Fig. 6.13 There are six ways to add the first RR chain.

second RR constraint is attached to new link of the first RR constraint. In the first case, we choose two of the six constraint to obtain $\binom{6}{2} = 15$ eight-bar structures. In addition to this, there are two cases where both RR constraints attached to the same bodies is possible, (i) B_1B_5, and (ii) B_2B_4. The 17 eight-bar structures are shown in Figure 6.14. We label these structures by denoting as Bij the RR constraint connecting B_i and B_j, so the structure with constraints Bij and Bkl is BijBkl. Using this notation, the eight-bar linkage B14B15 has RR constraints B_1B_4 and B_1B_5.

Let the link introduced by the first RR constraint be denoted B_7, then we have 15 eight-bar structures that arise from the attachment of the second RR constraint to B_7. These can be enumerated by considering that there are three ways to attach the second RR constraint, if the first constraint is one of the three, B_1B_4, B_1B_5, and B_2B_4; and two ways, if the first constraint is one of the three B_1B_3, B_2B_5 and B_3B_4. Thus, there are $3 \times 3 + 3 \times 2 = 15$ eight-bar structures, Figure 6.15.

6.4.2 The Number of Eight-Bar Linkage Designs

This design process yields 32 eight-bar structures available to a designer. Recall that the synthesis of an RR chain for a five-position task yields as many as four design candidates, so two RR constraints can yield as many as 16 candidates for a particular eight-bar structure. However, this occurs only for the case B15B24.

There are 17 cases where one of the existing links of the 6R loop arises in the synthesis of one of the RR constraints. For these structures, there are at most 12 design candidates. There are 12 cases in which the synthesis of both RR constraints includes two links of the 6R loop, which means that there are at most nine design candidates. Finally, the two cases B24B24, and B15B15 have RR chains attached to the same bodies, which means that there are at most $\binom{4}{2} = 6$ design candidates.

Thus, this design process can yield as many as 340 ($1 \times 16 + 17 \times 12 + 12 \times 9 + 2 \times 6$) eight-bar linkage design candidates. Comparing the linkage graphs in

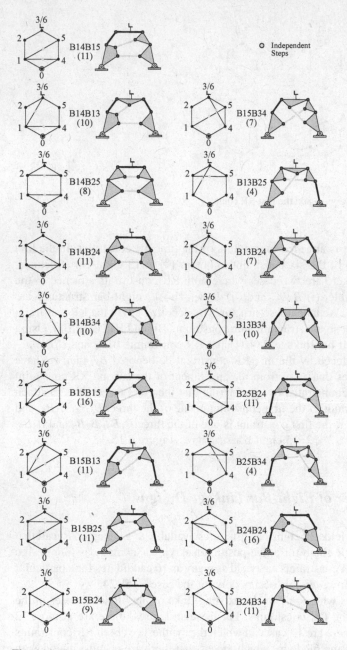

Fig. 6.14 The 17 eight-bar linkage structures obtained from the independent attachment of two RR constraints to a 6R loop. The number in parentheses identifies the associated eight-bar topology.

Figures 6.14 and 6.15 to the 16 eight-bar topologies in Figure 6.16, we see that these structures are inversions of the topologies 3, 4, 7, 8, 9, 10, 11, and 16.

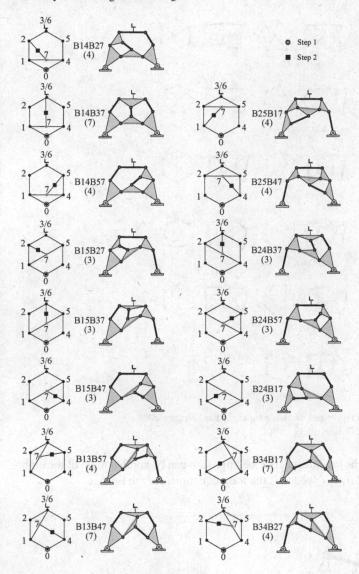

Fig. 6.15 The 15 eight-bar linkage structures obtained when the second RR constraint is attached to the link of the first RR constraint in a 6R loop. The number in parentheses identifies the associated eight-bar topology.

6.5 Analysis of an Eight-bar Linkage

The analysis of an eight-bar linkage to determine its configurations as a function of an input angle follows the analysis presented for a six-bar linkage using complex vectors and the Dixon determinant. For the purposes of this presentation, we focus

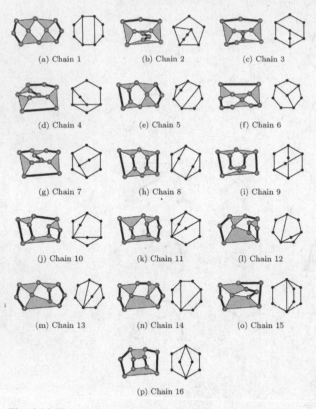

(a) Chain 1   (b) Chain 2   (c) Chain 3

(d) Chain 4   (e) Chain 5   (f) Chain 6

(g) Chain 7   (h) Chain 8   (i) Chain 9

(j) Chain 10   (k) Chain 11   (l) Chain 12

(m) Chain 13   (n) Chain 14   (o) Chain 15

(p) Chain 16

Fig. 6.16 The sixteen topologies available for eight-bar linkages.

on the B14B25 eight-bar linkage, but the approach can be applied to all of the eight-bar topologies. All that is needed is the loop equations for the linkage.

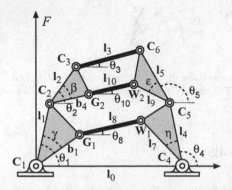

Fig. 6.17 The joint angle and link length parameters for the B14B25 eight-bar linkage.

6.5.1 Complex Loop Equations

Consider the B14B25 linkage shown in Figure 6.17. Introduce a coordinate frame F such that the base pivot of the 6R loop, $\mathbf{C}_1$, is the origin with its x-axis directed toward $\mathbf{C}_4$. The input angle is θ_1 and the output configuration angles are $\vec{\theta} = (\theta_2, \theta_3, \theta_4, \theta_5, \theta_8, \theta_{10})$ shown in Figure 6.17.

The vector loop equations for the three loops (i)$\mathbf{C}_1\mathbf{G}_1\mathbf{W}_1\mathbf{C}_4$, (ii) $\mathbf{C}_1\mathbf{C}_2\mathbf{G}_2\mathbf{W}_2\mathbf{C}_5\mathbf{C}_4$, and (iii) $\mathbf{C}_1\mathbf{C}_2\mathbf{C}_3\mathbf{C}_6\mathbf{C}_5\mathbf{C}_4$ are given by

$$
\begin{aligned}
\mathscr{F}_1 &: b_1\cos(\theta_1 - \gamma) + l_8\cos\theta_8 - l_7\cos(\theta_4 + \eta) - l_0 = 0, \\
\mathscr{F}_2 &: b_1\sin(\theta_1 - \gamma) + l_8\sin\theta_8 - l_7\sin(\theta_4 + \eta) = 0, \\
\mathscr{F}_3 &: l_1\cos\theta_1 + b_4\cos(\theta_2 - \beta) + l_{10}\cos\theta_{10} - l_9\cos(\theta_5 + \varepsilon) - l_4\cos\theta_4 - l_0 = 0, \\
\mathscr{F}_4 &: l_1\sin\theta_1 + b_4\sin(\theta_2 - \beta) + l_{10}\sin\theta_{10} - l_9\sin(\theta_5 + \varepsilon) - l_4\sin\theta_4 = 0, \\
\mathscr{F}_5 &: l_1\cos\theta_1 + l_2\cos\theta_2 + l_3\cos\theta_3 - l_4\cos\theta_4 - l_5\cos\theta_5 - l_0 = 0, \\
\mathscr{F}_6 &: l_1\sin\theta_1 + l_2\sin\theta_2 + l_3\sin\theta_3 - l_4\sin\theta_4 - l_5\sin\theta_5 = 0.
\end{aligned}
$$

$$(6.34)$$

For each value of the input angle these six equations define the six configuration angles.

Now introduce the complex vectors $\Theta_j = e^{i\theta_j}$, so the six equations (6.34) become three complex loop equations,

$$
\begin{aligned}
\mathscr{C}_1 &= b_1\Theta_1 e^{-i\gamma} + l_8\Theta_8 - l_7\Theta_4 e^{i\eta} - l_0, \\
\mathscr{C}_2 &= l_1\Theta_1 + b_4\Theta_2 e^{-i\beta} + l_{10}\Theta_{10} - l_9\Theta_5 e^{i\varepsilon} - l_4\Theta_4 - l_0, \\
\mathscr{C}_3 &= l_1\Theta_1 + l_2\Theta_2 + l_3\Theta_3 - l_4\Theta_4 - l_5\Theta_5 - l_0.
\end{aligned}
$$

$$(6.35)$$

The complex conjugate of these equations yields

$$
\begin{aligned}
\mathscr{C}_1^* &= b_1\Theta_1^{-1} e^{i\gamma} + l_8\Theta_8^{-1} - l_7\Theta_4^{-1} e^{-i\eta} - l_0, \\
\mathscr{C}_2^* &= l_1\Theta_1^{-1} + b_4\Theta_2^{-1} e^{i\beta} + l_{10}\Theta_{10}^{-1} - l_9\Theta_5^{-1} e^{-i\varepsilon} - l_4\Theta_4^{-1} - l_0, \\
\mathscr{C}_3^* &= l_1\Theta_1^{-1} + l_2\Theta_2^{-1} + l_3\Theta_3^{-1} - l_4\Theta_4^{-1} - l_5\Theta_5^{-1} - l_0.
\end{aligned}
$$

$$(6.36)$$

We solve the six complex loop equations (6.35) and (6.36) for the complex configuration angles $\vec{\Theta} = (\Theta_2, \Theta_3, \Theta_4, \Theta_5, \Theta_8, \Theta_{10})$, using the Dixon determinant.

6.5.2 The Dixon Determinant

Suppress Θ_3, so we have six complex equations in the five variables $\Theta_2, \Theta_4, \Theta_5, \Theta_8$, and Θ_{10}. The Dixon determinant constructed with the six functions $\mathscr{C}_1, \mathscr{C}_1^*, \mathscr{C}_2, \mathscr{C}_2^*$, $\mathscr{C}_3, \mathscr{C}_3^*$ as the first row. For the successive rows replace the five configuration angles by the variables α_i, to obtain

$$\Delta(\mathscr{C}_1, \mathscr{C}_1^*, \mathscr{C}_2, \mathscr{C}_2^*, \mathscr{C}_3, \mathscr{C}_3^*)$$

$$= \begin{vmatrix} \mathscr{C}_1(\Theta_2, \Theta_4, \Theta_5, \Theta_8, \Theta_{10}) & \cdots & \mathscr{C}_3^*(\Theta_2, \Theta_4, \Theta_5, \Theta_8, \Theta_{10}) \\ \mathscr{C}_1(\alpha_2, \Theta_4, \Theta_5, \Theta_8, \Theta_{10}) & \cdots & \mathscr{C}_3^*(\alpha_2, \Theta_4, \Theta_5, \Theta_8, \Theta_{10}) \\ \vdots & & \vdots \\ \mathscr{C}_1(\alpha_2, \alpha_4, \alpha_5, \alpha_8, \alpha_{10}) & \cdots & \mathscr{C}_3^*(\alpha_2, \alpha_4, \alpha_5, \alpha_8, \alpha_{10}) \end{vmatrix}. \quad (6.37)$$

This determinant is zero when Θ_j, $j = 2, 4, 5, 8, 10$ satisfy the loop equations, because the elements of the first row become zero.

As we did in the analysis of the six-bar linkage, we use the properties of the general form of the loop equations and row reduce the Dixon determinant, to obtain

$$\begin{vmatrix} c_{12}(\Theta_2 - \alpha_2) & \cdots & c_{32}^*(\Theta_2^{-1} - \alpha_2^{-1}) \\ c_{14}(\Theta_4 - \alpha_4) & \cdots & c_{34}^*(\Theta_4^{-1} - \alpha_4^{-1}) \\ \vdots & & \vdots \\ \mathscr{C}_1(\alpha_2, \alpha_4, \alpha_5, \alpha_8, \alpha_{10}) & \cdots & \mathscr{C}_3^*(\alpha_2, \alpha_4, \alpha_5, \alpha_8, \alpha_{10}) \end{vmatrix}. \quad (6.38)$$

This determinant contains extraneous roots of the form $\Theta_j = \alpha_j$, which we can remove by dividing out the factor $(\Theta_j^{-1} - \alpha_j^{-1})$ using $\Theta_j - \alpha_j = -\Theta_j \alpha_j (\Theta_j^{-1} - \alpha_j^{-1})$, in order to define

$$\delta = \frac{\Delta(\mathscr{C}_1, \mathscr{C}_1^*, \mathscr{C}_2, \mathscr{C}_2^*, \mathscr{C}_3, \mathscr{C}_3^*)}{\prod_{i=2,4,5,8,10}(\Theta_i^{-1} - \alpha_i^{-1})}. \quad (6.39)$$

This determinant expands to define a polynomial of the form

$$\delta = \mathbf{a}[W]\mathbf{t} = 0, \quad (6.40)$$

where the vectors $\mathbf{a}$ and $\mathbf{t}$ are defined as follows.

The vector $\mathbf{a} = \{\mathbf{a}_1\ \mathbf{a}_2\}$ is constructed from monomials of α_i, such that $\mathbf{a}_1$ is combinations $\alpha_i \alpha_j$ of the variables $(\alpha_2, \alpha_4, \alpha_5, \alpha_8, \alpha_{10})$, and $\mathbf{a}_2$ is its complement $\alpha_k \alpha_l \alpha_m$, given by

$$\mathbf{a}_1^T = \left\{ \begin{array}{c} \alpha_2 \alpha_4 \\ \alpha_2 \alpha_5 \\ \alpha_2 \alpha_8 \\ \alpha_2 \alpha_{10} \\ \alpha_4 \alpha_5 \\ \alpha_4 \alpha_{10} \\ \alpha_5 \alpha_8 \\ \alpha_5 \alpha_{10} \\ \alpha_8 \alpha_{10} \end{array} \right\}, \quad \mathbf{a}_2^T = \left\{ \begin{array}{c} \alpha_5 \alpha_8 \alpha_{10} \\ \alpha_4 \alpha_8 \alpha_{10} \\ \alpha_4 \alpha_5 \alpha_{10} \\ \alpha_4 \alpha_5 \alpha_8 \\ \alpha_2 \alpha_8 \alpha_{10} \\ \alpha_2 \alpha_5 \alpha_8 \\ \alpha_2 \alpha_4 \alpha_{10} \\ \alpha_2 \alpha_4 \alpha_8 \\ \alpha_2 \alpha_4 \alpha_5 \end{array} \right\}. \quad (6.41)$$

Similarly the $\mathbf{t} = (\mathbf{t}_1, \mathbf{t}_2)^T$ is constructed from the monomials Θ_i, such that

$$
\mathbf{t}_1 = \left\{\begin{array}{c} \Theta_2\Theta_4 \\ \Theta_2\Theta_5 \\ \Theta_2\Theta_8 \\ \Theta_2\Theta_{10} \\ \Theta_4\Theta_5 \\ \Theta_4\Theta_{10} \\ \Theta_5\Theta_8 \\ \Theta_5\Theta_{10} \\ \Theta_8\Theta_{10} \end{array}\right\}, \quad
\mathbf{t}_2 = \left\{\begin{array}{c} \Theta_5\Theta_8\Theta_{10} \\ \Theta_4\Theta_8\Theta_{10} \\ \Theta_4\Theta_5\Theta_{10} \\ \Theta_4\Theta_5\Theta_8 \\ \Theta_2\Theta_8\Theta_{10} \\ \Theta_2\Theta_5\Theta_8 \\ \Theta_2\Theta_4\Theta_{10} \\ \Theta_2\Theta_4\Theta_8 \\ \Theta_2\Theta_4\Theta_5 \end{array}\right\}. \tag{6.42}
$$

The result is that (6.40) takes the form

$$
\{\mathbf{a}_1 \ \mathbf{a}_2\} \begin{bmatrix} D_1x + D_2 & A^T \\ A & D_1^*x^{-1} + D_2^* \end{bmatrix} \begin{Bmatrix} \mathbf{t}_1 \\ \mathbf{t}_2 \end{Bmatrix} = 0, \tag{6.43}
$$

where matrices $[D_1]$ and $[D_2[$ are diagonal, and $[A]$ is Hermitian, which means $a_{ij} = a_{ji}^*$.

There are 10 ways to form the products $\alpha_i\alpha_j$ from the variables $(\alpha_2, \alpha_4, \alpha_5, \alpha_8, \alpha_{10})$, which means there are many as 20 monomials in $\mathbf{a} = (\mathbf{a}_1, \mathbf{a}_2)$. However, for our equations, $\alpha_4\alpha_8$ and its complementary $\alpha_2\alpha_5\alpha_{10}$ are not in the monomial list because they have zero coefficients. Thus, we have 18 monomials instead of 20.

A set of values Θ_j that satisfy the loop equations (6.35) and (6.36) also yield $\delta = 0$, which will be true for arbitrary values of the auxiliary variables α_j. Thus solutions for these loop equations must also satisfy

$$
[W]\mathbf{t} = \begin{bmatrix} D_1x + D_2 & A^T \\ A & D_1^*x^{-1} + D_2^* \end{bmatrix} \begin{Bmatrix} \mathbf{t}_1 \\ \mathbf{t}_2 \end{Bmatrix} = 0. \tag{6.44}
$$

Because $[W]$ is a square matrix, this equation has non-zero solutions only if $\det[W] = 0$, which is a polynomial in $\lambda = \Theta_3$.

The structure $[W]$ allows us to write it in the form of a generalized eigenvalue problem

$$
\left[\begin{pmatrix} D_1 & 0 \\ A & D_2^* \end{pmatrix} \lambda - \begin{pmatrix} -D_2 & -A^T \\ 0 & -D_1^* \end{pmatrix} \right] \begin{Bmatrix} \mathbf{t}_1 \\ \mathbf{t}_2 \end{Bmatrix} = [M\lambda - N]\mathbf{t} = 0. \tag{6.45}
$$

The eigenvalues of this system are the solutions of the characteristic polynomial $p(x) = \det[M\lambda - N]$. For each eigenvalue $\lambda = \Theta_3$, we can solve (6.45) to obtain an associated eigenvector $\mathbf{t}$ that defines the values of the remaining joint angles $\Theta_j, j = 2, 4, 5, 8, 10$.

Because the eigenvector $\mathbf{t} = (t_1, t_2, \ldots, t_{18})^T$ is defined up to a constant multiple μ, we obtain the values of Θ_j by computing the ratios

$$
\Theta_2 = \frac{t_{18}}{t_5} = \frac{\mu\Theta_2\Theta_4\Theta_5}{\mu\Theta_4\Theta_5}, \quad \Theta_4 = \frac{t_{18}}{t_2} = \frac{\mu\Theta_2\Theta_4\Theta_5}{\mu\Theta_2\Theta_5}, \quad \Theta_5 = \frac{t_{18}}{t_1} = \frac{\mu\Theta_2\Theta_4\Theta_5}{\mu\Theta_2\Theta_4},
$$

$$
\Theta_8 = \frac{t_{17}}{t_1} = \frac{\mu\Theta_2\Theta_4\Theta_8}{\mu\Theta_2\Theta_4}, \quad \Theta_{10} = \frac{t_{16}}{t_1} = \frac{\mu\Theta_2\Theta_4\Theta_{10}}{\mu\Theta_2\Theta_4}. \tag{6.46}
$$

For a given value of the input angle Θ_1, there are as many as 18 roots for the configuration variables $\vec{\Theta} = (\Theta_2, \Theta_3, \Theta_4, \Theta_5, \Theta_8, \Theta_{10})$. Each of these roots defines an assembly for the eight-bar linkage.

6.5.3 Sorting Assemblies

For a sequence of input angles Θ_1^k, we can obtain a set of as many as 18 roots $\vec{\Theta}_i^k$, $i = 1, \ldots, 18$, for each input angle. In order to sort the solutions of the Dixon determinant among the assemblies of the eight-bar linkage, construct an approximation to the loop equations in the same way as discussed previously for the analysis of a six-bar linkage.

Compute the derivatives of the loop equations (6.35) and (6.36), and construct the Jacobian

$$[\nabla \mathscr{C}(\vec{\Theta}^k)] =$$

$$\begin{bmatrix} b_1 e^{-i\gamma} & 0 & 0 & -l_7 e^{i\eta} & 0 & l_8 & 0 \\ l_1 & b_4 e^{-i\beta} & 0 & -l_4 & -l_9 e^{i\varepsilon} & 0 & l_{10} \\ l_1 & l_2 & l_3 & -l_4 & -l_5 & 0 & 0 \\ -b_1 e^{i\gamma}\Theta_1^{-2} & 0 & 0 & l_7 e^{-i\eta}\Theta_4^{-2} & 0 & -l_8\Theta_8^{-2} & 0 \\ -l_1\Theta_1^{-2} & -b_4 e^{i\beta}\Theta_2^{-2} & 0 & l_4\Theta_4^{-2} & l_9 e^{-i\varepsilon}\Theta_5^{-2} & 0 & -l_{10}\Theta_{10}^{-2} \\ -l_1\Theta_1^{-2} & -l_2\Theta_2^{-2} & -l_3\Theta_3^{-2} & l_4\Theta_4^{-2} & l_5\Theta_5^{-2} & 0 & 0 \end{bmatrix}.$$

$$(6.47)$$

The approximation to the loop equations of the eight-bar linkage in the ith is given by

$$[\nabla \mathscr{C}(\vec{\Theta}_i^k)](\vec{\Psi} - \vec{\Theta}_i^k) = 0. \tag{6.48}$$

The solution of these equations yields Ψ that is on the tangent to the ith circuit at $\vec{\Theta}_i^k$. The root $\vec{\Theta}_i^{k+1}$ that is closest to Ψ is the root that matches the ith assembly.

6.6 Example: A Convertible Sofa

As an example of our planar eight-bar linkage design methodology, we consider the deployment linkage for a convertible sofa. The current deployment linage may pull the user toward the bed causing back strain. Our goal is a new design that deploys smoothly without this reverse movement.

For this design, we use a B14B25 linkage, see Figure 6.18. The backbone 6R loop is defined by $\mathbf{C}_1\mathbf{C}_2\mathbf{C}_3\mathbf{C}_6\mathbf{C}_5\mathbf{C}_4$. We add the RR constraint $\mathbf{G}_1\mathbf{W}_1$ that connects links B_1B_4 and a second constraint $\mathbf{G}_2\mathbf{W}_2$ that connects B_2B_5.

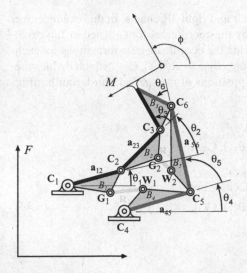

Fig. 6.18 The geometry of the B14B15 eight-bar linkage to be designed as the deployment linkage for convertible sofa .

Step 1.

The task positions that define the desired deployed positions of the bed, $[T_i]$, $i = 1,\ldots,5$, are given in Table 6.4. These positions define movement of the bed as it unfolds from the base of the sofa.

Table 6.4 Five task positions for the design of the convertible sofa (dimensions in cm)

Task Position (ϕ,x,y)
1    $(0°,-4,6.1)$
2    $(-49°,-5.4,18.5)$
3    $(-53°,-14,24)$
4    $(-42°,-20,22.4)$
5    $(0°,-28.6,13.6)$

The dimensions of the left 3R chain are chosen to be $a_{12} = a_{23} = 16.28cm$. The location of the base and moving pivots of this 3R chain, $[G]$ and $[H]$, are specified by the translation vectors $(0°,-6,0)$ and $(0°,4,0)$ respectively. The dimensions of the right 3R chain are chosen to be $a_{45} = a_{56} = 18.00cm$, and base and moving pivots are defined by the translation vectors $(0°,2,0)$ and $(0°,0,0)$ respectively.

Use this data to formulate the inverse kinematics equations of the 6R loop and solve for the configuration angles $q_j = (\theta_1,\theta_2,\theta_3,\theta_4,\theta_5,\theta_6)$, $j = 1,\ldots,5$, that reach the specified task positions. There are four solutions corresponding to the elbow up and elbow down configurations for the left and right 3R chains. In this example, we choose the elbow configuration defined by the coordinates for C_i listed in Table 6.5.

The five positions of the 6R loop provide the coordinate transformations for each link relative to the ground frame. Compute $[B_{1j}] = [R(\theta_{1j}), C_{1j}]$, which defines the five positions of the first link in F. The positions of the second, third, fourth, fifth, and sixth links are given by

$$
\begin{aligned}
[B_{2j}] &= [R(\theta_{1j} + \theta_{2j}), C_{2j}], \\
[B_{3j}] &= [R(\theta_{1j} + \theta_{2j} + \theta_{3j}), C_{3j}], \\
[B_{4j}] &= [R(\theta_{4j}), C_{4j}], \\
[B_{5j}] &= [R(\theta_{4j} + \theta_{5j}), C_{5j}], \quad \text{and} \\
[B_{6j}] &= [R(\theta_{4j} + \theta_{5j} + \theta_{6j}), C_{6j}], \quad j = 1, \ldots, 5.
\end{aligned}
\tag{6.49}
$$

Step 2.

Using the five positions $B_{1j}, j = 1, \ldots, 5$ of the link B_1 and $B_{4j}, j = 1, \ldots, 5$ of the link B_4, assemble the design equations for an RR chain, denoted $G_1 W_1$ in Figure 6.18. In this case, we obtain (6.6) with $[D_{1j}] = [B_{1,1}][B_{1,j}]^{-1}[B_{4,j}][B_{4,1}]^{-1}$, $i = 1, \ldots, 5$. The solution yields the values listed in Table 6.5.

Notice that the RR chain formed by $C_1 C_4$ is among the computed RR constraints, therefore only one solution is available for this design.

Step 3.

The synthesis of $G_2 W_2$ can occur in parallel with Step 2 because it does not rely on the result of the new link B_7 formed by the RR chain $G_1 W_1$.

Using the positions of the link B_2 and the positions of the link B_5, we assemble the design equations for an RR constraint $G_2 W_2$. The solution to these equations are shown in Table 6.5. In this case, the chain $C_3 C_6$ appears among the solutions, which guarantees a second real solution. Thus, this design process also results in at least one real B14B25 eight-bar linkage.

Analysis

In order to animate the movement of this system, solve the loop equations of the eight-bar linkage for a sequence of input angles interpolated between those defined by the task positions. Let these input angles $\theta_1^k, k = 1 \ldots, N$, then, for each value, θ_1^k, solve the loop equations (6.35) and (6.36) to determine the remaining configuration angles $\Theta^k = (\theta_2^k, \theta_3^k, \theta_4^k, \theta_5^k, \theta_8^k, \theta_{10}^k)^T$.

There can be as many as eighteen roots for each Θ^k because the $\det[M\lambda - N]$ for this linkage is of dimension 18, only one of which corresponds the assembly of the

Table 6.5 Step 1. Select the joint coordinates for the 6R loop; Step 2. Solve for the first RR chain; Step 3. Solve for the second RR chain; The selected values are highlighted in bold (dimensions in cm)

Step 1	C_1	C_2	C_3	C_4	C_5	C_6
	$(-6,0)$	$(8.2,8.0)$	$(-8,6.1)$	$(2,0)$	$(11.5,15.3)$	$(-4,6.1)$

Step 2					G_1	W_1
1					*Complex*	*Solution*
2					*Complex*	*Solution*
3					$(-6,0)$	$(2,0)$
4					$(0.84,1.02)$	$(6.85,4.55)$

Step 3					G_2	W_2
1					$(-8,6.1)$	$(-4,6.1)$
2					*Complex*	*Solution*
3					*Complex*	*Solution*
4					$(0.87,5.37)$	$(4.35,5.38)$

eight-bar specified in each of the task positions. Identify the root Θ^k for the correct assembly by formulating and solving the approximation to the loop equations (6.48). This calculation can be checked against the known configurations at each of the task positions. The results for our design are shown in Table 6.6. Figure 6.19 shows an image sequence of the resulting eight bar linkage reaching each of the task positions.

Table 6.6 Analysis solution for the sofabed mechanism.

	θ_1	θ_2	θ_3	θ_4	θ_5	θ_8	θ_{10}
1	29.52°	186.78°	0°	58.27°	210.78°	30.44°	0.23°
2	46.97°	143.78°	−49°	55.40°	168.20°	10.57°	−48.64°
3	84.36°	137.52°	−53°	86.92°	160.46°	1.14°	−55.89°
4	102.53°	145.66°	−42°	105.17°	163.80°	−1.57°	−56.03°
5	129.49°	176.35°	0°	134.47°	177.60°	−6.32°	−54.77°

6.7 Summary

In this chapter, we present a design procedure for single degree-of-freedom planar six-bar and eight-bar linkages that guide a workpiece through five task positions. This was done by adding two RR chains to an existing backbone chains. Starting with a 3R serial chain, the addition of two RR constraints yields a planar six-bar linkage, which yields seven different six-bar structures and as many as 72 design

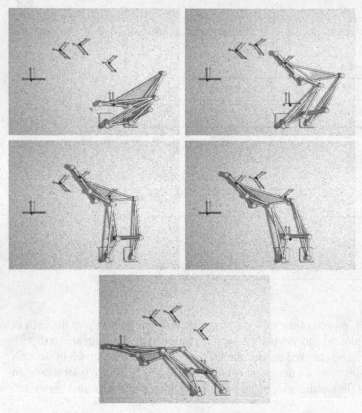

Fig. 6.19 The image sequence of the chosen planar eight-bar linkage design reaching each of the five task positions.

candidates. The planar eight-bar linkage is obtained by adding two RR constraints to a 6R loop, and yields thirty-two different eight-bar structures with as many 340 design candidates.

The design and analysis of a six-bar linkage for use as a steering linkage is presented as an example. Similarly, the design and analysis of an eight-bar linkage for use in the deployment of a convertible sofa is presented in detail.

6.8 References

The dimensional synthesis for planar multiloop mechanism can be found in Soh et al. [116], and Soh and McCarthy [117, 118, 119, 120]. Erdman [29], Erdman and Sandor [30], and Chase [9] used the complex vector formulation to synthesize different types of planar six-bar linkages. Tsai [138] describes the use of graph theory on planar multiloop mechanisms. Mirth and Chase [83], and Foster and Cipra [35] examine circuits and branches of planar multiloop mechanisms.

Exercises

1. Design a Watt Ia linkage that moves a foot pedal through the five positions in Table 6.7 using the given 3R chain. How many design candidates are obtained?

Table 6.7 Five task positions and the 3R chain for an adjustable foot pedal linkage (dimensions in mm)

Task	Position (ϕ, x, y)	Pivots	Coordinates
1	$(40°, 79, -216)$	C_1	$(0, 0)$
2	$(42°, 104, -208)$	C_2	$(43.2, -165.1)$
3	$(43.5°, 130, -203)$	C_3	$(12.7, -175.3)$
4	$(45°, 152, -198)$		
5	$(48°, 180, -191)$		

2. Analyze the the various design candidates for the above problem. How many of the design candidates pass smoothly through all five task positions?

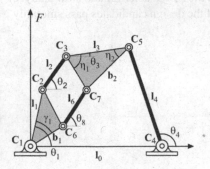

Fig. 6.20 The Stephenson IIb Linkage.

3. Formulate the closed-form solution to the Stephenson IIb linkage as shown in Figure 6.20 using complex number coordinates and the Dixon determinant.
4. Compare the computation time for the two methods for the algebraic solution of the design equations of an RR crank. Use the data in Table 6.7.
5. Design a B14B24 eight-bar linkage that moves a surgical tool around a remote center at each of the five positions listed in Table 6.8. You are free to choose your own 6R chain. How many design candidates are obtained?

Table 6.8 Five task positions for a remote center surgical tool (dimensions in mm)

Task	Position (ϕ, x, y)
1	$(55°, -77, 54)$
2	$(30°, -47, 81)$
3	$(0°, 0, 94)$
4	$(-30°, 47, 81)$
5	$(-55°, 77, 54)$

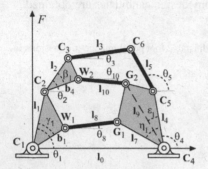

Fig. 6.21 The B14B24 eight-bar linkage.

6. Formulate the closed-form solution to the B14B24 linkage as shown in Figure 6.21 using complex number coordinates and the Dixon determinant. Now analyze the various design candidates. How many of the design candidates pass smoothly through all five task positions?

Chapter 7
Analysis of Spherical Linkages

In this chapter we examine spherical linkages. These linkages have the property that every link in the system rotates about the same fixed point. Thus, trajectories of points in each link lie on concentric spheres with this point as the center. Only the revolute joint is compatible with this rotational movement and its axis must pass through the fixed point. We study the spherical RR and 3R open chains and the 4R closed chain and determine their configuration as a function of the joint variables and the dimensions of the links.

7.1 Coordinate Rotations

A revolute joint in a spherical linkage allows spatial rotation about its axis. To define this rotation, we introduce a fixed frame F and a moving frame M attached to the moving link. The coordinate transformation between these frames defines the rotation of the link.

Consider a link connected to ground by one revolute joint. Let the $\mathbf{O}$ be directed along the axis of this joint and choose $\mathbf{A}$ to define the other end of the link. Both $\mathbf{O}$ and $\mathbf{A}$ are unit vectors that originate at the center $\mathbf{c}$. The angle α between these vectors defines the size of this link.

Choose an initial configuration and locate the fixed frame F so its origin is at $\mathbf{c}$, its z-axis directed along $\mathbf{O}$, and its y-axis directed along the vector $\mathbf{O} \times \mathbf{A}$. This convention ensures that $\mathbf{A}$ has $\sin \alpha$ as its positive x-component. Attach the moving frame M to $\mathbf{OA}$ so that in the initial configuration it is aligned with F. As the crank rotates, the angle θ measured counterclockwise about $\mathbf{O}$ from the x-axis of F to the x-axis of M defines the rotation of the link.

The orientation of $\mathbf{OA}$ is defined by transformation between coordinates $\mathbf{x} = (x,y,z)^T$ in M to $\mathbf{X} = (X,Y,Z)^T$ in F, given by the matrix equation

J.M. McCarthy and G.S. Soh, *Geometric Design of Linkages*, Interdisciplinary Applied Mathematics 11, DOI 10.1007/978-1-4419-7892-9_7,
© Springer Science+Business Media, LLC 2011

$$\begin{Bmatrix} X \\ Y \\ Z \end{Bmatrix} = \begin{bmatrix} \cos\theta & -\sin\theta & 0 \\ \sin\theta & \cos\theta & 0 \\ 0 & 0 & 1 \end{bmatrix} \begin{Bmatrix} x \\ y \\ z \end{Bmatrix}, \tag{7.1}$$

or

$$\mathbf{X} = [Z(\theta)]\mathbf{x}. \tag{7.2}$$

The notation $[Z(\cdot)]$ represents a rotation about the z-axis.

We can define similar matrices $[X(\cdot)]$ and $[Y(\cdot)]$ to represent rotations about the x-axis and y-axis, given by

$$[X(\alpha)] = \begin{bmatrix} 1 & 0 & 0 \\ 0 & \cos\alpha & -\sin\alpha \\ 0 & \sin\alpha & \cos\alpha \end{bmatrix} \quad \text{and} \quad [Y(\alpha)] = \begin{bmatrix} \cos\alpha & 0 & \sin\alpha \\ 0 & 1 & 0 \\ -\sin\alpha & 0 & \cos\alpha \end{bmatrix}. \tag{7.3}$$

These coordinate rotation matrices are useful in the analysis of spherical linkages.

An important property of rotation matrices is that their inverse is obtained by computing the matrix transpose. This means that

$$[Z(\theta)^{-1}] = [Z(\theta)^{T}] = \begin{bmatrix} \cos\theta & \sin\theta & 0 \\ -\sin\theta & \cos\theta & 0 \\ 0 & 0 & 1 \end{bmatrix}. \tag{7.4}$$

For the coordinate rotations $[X(\cdot)]$, $[Y(\cdot)]$, and $[Z(\cdot)]$ this transpose operation simply moves the negative sign from one sine element to the other. In fact, the inverse is the rotation in the negative angular direction, that is, $[Z(\theta)^{-1}] = [Z(-\theta)]$.

7.1.1 The Composition of Coordinate Rotations

Let $\mathbf{a} = (\sin\alpha, 0, \cos\alpha)^{T}$ be the coordinates in M of the vector $\mathbf{A}$ in the link $\mathbf{OA}$ above. Notice that $\mathbf{a}$ can be defined by rotating the unit vector $\vec{k} = (0,0,1)^{T}$ about the y-axis, so we have

$$\mathbf{a} = [Y(\alpha)]\vec{k}. \tag{7.5}$$

Substitute this into (7.2) to obtain $\mathbf{A}$ after a rotation by θ, which is given by the composition of two coordinate rotations

$$\mathbf{A} = [Z(\theta)][Y(\alpha)]\vec{k}. \tag{7.6}$$

This equation can be read from right to left as the rotation of $\vec{k}$ by the angle α around y followed by a rotation by θ around z.

7.1.2 The RR Open Chain

A spherical RR chain consists of a crank **OA** with fixed axis **O** and moving axis **A** connected to a floating link. Define the initial configuration of this chain so the fixed frame F has its z-axis is aligned with **O** and its y-axis directed along $\mathbf{O} \times \mathbf{A}$. Attach the moving frame M to the floating link so that its z-axis aligned with **A** and its y-axis is aligned with F in the initial configuration.

The rotation of the crank θ_1 about **O** is measured from the x-axis of F to the plane of the crank **OA**. The rotation angle θ_2 of the floating link is measured about **A** from the plane of **OA** to the x-axis of M. We now have the orientation of M relative to F defined by the matrix transformation

$$\mathbf{X} = [Z(\theta_1)][Y(\alpha)][Z(\theta_2)]\mathbf{x}. \tag{7.7}$$

The set of rotations reachable by the end-link of the RR chain is defined by

$$[R] = [Z(\theta_1)][Y(\alpha)][Z(\theta_2)], \tag{7.8}$$

or

$$[R] = \begin{bmatrix} c\theta_1 c\alpha c\theta_2 - s\theta_1 s\theta_2 & -c\theta_1 c\alpha s\theta_2 - s\theta_1 c\theta_2 & c\theta_1 s\alpha \\ s\theta_1 c\alpha c\theta_2 + c\theta_1 s\theta_2 & -s\theta_1 c\alpha s\theta_2 + c\theta_1 c\theta_2 & s\theta_1 s\alpha \\ -s\alpha c\theta_2 & s\alpha s\theta_2 & c\alpha \end{bmatrix}, \tag{7.9}$$

where s and c denote the sine and cosine functions. This matrix equation forms the kinematics equations of the RR chain and defines its *workspace*.

We now consider whether a specified rotation $[R]$ is in the workspace of this RR chain. Suppose that we know the elements of $[R]$, which we denote by

$$[R] = \begin{bmatrix} a_{11} & a_{12} & a_{13} \\ a_{21} & a_{22} & a_{23} \\ a_{31} & a_{32} & a_{33} \end{bmatrix}. \tag{7.10}$$

Equate this to (7.9) and examine the third column and third row to see that

$$\theta_1 = \arctan \frac{a_{23}}{a_{13}} \quad \text{and} \quad \theta_2 = \arctan \frac{a_{32}}{-a_{31}}. \tag{7.11}$$

Notice that the elements of the given orientation $[R]$ must satisfy the condition

$$\alpha = \arctan \frac{\sqrt{a_{31}^2 + a_{32}^2}}{a_{33}}, \tag{7.12}$$

where α is the angular length of **OA**. This characterizes the workspace of the spherical RR chain.

7.1.3 The 3R Open Chain

If the angular length α of an RR chain is allowed to vary, then we obtain a three-degree-of-freedom spherical robot. We can control this change in length by a revolute joint **E** inserted between **O** and **A**. The result is a spherical 3R open chain.

Let the angular lengths of the links **OE** and **EA** be α_1 and α_2, respectively. If ϕ is the dihedral angle measured about the **E** from the plane of **OE** to **EA**, then α is given by the spherical cosine law (C.5) as

$$\cos \alpha = \cos \alpha_1 \cos \alpha_2 - \sin \alpha_1 \sin \alpha_2 \cos \phi. \tag{7.13}$$

Notice that α lies between $|\alpha_2 - \alpha_1|$ and $\alpha_1 + \alpha_2$.

For a given rotation $[R]$ that defines the orientation of the end-link M, we can compute α using equation (7.12). We can also use (7.11) to determine the angle θ_1 measured to the plane of **OA** and the angle θ_2 from this plane to M.

Let ψ_1 be the rotation of the crank **OE** and let ψ_2 be the angle measured from **EA** to the x-axis of M. The interior angle of the spherical triangle $\triangle$**OEA** at **O** is $\theta_1 - \psi_1$, and the interior angle at **A** is $\theta_2 - \psi_2$. Use the sine and sine–cosine laws in (C.5) to compute

$$\sin \sigma \sin \left(\pi - (\theta_1 - \psi_1) \right) = \sin \alpha_2 \sin \phi,$$
$$\sin \sigma \cos \left(\pi - (\theta_1 - \psi_1) \right) = -(\sin \alpha_1 \cos \alpha_2 + \cos \alpha_1 \sin \alpha_2 \cos \phi). \tag{7.14}$$

Notice that π appears in these equations because $\theta_1 - \psi_1$ is the interior, not exterior, angle at **O**. However, because $\sin(\pi - \theta) = \sin \theta$ and $\cos(\pi - \theta) = -\cos \theta$, we have

$$\theta_1 - \psi_1 = \arctan \left(\frac{\sin \alpha_2 \sin \phi}{\sin \alpha_1 \cos \alpha_2 + \cos \alpha_1 \sin \alpha_2 \cos \phi} \right). \tag{7.15}$$

To determine $\theta_2 - \psi_2$, we use the sine law and sine–cosine law in (C.18) to obtain

$$\sin \sigma \sin \left(\pi - (\theta_2 - \psi_2) \right) = \sin \alpha_1 \sin \phi,$$
$$-\sin \sigma \cos \left(\pi - (\theta_2 - \psi_2) \right) = \sin \alpha_2 \cos \alpha_1 + \cos \alpha_2 \sin \alpha_1 \cos \phi. \tag{7.16}$$

Solve these equations to obtain

$$\theta_2 - \psi_2 = \arctan \left(\frac{\sin \alpha_1 \sin \phi}{\sin \alpha_2 \cos \alpha_1 + \cos \alpha_2 \sin \alpha_1 \cos \phi} \right). \tag{7.17}$$

The joint angle ϕ at **E** is that provides the desired angular length α is found from (7.13) to be

$$\phi = \arccos \left(\frac{\cos \alpha_1 \cos \alpha_2 - \cos \sigma}{\sin \alpha_1 \sin \alpha_2} \right). \tag{7.18}$$

The result is a set of values for ψ_1 and ψ_2 associated with each of the solutions $\pm\phi$. These are known as the elbow-down and elbow-up solutions for the 3R chain.

7.2 Position Analysis of a 4R Linkage

A spherical 4R linkage is constructed by connecting the end-links of two spherical RR chains, Figure 7.1. This linkage is defined by the axes of the revolute joints, which are lines through the origin **c**. We choose one of the two directions along each of these lines to be the unit vectors that define the linkage. Denote by **O** and **A** the unit vectors along fixed and moving axes of the input crank, and by **C** and **B** the fixed or moving axes of the output crank. The quadrilateral **OABC** on the unit sphere characterizes the spherical 4R linkage. Notice that there are actually sixteen spherical quadrilaterals that define the same spherical 4R linkage. In what follows, we consider each axis to be directed toward the $z > 0$ half-plane of the fixed frame F. However, the derivations apply to all sixteen cases.

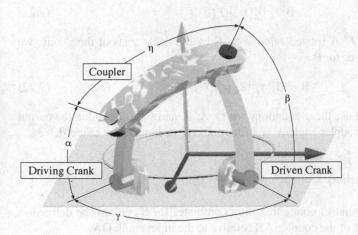

Fig. 7.1 A spherical 4R linkage is a spherical quadrilateral with angular dimensions α, β, γ, and η.

7.2.1 The Coordinates of O, A, B, and C

Given the spherical quadrilateral **OABC** that defines a spherical 4R linkage, we can compute the angular lengths of each of the links to be

$$\alpha = \arccos(\mathbf{O} \cdot \mathbf{A}), \qquad\qquad \beta = \arccos(\mathbf{C} \cdot \mathbf{B}),$$
$$\gamma = \arccos(\mathbf{O} \cdot \mathbf{C}), \qquad\qquad \eta = \arccos(\mathbf{A} \cdot \mathbf{B}). \qquad (7.19)$$

These dimensions completely define the movement of the linkage.

Attach a fixed frame F with its origin at $\mathbf{c}$, oriented so its z-axis is along $\mathbf{O}$ and its y-axis is directed along the vector $\mathbf{O} \times \mathbf{C}$. This ensures that the fixed axis $\mathbf{C}$ lies in the positive x direction of the xz coordinate plane. Let θ define the angle of the input crank about $\mathbf{O}$, and ψ the angle of the output crank around $\mathbf{C}$, both measured counterclockwise from the xz plane.

In F, we have the coordinates of $\mathbf{O}$ and $\mathbf{C}$ defined by

$$\mathbf{O} = \vec{k} \quad \text{and} \quad \mathbf{C} = [Y(\gamma)]\vec{k}. \tag{7.20}$$

The coordinates of $\mathbf{A}$ are the same as those given in (7.6). It is the coordinates of $\mathbf{B}$ that require some consideration.

Introduce a frame F' with its z-axis along $\mathbf{C}$. Then the coordinates $\mathbf{B}'$ in F' of the axis $\mathbf{B}$ are defined by an expression similar to (7.6), that is,

$$\mathbf{B}' = [Z(\psi)][Y(\beta)]\vec{k}. \tag{7.21}$$

Because the frame F' is rotated relative to F by the angle γ about the y-axis, we obtain the coordinates for $\mathbf{B}$ as

$$\mathbf{B} = [Y(\gamma)][Z(\psi)][Y(\beta)]\vec{k}. \tag{7.22}$$

In what follows, we use these equations for $\mathbf{O}$, $\mathbf{A}$, $\mathbf{B}$, and $\mathbf{C}$ to determine the output crank angle ψ and coupler angle ϕ as a functions of the input crank angle θ.

7.2.1.1 Loop Equations

It is useful at this point to notice that the coordinates for $\mathbf{B}$ can also be defined in terms of the angle ϕ of the coupler $\mathbf{AB}$ relative to the input crank $\mathbf{OA}$.

Introduce a frame M' with its z-axis along $\mathbf{A}$ and its y axis in the direction $\mathbf{O} \times \mathbf{A}$. The coordinates $\mathbf{B}'$ are defined by an equation similar to (7.21), given by

$$\mathbf{B}' = [Z(\phi)][Y(\eta)]\vec{k}. \tag{7.23}$$

The transformation from F to the frame M' is defined by the composite rotation $[Z(\theta)][Y(\alpha)]$, given in (7.6). Thus,

$$\mathbf{B} = [Z(\theta)][Y(\alpha)][Z(\phi)][Y(\eta)]\vec{k}. \tag{7.24}$$

The *loop equations* for the 4R linkage are obtained by equating (7.22) and (7.24).

7.2.2 The Output Angle

The angular dimension η between the moving axes $\mathbf{A}$ and $\mathbf{B}$ of a spherical 4R linkage is constant throughout the movement of the linkage. This provides the constraint equation

$$\mathbf{A} \cdot \mathbf{B} = \cos \eta, \tag{7.25}$$

which we use to determine the output angle ψ as a function of the input angle θ. The coordinates of $\mathbf{A}$ and $\mathbf{B}$, defined in (7.6) and (7.22), are

$$\mathbf{A} = \left\{ \begin{array}{c} \cos \theta \sin \alpha \\ \sin \theta \sin \alpha \\ \cos \alpha \end{array} \right\}, \quad \mathbf{B} = \left\{ \begin{array}{c} \cos \gamma \cos \psi \sin \beta + \sin \gamma \cos \beta \\ \sin \psi \sin \beta \\ -\sin \gamma \cos \psi \sin \beta + \cos \gamma \cos \beta \end{array} \right\}. \tag{7.26}$$

Substitute this into (7.25) to obtain

$$c\theta s\alpha(c\gamma c\psi s\beta + s\gamma c\beta) + s\theta s\alpha s\psi s\beta + c\alpha(-s\gamma c\psi s\beta + c\gamma c\beta) = c\eta, \tag{7.27}$$

where s and c denote the sine and cosine functions.

Collect the coefficients of $\cos \psi$ and $\sin \psi$ in this equation, so this constraint equation takes the form

$$A(\theta)\cos \psi + B(\theta)\sin \psi = C(\theta), \tag{7.28}$$

where

$$A(\theta) = \cos \theta \sin \alpha \cos \gamma \sin \beta - \cos \alpha \sin \gamma \sin \beta,$$
$$B(\theta) = \sin \theta \sin \alpha \sin \beta,$$
$$C(\theta) = \cos \eta - \cos \theta \sin \alpha \sin \gamma \cos \beta - \cos \alpha \cos \gamma \cos \beta. \tag{7.29}$$

This equation has the solution given by

$$\psi(\theta) = \arctan\left(\frac{B}{A}\right) \pm \arccos\left(\frac{C}{\sqrt{A^2 + B^2}}\right). \tag{7.30}$$

For reference, see (A.1). Notice that there are two output angles ψ associated with each input angle θ, Figure 7.2.

The two output angles ψ result from the fact that the spherical triangle $\triangle \mathbf{ABC}$ can be assembled with $\mathbf{B}$ on either side of the diagonal $\mathbf{AC}$. The angle $\delta = \arctan(B/A)$ locates the diagonal $\mathbf{AC}$, and $\kappa = \arccos(C/\sqrt{A^2 + B^2})$ is the angle above and below this diagonal that locates the driven crank.

The argument of the arc-cosine function must be in the range -1 to $+1$, which means that the link lengths and the input angle θ must combine so

$$A(\theta)^2 + B(\theta)^2 - C(\theta)^2 \geq 0. \tag{7.31}$$

If this constraint is not satisfied, then the linkage cannot be assembled for the specified value of θ.

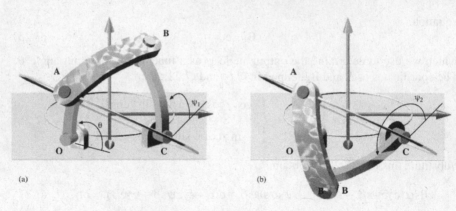

Fig. 7.2 A spherical 4R linkage has two solutions for the output angle ψ for each value of the input angle θ.

7.2.2.1 Hooke's Coupling

A special case of a spherical linkage known as *Hooke's coupling* is found in most automobiles connecting the output shaft of the transmission to the drive shaft. This linkage introduces an angle γ between the input and output axes. The dimensions of the input and output cranks and the coupler are $\alpha = \beta = \eta = \pi/2$. In this case the constraint equation (7.27) simplifies to become

$$\cos\theta\cos\gamma\cos\psi + \sin\theta\sin\psi = 0. \tag{7.32}$$

This can be written in the form

$$\tan\theta\tan\psi = -\sec\gamma, \tag{7.33}$$

which is the input-output equation for Hooke's coupling.

7.2.3 The Coupler Angle

We now determine the coupler angle ϕ, using the loop equations of the 4R chain. Equating the two ways (7.22) and (7.24) to define the coordinates of the moving axis **B**, we have

$$[Z(\theta)][Y(\alpha)][Z(\phi)][Y(\eta)]\vec{k} = [Y(\gamma)][Z(\psi)][Y(\beta)]\vec{k}. \tag{7.34}$$

Multiply both sides of this equation by $[Z(\theta)^{-1}] = [Z(\theta)^T]$ and expand to obtain

$$\left\{\begin{array}{c} c\alpha c\phi s\eta + s\alpha c\eta \\ s\phi s\eta \\ -s\alpha c\phi s\eta + c\alpha c\eta \end{array}\right\} = \left\{\begin{array}{c} c\theta c\gamma c\psi s\beta + c\theta s\gamma c\beta + s\theta s\psi s\beta \\ -s\theta c\gamma c\psi s\beta - s\theta s\gamma c\beta + c\theta s\psi s\beta \\ -s\gamma c\psi s\beta + c\gamma c\beta \end{array}\right\} \qquad (7.35)$$

The first two components of this equation allow us to determine

$$c\phi = \frac{s\beta c\theta c\gamma c\psi + c\theta s\gamma c\beta + s\beta s\theta s\psi - s\alpha c\eta}{c\alpha s\eta},$$

$$s\phi = \frac{-s\beta s\theta c\gamma c\psi - s\theta s\gamma c\beta + s\beta c\theta s\psi}{s\eta}. \qquad (7.36)$$

The arctan function yields a unique angle ϕ given values for the angles θ and ψ. This provides a reliable way to determine the coupler angle ϕ associated with each of the two output angles ψ for a given input crank angle θ.

7.2.3.1 An Alternative Derivation

A formula for ϕ can be obtained directly in terms of the crank angle θ independent of the output angle ψ. To do this we use the fact that the angular dimension β of the output crank **CB** is constant during the movement of the linkage.

Consider the frame F' positioned so its z-axis is along **A** and its y-axis is along the vector $\mathbf{A} \times \mathbf{O}$. In this frame, the axes **B** and **C** have the coordinates

$$^{F'}\mathbf{B} = \left\{\begin{array}{c} s\eta c(\phi - \pi) \\ s\eta s(\phi - \pi) \\ c\gamma \end{array}\right\} \quad \text{and} \quad ^{F'}\mathbf{C} = \left\{\begin{array}{c} c\alpha s\gamma c(\pi - \theta) + s\alpha c\gamma \\ s\gamma s(\pi - \theta) \\ -s\alpha s\gamma c(\pi - \theta) + c\alpha c\gamma \end{array}\right\}. \qquad (7.37)$$

The condition $\mathbf{B} \cdot \mathbf{C} = \cos\beta$ yields the equation

$$A(\theta)\cos\phi + B(\theta)\sin\phi = C(\theta), \qquad (7.38)$$

where

$$A(\theta) = \sin\eta\,(\cos\alpha\sin\gamma\cos\theta - \sin\alpha\cos\gamma),$$
$$B(\theta) = -\sin\eta\sin\gamma\sin\theta,$$
$$C(\theta) = \cos\beta - \cos\eta\,(\sin\alpha\sin\gamma\cos\theta - \cos\alpha\cos\gamma). \qquad (7.39)$$

This is solved in the same way as before to define

$$\phi(\theta) = \arctan\left(\frac{B}{A}\right) \pm \arccos\left(\frac{C}{\sqrt{A^2 + B^2}}\right). \qquad (7.40)$$

The result is two values for ϕ for each value of the input angle θ. We use this equation to determine the coupler angle in our solutions for four- and five-position synthesis for a spherical 4R linkage.

7.2.4 Coupler Curves

As the linkage moves, a point in the coupler traces a curve on a sphere in the fixed frame. This curve can be generated as a function of the input angle θ as follows. Let M be a reference frame attached to the coupler so its z-axis is aligned with the moving pivot $\mathbf{A}$ and its y-axis is along $\mathbf{A} \times \mathbf{B}$. Let a point in the coupler have the coordinate vector $\mathbf{x}$ measured in M. Its coordinates in F are given by the matrix equation

$$\mathbf{X}(\theta) = [Z(\theta)][Y(\alpha)][Z(\phi)]\mathbf{x}. \tag{7.41}$$

Recall that the coupler rotation angle ϕ depends on θ. To trace the coupler curve, compute $\mathbf{X}$ as θ varies through its range of movement. The points on this curve separate into sets that are given by the two different solutions for ϕ.

See Chiang [12] for the derivation and analysis of the algebraic equation of the spherical coupler curve, which when projected on a plane is a curve of eighth degree.

7.2.5 The Transmission Angle

The exterior angle ζ between the coupler and the output crank at the moving axis $\mathbf{B}$ is called the *transmission angle*. To determine ζ in terms of the input angle θ, we equate the spherical cosine laws for the triangles $\triangle \mathbf{COA}$ and $\triangle \mathbf{ABC}$. Let the diagonal shared by these triangles have the angular length δ, so that $\cos \delta = \mathbf{A} \cdot \mathbf{C}$. Notice that θ is the interior angle at $\mathbf{O}$, so we have the identities

$$\begin{aligned}\cos \delta &= \cos \gamma \cos \alpha + \sin \gamma \sin \alpha \cos \theta, \\ &= \cos \beta \cos \eta - \sin \beta \sin \eta \cos \zeta. \end{aligned} \tag{7.42}$$

Solving for $\cos \zeta$, we obtain the formula

$$\cos \zeta = \frac{\cos \beta \cos \eta - \cos \gamma \cos \alpha - \sin \gamma \sin \alpha \cos \theta}{\sin \beta \sin \eta}. \tag{7.43}$$

If the only external loads on the linkage are an input torque applied to the crank $\mathbf{OA}$ and an output torque applied by the crank $\mathbf{CB}$, then the joint reaction forces $\mathbf{F}_A$ and $\mathbf{F}_B$ acting on the coupler must act along the segment $\mathbf{AB}$. The usual assumption in the static analysis of a spherical linkage is that the joints $\mathbf{A}$ and $\mathbf{B}$ do not support forces along their axes, which means that $\mathbf{F}_A$ and $\mathbf{F}_B$ must be tangent to the unit sphere. Because these forces cannot be directed along $\mathbf{AB}$ and tangent to the sphere at the same time, they must both be zero. The conclusion is that under these conditions there are no internal reaction forces only reaction moments.

Let the internal reaction moments at joints $\mathbf{A}$ and $\mathbf{B}$ be denoted by $\mathbf{M}_A$ and $\mathbf{M}_B$. In order to be in static equilibrium these moments must be equal and opposite in sign, that is, $\mathbf{M}_A = -\mathbf{M}_B$. Since a hinge does not support a moment along its axis, these vectors must be perpendicular to the plane through **AB**, which means that they are perpendicular to the plane of the coupler.

Now consider the components of the moment $\mathbf{M_B}$ in the frame M' that has its z-axis along **B** and its y-axis in the direction $\mathbf{B} \times \mathbf{C}$. Because $\mathbf{M_B}$ must lie in the xy-plane of this frame and be perpendicular to the plane of the coupler, we have

$$\mathbf{M_B} = \left\{ \begin{array}{c} M_B \cos \zeta \\ M_B \sin \zeta \\ 0 \end{array} \right\}. \tag{7.44}$$

Only the component $M_B \sin \zeta$ contributes to the external torque on the driven crank. Thus, $\sin \zeta$ determines the magnitude of the moment on the coupler that is transmitted to the output crank. The component $M_B \cos \zeta$ is absorbed by the reaction moment at the fixed axis of the output crank.

7.3 Range of Movement

The condition that defines whether or not a spherical linkage can be assembled for a given input crank angle yields a formula for the range of movement of this crank in terms of the link dimensions. A similar analysis yields the range of movement for the output crank. This results in a classification of spherical 4R linkages based on the angular lengths of its links.

7.3.1 Limits on the Input Crank Angle

A solution exists for an output angle ψ for a given input θ only if the condition (7.31) is satisfied. The extreme values available for θ are defined by $A^2 + B^2 - C^2 = 0$, where A, B, and C are given by (7.29). This is a quadratic equation in $\cos \theta$ that has the solutions

$$\cos \theta_{\min} = \frac{\cos(\eta - \beta) - \cos \alpha \cos \gamma}{\sin \alpha \sin \gamma}, \quad \cos \theta_{\max} = \frac{\cos(\eta + \beta) - \cos \alpha \cos \gamma}{\sin \alpha \sin \gamma}. \tag{7.45}$$

These equations are the spherical cosine laws for the triangles formed by the two ways that the coupler link can align with the output link. These configurations define the range of movement of the input crank. Because the cosine function does not distinguish between positive and negative angles, an additional pair of limits exist reflected through the xz-plane of F. These negative limits apply when the crank passes through 0 or π into the lower half-plane.

The angular limits $\theta_{\min}$ and $\theta_{\max}$ exist only if the dimensions of the links combine so (7.45) takes values between -1 and 1. This provides conditions that define whether or not limits exist to the rotation of the input crank.

7.3.1.1 The Lower Limit: $\theta_{\min}$

If $\cos\theta_{\min} > 1$, then the limiting angle $\theta_{\min}$ does not exist, and the input crank can rotate through $\theta = 0$ into the lower half-plane. Thus, the condition that there is no lower limit is

$$\frac{\cos(\eta - \beta) - \cos\alpha\cos\gamma}{\sin\alpha\sin\gamma} > 1,$$

or

$$\cos(\eta - \beta) > \cos(\gamma - \alpha). \tag{7.46}$$

Because the cosine function decreases as the absolute value of its argument increases, we can replace this by the equivalent condition

$$(\gamma - \alpha)^2 > (\eta - \beta)^2. \tag{7.47}$$

Subtract the right side of this inequality from the left side and factor the difference of two squares to obtain

$$(\gamma - \alpha + \eta - \beta)(\gamma - \alpha - \eta + \beta) > 0,$$
$$T_1 T_2 > 0, \tag{7.48}$$

where

$$T_1 = \gamma - \alpha + \eta - \beta \quad \text{and} \quad T_2 = \gamma - \alpha - \eta + \beta. \tag{7.49}$$

Thus, the condition that there is no lower limit to the range of movement of the input crank is simply that both T_1 and T_2 have the same sign, either positive or negative. If the signs of these parameters are opposite to each other, then the input crank cannot reach the value $\theta = 0$.

7.3.1.2 The Upper Limit: $\theta_{\max}$

If $\cos\theta_{\max} < -1$, then the upper limit $\theta_{\max}$ does not exist, and the input crank can rotate smoothly through $\theta = \pi$ into the lower half-plane. Therefore, the condition that there is no upper limit to the crank rotation is

$$\frac{\cos(\eta + \beta) - \cos\alpha\cos\gamma}{\sin\alpha\sin\gamma} < -1, \tag{7.50}$$

which simplifies to the relation

$$\cos(\eta + \beta) < \cos(\gamma + \alpha). \tag{7.51}$$

The angular distances that characterize the spherical quadrilateral are always positive. However, the sum of any two can be greater than π, though not greater than 2π. We ensure that the absolute value of the argument to cosine is in the range from 0 to π by replacing $\cos(\eta + \beta)$ by $-\cos(\pi - (\eta + \beta))$ and $\cos(\gamma + \alpha)$ by $-\cos(\pi - (\gamma + \alpha))$. The inequality becomes

$$\cos(\pi - (\eta + \beta)) > \cos(\pi - (\gamma + \alpha)), \tag{7.52}$$

which allows us to define the equivalent relationship

$$(\pi - \gamma - \alpha)^2 > (\pi - \eta - \beta)^2. \tag{7.53}$$

This can be simplified to become

$$(\eta + \beta - \gamma - \alpha)(2\pi - \eta - \beta - \gamma - \alpha) < 0,$$
$$T_3 T_4 < 0, \tag{7.54}$$

where

$$T_3 = \eta + \beta - \gamma - \alpha \quad \text{and} \quad T_4 = 2\pi - \eta - \beta - \gamma - \alpha. \tag{7.55}$$

The range of movement of the input crank does not have an upper limit if T_3 and T_4 have the same sign. If these signs are opposite to each other, then the crank cannot reach the angular value $\theta = \pi$.

7.3.1.3 Input Crank Types

We can now identify four types of movement available to the input crank of a spherical four-bar linkage:

1. **A crank:** $T_1 T_2 > 0$ and $T_3 T_4 > 0$, in which case neither limit θ_{min} nor θ_{max} exists, and the input crank can fully rotate.
2. **A 0-rocker:** $T_1 T_2 > 0$ and $T_3 T_4 < 0$, for which θ_{max} exists but not θ_{min}, and the input crank passes through $\theta = 0$ rocking between the values $\pm\theta_{max}$.
3. **A π-rocker:** $T_1 T_2 < 0$ and $T_3 T_4 > 0$, which means that θ_{min} exists but not θ_{max}, and the input crank rocks through $\theta = \pi$ between the values $\pm\theta_{min}$.
4. **A rocker:** $T_1 T_2 < 0$ and $T_3 T_4 < 0$, which means that both upper and lower limit angles exist, and the crank cannot pass through either 0 or π. Instead, it rocks in one of two separate ranges: (i) $\theta_{min} \leq \theta \leq \theta_{max}$, or (ii) $-\theta_{max} \leq \theta \leq -\theta_{min}$.

7.3.2 Limits on the Output Crank Angle

The range of movement of the output crank is defined by the two configurations in which the input crank and coupler link can become aligned. The limits ψ_{min} and ψ_{max} are obtained by applying the spherical cosine law to the triangles formed by the linkage in these two configurations:

$$\cos \psi_{min} = \frac{\cos \gamma \cos \beta - \cos(\eta + \alpha)}{\sin \gamma \sin \beta}, \quad \cos \psi_{max} = \frac{\cos \gamma \cos \beta - \cos(\eta - \alpha)}{\sin \gamma \sin \beta}.$$

(7.56)

As we saw previously, the cosine does not distinguish between positive and negative angles, so we obtain two sets of limits. One set limits the angular range of the output crank when it is above the half-plane defined by **OC**, and the other set limits rotation below this half-plane.

The limiting angles (7.56) exist only if these formulas yield values that are between -1 and 1. This allows us to characterize the movement of the output crank in terms of the dimensions of the linkage.

7.3.2.1 The Lower Limit: ψ_{min}

If $\cos \psi_{min} > 1$, then the limiting angle ψ_{min} does not exist, and the output crank can move smoothly through $\psi = 0$. Thus, the condition that there is no lower limit to the movement of the output crank is

$$\frac{\cos \gamma \cos \beta - \cos(\eta + \alpha)}{\sin \gamma \sin \beta} > 1,$$

(7.57)

which can be written as

$$-\cos(\eta + \alpha) > -\cos(\gamma + \beta).$$

(7.58)

Replace $-\cos(\eta + \alpha)$ and $-\cos(\gamma + \beta)$ by $\cos(\pi - (\eta + \alpha))$ and $\cos(\pi - (\gamma + \beta))$ to obtain

$$\cos(\pi - (\eta + \alpha)) > \cos(\pi - (\gamma + \beta)).$$

(7.59)

This is equivalent to the inequality

$$(\pi - (\gamma + \beta))^2 > (\pi - (\eta + \alpha))^2,$$

(7.60)

which simplifies to yield

$$(\eta + \alpha - \gamma - \beta)(2\pi - \eta - \alpha - \gamma - \beta) > 0,$$
$$(-T_2)(T_4) > 0.$$

(7.61)

Thus, the parameters defined above also provide insight to the rotation of the output crank. If the T_2 and T_4 have opposite signs, then there is no lower limit to the rotation of this crank. This lower limit exists if $T_2 T_4 > 0$.

7.3.2.2 The Upper Limit: ψ_{max}

If $\cos\psi_{max} < -1$, then there is no upper limit ψ_{max} to the movement of the output crank. Thus, we have the condition

$$\frac{\cos\gamma\cos\beta - \cos(\eta - \alpha)}{\sin\gamma\sin\beta} < -1, \tag{7.62}$$

which ensures that output crank can move smoothly through $\psi = \pi$. Simplify this equation to obtain

$$\cos(\eta - \alpha) > \cos(\gamma - \beta). \tag{7.63}$$

As we have seen above, this is equivalent to the condition

$$(\gamma - \beta)^2 - (\eta - \alpha)^2 > 0, \tag{7.64}$$

which becomes

$$(\gamma - \beta + \eta - \alpha)(\gamma - \beta - \eta + \alpha) > 0,$$
$$(T_1)(-T_3) > 0. \tag{7.65}$$

The result is that the output crank passes through $\psi = \pi$ when T_1 and T_3 have opposite signs. The upper limit ψ_{max} exists when $T_1 T_3 > 0$.

7.3.2.3 Output Crank Types

We can identify the four types of output cranks:

1. **A rocker:** $T_1 T_3 > 0$ and $T_2 T_4 > 0$, in which case both limits ψ_{min} and ψ_{max} exist, and the output crank rocks in one of two separate ranges: (i) $\psi_{min} \leq \psi \leq \psi_{max}$, or (ii) $-\psi_{max} \leq \psi \leq -\psi_{min}$.
2. **A 0-rocker:** $T_1 T_3 > 0$ and $T_2 T_4 < 0$, in which case ψ_{max} exists but not ψ_{min}, and the output crank passes through $\theta = 0$ rocking between the values $\pm\psi_{max}$.
3. **A π-rocker:** $T_1 T_3 < 0$ and $T_3 T_4 > 0$, which means that ψ_{min} exists but not ψ_{max}, and the output crank rocks through $\psi = \pi$ between the values $\pm\psi_{min}$.
4. **A crank:** $T_1 T_2 \leq 0$ and $T_3 T_4 \leq 0$, which means that neither ψ_{min} nor ψ_{max} exists, and the output crank can fully rotate.

7.3.3 Classification of Spherical 4R Linkages

A spherical linkage is described in terms of the movement of its two cranks in the same way as a planar linkage. For example, a crank-rocker has a fully rotatable input crank and a rocker as the output crank. By assigning positive and negative signs to the four parameters T_i, $i = 1, 2, 3, 4$, we obtain 16 types of spherical four-bar linkage.

We separate these linkage types into those with $T_4 > 0$ and those with $T_4 < 0$. In the first case, the sum of the angular dimensions of the four links is less than 2π. Spherical linkages in this class lie on one side of the sphere. If $T_4 > 0$, then the link dimensions to add up to greater than 2π and the linkage wraps around the sphere. Within these two general classes we can identify spherical versions of the eight basic types found in the plane.

7.3.3.1 The Eight Basic Types, $T_4 > 0$

Given a spherical four-bar linkage with angular dimensions α, β, γ, and η, we compute the four parameters T_i, $i = 1, 2, 3, 4$. Assume for the moment that $T_4 > 0$. Then the signs of the three parameters T_1, T_2, and T_3 define the same linkage type on the sphere as they do on the plane. In particular, the combination $T_1 T_2 T_3 T_4 > 0$ defines *Grashof* linkages, while $T_1 T_2 T_3 T_4 < 0$ defines *nonGrashof* linkages, and there are four Grashof and four nonGrashof linkage types.

We consider the Grashof cases first:

1. $(+, +, +, +)$: Because $T_1 T_2 > 0$ and $T_3 T_4 > 0$ the input link can fully rotate. Similarly, because $T_1 T_3 > 0$ and $T_2 T_4 > 0$ the output link is a rocker with two output ranges. This linkage is the *spherical crank-rocker*.
2. $(+, -, -, +)$: With $T_1 T_2 < 0$ and $T_3 T_4 < 0$ the input is a rocker, and with $T_1 T_3 < 0$ and $T_2 T_4 < 0$ the output is a crank. Thus, this is a *spherical rocker-crank* linkage.
3. $(-, -, +, +)$: In this case, $T_1 T_2 > 0$ and $T_3 T_4 > 0$, so that the input link is a crank, and $T_1 T_3 < 0$ and $T_2 T_4 < 0$, which means that the output link is also a crank. This defines the *spherical double-crank* linkage.
4. $(-, +, -, +)$: $T_1 T_2 < 0$ and $T_3 T_4 < 0$ define the input as a rocker, and with $T_1 T_3 > 0$ and $T_2 T_4 > 0$ the output is a rocker as well. Thus, this defines the *spherical Grashof double-rocker* linkage.

And the following are the remaining nonGrashof cases:

5. $(-, -, -, +)$: Here we have $T_1 T_2 > 0$ and $T_3 T_4 < 0$. Therefore, the input link rocks through the value $\theta = 0$. With $T_1 T_3 > 0$ and $T_2 T_4 < 0$, we see that the output link also rocks through the value $\psi = 0$. This type of linkage is termed a 00 *spherical double rocker*.
6. $(+, +, -, +)$: In this case, the input again rocks through $\theta = 0$. However, with $T_1 T_3 < 0$ and $T_2 T_4 > 0$ the output rocks through $\psi = \pi$. This linkage is called a 0π *spherical double rocker*.
7. $(+, -, +, +)$: With $T_1 T_2 > 0$ and $T_3 T_4 > 0$ we see that the input link rocks through π, and because $T_1 T_3 < 0$ and $T_2 T_4 < 0$ the output link rocks through 0. This is the $\pi 0$ *spherical double rocker*.
8. $(-, +, +, +)$: Finally, the input again rocks through π, as does the output, which we term the $\pi\pi$ *spherical double rocker*.

This classification is summarized in Table 7.1.

7.3.3.2 The Linkages Types for $T_4 < 0$

If the parameter T_4 is negative, then we have eight linkage types that wrap around the sphere. However, the movement of the input and output cranks for each of these cases corresponds to one of the eight basic types presented above. This correspondence is identified by simply negating each of the parameters T_1, T_2, and T_3. For example, the spherical linkage of type $(+,+,+,-)$ has the same input and output crank movement as that given by $(-,-,-,+)$, which is a 00 spherical double rocker.

To see that this is true, notice that the input crank movement is defined by the signs of the products $T_1 T_2$ and $T_3 T_4$, and the output crank movement by the signs of $T_1 T_3$ and $T_2 T_4$. It is easy to see that if $T_4 < 0$ then both T_3 and T_2 must be negated in order for $T_3 T_4$ and $T_2 T_4$ to maintain the same signs. We can ensure that the signs of the products $T_1 T_2$ and $T_1 T_3$ are unchanged by also negating T_1. Thus, the linkage type $(\text{sgn}\, T_1, \text{sgn}\, T_2, \text{sgn}\, T_3, +)$ has the same crank movement as the type $(-\text{sgn}\, T_1, -\text{sgn}\, T_2, -\text{sgn}\, T_3, -)$.

Table 7.1 Basic Spherical 4R Linkage types

	Linkage type	T_1	T_2	T_3	T_4
1	Crank-rocker	+	+	+	+
2	Rocker-crank	+	−	−	+
3	Double-crank	−	−	+	+
4	Grashof double rocker	−	+	−	+
5	00+ double rocker	−	−	−	+
6	0π+ double rocker	+	+	−	+
7	π0+ double rocker	+	−	+	+
8	$\pi\pi$+ double rocker	−	+	+	+
9	Crank-rocker	−	−	−	−
10	Rocker-crank	−	+	+	−
11	Double-crank	+	+	−	−
12	Grashof double rocker	+	−	+	−
13	00− double rocker	+	+	+	−
14	0π− double rocker	−	−	+	−
15	π0− double rocker	−	+	−	−
16	$\pi\pi$− double rocker	+	−	−	−

7.3.4 Grashof Linkages

Grashof's criterion for planar linkages can be extended to spherical linkages. However, in order to uniquely identify the longest and shortest links, Chiang [12] defines a *model linkage* that can be selected from the 16 equivalent spherical linkages. It has the property that the sum of each pair of consecutive link angles is less than or equal to π. Using this model linkage, we have a spherical version of Grashof's criterion

$$s + l < p + q, \tag{7.66}$$

where s and l are the angular lengths of the shortest and longest links, and p and q are the angles of the other two links. If this condition is satisfied, then the shortest link will fully rotate relative to its neighbors.

As for planar linkages, any of the four links of the spherical linkage may be the shortest link. If the input or output link is the shortest, then we have a crank-rocker or rocker-crank. If the ground link is the shortest, then both the input and output links will fully rotate relative to the ground. This is the double-crank linkage. Finally, if the floating link is the shortest link, then the input and output links are rockers, which is the Grashof double rocker. Examining Table 7.1, it is easy to see that these four linkage types satisfy the condition

$$T_1 T_2 T_3 T_4 > 0. \tag{7.67}$$

This is equivalent to Grashof's criterion.

Like planar Grashof linkages, the two solutions for the output angle ϕ for a given input angle θ define independent assemblies for the spherical Grashof linkage. The configurations reachable in one assembly are separate from those reachable in the other assembly. To move from one set to the other the linkage must be disassembled.

7.3.5 Folding Linkages

The classification above considers only positive and negative values for the parameters T_i, $i = 1, 2, 3, 4$. If any of these parameters is zero, then a configuration exists with all four joints of the linkage **OABC** aligned in a plane, and the spherical linkage is said to fold. The number of parameters T_i that are zero is the number of folding configurations of the linkage.

If we consider that the parameters T_i, $i = 1, 2, 3, 4$, can take the values $(+, 0, -)$, then there are 81 types of spherical 4R linkages, 65 of which fold. The ability to fold can be useful, however, the position analysis equations often break down in these configurations.

7.4 Angular Velocity

The velocities of points in a spherical linkage are generated by the spatial rotation of the links. Thus, the time derivative of a rotation matrix defines these velocities and the *angular velocity* of the link.

The angular velocity of a rotating body can be visualized as the rate of a continuous rotation about an axis. Because this axis may also move in space, we must develop the definition of angular velocity with some care.

Let the orientation of a body M be defined by the rotation matrix $[A(t)]$, so that we have the trajectory $\mathbf{X}(t) = [A(t)]\mathbf{x}$ for any point $\mathbf{x}$ in M. The velocity of this point is $\mathbf{V} = \dot{\mathbf{X}} = [\dot{A}(t)]\mathbf{x}$. In order to focus on coordinates in F we make the substitution

$$\mathbf{V} = \dot{\mathbf{X}} = [\dot{A}(t)][A(t)^T]\mathbf{X} = [\Omega(t)]\mathbf{X}. \tag{7.68}$$

The matrix $[\Omega]$ is called as the *angular velocity matrix* and defines the rate of change of orientation of the moving body. It can be viewed as an operator that computes the velocity $\mathbf{V}$ by operating on a trajectory $\mathbf{X}(t)$.

The angular velocity matrix is skew-symmetric, that is, $[\Omega^T] = -[\Omega]$. Therefore, we can identify a vector $\mathbf{w}$ such that for any other vector $\mathbf{y}$, $[\Omega]\mathbf{y} = \mathbf{w} \times \mathbf{y}$. This vector $\mathbf{w}$ is the *angular velocity vector* of the body. For a link connected to ground by a revolute joint, this vector is directed along the joint axis, as we would expect.

For a link undergoing a general rotation $[A(t)]$, we can find the points $\mathbf{I}$ that have zero velocity by solving the equation

$$[\Omega(t)]\mathbf{I} = 0. \tag{7.69}$$

Clearly, these points lie on the line along the angular velocity vector $\mathbf{w}$, which is called the *instantaneous rotation axis*.

7.5 Velocity Analysis of an RR Chain

The velocity of trajectories traced by the end-link of a spherical RR chain are computed using the angular velocity matrix $[\Omega]$ constructed from its kinematics equations

$$
\begin{aligned}
[\Omega(\theta_1, \theta_2)] &= \frac{d}{dt}([Z(\theta_1)][Y(\alpha)][Z(\theta_2)])([Z(\theta_1)][Y(\alpha)][Z(\theta_2)])^T \\
&= \dot{\theta}_1[K] + \dot{\theta}_2[Z(\theta_1)[Y(\alpha)][K][Y(\alpha)^T][Z(\theta_1)^T], \tag{7.70}
\end{aligned}
$$

where $[K]$ is the skew-symmetric matrix defined so $[K]\mathbf{y} = \vec{k} \times \mathbf{y}$. Thus, the angular velocity vector of the end-link is

$$\mathbf{w} = \dot{\theta}_1\vec{k} + \dot{\theta}_2[Z(\theta_1)][Y(\alpha)]\vec{k}. \tag{7.71}$$

This vector can be written in the form of a matrix equation

$$\mathbf{w} = \begin{Bmatrix} w_x \\ w_y \\ w_z \end{Bmatrix} = \begin{bmatrix} 0 & \cos\theta_1\sin\alpha \\ 0 & \sin\theta_1\sin\alpha \\ 1 & \cos\alpha \end{bmatrix} \begin{Bmatrix} \dot{\theta}_1 \\ \dot{\theta}_2 \end{Bmatrix}. \tag{7.72}$$

This matrix is called the *Jacobian* of the spherical RR chain. It relates the angular velocity of the end-link to the joint rates, $\dot{\theta}_1$ and $\dot{\theta}_2$.

7.6 Velocity Analysis of the 4R Linkage

The time derivative of the loop equations (7.35) provides relationships among the angular velocities $\dot{\psi}$, $\dot{\phi}$, and $\dot{\theta}$. Expanding this derivative, we obtain

$$
\dot{\phi}\left\{\begin{array}{c} -c\alpha s\eta s\phi \\ s\eta c\phi \\ s\alpha s\eta s\phi \end{array}\right\} = \dot{\theta}\left\{\begin{array}{c} -s\theta(c\gamma c\psi s\beta + s\gamma c\beta) + c\theta(s\psi s\beta) \\ -c\theta(c\gamma c\psi s\beta + s\gamma c\beta) - s\theta(s\psi s\beta) \\ 0 \end{array}\right\}
$$
$$
+ \dot{\psi}\left\{\begin{array}{c} -c\theta c\gamma s\psi s\beta + s\theta c\psi s\beta \\ s\theta c\gamma s\psi s\beta + c\theta c\psi s\beta \\ s\gamma s\psi s\beta \end{array}\right\}. \tag{7.73}
$$

The first two components of these equations provide linear equations that define the angular velocities of the coupler and output crank, $\dot{\phi}$ and $\dot{\psi}$, in terms of the angular velocity of the input crank $\dot{\theta}$.

7.6.1 The Output Velocity Ratio

To obtain further insight to the relationship between the angular velocities $\dot{\theta}$ and $\dot{\psi}$, we derive it in another way. Compute the derivative of the constraint equation $\mathbf{A} \cdot \mathbf{B} = \cos \eta$ to obtain

$$
\dot{\mathbf{A}} \cdot \mathbf{B} + \mathbf{A} \cdot \dot{\mathbf{B}} = 0. \tag{7.74}
$$

In order to determine $\dot{\mathbf{A}}$ and $\dot{\mathbf{B}}$ we use the angular velocity vectors $\mathbf{w}_O$ and $\mathbf{w}_C$ of the input and output crank, which are given by

$$
\mathbf{w}_O = \dot{\theta}\mathbf{O} \quad \text{and} \quad \mathbf{w}_C = \dot{\psi}\mathbf{C}. \tag{7.75}
$$

Recall that $\mathbf{O}$ and $\mathbf{C}$ are unit vectors.

Now substitute $\dot{\mathbf{A}} = \mathbf{w}_O \times \mathbf{A}$ and $\dot{\mathbf{B}} = \mathbf{w}_C \times \mathbf{B}$ into (7.74) to obtain the relation

$$
(\dot{\theta}\mathbf{O} \times \mathbf{A}) \cdot \mathbf{B} + \mathbf{A} \cdot (\dot{\psi}\mathbf{C} \times \mathbf{B}) = 0. \tag{7.76}
$$

Interchange the dot and cross operations in this equation and collect terms to obtain

$$
(\dot{\theta}\mathbf{O} - \dot{\psi}\mathbf{C}) \cdot \mathbf{A} \times \mathbf{B} = 0. \tag{7.77}
$$

Substitute $\mathbf{O} = \vec{k}$ and $\mathbf{C} = \cos \gamma \vec{k} + \sin \gamma \vec{i}$ so we have

$$\left(\dot{\theta}\vec{k} - \psi(\cos\gamma\vec{k} + \sin\gamma\vec{i})\right) \cdot \mathbf{A} \times \mathbf{B} = 0. \tag{7.78}$$

Finally, using the relations $\vec{k} = \vec{i} \times \vec{j}$ and $\vec{i} = \vec{j} \times \vec{k}$, we write this equation in the form

$$\left(\dot{\theta}\vec{i} - \psi(\cos\gamma\vec{i} - \sin\gamma\vec{k})\right) \cdot \vec{j} \times (\mathbf{A} \times \mathbf{B}) = 0. \tag{7.79}$$

The vector $k\mathbf{I} = \vec{j} \times (\mathbf{A} \times \mathbf{B})$ is the intersection of the coupler plane $\mathbf{AB}$ and the xz coordinate plane. Let the coordinates of this vector be $\mathbf{I} = (\sin\rho, 0, \cos\rho)^T$, then from (7.79) we have

$$\frac{\psi}{\dot{\theta}} = \frac{\sin\rho}{\sin(\rho - \gamma)}. \tag{7.80}$$

7.6.2 The Coupler Velocity Ratio

A similar relationship can be derived for the velocity ratio of the coupler. To do this, we compute the velocity of $\dot{\mathbf{B}}$ using the relation

$$\dot{\mathbf{B}} = \dot{\mathbf{A}} + \mathbf{w}_{AB} \times (\mathbf{B} - \mathbf{A}), \tag{7.81}$$

where $\mathbf{w}_{AB}$ is the angular velocity of the coupler in F. This angular velocity is the sum of the angular velocity $\mathbf{w}_A = \dot{\phi}\mathbf{A}$ of the coupler relative to the crank $\mathbf{OA}$ and the angular velocity $\mathbf{w}_O = \dot{\theta}\mathbf{O}$ of the crank relative to F, that is,

$$\mathbf{w}_{AB} = \dot{\theta}\mathbf{O} + \dot{\phi}\mathbf{A} \tag{7.82}$$

Now, from the fact that $\dot{\mathbf{B}} = \mathbf{w}_C \times \mathbf{B}$, we have the condition

$$\dot{\mathbf{B}} \cdot \mathbf{C} = \left(\dot{\mathbf{A}} + \mathbf{w}_{AB} \times (\mathbf{B} - \mathbf{A})\right) \cdot \mathbf{C} = 0. \tag{7.83}$$

Recall $\mathbf{O} = \vec{k}$ and introduce the unit vector $\vec{e}$ in the plane of the input crank, so $\mathbf{A} = \cos\alpha\vec{k} + \sin\alpha\vec{e}$, then this equation becomes

$$\left(\dot{\theta}\vec{k} \times \mathbf{A} + (\dot{\theta}\vec{k} + \dot{\phi}(\cos\alpha\vec{k} + \sin\alpha\vec{e})) \times (\mathbf{B} - \mathbf{A})\right) \cdot \mathbf{C} = 0. \tag{7.84}$$

Cancel the $\vec{k} \times \mathbf{A}$ terms and interchange the dot and cross products to obtain

$$\left((\dot{\theta} + \dot{\phi}\cos\alpha)\vec{k} + \dot{\phi}\sin\alpha\vec{e}\right) \cdot \mathbf{B} \times \mathbf{C} = 0. \tag{7.85}$$

Finally, define the unit vector $\vec{e}^{\perp} = \vec{k} \times \vec{e}$ and use the relations $\vec{k} = \vec{e} \times \vec{e}^{\perp}$ and $\vec{e} = -\vec{k} \times \vec{e}^{\perp}$ to write this equation in the form

$$\left((\dot{\theta} + \dot{\phi}\cos\alpha)\vec{e} - \dot{\phi}\sin\alpha\vec{k}\right) \cdot \vec{e}^{\perp} \times (\mathbf{B} \times \mathbf{C}) = 0. \tag{7.86}$$

The vector $k\mathbf{J} = \vec{e}^{\perp} \times (\mathbf{B} \times \mathbf{C})$ defines the line of intersection of the plane of the input crank $\mathbf{OA}$ with the plane of the output crank $\mathbf{BC}$. If the coordinates of the unit vector along this line are $\mathbf{J} = \cos\rho\vec{k} + \sin\rho\vec{e}$, then from (7.86) we obtain

$$\frac{\dot{\phi}}{\dot{\theta}} = \frac{\sin\rho}{\sin(\alpha - \rho)}. \tag{7.87}$$

7.6.3 Instantaneous Rotation Axis

The output velocity ratio of the spherical four-bar linkage can be viewed as generated instantaneously by a pair of bevel gears connecting the input and output cranks. The gears have axes $\mathbf{O}$ and $\mathbf{C}$ and are in contact at the point $\mathbf{I}$ on the spherical section of their pitch cones. We now show that this is a general characteristic for two bodies rotating about the same point in a fixed frame F.

Let the two bodies have the instantaneous rotation axes $\mathbf{O}$ and $\mathbf{C}$ that are separated by the angle γ. We ask if there is a point $\mathbf{Q}$ in one and $\mathbf{P}$ in the other, that coincide with the same point $\mathbf{I} = (X, Y, Z)^T$ and have the same velocity in F. This is the property of points in contact on the pitch circle of the two gears.

Let ρ be the angle from $\mathbf{O}$ to $\mathbf{Q}$ and κ the angle from $\mathbf{C}$ to $\mathbf{P}$, so we have

$$\mathbf{Q} = \begin{Bmatrix} \sin\rho\cos\theta \\ \sin\rho\sin\theta \\ \cos\rho \end{Bmatrix}, \quad \mathbf{P} = \begin{Bmatrix} \cos\gamma\sin\kappa\cos\psi + \sin\gamma\cos\kappa \\ \sin\kappa\sin\psi \\ -\sin\gamma\sin\kappa\cos\psi + \cos\gamma\sin\kappa \end{Bmatrix}. \tag{7.88}$$

The velocities of these points are obtained by computing the derivatives

$$\dot{\mathbf{Q}} = \dot{\theta}\begin{Bmatrix} -\sin\rho\sin\theta \\ \sin\rho\cos\theta \\ 0 \end{Bmatrix}, \quad \dot{\mathbf{P}} = \dot{\psi}\begin{Bmatrix} -\cos\gamma\sin\kappa\sin\psi \\ \sin\kappa\cos\psi \\ -\sin\gamma\sin\kappa\sin\psi \end{Bmatrix}. \tag{7.89}$$

If $\mathbf{P}$ and $\mathbf{Q}$ are to be in contact, then $Y = \sin\kappa\sin\theta = \sin\rho\sin\psi$, and if they are to have the same velocity, then

$$\dot{\mathbf{P}} - \dot{\mathbf{Q}} = \begin{Bmatrix} -\dot{\psi}\cos\gamma Y + \dot{\theta}Y \\ \dot{\psi}\sin\kappa\cos\psi - \dot{\theta}\sin\rho\cos\theta \\ -\dot{\psi}\cos\gamma Y \end{Bmatrix} = \begin{Bmatrix} 0 \\ 0 \\ 0 \end{Bmatrix}. \tag{7.90}$$

The first and third components of this equation show $Y = 0$, which means that $\theta = \psi = 0$ or π. Let $\theta = \psi = 0$ and allow the angles ρ and κ to take positive and negative values, then (7.88) shows that $\rho - \gamma = \kappa$.

Finally, the second component of (7.90) yields the relation that defines ρ,

$$\frac{\dot{\psi}}{\dot{\theta}} = \frac{\sin\rho}{\sin(\rho - \gamma)}. \tag{7.91}$$

Thus, $\mathbf{I}$ derived in the previous section is the point that has zero relative velocity in both the input and output cranks. The fact that this instantaneous relative rotation axis must lie in the plane defined by the two instantaneous rotation axes $\mathbf{O}$ and $\mathbf{C}$ is the spherical version of *Kennedy's theorem*.

A similar analysis of the coupler velocity ratio shows that the axis $\mathbf{J}$ is the instantaneous relative rotation axis of the coupler relative to ground.

7.6.4 Mechanical Advantage

The relationship between the torque applied to the input crank and the torque at output crank of a spherical four-bar linkage is obtained using the principle of virtual work. In the absence of gravity and frictional loads, this principle equates the virtual work of torque applied to the input crank to the virtual work done by the output crank.

A virtual displacement of the input or output crank is defined to be the angular displacement that occurs when it rotates at a constant angular velocity for a virtual time period δt. This yields the virtual displacements

$$\delta\vec{\theta} = \dot{\theta}\mathbf{O}\delta t \quad \text{and} \quad \delta\vec{\psi} = \dot{\psi}\mathbf{C}\delta t. \tag{7.92}$$

Let $\mathbf{T_O} = T_O\mathbf{O}$ be the torque applied to the driving crank and $\mathbf{T_C} = T_C\mathbf{C}$ be the output torque at the driven crank. Then the principle of virtual work yields the relation

$$\mathbf{T_O} \cdot \delta\vec{\theta} = \mathbf{T_C} \cdot \delta\vec{\psi}. \tag{7.93}$$

Substitute (7.92) into this equation to obtain

$$T_O\dot{\theta}\delta t = T_C\dot{\psi}\delta t. \tag{7.94}$$

Since the virtual time increment is nonzero, we can cancel δt to obtain the torque ratio

$$\frac{T_C}{T_O} = \frac{\dot{\theta}}{\dot{\psi}} = \frac{\sin(\rho - \gamma)}{\sin\rho}. \tag{7.95}$$

The second equality is the output velocity ratio (7.80). Thus, the torque ratio of a spherical four-bar linkage is the inverse of its velocity ratio. Notice that this ratio changes as the configuration of the linkage changes.

Let the input torque be generated by a *couple* defined such that $T_O = aF_O$. Similarly, let the output torque be a couple such that $T_C = bF_C$. Then the ratio of output force F_C to input force F_O is given by

$$\frac{F_C}{F_O} = \frac{a}{b}\frac{\sin(\rho - \gamma)}{\sin\rho}. \tag{7.96}$$

This ratio is called the *mechanical advantage* of the linkage. For a given set of dimensions a and b the mechanical advantage is proportional to the velocity ratio of the input and output cranks.

7.7 Analysis of Multiloop Spherical Linkages

The analysis of multiloop spherical linkages is covered in detail in a later chapter. Here we note that our analysis procedure follows Wampler [149] and consists of decomposing the spherical linkage into a sequence of structures that are analyzed separately. This reduces the analysis of a multiloop spherical linkage to the analysis of specific types of spherical component structures.

7.8 Summary

This chapter has presented the position and velocity analysis of the spherical 3R open chain that forms a robot wrist and spherical 4R closed chain. The results follow closely those developed in the earlier chapter on planar linkages. The instantaneous rotation axis and Kennedy's theorem provide a convenient way to determine the mechanical advantage of a spherical linkage.

7.9 References

The fundamental reference for spherical linkages is Chiang [12]. Craig [15] and Tsai [136] present the analysis of robot wrists and the Hunt [50] and Crane and Duffy [16] examine spherical 4R linkages. The classification of spherical linkages is taken from Murray [86].

Exercises

1. Use the equations of spherical trigonometry to determine the joint angles of a spherical triangle with sides $\alpha_{12} = 120°$, $\alpha_{23} = 80°$, and $\alpha_{13} = 135°$ (Crane and Duffy [16]).
2. Analyze the spherical linkage $\alpha = 40°$, $\eta = 70°$, $\beta = 85°$, and $\gamma = 70°$. Let the input crank angle be $\theta = 75°$ (Crane and Duffy [16]).
3. A 4R linkage with an output link angle $\beta = \pi/2$ is considered the spherical version of a slider-crank, since the moving pivot moves on a great circle of the sphere. Derive the input/output equations for this linkage.
4. Derive the algebraic equation of the coupler curve of a spherical 4R linkage.
5. Determine the output velocity ratio for Hooke's coupling. Under what condition is this ratio constant.
6. Prove the general version of Kennedy's theorem that the three instantaneous rotation axes for three rotating bodies must lie in a plane.

Chapter 8
Spherical Kinematics

In this chapter we consider spatial displacements that are pure rotations in three-dimensional space. These are transformations that have the property that one point of the moving body M has the same coordinates in F before and after the displacement. Because the distance between this fixed point and points in M are constant, each point in the moving body moves on a sphere about this point. If the origins for both the fixed and moving frames are located at this fixed point, then the spatial displacement is defined by a 3×3 rotation matrix. The study of spherical kinematics benefits from both the properties of linear transformations and the geometry of a sphere.

8.1 Isometry

A *spatial displacement* preserves the distance between every pair of points in the moving body and is an isometry of three-dimensional space. As in the plane, this displacement is the composition of a translation and a rotation.

Let $\mathbf{P} = (P_x, P_y, P_z)^T$ and $\mathbf{Q} = (Q_x, Q_y, Q_z)^T$ be the coordinate vectors of two points in three-dimensional space. The distance between these points is the magnitude of their relative position vector $\mathbf{Q} - \mathbf{P}$,

$$|\mathbf{Q} - \mathbf{P}| = \sqrt{(Q_x - P_x)^2 + (Q_y - P_y)^2 + (Q_z - P_z)^2}, \qquad (8.1)$$

which is also called the *Euclidean metric*. Using vector notation, this formula takes the same form as that used for planar kinematics, that is,

$$|\mathbf{Q} - \mathbf{P}|^2 = (\mathbf{Q} - \mathbf{P}) \cdot (\mathbf{Q} - \mathbf{P}) = (\mathbf{Q} - \mathbf{P})^T (\mathbf{Q} - \mathbf{P}). \qquad (8.2)$$

The second equality is the matrix form of the vector scalar product.

J.M. McCarthy and G.S. Soh, *Geometric Design of Linkages*, Interdisciplinary Applied Mathematics 11, DOI 10.1007/978-1-4419-7892-9_8,
© Springer Science+Business Media, LLC 2011

8.1.1 Spatial Translations

As we saw in the plane, the addition of a vector $\mathbf{d} = (d_x, d_y, d_z)^T$ to the coordinates of all the points in a body, such that $\mathbf{X} = \mathbf{x} + \mathbf{d}$, is called a *translation*. Let the points $\mathbf{p}$ and $\mathbf{q}$ be translated so $\mathbf{P} = \mathbf{p} + \mathbf{d}$ and $\mathbf{Q} = \mathbf{q} + \mathbf{d}$, then we can compute

$$|\mathbf{Q} - \mathbf{P}| = |(\mathbf{q} + \mathbf{d}) - (\mathbf{p} + \mathbf{d})| = |\mathbf{q} - \mathbf{p}|. \tag{8.3}$$

Thus, translations preserve the distance between points.

8.1.2 Spatial Rotations

A spatial rotation has the same basic properties as a planar rotation, though now applied to three-dimensional vectors. A rotation takes M from a position initially aligned with F and reorients it, while keeping the origins of the two frames located at the same point $\mathbf{c}$. Let $\vec{\imath}$, $\vec{\jmath}$, and $\vec{k}$ be the unit vectors along the coordinate axes of F. The rotation changes the direction of each of these vectors. Let these new directions be given by the orthogonal unit vectors $\mathbf{e}_x$, $\mathbf{e}_y$, and $\mathbf{e}_z$, such that $\mathbf{e}_x \times \mathbf{e}_y = \mathbf{e}_z$. This last condition ensures that $\mathbf{e}_x$, $\mathbf{e}_y$, and $\mathbf{e}_z$ form a right-handed frame like $\vec{\imath}$, $\vec{\jmath}$, and $\vec{k}$. A point with coordinates $\mathbf{x} = (x, y, z)^T = x\vec{\imath} + y\vec{\jmath} + z\vec{k}$ before the rotation will have coordinates $\mathbf{X}$ after the rotation, given by

$$\mathbf{X} = X\vec{\imath} + Y\vec{\jmath} + Z\vec{k} = x\mathbf{e}_x + y\mathbf{e}_y + z\mathbf{e}_z. \tag{8.4}$$

Let the components of $\mathbf{e}_x$ be $(e_{x,1}, e_{x,2}, e_{x,3})^T$. A similar definition for $\mathbf{e}_y$ and $\mathbf{e}_z$ allows us to form the matrix equation

$$\begin{Bmatrix} X \\ Y \\ Z \end{Bmatrix} = \begin{bmatrix} e_{x,1} & e_{y,1} & e_{z,1} \\ e_{x,2} & e_{y,2} & e_{z,2} \\ e_{x,3} & e_{y,3} & e_{z,3} \end{bmatrix} \begin{Bmatrix} x \\ y \\ z \end{Bmatrix}, \tag{8.5}$$

or

$$\mathbf{X} = \begin{bmatrix} \mathbf{e}_x, & \mathbf{e}_y, & \mathbf{e}_z \end{bmatrix} \mathbf{x} = [A]\mathbf{x}. \tag{8.6}$$

All spatial rotations are represented by matrices constructed in this way, which are known as *rotation matrices*.

8.1.2.1 Distances

We now show that rotations preserve distances between points. Let $\mathbf{p}$ and $\mathbf{q}$ be the coordinates of two points before the rotation, and let $\mathbf{P} = [A]\mathbf{p}$ and $\mathbf{Q} = [A]\mathbf{q}$ be their coordinates after the rotation. We compute

$$|\mathbf{Q} - \mathbf{P}|^2 = (\mathbf{p} - \mathbf{q})^T [A^T][A](\mathbf{p} - \mathbf{q})^T. \qquad (8.7)$$

Notice that this equality is satisfied only if $[A^T][A] = [I]$.

This condition is always true for matrices constructed from orthogonal unit vectors as in (8.6), which can be seen from the computation

$$[A^T][A] = \begin{bmatrix} \mathbf{e}_x^T \\ \mathbf{e}_y^T \\ \mathbf{e}_z^T \end{bmatrix} [\mathbf{e}_x, \mathbf{e}_y, \mathbf{e}_z] = \begin{bmatrix} \mathbf{e}_x^T \mathbf{e}_x & \mathbf{e}_x^T \mathbf{e}_y & \mathbf{e}_x^T \mathbf{e}_z \\ \mathbf{e}_y^T \mathbf{e}_x & \mathbf{e}_y^T \mathbf{e}_y & \mathbf{e}_y^T \mathbf{e}_z \\ \mathbf{e}_z^T \mathbf{e}_x & \mathbf{e}_z^T \mathbf{e}_y & \mathbf{e}_z^T \mathbf{e}_z \end{bmatrix} = [I]. \qquad (8.8)$$

Important examples are the coordinate rotations $[X(\cdot)]$, $[Y(\cdot)]$, and $[Z(\cdot)]$ presented in (7.3) and (7.1) in the previous chapter.

In general, a spatial rotation is a linear transformation that preserves the distances between points and the orientation of the reference frames. Matrices $[A]$ that satisfy the condition (8.8) are termed *orthogonal*. However, in order to preserve the orientation of the coordinate frame we must add the requirement that the determinant $|A|$ be positive. From the calculation

$$\det([A^T][A]) = |A|^2 = 1 \qquad (8.9)$$

we see that an orthogonal matrix can have a determinant of either $+1$ or -1. Those with $|A| = +1$ are *rotations*. Those with $|A| = -1$ are *reflections*. An example of a reflection is the matrix

$$\begin{bmatrix} 1 & 0 & 0 \\ 0 & 1 & 0 \\ 0 & 0 & -1 \end{bmatrix}. \qquad (8.10)$$

The columns of this matrix are orthogonal unit vectors, and its transpose is its inverse. However, it changes the orientation of the frame by reversing the direction of the z-axis relative to the xy-plane.

8.1.2.2 Angles

A spatial rotation preserves the relative distances between three points $\mathbf{P}$, $\mathbf{Q}$, and $\mathbf{R}$. Therefore, it preserves the angle $\kappa = \angle \mathbf{QRP}$. In order to show this, it is useful to recall that the sine and cosine of the angle about $\mathbf{R}$ from $\mathbf{P}$ to $\mathbf{Q}$ can be computed from the relative vectors $\mathbf{P} - \mathbf{R}$ and $\mathbf{Q} - \mathbf{R}$ by the formulas

$$\sin \kappa = \frac{(\mathbf{P} - \mathbf{R}) \times (\mathbf{Q} - \mathbf{R}) \cdot \mathbf{N}}{|\mathbf{P} - \mathbf{R}||\mathbf{Q} - \mathbf{R}|}, \quad \cos \kappa = \frac{(\mathbf{P} - \mathbf{R}) \cdot (\mathbf{Q} - \mathbf{R})}{|\mathbf{P} - \mathbf{R}||\mathbf{Q} - \mathbf{R}|}, \qquad (8.11)$$

where $\mathbf{N}$ is the unit vector in the direction of $(\mathbf{P} - \mathbf{R}) \times (\mathbf{Q} - \mathbf{R})$.

Now let the triangle $\triangle \mathbf{QRP}$ be the result of a rotation by $[A]$, so we can make the substitutions $\mathbf{P} - \mathbf{R} = [A](\mathbf{p} - \mathbf{r})$, $\mathbf{Q} - \mathbf{R} = [A](\mathbf{q} - \mathbf{r})$, and $\mathbf{N} = [A]\mathbf{n}$ in (8.11). Note that for rotation matrices only, we have the identity

$$([A]\mathbf{x}) \times ([A]\mathbf{y}) = [A](\mathbf{x} \times \mathbf{y}). \tag{8.12}$$

This allows us to factor $[A]$ from the vector product in $\sin \kappa$ and obtain

$$(\mathbf{P} - \mathbf{R}) \times (\mathbf{Q} - \mathbf{R}) \cdot \mathbf{N} = (\mathbf{p} - \mathbf{r}) \times (\mathbf{q} - \mathbf{r}) \cdot \mathbf{n}, \tag{8.13}$$

where $[A]$ cancels in the scalar product because $[A^T][A] = [I]$. For the same reason $[A]$ cancels in the expression for $\cos \kappa$. The conclusion is that these formulas apply without change to coordinates in M and in F, and therefore the angle κ is the same before and after the rotation.

If the point $\mathbf{R} = (0,0,0)^T = \mathbf{c}$ is the origin of F, then (8.11) simplifies to define the angle between the vectors $\mathbf{P}$ and $\mathbf{Q}$ as

$$\sin \kappa = \frac{\mathbf{P} \times \mathbf{Q} \cdot \mathbf{N}}{|\mathbf{P}||\mathbf{Q}|}, \quad \cos \kappa = \frac{\mathbf{P} \cdot \mathbf{Q}}{|\mathbf{P}||\mathbf{Q}|}. \tag{8.14}$$

And we have that $\angle\mathbf{PcQ}=\angle\mathbf{pcq}$ for any two points $\mathbf{p}$ and $\mathbf{q}$ in the moving body.

8.1.3 Spatial Displacements

A spatial displacement consists of a spatial rotation $[A]$ of the moving frame M from its initial position to a new orientation M' followed by a translation $\mathbf{d}$ to M''. The initial position of M is aligned with F, and its final position is aligned with M''. As we did in the plane, we let F and M be the initial and final positions, and define the *spatial displacement* of M relative to F by the transformation $[T] = [A, \mathbf{d}]$.

Clearly, if $\mathbf{d} = 0$, then the displacement is a spatial rotation about the origin of F. Therefore, the *orientation* of M relative to F is defined by the rotation matrix $[A]$. Below we consider spatial displacements that are equivalent to pure rotations about other points in F. The properties of these rotational displacements are the same as for rotations about the origin of F, which are our focus of study in what follows.

8.1.4 Composition of Rotations

Consider two rotation matrices $[A_1]$ and $[A_2]$. Their product $[A_1][A_2]$ is an orthogonal matrix, as can be determined from the computation

$$([A_1][A_2])^T ([A_1][A_2]) = [A_2^T][A_1^T][A_1][A_2] = [I]. \tag{8.15}$$

This is a rotation matrix because its determinant is the product of the determinants of $[A_1]$ and $[A_2]$.

The orientation of M defined by this product results from the orientation of a frame M' relative to F defined by the equation $\mathbf{X} = [A_1]\mathbf{y}$, combined with the orientation of M relative to M' given by $\mathbf{y} = [A_2]\mathbf{x}$. Compose these two rotations by direct substitution for $\mathbf{y}$, and the result is the orientation of M relative to F, given by

$$\mathbf{X} = [A_3]\mathbf{x} = [A_1][A_2]\mathbf{x}. \tag{8.16}$$

The composite rotation is obtained from the matrix product.

The *inverse* $[A^{-1}]$ of a rotation $[A]$ is the rotation defined such that the composition $[A^{-1}][A]$ is the identity. Therefore, $[A^T] = [A^{-1}]$ is the inverse rotation.

8.1.4.1 Changing Coordinates of a Rotation Matrix

Consider the rotation $\mathbf{X} = [A]\mathbf{x}$ of M relative to F. We now determine the rotation matrix $[A']$ between a pair of fixed and moving frames F' and M' that are rotated by the same matrix $[R]$ relative to the original frames. Coordinates $\mathbf{Y}$ and $\mathbf{y}$ in the new frames are related to the original coordinates by $\mathbf{Y} = [R]\mathbf{X}$ and $\mathbf{y} = [R]\mathbf{x}$. Therefore, we have

$$\mathbf{X} = [A]\mathbf{x} = [R^T]\mathbf{Y} = [A][R^T]\mathbf{y},$$

or

$$\mathbf{Y} = [R][A][R^T]\mathbf{y}. \tag{8.17}$$

Thus, the original matrix $[A]$ is transformed by the change of coordinates into $[A'] = [R][A][R^T]$.

8.1.5 Relative Rotations

Consider two orientations M_1 and M_2 of a body relative to F defined by the rotation matrices $[A_1]$ and $[A_2]$. Let $\mathbf{X}$ be the coordinates in F of a point $\mathbf{x}$ in M when in the orientation M_1. Similarly, let $\mathbf{Y}$ be the coordinates of the same point when M is in orientation M_2. Then we have $\mathbf{X} = [A_1]\mathbf{x}$ and $\mathbf{Y} = [A_2]\mathbf{x}$, respectively. The *relative rotation* matrix $[A_{12}]$ that transforms the coordinates $\mathbf{X}$ into $\mathbf{Y}$ is defined by

$$\mathbf{Y} = [A_{12}]\mathbf{X}. \tag{8.18}$$

Substitute for $\mathbf{X}$ and $\mathbf{Y}$ in order to obtain

$$[A_2]\mathbf{x} = [A_{12}][A_1]\mathbf{x}. \tag{8.19}$$

Equating the matrices on both sides of this equation, we see that $[A_{12}]$ is given by

$$[A_{12}] = [A_2][A_1^T]. \tag{8.20}$$

This defines the orientation of M_2 relative to M_1 measured in the frame F.

Relative rotations are easy to compute for coordinate rotations $[X(\cdot)]$, $[Y(\cdot)]$, and $[Z(\cdot)]$. Consider, for example, two orientations of M defined by the z-rotations $[Z(\theta_1)]$ and $[Z(\theta_2)]$. The relative rotation is given by

$$[Z(\theta_2)][Z(\theta_1)^T] = [Z(\theta_2)][Z(-\theta_1)] = [Z(\theta_2 - \theta_1)] = [Z(\theta_{12})], \qquad (8.21)$$

where $\theta_{12} = \theta_2 - \theta_1$. This calculation uses the fact that the inverse of a coordinate rotation is just the rotation by the negative value of the rotation angle.

In general, given a set of orientations M_i, $i = 1, \ldots, n$, we have the relative rotations defined by

$$[A_{ij}] = [A_j][A_i^T]. \qquad (8.22)$$

8.1.6 Relative Inverse Rotations

For two orientations M_1 and M_2 of a body, we can determine the inverse orientations F_1 and F_2 of the fixed frame F as viewed from M. These are defined by the inverse rotations $[A_1^{-1}]$ and $[A_2^{-1}]$. Let $\mathbf{X}$ be a point in F that corresponds to a point $\mathbf{x}$ when M is in orientation M_1, or equivalently, when F coincides with F_1. Let this $\mathbf{X}$ correspond to $\mathbf{y}$ in M when in orientation M_2, which is the same as when F aligns with F_2 as viewed from M. These coordinates are related by the equations $\mathbf{x} = [A_1^{-1}]\mathbf{X}$ and $\mathbf{y} = [A_2^{-1}]\mathbf{X}$. The *relative inverse rotation* $[A_{12}^\dagger]$ transforms the coordinates $\mathbf{x}$ into $\mathbf{y}$ by

$$\mathbf{y} = [A_{12}^\dagger]\mathbf{x},$$

or

$$[A_2^{-1}]\mathbf{X} = [A_{12}^\dagger][A_1^{-1}]\mathbf{X}. \qquad (8.23)$$

Thus, we have

$$[A_{12}^\dagger] = [A_2^T][A_1]. \qquad (8.24)$$

This rotation defines the rotation of F from F_1 into F_2 as viewed from the moving frame M. Notice that this is not the inverse of the relative rotation matrix $[A_{12}]$, which is $[A_{12}^{-1}] = [A_1][A_2^T]$.

For a general set of orientations M_i, $i = 1, \ldots, n$, we have the relative inverse rotations

$$[A_{ij}^\dagger] = [A_j^T][A_i]. \qquad (8.25)$$

The relative inverse rotation is defined from the point of view of the moving frame M. We can choose a specific orientation M_j and transform its coordinates to the fixed frame. In particular, transform the relative inverse rotation $[A_{ik}^\dagger]$ to M_j in F by the rotation $[A_j]$, to obtain

$$[A_{ik}^j] = [A_j][A_{ik}^\dagger][A_j^T].$$ (8.26)

This is known as the image of the relative inverse rotation for position M_j in F.

Notice that if M_j is one of the orientations of the relative inverse rotation, say $j = i$, then

$$[A_{ik}^i] = [A_i][A_{ik}^\dagger][A_i^T] = [A_i][A_k^T][A_i][A_i^T] = [A_i][A_k^T] = [A_{ik}^T].$$ (8.27)

This result is also obtained for $j = k$. Thus, in these cases the image of the relative inverse rotation is the inverse of the relative rotation.

8.2 The Geometry of Rotation Axes

Every rotation has an axis, which is the set of points that are invariant under the transformation. The geometric properties of these axes are fundamental tools in the synthesis of spherical RR chains.

8.2.1 The Rotation Axis

The points that remain fixed during a rotation $[A]$ form its *rotation axis*. To find these points we consider the transformation equation

$$\mathbf{X} = [A]\mathbf{X}.$$ (8.28)

This shows that a fixed point $\mathbf{X}$ is the solution to

$$[I - A]\mathbf{X} = 0.$$ (8.29)

This equation has the solution $\mathbf{X} = 0$, which tells us that origin is a fixed point, as expected. For there to be other fixed points, the determinant of the coefficient matrix must be zero, that is, $|I - A| = 0$.

It happens that this condition is satisfied for all spatial rotation matrices. Another way of saying this is that these matrices always have $\lambda = 1$ as an eigenvalue. Notice that if $\mathbf{S}$ is a nonzero solution, then every point $\mathbf{P} = t\mathbf{S}$ on the line through the origin and $\mathbf{S}$ is also a solution. This line of points is the rotation axis.

8.2.1.1 Cayley's Formula

In order to obtain an explicit equation for the rotation axis we first derive *Cayley's formula* for a spatial rotation matrix. Consider the points $\mathbf{x}$ and $\mathbf{X}$ in F that represent the initial and final positions obtained from the rotation $\mathbf{X} = [A]\mathbf{x}$. Using the fact that

$|\mathbf{x}| = |\mathbf{X}|$, we compute

$$\mathbf{X} \cdot \mathbf{X} - \mathbf{x} \cdot \mathbf{x} = (\mathbf{X} - \mathbf{x}) \cdot (\mathbf{X} + \mathbf{x}) = 0. \tag{8.30}$$

This equation states that the diagonals $\mathbf{X} - \mathbf{x}$ and $\mathbf{X} + \mathbf{x}$ of the rhombus formed by the vertices $\mathbf{O}$, $\mathbf{x}$, $\mathbf{X}$, and $\mathbf{x} + \mathbf{X}$ must be perpendicular.

The diagonals $\mathbf{X} - \mathbf{x}$ and $\mathbf{X} + \mathbf{x}$ are also given by the equations

$$\mathbf{X} - \mathbf{x} = [A - I]\mathbf{x} \quad \text{and} \quad \mathbf{X} + \mathbf{x} = [A + I]\mathbf{x}. \tag{8.31}$$

Substitute for $\mathbf{x}$ on the right side of these two equations in order obtain

$$\mathbf{X} - \mathbf{x} = [A - I][A + I]^{-1}(\mathbf{X} + \mathbf{x}) = [B](\mathbf{X} + \mathbf{x}). \tag{8.32}$$

The matrix $[B]$ operates on the diagonal $\mathbf{X} + \mathbf{x}$ to rotate it $90°$ and change its length. The result is the other diagonal $\mathbf{X} - \mathbf{x}$. From the fact that

$$(\mathbf{X} + \mathbf{x})^T [B](\mathbf{X} + \mathbf{x}) = 0, \tag{8.33}$$

we can see that $[B]$ must have the form

$$[B] = \begin{bmatrix} 0 & -b_z & b_y \\ b_z & 0 & -b_x \\ -b_y & b_x & 0 \end{bmatrix}. \tag{8.34}$$

This matrix is *skew-symmetric*, which means that $[B^T] = -[B]$.

The elements of $[B]$ can be assembled into the vector $\mathbf{b} = (b_x, b_y, b_z)^T$ that has the property that for any vector $\mathbf{y}$,

$$[B]\mathbf{y} = \mathbf{b} \times \mathbf{y}, \tag{8.35}$$

where $\times$ is the usual vector product. The vector $\mathbf{b}$ is called *Rodrigues's vector* and (8.32) is often written as

$$\mathbf{X} - \mathbf{x} = \mathbf{b} \times (\mathbf{X} + \mathbf{x}). \tag{8.36}$$

The equation $[B] = [A - I][A + I]^{-1}$ can be solved to obtain Cayley's formula for rotation matrices

$$[A] = [I - B]^{-1}[I + B]. \tag{8.37}$$

This shows that the nine elements of a 3×3 rotation matrix depend on three independent parameters. Another way to say this is that the set of rotation matrices $SO(3)$ is three-dimensional.

We now solve (8.29) explicitly to determine a nonzero point $\mathbf{S}$ on the rotation axis. Substitute Cayley's formula for $[A]$ into this equation to obtain

$$\left[I - [I - B]^{-1}[I + B] \right] \mathbf{X} = 0, \tag{8.38}$$

which simplifies to

$$[B]\mathbf{X} = 0. \tag{8.39}$$

Since $[B]\mathbf{X} = \mathbf{b} \times \mathbf{X}$, it is clear that $\mathbf{X} = \mathbf{b}$ is a solution. Thus, Rodrigues's vector defines the rotation axis.

We denote by $\mathbf{S}$ the unit vector in the direction of Rodrigues's vector $\mathbf{b}$, and use it to identify the rotation axis of $[A]$.

8.2.2 Perpendicular Bisectors and the Rotation Axis

The angle between the rotation axis $\mathbf{S}$ and vectors through the origin $\mathbf{c}$ to any point $\mathbf{x}$ in the moving body is preserved by the rotation. This means that $\angle\mathbf{xcS}$ equals $\angle\mathbf{XcS}$ and we have

$$\mathbf{S} \cdot \mathbf{X} = \mathbf{S} \cdot \mathbf{x}. \tag{8.40}$$

This equation states that the component of any point $\mathbf{x}$ in the direction of the rotation axis $\mathbf{S}$ is unchanged by the rotation $[A]$. This can be made explicit by computing

$$\mathbf{S} \cdot \mathbf{X} - \mathbf{S} \cdot \mathbf{x} = \mathbf{S} \cdot (\mathbf{X} - \mathbf{x}) = 0. \tag{8.41}$$

Thus, the vector joining the initial position of a point $\mathbf{x}$ to its final position $\mathbf{X}$ is perpendicular to the direction of the rotation axis.

In order to examine this relation (8.41) in more detail, choose the pair of corresponding points $\mathbf{p}$ and $\mathbf{P}$ and consider the set of points $\mathbf{Y}$ that satisfy the equation

$$\mathbf{Y} \cdot (\mathbf{P} - \mathbf{p}) = 0. \tag{8.42}$$

This defines a plane through the origin, because $\mathbf{Y} = 0$ satisfies this equation. Furthermore, the midpoint $\mathbf{V}$ of the segment $\mathbf{P} - \mathbf{p}$ lies on this plane because

$$\mathbf{V} \cdot (\mathbf{P} - \mathbf{p}) = \frac{\mathbf{P} + \mathbf{p}}{2} \cdot (\mathbf{P} - \mathbf{p}) = \frac{\mathbf{P} \cdot \mathbf{P} - \mathbf{p} \cdot \mathbf{p}}{2} = 0. \tag{8.43}$$

The last equality simply restates that $|\mathbf{P}| = |\mathbf{p}|$.

This shows that the plane defined by (8.42) is perpendicular to the segment $\mathbf{pP}$ and passes through its mid-point $\mathbf{V}$. So it is the *perpendicular bisector* of the vector $\mathbf{P} - \mathbf{p}$. Thus, we see that the rotation axis lies on the perpendicular bisector of all vectors $\mathbf{X} - \mathbf{x}$ for every point $\mathbf{x}$ in the moving body.

8.2.2.1 Constructing the Rotation Axis

This result provides a convenient way to determine the rotation axis. Choose two points $\mathbf{p}$ and $\mathbf{q}$ and determine their transformed positions $\mathbf{P} = [A]\mathbf{p}$ and $\mathbf{Q} = [A]\mathbf{q}$. The perpendicular bisectors of the segments $\mathbf{P} - \mathbf{p}$ and $\mathbf{Q} - \mathbf{q}$ are

$$Y \cdot (P - p) = 0,$$
$$Y \cdot (Q - q) = 0. \tag{8.44}$$

The rotation axis is the line of intersection of these two planes, and we have the solution $Y = S$ given by

$$S = \frac{(P - p) \times (Q - q)}{|(P - p) \times (Q - q)|}. \tag{8.45}$$

8.2.3 The Rotation Angle

The plane defined by $\triangle \mathbf{x c S}$ rotates about $\mathbf{S}$ in order to reach its final position containing $\triangle \mathbf{X c S}$, Figure 8.1. The dihedral angle ϕ between these two planes is called the *rotation angle* of $[A]$. It can be determined from the two vectors $\mathbf{S} \times \mathbf{x}$ and $\mathbf{S} \times \mathbf{X}$ that are perpendicular to these respective planes, as well as perpendicular to $\mathbf{S}$. Compute the sine and cosine of the angle between these vectors using (8.14),

$$\sin \phi = \frac{(S \times x) \times (S \times X) \cdot S}{|S \times x||S \times X|}, \quad \cos \phi = \frac{(S \times x) \cdot (S \times X)}{|S \times x||S \times X|}. \tag{8.46}$$

The numerator of $\sin \phi$ simplifies to $(S \times x) \cdot X$, so we have

$$\phi = \arctan \left(\frac{(S \times x) \cdot X}{(S \times x) \cdot (S \times X)} \right). \tag{8.47}$$

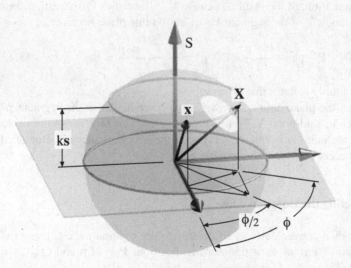

Fig. 8.1 The rotation axis and two positions of a general point.

8.2.3.1 Rodrigues's Equation

Consider the projections of the points $\mathbf{x}$ and $\mathbf{X}$ onto the plane perpendicular to the rotation axis $\mathbf{S}$ through $\mathbf{c}$, which we denote by $\mathbf{x}^*$ and $\mathbf{X}^*$, respectively. The isosceles triangle $\triangle \mathbf{x}^*\mathbf{c}\mathbf{X}^*$ has the rotation angle ϕ as its vertex angle. The altitude $\mathbf{V}^*$ of this triangle is the projection of the midpoint of the segment $\mathbf{X} - \mathbf{x}$. Therefore, we have

$$\tan\frac{\phi}{2} = \frac{|\mathbf{X}^* - \mathbf{V}^*|}{|\mathbf{V}^*|}. \tag{8.48}$$

Notice that the vectors $\mathbf{X}^* - \mathbf{V}^*$ and $\mathbf{V}^*$ are related by the equation

$$\mathbf{X}^* - \mathbf{V}^* = \tan\frac{\phi}{2}\mathbf{S} \times \mathbf{V}^*, \tag{8.49}$$

where the vector product by $\mathbf{S}$ rotates $\mathbf{V}^*$ by $90°$.

We expand this relation to obtain Rodrigues's equation. First, notice that

$$\mathbf{S} \times \mathbf{V}^* = \mathbf{S} \times \frac{\mathbf{X}^* + \mathbf{x}^*}{2} = \mathbf{S} \times \frac{\mathbf{X} + \mathbf{x}}{2}. \tag{8.50}$$

The components of $\mathbf{x}$ and $\mathbf{X}$ in the direction $\mathbf{S}$ are canceled by the vector product with $\mathbf{S}$. Next, we have

$$\mathbf{X}^* - \mathbf{V}^* = \mathbf{X}^* - \frac{\mathbf{X}^* + \mathbf{x}^*}{2} = \frac{\mathbf{X}^* - \mathbf{x}^*}{2} = \frac{\mathbf{X} - \mathbf{x}}{2}, \tag{8.51}$$

because the components of $\mathbf{x}$ and $\mathbf{X}$ along $\mathbf{S}$ cancel. Combining these results, we obtain

$$\mathbf{X} - \mathbf{x} = \tan\frac{\phi}{2}\mathbf{S} \times (\mathbf{X} + \mathbf{x}). \tag{8.52}$$

This is another derivation of *Rodrigues's equation*. However, in this case, we see that the magnitude of Rodrigues's vector is $\tan(\phi/2)$, that is, $\mathbf{b} = \tan(\phi/2)\mathbf{S}$.

8.2.4 The Rotation Defined by ϕ and $\mathbf{S}$

A rotation matrix $[A]$ is characterized by its rotation axis $\mathbf{S}$ and rotation angle ϕ. Cayley's formula combines with the definition of Rodrigues's vector to yield an explicit formula for $[A(\phi,\mathbf{S})]$. Because $\mathbf{b} = \tan(\phi/2)\mathbf{S}$, we have that the matrix $[B]$ is given by

$$[B] = \tan\frac{\phi}{2}\begin{bmatrix} 0 & -s_z & s_y \\ s_z & 0 & -s_x \\ -s_y & s_x & 0 \end{bmatrix}, \tag{8.53}$$

where $\mathbf{S} = (s_x, s_y, s_z)^T$. Thus, Cayley's formula yields

$$[A(\phi, \mathbf{S})] = [I - \tan\frac{\phi}{2}[S]]^{-1}[I + \tan\frac{\phi}{2}[S]], \tag{8.54}$$

which can be expanded to obtain

$$[A(\phi, \mathbf{S})] = [I] + \sin\phi[S] + (1 - \cos\phi)[S^2]. \tag{8.55}$$

This equation defines the rotation matrix in terms of its rotation axis and the angle of rotation about this axis.

Important examples of (8.55) are the coordinate rotations $[X(\theta)]$, $[Y(\theta)]$, and $[Z(\theta)]$, which are rotations by the angle θ about the axes $\vec{\imath} = (1,0,0)^T$, $\vec{\jmath} = (0,1,0)^T$, and $\vec{k} = (0,0,1)^T$, respectively.

8.2.4.1 Inverse Rotations

The rotation matrix $[A]$ and its inverse $[A^T]$ have the same rotation axis $\mathbf{S}$. This is easily seen by multiplying $[A]\mathbf{S} = \mathbf{S}$ by $[A^T]$ to obtain $\mathbf{S} = [A^T]\mathbf{S}$. We now compute $[A(\phi, \mathbf{S})^T]$ using (8.55). Notice that $[S^T] = -[S]$ and $[S^2]^T = [S^2]$, so we have

$$[A^T] = [I] - \sin\phi[S] + (1 - \cos\phi)[S^2], \tag{8.56}$$

where ϕ is the rotation angle of $[A]$. Let ϕ' be the rotation angle of $[A^T]$. Then we see from $\sin\phi' = -\sin\phi$ and $\cos\phi' = \cos\phi$ that $\phi' = -\phi$. Thus, the inverse rotation is simply the rotation by the negative angle around the same axis.

8.2.4.2 A Change of Coordinates

Equation (8.55) provides a convenient way to understand the change of coordinates of a rotation matrix. Consider the transformation $[A'] = [R][A(\phi, \mathbf{S})][R^T]$, where $[A]$ is defined in terms of its rotation angle and axis. Then we have

$$\begin{aligned}[A'] &= [R]\big([I] + \sin\phi[S] + (1 - \cos\phi)[S^2]\big)[R^T] \\ &= [I] + \sin\phi([R][S][R^T]) + (1 - \cos\phi)([R][S^2][R^T]). \end{aligned} \tag{8.57}$$

It is easy to show that $[S'] = [R][S][R^T]$ is the skew symmetric matrix associated with the vector $\mathbf{S}' = [R]\mathbf{S}$, and we have

$$[A'] = [I] + \sin\phi[S'] + (1 - \cos\phi)[S'^2]. \tag{8.58}$$

Thus, a change of coordinates $[R][A(\phi, \mathbf{S})][R^T]$ leaves the rotation angle unchanged, and transforms the rotation axis by $[R]$, so $[A'] = [A(\phi, [R]\mathbf{S})]$.

8.2.5 Eigenvalues of a Rotation Matrix

In this section we consider the matrix equation (8.29) in more detail. The matrix $[I - A]$ can be considered to be $[\lambda I - A]$, where $\lambda = 1$. This leads us to consider the eigenvalue equation

$$[A]\mathbf{X} = \lambda \mathbf{X}, \tag{8.59}$$

which has a solution only if the determinant $|\lambda I - A|$ is zero. This yields the characteristic polynomial

$$\lambda^3 - (a_{11} + a_{22} + a_{33})\lambda^2 + (M_{11} + M_{22} + M_{33})\lambda - 1 = 0, \tag{8.60}$$

where M_{ij} is the minor of the submatrix of $[A]$ obtained by removing row i and column j.

Rotation matrices have the property that each element is equal to its associated minor, that is,

$$M_{ij} = a_{ij}. \tag{8.61}$$

This follows directly from the fact that the inverse of a rotation matrix is its transpose.

We use this to simplify (8.60) to obtain

$$\lambda^3 - (a_{11} + a_{22} + a_{33})\lambda^2 + (a_{11} + a_{22} + a_{33})\lambda - 1 = 0. \tag{8.62}$$

It is now easy to check that $\lambda = 1$ is a root of this polynomial for all rotation matrices. This means that $|I - A| = 0$, so (8.29) always has solutions other than $\mathbf{X} = 0$.

To obtain the other two roots of (8.62), divide by $(\lambda - 1)$ to obtain

$$\lambda^2 - (a_{11} + a_{22} + a_{33} - 1)\lambda + 1 = 0. \tag{8.63}$$

The roots of this equation are $\lambda = e^{i\phi}$ and $\lambda = e^{-i\phi}$, where the angle ϕ is given by

$$\phi = \arccos\left(\frac{a_{11} + a_{22} + a_{33} - 1}{2}\right). \tag{8.64}$$

8.2.6 Rotation Axis of a Relative Rotation

For two orientations defined by the rotations $[A_1]$ and $[A_2]$ we have the relative rotation matrix $[A_{12}] = [A_2][A_1^T]$. This matrix transforms the coordinates $\mathbf{X}^1$ of point $\mathbf{x}$ in M in orientation M_1 into the coordinates $\mathbf{X}^2$ when the body is in M_2, that is,

$$\mathbf{X}^2 = [A_{12}]\mathbf{X}^1. \tag{8.65}$$

The axis of rotation $\mathbf{S}_{12}$ satisfies the condition (8.40), which in this case becomes

$$\mathbf{S}_{12} \cdot \mathbf{X}^2 = \mathbf{S}_{12} \cdot \mathbf{X}^1, \tag{8.66}$$

or

$$\mathbf{S}_{12} \cdot (\mathbf{X}^2 - \mathbf{X}^1) = 0. \tag{8.67}$$

Thus, the relative rotation axis $\mathbf{S}_{12}$ lies on the perpendicular bisector of all segments joining corresponding points in orientations M_1 and M_2.

Let $\mathbf{P}^1$ and $\mathbf{P}^2$ and $\mathbf{Q}^1$ and $\mathbf{Q}^2$ be a pair of corresponding points in orientations M_1 and M_2. Then $\mathbf{S}_{12}$ is the solution to the pair of equations

$$\mathbf{S}_{12} \cdot (\mathbf{P}^2 - \mathbf{P}^1) = 0,$$
$$\mathbf{S}_{12} \cdot (\mathbf{Q}^2 - \mathbf{Q}^1) = 0, \tag{8.68}$$

given by

$$\mathbf{S}_{12} = \frac{(\mathbf{P}^2 - \mathbf{P}^1) \times (\mathbf{Q}^2 - \mathbf{Q}^1)}{|(\mathbf{P}^2 - \mathbf{P}^1) \times (\mathbf{Q}^2 - \mathbf{Q}^1)|}. \tag{8.69}$$

The relative rotation angle ϕ_{12} about $\mathbf{S}_{12}$ is the dihedral angle between the planes containing $\triangle \mathbf{X}^1 O \mathbf{S}_{12}$ and $\triangle \mathbf{X}^1 O \mathbf{S}_{12}$, which by (8.47) is

$$\phi_{12} = \arctan\left(\frac{(\mathbf{S}_{12} \times \mathbf{X}^1) \cdot \mathbf{X}^2}{(\mathbf{S} \times \mathbf{X}^1) \cdot (\mathbf{S} \times \mathbf{X}^2)}\right). \tag{8.70}$$

Finally, we see from (8.55) that the relative rotation $[A_{12}]$ is defined in terms of its rotation angle ϕ_{12} and axis $\mathbf{S}_{12}$ by the formula

$$[A(\phi_{12}, \mathbf{S}_{12})] = [I] + \sin\phi_{12}[S_{12}] + (1 - \cos\phi_{12})[S_{12}^2]. \tag{8.71}$$

8.2.7 Rotation Axis of a Relative Inverse Rotation

Given the two orientations M_1 and M_2, we can compute the inverse rotations $[A_1^T]$ and $[A_2^T]$ that define the orientations F_1 and F_2 of the fixed frame relative to M. The *relative inverse rotation* is given by $[A_{12}^\dagger] = [A_2^T][A_1]$. The rotation axis $\mathbf{s}_{12}$ of $[A_{12}^\dagger]$ is computed using the formulas above, but it now lies in M. Let $\phi_{12}^\dagger$ be the relative rotation angle.

In general, we can transform the relative inverse rotation $[A_{ik}^\dagger]$ to the fixed frame F when M is in orientation M_j by the computation

$$[A_{ik}^j] = [A_j][A_{ik}^\dagger][A_j^T]. \tag{8.72}$$

This transforms the coordinates of $\mathbf{s}_{ik}$ to

$$\mathbf{S}_{ik}^j = [A_j]\mathbf{s}_{ik}, \tag{8.73}$$

which is the *image* of the relative inverse rotation axis.

For the two cases $j = 1$ and $j = 2$ we have

$$[A_{12}^1] = [A_{12}^2] = [A_{12}^T].$$ (8.74)

Thus, the relative inverse rotation angle is $\phi_{12}^\dagger = -\phi_{12}$. The rotation axis $\mathbf{s}_{12}$ is transformed to the fixed frame such that

$$\mathbf{S}_{12} = [A_1]\mathbf{s}_{12} = [A_2]\mathbf{s}_{12},$$ (8.75)

which is what we expect for a relative rotation axis.

8.2.8 Rotational Displacements

We now consider spatial displacements that are rotations but around points other than the origin of F. In particular, we determine the condition under which the displacement $[T] = [A, \mathbf{d}]$ has a nonzero fixed point $\mathbf{c}$. We seek the points $\mathbf{X}$ such that

$$\mathbf{X} = [A]\mathbf{X} + \mathbf{d},$$ (8.76)

or

$$[I - A]\mathbf{X} = \mathbf{d}.$$ (8.77)

We have already seen that $|I - A| = 0$. Therefore, this equation does not have a solution, in general, and there are no fixed points. However, we can determine a condition that the translation vector $\mathbf{d}$ must satisfy in order for a solution to exist.

Substitute Cayley's formula for $[A]$, as was done in (8.39). The result can be simplified to the form

$$\mathbf{b} \times \mathbf{X} = \frac{1}{2}(\mathbf{b} \times \mathbf{d} - \mathbf{d}).$$ (8.78)

The left side of this equation is orthogonal to the Rodrigues's vector $\mathbf{b}$. This means that the right side must be orthogonal to $\mathbf{b}$ as well. Thus, we obtain the condition

$$\mathbf{b} \cdot (\mathbf{b} \times \mathbf{d} - \mathbf{d}) = 0, \quad \text{or} \quad \mathbf{b} \cdot \mathbf{d} = 0.$$ (8.79)

This is clearly satisfied when $\mathbf{d} = 0$. However, we now see that a spatial displacement has a fixed point when the translation vector is orthogonal to the rotation axis.

In this case, we can solve (8.78) by computing

$$\mathbf{b} \times (\mathbf{b} \times \mathbf{X}) = \mathbf{b} \times \frac{1}{2}(\mathbf{b} \times \mathbf{d} - \mathbf{d}),$$ (8.80)

which simplifies to yield

$$\mathbf{X} = \frac{\mathbf{b} \times (\mathbf{d} - \mathbf{b} \times \mathbf{d})}{2\mathbf{b} \cdot \mathbf{b}}.$$ (8.81)

Not only is the point $\mathbf{X} = \mathbf{c}$ fixed under this displacement, but every point on the line $\mathsf{L} : \mathbf{Y} = \mathbf{c} + t\mathbf{S}$ is fixed as well. Therefore, the displacement is a pure rotation about the axis L, and we call it a *rotational displacement*.

8.3 The Spherical Pole Triangle

8.3.1 The Axis of Composite Rotation

There is an important geometric relationship between the axes of two rotations $[A]$ and $[B]$ and the axis of their product $[C] = [B][A]$. This is easily derived by using Rodrigues's equation (8.52) to represent each of the rotations. Let $[C]$ have the rotation axis $\mathbf{C}$ and rotation angle γ, so we have Rodrigues's vector $\tan(\gamma/2)\mathbf{C}$. Similarly, let Rodrigues's vectors for $[A]$ and $[B]$ be $\tan(\alpha/2)\mathbf{A}$ and $\tan(\beta/2)\mathbf{B}$, respectively.

We now consider the composite rotation $[B][A]$ as the transformation $\mathbf{y} = [A]\mathbf{x}$, followed by the transformation $\mathbf{X} = [B]\mathbf{y}$. Thus, for $[A]$ we have

$$\mathbf{y} - \mathbf{x} = \tan\frac{\alpha}{2}\mathbf{A} \times (\mathbf{y} + \mathbf{x}). \tag{8.82}$$

And for the rotation $[B]$ we have

$$\mathbf{X} - \mathbf{y} = \tan\frac{\beta}{2}\mathbf{B} \times (\mathbf{X} + \mathbf{y}). \tag{8.83}$$

The vector $\mathbf{y}$ can be eliminated between these two equations to yield

$$\mathbf{X} - \mathbf{x} = \tan\frac{\gamma}{2}\mathbf{C} \times (\mathbf{X} + \mathbf{x}),$$

where

$$\tan\frac{\gamma}{2}\mathbf{C} = \frac{\tan\frac{\beta}{2}\mathbf{B} + \tan\frac{\alpha}{2}\mathbf{A} + \tan\frac{\beta}{2}\tan\frac{\alpha}{2}\mathbf{B} \times \mathbf{A}}{1 - \tan\frac{\beta}{2}\tan\frac{\alpha}{2}\mathbf{B} \cdot \mathbf{A}}. \tag{8.84}$$

This result is known as *Rodrigues's formula* for the composition of rotations.

8.3.1.1 A Spherical Triangle

We now show that (8.84) is the equation of the spherical triangle formed by $\mathbf{A}$, $\mathbf{B}$, and $\mathbf{C}$, with interior angles $\alpha/2$ and $\beta/2$ at $\mathbf{A}$ and $\mathbf{B}$, and the exterior angle $\gamma/2$ at $\mathbf{C}$, Figure 8.2.

Introduce the planes E_A and E_B through the center of the sphere, which define the sides $\mathbf{AC}$ and $\mathbf{BC}$ of the spherical triangle. These planes intersect along the vector $\mathbf{C}$ and lie at the dihedral angle $\gamma/2$ relative to each other. Let $\mathbf{n}_A$ be the unit vector in the direction $\mathbf{C} \times \mathbf{A}$ normal to E_A, and let $\mathbf{n}_B$ be the unit vector along $\mathbf{B} \times \mathbf{C}$ normal

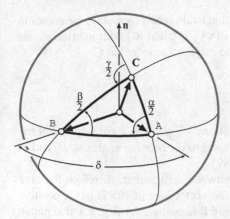

Fig. 8.2 The spherical triangle $\triangle ABC$ with interior angles $\alpha/2$ and $\beta/2$ at **A** and **B**, and $\gamma/2$ as its exterior angle at **C**.

to E_B. Using these conventions we have

$$\mathbf{n}_A \times \mathbf{n}_B = \sin\frac{\gamma}{2}\mathbf{C} \quad \text{and} \quad \mathbf{n}_B \cdot \mathbf{n}_A = \cos\frac{\gamma}{2}, \tag{8.85}$$

where $\gamma/2$ is the exterior angle at the vertex **C**. We now compute **C** in terms of the vertices **A** and **B** and their interior angles.

We can expand $\mathbf{n}_A$ and $\mathbf{n}_B$ in terms of the unit vectors **B**, **v**, and **n**, where **n** is the unit vector in the direction $\mathbf{B} \times \mathbf{A}$ and $\mathbf{v} = \mathbf{n} \times \mathbf{B}$. If δ is the angle measured from **B** to **A** in the **AB** plane, that is, $\cos\delta = \mathbf{B} \cdot \mathbf{A}$, then we have

$$\mathbf{n}_A = \sin\frac{\alpha}{2}(\cos\delta\,\mathbf{v} - \sin\delta\,\mathbf{B}) + \cos\frac{\alpha}{2}\mathbf{n},$$

$$\mathbf{n}_B = -\sin\frac{\beta}{2}\mathbf{v} + \cos\frac{\beta}{2}\mathbf{n}. \tag{8.86}$$

Computing the scalar and vector products in (8.85) we obtain

$$\sin\frac{\gamma}{2}\mathbf{C} = \sin\frac{\beta}{2}\cos\frac{\alpha}{2}\mathbf{B} + \sin\frac{\alpha}{2}\cos\frac{\beta}{2}\mathbf{A} + \sin\frac{\beta}{2}\sin\frac{\alpha}{2}\mathbf{B}\times\mathbf{A},$$

$$\cos\frac{\gamma}{2} = \cos\frac{\alpha}{2}\cos\frac{\beta}{2} - \sin\frac{\alpha}{2}\sin\frac{\beta}{2}\mathbf{B}\cdot\mathbf{A}. \tag{8.87}$$

Divide these two equations to obtain

$$\tan\frac{\gamma}{2}\mathbf{C} = \frac{\tan\frac{\beta}{2}\mathbf{B} + \tan\frac{\alpha}{2}\mathbf{A} + \tan\frac{\beta}{2}\tan\frac{\alpha}{2}\mathbf{B}\times\mathbf{A}}{1 - \tan\frac{\beta}{2}\tan\frac{\alpha}{2}\mathbf{B}\cdot\mathbf{A}}. \tag{8.88}$$

Compare this equation with (8.84) to see that Rodrigues's formula is the equation of the triangle formed by the rotation axes of $[A]$, $[B]$, and $[C]$ with interior angles $\alpha/2$ and $\beta/2$ at the vertices $\mathbf{A}$ and $\mathbf{B}$, and the exterior angle $\gamma/2$ at $\mathbf{C}$.

8.3.1.2 The Composite Axis Theorem

The rotation axes of the composition of rotations $[C] = [B][A]$ form a spherical triangle $\triangle \mathbf{ABC}$ with vertex angles directly related to the rotation angles α, β, and γ. We examine this triangle using equation (8.87).

Notice that $\alpha/2$ and $\beta/2$ take values between zero and π, therefore the sine of these angles are always positive. Thus, the vector part of (8.87) has a positive component along $\mathbf{B} \times \mathbf{A}$. The component along $\mathbf{B}$ is positive for $\beta < \pi$ and negative for $\beta > \pi$. We now introduce the convention that $\mathbf{C}$ is directed so it always has a positive component along $\mathbf{B}$. This allows $\sin(\gamma/2)$ to take positive and negative values. Notice that if $\sin(\gamma/2)$ is negative, then $\gamma/2 > \pi$.

Because $\mathbf{A} = \cos\delta \mathbf{B} + \sin\delta \mathbf{v}$, we have

$$\sin\frac{\gamma}{2}\mathbf{C} = \left(\sin\frac{\beta}{2}\cos\frac{\alpha}{2} + \sin\frac{\alpha}{2}\cos\frac{\beta}{2}\cos\delta\right)\mathbf{B} + \sin\frac{\alpha}{2}\cos\frac{\beta}{2}\sin\delta\mathbf{v}$$
$$+ \sin\frac{\beta}{2}\sin\frac{\alpha}{2}\mathbf{B} \times \mathbf{A}. \tag{8.89}$$

Introduce the angle τ so that $\cos\tau = \mathbf{B} \cdot \mathbf{C}$. Then we have

$$\sin\frac{\gamma}{2}\cos\tau = \sin\frac{\beta}{2}\cos\frac{\alpha}{2} + \sin\frac{\alpha}{2}\cos\frac{\beta}{2}\cos\delta. \tag{8.90}$$

Our convention for the direction of $\mathbf{C}$ ensures that $\cos\tau$ is always positive, so we have the two cases:

Case 1. $\sin(\gamma/2) > 0$, that is, $\gamma < 2\pi$.

In this case the vertex $\mathbf{C}$ has a positive component along $\mathbf{B} \times \mathbf{A}$. The derivation above shows that $\alpha/2$ and $\beta/2$ are the interior angles of $\triangle \mathbf{ABC}$ at the vertices $\mathbf{A}$ and $\mathbf{B}$. The angle $\gamma/2$ is the exterior angle at $\mathbf{C}$.

Case 2. $\sin(\gamma/2) < 0$, that is, $\gamma > 2\pi$.

In this case the vertex $\mathbf{C}$ lies below the $\mathbf{AB}$ plane and has a component directed opposite to $\mathbf{B} \times \mathbf{A}$. The angles $\alpha/2$ and $\beta/2$ are the exterior angles of $\triangle \mathbf{ABC}$ at $\mathbf{A}$ and $\mathbf{B}$, respectively. If the angle κ is the interior angle at $\mathbf{C}$, then $\gamma/2 = \kappa + \pi$.

We collect these results in the following theorem:

Theorem 9 (The Composite Axis Theorem). *The axis* **C** *of a composite rotation* [C] = [B][A] *forms a triangle with the axes* **B** *and* **A** *of the rotations* [B] *and* [A], *respectively. If* $\sin(\gamma/2) > 0$, *then the interior angles of this triangle at* **A** *and* **B** *are* $\alpha/2$ *and* $\beta/2$, *respectively. If* $\sin(\gamma/2) < 0$, *then* $\alpha/2$ *and* $\beta/2$ *are the exterior angles at these vertices. In this case, if* κ *is the interior angle at* **C**, *then* $\gamma/2 = \kappa + \pi$.

8.3.1.3 Quaternions and the Spherical Triangle

W. R. Hamilton [43] introduced quaternions to generalize to three dimensions the geometric properties of complex numbers. A quaternion is the formal sum of a scalar q_0 and a vector $\mathbf{q} = (q_x, q_y, q_z)^T$, written as $Q = q_0 + \mathbf{q}$.

Quaternions can be added together, and multiplied by a scalar, componentwise like four-dimensional vectors. A new operation, invented by Hamilton, defines the product of two quaternions $P = p_0 + \mathbf{p}$ and $Q = q_0 + \mathbf{q}$ by the rule

$$R = PQ = (p_0 + \mathbf{p})(q_0 + \mathbf{q}) = (p_0 q_0 - \mathbf{p} \cdot \mathbf{q}) + (q_0 \mathbf{p} + p_0 \mathbf{q} + \mathbf{p} \times \mathbf{q}), \qquad (8.91)$$

where the dot and cross denote the usual vector operations.

The conjugate of the quaternion $Q = q_0 + \mathbf{q}$ is $Q^* = q_0 - \mathbf{q}$, and the product QQ^* is the positive real number.

$$QQ^* = (q_0 + \mathbf{q})(q_0 - \mathbf{q}) = q_0^2 + \mathbf{q} \cdot \mathbf{q} = |Q|^2. \qquad (8.92)$$

The scalar $|Q|$ is called the *norm* of the quaternion.

We are interested in quaternions of norm equal to 1. These so-called unit quaternions can be written in the form

$$Q = \cos\frac{\theta}{2} + \sin\frac{\theta}{2}\mathbf{S}, \qquad (8.93)$$

where $\mathbf{S} = (s_x, s_y, s_z)^T$ is a unit vector. The quaternion product of $A(\alpha/2) = \cos(\alpha/2) + \sin(\alpha/2)\mathbf{A}$ and $B(\beta/2) = \cos(\beta/2) + \sin(\beta/2)\mathbf{B}$ yields the unit quaternion $C(\gamma/2) = B(\beta/2)A(\alpha/2)$, given by

$$\cos\frac{\gamma}{2} + \sin\frac{\gamma}{2}\mathbf{C} = \left(\cos\frac{\beta}{2}\cos\frac{\alpha}{2} - \sin\frac{\beta}{2}\sin\frac{\alpha}{2}\mathbf{B} \cdot \mathbf{A} \right)$$
$$+ \left(\sin\frac{\beta}{2}\cos\frac{\alpha}{2}\mathbf{B} + \sin\frac{\alpha}{2}\cos\frac{\beta}{2}\mathbf{A} + \sin\frac{\beta}{2}\sin\frac{\alpha}{2}\mathbf{B} \times \mathbf{A} \right). \quad (8.94)$$

Compare this equation to (8.87) to see that quaternion multiplication yields one vertex of a spherical triangle from the other two. We conclude that each rotation $[A(\phi, \mathbf{S})]$ can be identified with a quaternion $S(\phi/2) = \cos(\phi/2) + \sin(\phi/2)\mathbf{S}$.

8.3.2 The Triangle of Relative Rotation Axes

For three orientations M_i, M_j, and M_k of a moving body, we can construct the relative rotations $[A_{ij}]$, $[A_{jk}]$, and $[A_{ik}]$. Notice that the relative rotation $[A_{ik}]$ is given by the product $[A_{jk}][A_{ij}]$, as is seen from

$$[A_{ik}] = [A_k][A_i^T] = ([A_k][A_j^T])([A_j][A_i^T]) = [A_{jk}][A_{ij}]. \qquad (8.95)$$

Rodrigues's formula for this composition of rotations yields

$$\tan\frac{\phi_{ik}}{2}\mathbf{S}_{ik} = \frac{\tan\frac{\phi_{jk}}{2}\mathbf{S}_{jk} + \tan\frac{\phi_{ij}}{2}\mathbf{S}_{ij} + \tan\frac{\phi_{jk}}{2}\tan\frac{\phi_{ij}}{2}\mathbf{S}_{jk}\times\mathbf{S}_{ij}}{1 - \tan\frac{\phi_{jk}}{2}\tan\frac{\phi_{ij}}{2}\mathbf{S}_{jk}\cdot\mathbf{S}_{ij}}. \qquad (8.96)$$

This is the equation of the spherical triangle formed by the relative rotation axes $\triangle\mathbf{S}_{ij}\mathbf{S}_{jk}\mathbf{S}_{ik}$. The composite-axis theorem defines the relationship between the vertex angles of this triangle and the relative rotation angles $\phi_{ij}/2$, $\phi_{jk}/2$, and $\phi_{ik}/2$. For example, if $\mathbf{S}_{ik}$ lies above the plane through $\mathbf{S}_{ij}\mathbf{S}_{jk}$, then the interior angles at $\mathbf{S}_{ij}$ and $\mathbf{S}_{jk}$ are $\phi_{ij}/2$ and $\phi_{jk}/2$, respectively, and the exterior angle at $\mathbf{S}_{ik}$ is $\phi_{ik}/2$, Figure 8.3. This triangle is analogous to the planar pole triangle and is called the *spherical pole triangle*.

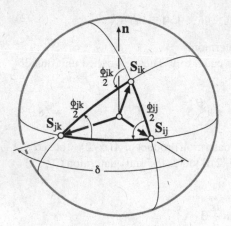

Fig. 8.3 The spherical pole triangle.

8.3.3 The Spherical Image Pole Triangle

We now consider the inverse rotations associated with the orientations M_i, M_j, M_k, given by $[A_i^T]$, $[A_j^T]$, and $[A_k^T]$. The relative inverse rotation $[A_{ik}^+]$ is the composition of the relative inverse rotations $[A_{jk}^{\dagger}][A_{ij}^{\dagger}]$, as can be seen from

$$[A_{ik}^{\dagger}] = [A_k^T][A_i] = \left([A_k^T][A_j]\right)\left([A_j^T][A_i]\right) = [A_{jk}^{\dagger}][A_{ij}^{\dagger}]. \tag{8.97}$$

We can transform each relative inverse rotation to the fixed frame F for M aligned with M_m, that is, we compute

$$[A_{ik}^m] = [A_m][A_{ik}^{\dagger}][A_m^T]. \tag{8.98}$$

We obtain the composition of the relative inverse rotations as seen from F,

$$[A_{ik}^m] = [A_{jk}^m][A_{ij}^m]. \tag{8.99}$$

Rodrigues's formula for this composition defines the triangle of image relative rotation axes, which is known as the *spherical image pole triangle*.

Let $m = i$, for example, and notice that $\mathbf{S}_{ij}^i = \mathbf{S}_{ij}$ and $\mathbf{S}_{ik}^i = \mathbf{S}_{ik}$, and we have the image pole triangle $\triangle\mathbf{S}_{ij}\mathbf{S}_{jk}^i\mathbf{S}_{ik}$. The relative inverse rotation angles are the negatives of the relative rotation angles. Therefore, $\phi_{ik}^{\dagger} = -\phi_{ik}$. Thus, if the spherical pole triangle has $\phi_{ij}/2$ as its interior angle at $\mathbf{S}_{ij}$, then the image pole triangle $\triangle\mathbf{S}_{ij}\mathbf{S}_{jk}^i\mathbf{S}_{ik}$ has $-\phi_{12}/2$ as its associated interior angle. The result is that the axis $\mathbf{S}_{jk}^i$ is the reflection of $\mathbf{S}_{jk}$ through the plane defined by $\mathbf{S}_{ij}\mathbf{S}_{ik}$, Figure 8.4.

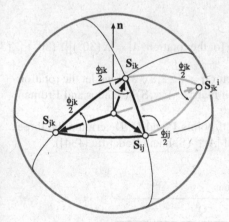

Fig. 8.4 The image pole $\mathbf{S}_{jk}^i$ is the reflection of $\mathbf{S}_{jk}$ through the side $\mathbf{S}_{ij}\mathbf{S}_{ik}$ of the spherical pole triangle.

8.4 Summary

This chapter has presented the geometric theory of spatial rotations. Of fundamental importance is the spherical pole triangle, which is the analog of the planar pole triangle. Notice that Hamilton's quaternions can be viewed as the generalization of complex vectors, and that they provide a convenient tool for computations using the spherical triangle. The similar form of the planar and spherical results provides an avenue for visualizing three-dimensional geometry using intuition drawn from plane geometry.

8.5 References

The kinematic theory of spatial rotations can be found in Bottema and Roth [5]. Crane and Duffy [16] present a detailed development of the trigonometric formulas for spherical triangles. Cheng and Gupta [11] discuss the history of the various representations of the rotation matrix. The interesting history surrounding Hamilton's quaternions and Rodrigues's formula is described by Altmann [1].

Exercises

1. Determine the rotation axis $\mathbf{S}$ and angle ϕ for the rotation $[A] = [X(30°)][Y(30°)][Z(30°)$ (Sandor and Erdman [112]).
2. Let the axis of a rotation be along the vector $\mathbf{q} = (2,2,2\sqrt{2})^T$ and let the rotation angle be $\phi = 30°$. Determine the rotation matrix $[A(\phi,\mathbf{S})]$ (Sandor and Erdman [112]).
3. Table 8.1 gives four locations of a pair of points $\mathbf{P}$ and $\mathbf{Q}$. Determine the three relative rotation matrices $[A_{12}]$, $[A_{13}]$, and $[A_{14}]$ (Suh and Radcliffe [134]).

Table 8.1 Point coordinates defining four orientations

M_i	$\mathbf{P}^i$	$\mathbf{Q}^i$
1	$(0.105040, 0.482820, 0.869397)^T$	$(-0.464640, -0.676760, 0.571057)^T$
2	$(0.090725, 0.541283, 0.835931)^T$	$(-0.133748, -0.751642, 0.645868)^T$
3	$(0.104155, 0.620000, 0.777658)^T$	$(0.161113, -0.702067, 0.693646)^T$
4	$(0.096772, 0.725698, 0.681173)^T$	$(0.400762, -0.564306, 0.721769)^T$

4. Prove that each element of a rotation matrix is equal to its associated minor.
5. Show that the change of coordinates $[R][A][R^T]$ of a rotation matrix $[A]$ has the rotation axis $[R]\mathbf{S}$, where $\mathbf{S}$ is the rotation axis of $[A]$.

6. Let $c_1 = s_x \sin(\phi/2)$, $c_2 = s_y \sin(\phi/2)$, $c_3 = s_z \sin(\phi/2)$ and $c_4 = \cos(\phi/2)$ denote the components of a unit quaternion. Use (8.54) to obtain a formula for $[A(\phi, \mathbf{S})]$ with each element quadratic in c_i.

7. Derive Rodrigues's formula for the composition of rotations (8.84).

8. Consider a spherical pole triangle $\triangle \mathbf{S}_{12}\mathbf{S}_{23}\mathbf{S}_{13}$. Show that the first position of a point $\mathbf{Q}^1$ reflects through the side $N_1 : \mathbf{S}_{12}\mathbf{S}_{13}$ to the cardinal point $\mathbf{Q}^*$ and that the corresponding points $\mathbf{Q}^i$ reflect through the sides N_i to the same point.

Chapter 9
Algebraic Synthesis of Spherical Chains

In this chapter we formulate the design theory for spherical RR chains. The axes of the two revolute joints must lie in the same plane, and therefore intersect in a point. The floating link of this system moves in pure rotation about this point.

Two RR chains can be connected to form a one-degree-of-freedom spherical 4R linkage. The result is that the coupler is guided along a general rotational movement. Notice that the axes of the four revolute joints must pass through the same point. While it would seem difficult to ensure that the four axes intersect in one point, in practice the internal forces of the system tend to align the axes so that the linkage moves smoothly.

The synthesis theory for spherical linkages follows the same geometric principles as the planar theory. Intuition gained from working in the plane can be used to guide the design process for spherical linkages.

9.1 A Single Revolute Joint

A revolute joint cannot rotate a body between two general positions in space. However, if the two positions have the property that their relative displacement $[T_{12}] = [A_{12}, \mathbf{d}_{12}]$ is a rotational displacement, then a revolute joint aligned with the rotation axis of this displacement provides the desired movement.

Let the two spatial positions M_1 and M_2 be defined by $[T_1]$ and $[T_2]$, so we have the relative displacement $[T_{12}] = [T_2][T_1^{-1}]$. If the relative translation vector $\mathbf{d}_{12}$ is orthogonal to the rotation axis $\mathbf{S}_{12}$, that is, $\mathbf{d}_{12} \cdot \mathbf{S}_{12} = 0$, then we can determine a fixed point $\mathbf{c}$ by (8.81)

$$\mathbf{c} = \frac{\mathbf{b}_{12} \times (\mathbf{d}_{12} - \mathbf{b}_{12} \times \mathbf{d}_{12})}{2\mathbf{b}_{12} \cdot \mathbf{b}_{12}}, \tag{9.1}$$

where $\mathbf{b}_{12} = \tan(\phi_{12}/2)\mathbf{S}_{12}$ is Rodrigues's vector for the relative rotation. The line $\mathsf{L}_{12}\colon \mathbf{Y}(t) = \mathbf{c} + t\mathbf{S}_{12}$ is the axis of the revolute joint.

J.M. McCarthy and G.S. Soh, *Geometric Design of Linkages*, Interdisciplinary Applied Mathematics 11, DOI 10.1007/978-1-4419-7892-9_9,

A point in the moving body that lies on this axis, as seen from the moving frame M, is found from the relative inverse rotational displacement $[T_{12}^\dagger] = [T_2^{-1}][T_1] = [A_{12}^\dagger, \mathbf{d}_{12}^\dagger]$. This point defines the line $\mathsf{L}_{12}^\dagger: \mathbf{y}(t) = \mathbf{c}^\dagger + t\mathbf{s}_{12}$ in M that coincides with the line L_{12} to form the revolute joint.

9.2 Spherical Displacements

If three or more spatial positions M_i, $i = 1,\ldots,n$, are to be reachable by the floating link of a spherical chain, then the relative displacements $[T_{1j}]$, $j = 2,\ldots,n$, must each be rotational displacements. Furthermore, the axes of these displacements must pass through the same point $\mathbf{c}$. If this occurs, then the axes of the relative displacements have the form $\mathsf{L}_{1j}: \mathbf{Y}(t) = \mathbf{c} + t\mathbf{S}_{1j}$, $j = 2,\ldots,n$. The relative translation vectors $\mathbf{d}_{1j}$ can now be written as

$$[I - A_{1j}]\mathbf{c} = \mathbf{d}_{1j}, j = 2,\ldots,n. \tag{9.2}$$

Substitute this into the transformation equation $\mathbf{Y}^j = [A_{1j}]\mathbf{Y}^1 + \mathbf{d}_{1j}$ to obtain

$$\mathbf{Y}^j - \mathbf{c} = [A_{1j}](\mathbf{Y}^1 - \mathbf{c}), j = 2,\ldots,n. \tag{9.3}$$

Spatial displacements that can be put into this form are called *spherical displacements*. The change of coordinates $\mathbf{X} = \mathbf{Y} - \mathbf{c}$ transforms these equations into pure rotations about the origin of F. In what follows, we assume this transformation and consider the positions M_i, $i = 1,\ldots,n$, to be specified by pure rotations $[A_i]$, $i = 1,\ldots,n$.

9.2.0.1 Longitude, Latitude, and Roll Angles

There are many ways to parameterize the set of rotation matrices. For example, almost any combination of three coordinate rotations can be used to define a general rotation. Here we introduce a set of parameters that is analogous to the x and y translation and rotation ϕ used to define planar positions. They are the longitude, latitude, and roll angles that define the orientation of M relative to F. These parameters allow us to use a globe to illustrate the orientation of each task position, Figure 9.1.

Locate the fixed frame F at the center of a globe so its y-axis is directed toward the north pole and its z-axis passes through intersection of the prime meridian, $0°$ longitude, and the equator, $0°$ latitude. Longitude and latitude coordinates α, β locate a point $\mathbf{X}$ on the surface of the globe such that its coordinates are

$$\mathbf{X} = R \left\{ \begin{array}{c} \sin\alpha\cos\beta \\ \sin\beta \\ \cos\alpha\cos\beta \end{array} \right\}, \tag{9.4}$$

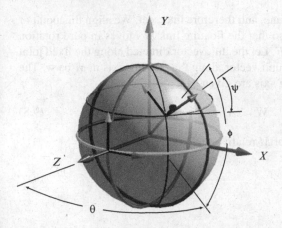

Fig. 9.1 Longitude, latitude, and roll coordinates defining orientation relative to a global frame.

where R is the distance from the origin $\mathbf{c}$. We can compute longitude and latitude angles for a vector $\mathbf{X} = (X, Y, Z)^T$ using the formulas

$$\alpha = \arctan \frac{X}{Z}, \quad \beta = \arctan \frac{Y}{\sqrt{X^2 + Z^2}}, \quad R = \sqrt{X^2 + Y^2 + Z^2}. \quad (9.5)$$

The orientation of the moving frame M can now be defined in terms of the longitude θ and latitude ϕ of its z-axis, and the roll ψ about this axis. The rotation matrix $[A]$ is given by the composition of coordinate rotations

$$[A(\theta, \phi, \psi)] = [Y(\theta)][X(-\phi)][Z(\psi)], \quad (9.6)$$

that is,

$$[A] = \begin{bmatrix} \cos\theta & 0 & \sin\theta \\ 0 & 1 & 0 \\ -\sin\theta & 0 & \cos\theta \end{bmatrix} \begin{bmatrix} 1 & 0 & 0 \\ 0 & \cos\phi & \sin\phi \\ 0 & -\sin\phi & \cos\phi \end{bmatrix} \begin{bmatrix} \cos\psi & -\sin\psi & 0 \\ \sin\psi & \cos\psi & 0 \\ 0 & 0 & 1 \end{bmatrix}. \quad (9.7)$$

Notice that we have used the identities $\cos(-\phi) = \cos\phi$ and $\sin(-\phi) = -\sin\phi$ in $[X(-\phi)]$.

We often draw the moving frame on the surface of the unit sphere, though it is understood that the origins of both the moving and fixed frames, M and F, are located at the center of the sphere.

9.3 The Geometry of Spherical RR Chains

A spherical RR chain consists of a floating link connected to a crank by a revolute joint, which in turn is connected to ground by a revolute joint, Figure 9.2. The axes of the two joints lie in the same plane, and therefore intersect. We align this point of intersection with the origin of F so that the floating link M moves in pure rotation with no translation component in F. Let the unit vector directed along the fixed joint axis be denoted by $\mathbf{G}$, and let the unit vector along the moving axis in M be $\mathbf{w}$. The coordinates $\mathbf{W}$ in F of the moving axis are given by

$$\mathbf{W} = [A]\mathbf{w}, \tag{9.8}$$

where $[A]$ represents the rotation of M relative to the fixed frame F.

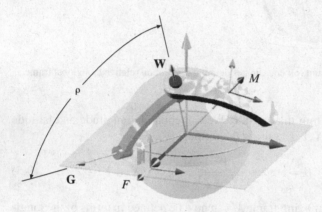

Fig. 9.2 The spherical RR chain with $\mathbf{G}$ as the fixed axis, $\mathbf{W}$ as the moving axis, and ρ is the angular dimension of the chain.

The angular length ρ of the RR chain is constant during the movement, so we have the condition

$$\mathbf{G} \cdot \mathbf{W} = |\mathbf{G}||\mathbf{W}| \cos\rho. \tag{9.9}$$

This must be true for any rotation $[A]$ of the end-link. This constraint characterizes the spherical RR chain.

9.3.1 Perpendicular Bisectors

Let n orientations of the frame M in the end-link of a spherical RR chain be defined by the rotations $[A_i]$, $i = 1, \ldots, n$. Each vector $\mathbf{W}^i$, which locates the moving axis in F for M in orientation M_i, satisfies (9.9). Therefore,

$$\mathbf{G} \cdot \mathbf{W}^i = |\mathbf{G}||\mathbf{W}^i| \cos\rho, \, i = 1, \ldots, n. \tag{9.10}$$

Because $|\mathbf{W}^1| = |\mathbf{W}^i|$, we can subtract the first equation from the remaining $n-1$ and obtain

$$\mathscr{P}_{1i}: \ \mathbf{G}\cdot(\mathbf{W}^i - \mathbf{W}^1) = 0, \ i = 2,\dots,n. \tag{9.11}$$

Recall from (8.30) that this is the equation of the perpendicular bisector $\mathscr{P}_{1i}$ to each segment $\mathbf{W}^i - \mathbf{W}^1$. Equation (9.11) states that all of these planes pass through $\mathbf{G}$. This is an algebraic expression of the fact that the moving axes $\mathbf{W}^i$ lie on right circular cone with $\mathbf{G}$ as its axis. The equations of these perpendicular bisectors form the *design equations* for a spherical RR chain.

9.3.2 The Spherical Dyad Triangle

The movement of a spherical RR chain can be viewed as a rotation about the moving pivot $\mathbf{W}^1$ by the angle α_{1i} that is followed by a rotation about $\mathbf{G}$ by the angle β_{1i}. The composition of these two rotations yields the rotation ϕ_{1i} about the relative rotation axis $\mathbf{S}_{1i}$ that moves the end-link M from orientation M_1 to M_i, that is,

$$[A(\phi_{1i}, \mathbf{S}_{1i})] = [A(\beta_{1i}, \mathbf{G})][A(\alpha_{1i}, \mathbf{W}^1)]. \tag{9.12}$$

Rodrigues's formula yields the equation of the *spherical dyad triangle* $\triangle \mathbf{W}^1 \mathbf{G} \mathbf{S}_{1i}$ as

$$\tan\frac{\phi_{1i}}{2}\mathbf{S}_{1i} = \frac{\tan\frac{\beta_{1i}}{2}\mathbf{G} + \tan\frac{\alpha_{1i}}{2}\mathbf{W}^1 + \tan\frac{\beta_{1i}}{2}\tan\frac{\alpha_{1i}}{2}\mathbf{G}\times\mathbf{W}^1}{1 - \tan\frac{\beta_{1i}}{2}\tan\frac{\alpha_{1i}}{2}\mathbf{G}\cdot\mathbf{W}^1}. \tag{9.13}$$

The spherical triangle $\triangle \mathbf{W}^1 \mathbf{G} \mathbf{S}_{1i}$ has two configurations relative to the plane containing the axes $\mathbf{G}$ and $\mathbf{W}^1$. In order to distinguish these configurations let $\mathbf{R}_{1i}$ be the point that coincides with the moving pivot $\mathbf{W}$ when the crank angle is $\beta_{1i}/2$. We can now view $\mathbf{S}_{1i}$ as being on the same, or opposite, side of $\mathbf{G}$ along the perpendicular bisector $\mathscr{P}_{1i}$ as the point $\mathbf{R}_{1i}$. If $\mathbf{S}_{1i}$ is on the same side as $\mathbf{R}_{1i}$, then it is above the plane through $\mathbf{G}\mathbf{W}^1$ and the sum of the joint angles $\alpha + \beta$ is less than 2π, see Figure 9.3. If $\mathbf{S}_{1i}$ is on the side opposite to $\mathbf{R}_{1i}$, then $\mathbf{S}_{1i}$ is below the plane through $\mathbf{G}\mathbf{W}^1$ and $\alpha + \beta$ is greater that 2π, Figure 9.4. In this case, if κ is the interior angle at $\mathbf{S}_{1i}$, then $\phi_{1i}/2 = \kappa + \pi$.

9.3.2.1 Quaternion Equations

We can use Hamilton's quaternions to define the spherical dyad triangle. Introduce the quaternions $G(\beta_{1i}/2) = \cos(\beta_{1i}/2) + \sin(\beta_{1i}/2)\mathbf{G}$ and $W^1(\alpha_{1i}/2) = \cos(\alpha_{1i}/2) + \sin(\alpha_{1i}/2)\mathbf{W}^1$ that represent rotations about the fixed and moving axes. The quaternion product yields

$$S_{1i}(\frac{\phi_{1i}}{2}) = G(\frac{\beta_{1i}}{2})W^1(\frac{\alpha_{1i}}{2}), \tag{9.14}$$

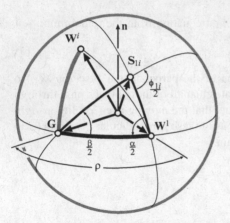

Fig. 9.3 The spherical dyad triangle $\triangle \mathbf{W}^1\mathbf{G}\mathbf{S}_{1i}$ formed by the fixed axis $\mathbf{G}$, the moving axis $\mathbf{W}^1$, and the rotation axis $\mathbf{S}_{1i}$.

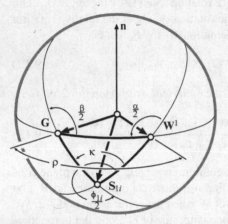

Fig. 9.4 The spherical dyad triangle $\triangle \mathbf{W}^1\mathbf{G}\mathbf{S}_{1i}$ with $\mathbf{S}_{1i}$ below the $\mathbf{G}\mathbf{W}^1$ plane.

where $S_{1i}(\phi_{1i}/2) = \cos(\phi_{1i}/2) + \sin\phi_{1i}/2\mathbf{S}_{1i}$ defines the relative rotation of the end-link. Separate the scalar and vector parts of this quaternion to obtain

$$\cos\frac{\phi_{1i}}{2} = \cos\frac{\beta_{1i}}{2}\cos\frac{\alpha_{1i}}{2} - \sin\frac{\beta_{1i}}{2}\sin\frac{\alpha_{1i}}{2}\mathbf{G}\cdot\mathbf{W}^1,$$

$$\sin\frac{\phi_{1i}}{2}\mathbf{S}_{1i} = \sin\frac{\beta_{1i}}{2}\cos\frac{\alpha_{1i}}{2}\mathbf{G} + \sin\frac{\alpha_{1i}}{2}\cos\frac{\beta_{1i}}{2}\mathbf{W}^1 + \sin\frac{\beta_{1i}}{2}\sin\frac{\alpha_{1i}}{2}\mathbf{G}\times\mathbf{W}^1.$$

$$(9.15)$$

Notice that (9.13) is obtained by dividing the vector equation by the scalar equation.

9.3.3 The Center-Axis Theorem

For three orientations of the moving body M_1, M_2, and M_3 we have three positions of the moving pivot $\mathbf{W}^i$, $i = 1, 2, 3$. The crank rotation angle about $\mathbf{G}$ between each of these positions is the dihedral angle β_{ij} between the planes defined by $\mathbf{GW}^i$ and $\mathbf{GW}^j$. Notice that $\beta_{13} = \beta_{23} + \beta_{12}$.

Recall that the plane $\mathscr{P}_{12}$ contains $\mathbf{GS}_{12}$ and bisects the segment $\mathbf{W}^1\mathbf{W}^2$. This means that the angle $\angle\mathbf{W}^1\mathbf{GS}_{12}$ equals $\beta_{12}/2$ or $\beta_{12}/2 + \pi$, depending on the location of $\mathbf{S}_{12}$ above or below the plane through $\mathbf{GW}^1$. Similarly, because $\mathscr{P}_{23}$ bisects the segment $\mathbf{W}^2\mathbf{W}^3$, we have that the angle $\angle\mathbf{W}^2\mathbf{GS}_{23}$ is either $\beta_{23}/2$ or $\beta_{23}/2 + \pi$. Notice that $\angle\mathbf{W}^1\mathbf{GS}_{12} = \angle\mathbf{S}_{12}\mathbf{GW}^2$. Considering each of the possible cases for $\angle\mathbf{S}_{12}\mathbf{GS}_{23} = \angle\mathbf{S}_{12}\mathbf{GW}^2 + \angle\mathbf{W}^2\mathbf{GS}_{23}$, we see that this angle must be either $\beta_{13}/2$ or $\beta_{13}/2 + \pi$. Thus, $\mathbf{G}$ views the two relative rotation axes $\mathbf{S}_{12}$ and $\mathbf{S}_{23}$ in the angle $\beta_{13}/2$ or $\beta_{13}/2 + \pi$, Figure 9.5.

This generalizes to yield a spherical version of the planar center-point theorem:

Theorem 10 (The Center-Axis Theorem). *The center axis $\mathbf{G}$ of an RR chain that reaches orientations M_i, M_j, and M_k views the relative rotation axes $\mathbf{S}_{ij}$ and $\mathbf{S}_{jk}$ in the dihedral angle $\beta_{ik}/2$ or $\beta_{ik}/2 + \pi$, where β_{ik} is the crank rotation angle from position M_i to M_k.*

This theorem provides the foundation for the generalization of Burmester's techniques to spherical RR chain design.

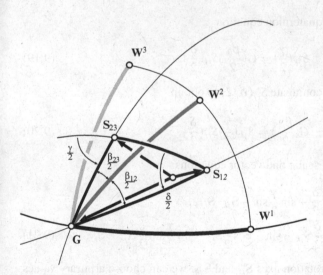

Fig. 9.5 The center axis $\mathbf{G}$ views the relative rotation axes $\mathbf{S}_{12}$ and $\mathbf{S}_{23}$ in the angle $\beta_{13}/2 = \beta_{12}/2 + \beta_{23}/2$.

9.3.3.1 The Dihedral Angle at G

The center-axis theorem provides a formula to compute the crank angle β_{ik} given the relative rotation axes $\mathbf{S}_{ij}$ and $\mathbf{S}_{jk}$. Consider the vectors $\mathbf{G} \times \mathbf{S}_{ij}$ and $\mathbf{G} \times \mathbf{S}_{jk}$ that are normal, respectively, to the planes containing $\mathbf{GS}_{ij}$ and $\mathbf{GS}_{jk}$. The angle between these vectors is given by (8.47) as

$$\frac{\beta_{ik}}{2} = \arctan\left(\frac{(\mathbf{G} \times \mathbf{S}_{ij}) \cdot \mathbf{S}_{jk}}{(\mathbf{G} \times \mathbf{S}_{ij}) \cdot (\mathbf{G} \times \mathbf{S}_{jk})}\right). \tag{9.16}$$

We use this equation to compute β_{ik} for a given set of relative rotation axes $\mathbf{S}_{ij}$ and $\mathbf{S}_{jk}$. This equation remains true whether $\mathbf{G}$ views these axes in $\beta_{ik}/2$ or $\beta_{ik}/2 + \pi$, because $\tan\kappa = \tan(\kappa + \pi)$.

Alternatively, we can specify β_{ik} in this equation and obtain

$$\mathscr{C}_{ik}: \ (\mathbf{G} \times \mathbf{S}_{ij}) \cdot (\mathbf{G} \times \mathbf{S}_{jk}) \tan\frac{\beta_{ik}}{2} - (\mathbf{G} \times \mathbf{S}_{ij}) \cdot \mathbf{S}_{jk} = 0, \tag{9.17}$$

which is a quadric cone on which the fixed axes $\mathbf{G}$ must lie.

Another approach to defining the crank rotation angle β_{ik} is to use the quaternion form of the spherical triangle. Assume that the spherical triangle $\triangle \mathbf{S}_{ij}\mathbf{GS}_{jk}$ is the pole triangle for the composite rotation

$$[A(\gamma, \mathbf{S}_{jk})] = [A(\beta_{ik}, \mathbf{G})][A(\delta, \mathbf{S}_{ij})]. \tag{9.18}$$

Associated with this is the quaternion equation

$$S_{jk}(\frac{\gamma}{2}) = G(\frac{\beta_{ik}}{2})S_{ij}(\frac{\delta}{2}). \tag{9.19}$$

Multiply both sides by the conjugate $S_{ij}^*(\delta/2)$ to obtain

$$G(\frac{\beta_{ik}}{2}) = S_{jk}(\frac{\gamma}{2})S_{ij}^*(\frac{\delta}{2}), \tag{9.20}$$

which expands to yield the scalar and vector equations

$$\cos\frac{\beta_{ik}}{2} = \cos\frac{\gamma}{2}\cos\frac{\delta}{2} + \sin\frac{\gamma}{2}\sin\frac{\delta}{2}\mathbf{S}_{jk}\cdot\mathbf{S}_{ij},$$
$$\sin\frac{\beta_{ik}}{2}\mathbf{G} = \sin\frac{\gamma}{2}\cos\frac{\delta}{2}\mathbf{S}_{jk} - \sin\frac{\delta}{2}\cos\frac{\gamma}{2}\mathbf{S}_{ij} - \sin\frac{\gamma}{2}\sin\frac{\delta}{2}\mathbf{S}_{jk}\times\mathbf{S}_{ij}. \tag{9.21}$$

For a given set of relative rotation axes $\mathbf{S}_{ij}$ and $\mathbf{S}_{jk}$, we can choose arbitrary values for γ and δ and obtain the axis $\mathbf{G}$ and the rotation angle β_{ik}.

On the other hand, if we specify β_{ik}, then we can determine δ as a function of γ from the scalar equation of (9.21). Notice that this equation has the form

$$A\cos\frac{\delta}{2} + B\sin\frac{\delta}{2} = C, \tag{9.22}$$

where

$$A = \cos\frac{\gamma}{2}, \quad B = \sin\frac{\gamma}{2}\mathbf{S}_{jk}\cdot\mathbf{S}_{ij}, \quad \text{and} \quad C = \cos\frac{\beta_{ik}}{2}. \tag{9.23}$$

The solution is

$$\delta(\gamma) = 2\arctan\frac{B}{A} \pm 2\arccos\frac{C}{\sqrt{A^2 + B^2}}. \tag{9.24}$$

Substitute this into the vector equation of (9.21) in order to obtain a cone of fixed axes $\mathbf{G}$ parameterized by γ. Each of these axes views $\mathbf{S}_{ij}\mathbf{S}_{jk}$ in the dihedral angle $\beta_{ik}/2$ or $\beta_{ik}/2 + \pi$.

9.4 Finite Position Synthesis of RR Chains

9.4.1 The Algebraic Design Equations

The equations of the perpendicular bisectors $\mathscr{P}_{1i}$ can be modified to provide a convenient set of algebraic design equations for a spherical RR chain. If n task orientations M_i, $i = 1,\dots,n$, are specified for the end-link of the chain, then we can determine the relative rotation matrices $[A_{1i}]$, $i = 2,\dots,n$, such that $\mathbf{W}^i = [A_{1i}]\mathbf{W}^1$. This allows us to write the equations of the perpendicular bisectors (9.11) as

$$\mathscr{P}_{1i}: \ \mathbf{G}\cdot[A_{1i} - I]\mathbf{W}^1 = 0, i = 2,\dots,n. \tag{9.25}$$

This is a homogeneous bilinear equation in the six unknown coordinates for $\mathbf{G} = (x,y,z)^T$ and $\mathbf{W}^1 = (\lambda,\mu,\nu)^T$. In what follows we solve these equations to obtain a spherical RR chain that reaches five task orientations.

To see the structure of these design equations in more detail, we use (8.55) to write $[A_{1i}]$ in terms of its rotation axis $\mathbf{S}_{1i}$ and rotation angle ϕ_{1i}. Introducing $\phi_{1i}/2$ into the resulting equation for $[A_{1i} - I]$, we obtain

$$[A_{1i} - I] = 2\sin\frac{\phi_{1i}}{2}\cos\frac{\phi_{1i}}{2}\left([S_{1i}] + \tan\frac{\phi_{1i}}{2}[S_{1i}]^2\right). \tag{9.26}$$

Write the components of the rotation axis such that $\mathbf{S}_{1i} = (p_i, q_i, r_i)^T$. Then a typical equation (9.25) is

$$\begin{Bmatrix}x\\y\\z\end{Bmatrix}^T\begin{bmatrix}-\tan\frac{\phi_{1i}}{2}(r_i^2+q_i^2) & -r_i+\tan\frac{\phi_{1i}}{2}q_ip_i & q_i+\tan\frac{\phi_{1i}}{2}r_ip_i\\ r_i+\tan\frac{\phi_{1i}}{2}q_ip_i & -\tan\frac{\phi_{1i}}{2}(r_i^2+p_i^2) & -p_i+\tan\frac{\phi_{1i}}{2}r_iq_i\\ -q_i+\tan\frac{\phi_{1i}}{2}r_ip_i & p_i+\tan\frac{\phi_{1i}}{2}r_iq_i & -\tan\frac{\phi_{1i}}{2}(q_i^2+p_i^2)\end{bmatrix}\begin{Bmatrix}\lambda\\\mu\\\nu\end{Bmatrix} = 0.$$
$$\tag{9.27}$$

If the vector $\mathbf{G} = (x, y, z)^T$ is considered to be known, then these equations are linear and homogeneous in the components of $\mathbf{W}^1 = (\lambda, \mu, \nu)^T$. In the same way, if $\mathbf{W}^1$ is known, then these equations are linear and homogeneous in the components of $\mathbf{G}$. This structure provides a convenient strategy for solving these equations.

9.4.1.1 The Bilinear Structure

The solution of the design equations for spherical RR chains follows closely the results for planar RR chains. As in the plane, these equations are linear separately in the coordinates of the fixed and moving axes. This bilinear structure allows us to consider selecting the fixed axis or the moving axis as part of the solution process.

If we select the fixed axis $\mathbf{G}$, then x, y, z are known, and we can collect coefficients of λ, μ, ν to obtain the design equations

$$A_i \lambda + B_i \mu + C_i \nu = 0, \, i = 2, \dots, n, \tag{9.28}$$

where

$$A_i = -\tan\frac{\phi_{1i}}{2}(r_i^2 + q_i^2)x + (r_i + \tan\frac{\phi_{1i}}{2}q_i p_i)y + (-q_i + \tan\frac{\phi_{1i}}{2}r_i p_i)z,$$

$$B_i = (-r_i + \tan\frac{\phi_{1i}}{2}q_i p_i)x - \tan\frac{\phi_{1i}}{2}(r_i^2 + p_i^2)y + (p_i + \tan\frac{\phi_{1i}}{2}r_i q_i)z,$$

$$C_i = (q_i + \tan\frac{\phi_{1i}}{2}r_i p_i)x + (-p_i + \tan\frac{\phi_{1i}}{2}r_i q_i)y - \tan\frac{\phi_{1i}}{2}(q_i^2 + p_i^2)z. \tag{9.29}$$

This is a set of $n - 1$ linear homogeneous equations in the components of the moving axis $\mathbf{W}^1$.

Rather than select the fixed axis, we can specify a moving axis $\mathbf{W}^1 = (\lambda, \mu, \nu)^T$ and collect the coefficients of (x, y, z) to obtain

$$A_i' x + B_i' y + C_i' z = 0, \, i = 2, \dots, n, \tag{9.30}$$

where

$$A_i' = -\tan\frac{\phi_{1i}}{2}(r_i^2 + q_i^2)\lambda + (-r_i + \tan\frac{\phi_{1i}}{2}q_i p_i)\mu + (q_i + \tan\frac{\phi_{1i}}{2}r_i p_i)\nu,$$

$$B_i' = (r_i + \tan\frac{\phi_{1i}}{2}q_i p_i)\lambda - \tan\frac{\phi_{1i}}{2}(r_i^2 + p_i^2)\mu + (-p_i + \tan\frac{\phi_{1i}}{2}r_i q_i)\nu,$$

$$C_i' = (-q_i + \tan\frac{\phi_{1i}}{2}r_i p_i)\lambda + (p_i + \tan\frac{\phi_{1i}}{2}r_i q_i)\mu - \tan\frac{\phi_{1i}}{2}(q_i^2 + p_i^2)\nu. \tag{9.31}$$

We solve these $n - 1$ equations to determine the coordinates of the fixed axis $\mathbf{G}$.

9.4.2 Parameterized Form of the Design Equations

The movement of the end-link of an RR chain from orientation M_1 to each of the orientations M_i, $i = 2, \ldots, n$, is defined by the composite rotations

$$[A(\phi_{1i}, \mathbf{S}_{1i})] = [A(\beta_{1i}, \mathbf{G})][A(\alpha_{1i}, \mathbf{W}^1)], i = 2, \ldots, n, \qquad (9.32)$$

where β_{1i} and α_{1i} are the rotation angles about the fixed and moving axes, respectively. The quaternion form of these equations are

$$S(\frac{\phi_{1i}}{2}) = G(\frac{\beta_{1i}}{2})W^1(\frac{\alpha_{1i}}{2}), i = 2, \ldots, n. \qquad (9.33)$$

The scalar and vector parts of these quaternions provide design equations in which the joint rotation angles β_{1i} and α_{1i} appear explicitly. They are used to design spherical RR chains for selected crank rotation angles.

9.4.3 Two Specified Orientations

If two orientations M_1 and M_2 are specified for the end-link of a spherical RR chain, then we have the rotations $[A_1]$ and $[A_2]$. From these rotations we construct the relative rotation matrix $[A_{12}]$ and obtain the relative rotation angle ϕ_{12} and axis $\mathbf{S}_{12}$. This is the information that we need to form the single design equation

$$\mathbf{G} \cdot [A(\phi_{12}, \mathbf{S}_{12}) - I]\mathbf{W}^1 = 0. \qquad (9.34)$$

We may choose either the fixed or moving axis and solve for the other.

9.4.3.1 Select the Fixed Axis

If we specify the coordinates $\mathbf{G} = (x, y, z)^T$, then the design equation (9.28) becomes

$$A_2\lambda + B_2\mu + C_2v = 0. \qquad (9.35)$$

This is the equation of a plane through the origin of F. Any vector in this plane can be used as a moving pivot $\mathbf{W}^1$. Set v equal to 1, so this plane defines a line in the $z = 1$ plane. We can then choose either λ or μ and solve for the other.

9.4.3.2 Select the Moving Axis

A similar result is obtained if we choose the moving axis $\mathbf{W}^1 = (\lambda, \mu, v)^T$. In this case the design equation (9.30) becomes

$$A'_2 x + B'_2 y + C'_2 z = 0. \tag{9.36}$$

This equation defines the perpendicular bisector to the segment $\mathbf{W}^1\mathbf{W}^2$. Set $z = 1$, so this equation defines a line in the $z = 1$ plane. Then choose either x or y and solve for the other.

9.4.3.3 Select the Crank Angle

The quaternion equation of the dyad triangle can be used to design a spherical RR chain with a specified crank rotation angle β_{12}. The parameterized design equation is

$$S(\frac{\phi_{12}}{2}) = G(\frac{\beta_{12}}{2})W^1(\frac{\alpha_{12}}{2}). \tag{9.37}$$

For the specified orientations, we can determine the quaternion $S(\phi_{12}/2)$. We now assume that we have specified the fixed axis $\mathbf{G}$ and the crank angle β_{12}, so we have the quaternion $G(\beta_{12}/2)$. Multiply this equation on the left by the conjugate G^* to obtain a formula defining the moving axis $\mathbf{W}^1$ and the rotation angle α_{12},

$$W^1(\frac{\alpha_{12}}{2}) = G^*(\frac{\beta_{12}}{2})S(\frac{\phi_{12}}{2}). \tag{9.38}$$

Recall that the conjugate G^* is obtained by negating its vector part, or equivalently, by negating the angle β_{12}.

We can select $\mathbf{W}^1$ and α_{12} instead. Then $\mathbf{G}$ and β_{12} are determined by multiplying (9.37) on the right by the conjugate W^{1*}.

9.4.4 Three Specified Orientations

Given three task orientations M_1, M_2, and M_3, we have three rotation matrices $[A_i]$, $i = 1,2,3$, which we use to construct the relative rotations $[A(\phi_{12},\mathbf{S}_{12})]$ and $[A(\phi_{13},\mathbf{S}_{13})]$. The result is the two design equations

$$\mathbf{G} \cdot [A(\phi_{12},\mathbf{S}_{12}) - I]\mathbf{W}^1 = 0,$$
$$\mathbf{G} \cdot [A(\phi_{13},\mathbf{S}_{13}) - I]\mathbf{W}^1 = 0. \tag{9.39}$$

Selecting one axis, these equations yield a unique solution for the other.

9.4.4.1 Select the Fixed Axis

For a selected direction of the fixed axis $\mathbf{G}$, the design equations (9.28) can be assembled into the matrix equation

$$\begin{bmatrix} A_2 & B_2 & C_2 \\ A_3 & B_3 & C_3 \end{bmatrix} \begin{Bmatrix} \lambda \\ \mu \\ \nu \end{Bmatrix} = \begin{Bmatrix} 0 \\ 0 \end{Bmatrix}. \tag{9.40}$$

This equation is solved by forming the two vectors $\mathbf{D}_1 = (A_1, B_1, C_1)^T$ and $\mathbf{D}_2 = (A_2, B_2, C_2)^T$. The result is

$$\mathbf{W}^1 = k\mathbf{D}_1 \times \mathbf{D}_2, \tag{9.41}$$

where k is used to normalize this to a unit vector.

9.4.4.2 Select the Moving Axis

For a selected moving axis $\mathbf{W}^1$, we have from (9.30) the equations for the fixed axis $\mathbf{G}$ given by

$$\begin{bmatrix} A_2' & B_2' & C_2' \\ A_3' & B_3' & C_3' \end{bmatrix} \begin{Bmatrix} x \\ y \\ z \end{Bmatrix} = \begin{Bmatrix} 0 \\ 0 \end{Bmatrix}. \tag{9.42}$$

These equations define the perpendicular bisectors to the segments $\mathbf{W}^1\mathbf{W}^2$ and $\mathbf{W}^1\mathbf{W}^3$. The two planes intersect in the axis $\mathbf{G}$. Solve these equations in the same way as shown above for the moving axis.

9.4.4.3 Select the Crank Angle

Given three orientations M_i, $i = 1, 2, 3$, we have the spherical pole triangle $\triangle S_{12}S_{23}S_{13}$. The center-axis theorem shows that $\mathbf{G}$ views the sides $S_{13}S_{23}$ and $S_{12}S_{23}$ in the angles $\beta_{12}/2$ and $\beta_{13}/2$, respectively. Given the crank angles β_{12} and β_{13}, we can use (9.17) to determine two quadric cones that intersect to define $\mathbf{G}$. They are

$$\mathscr{C}_{12}: \quad (\mathbf{G} \times \mathbf{S}_{13}) \cdot (\mathbf{G} \times \mathbf{S}_{23}) \tan\frac{\beta_{12}}{2} - (\mathbf{G} \times \mathbf{S}_{13}) \cdot \mathbf{S}_{23} = 0,$$

$$\mathscr{C}_{13}: \quad (\mathbf{G} \times \mathbf{S}_{12}) \cdot (\mathbf{G} \times \mathbf{S}_{23}) \tan\frac{\beta_{13}}{2} - (\mathbf{G} \times \mathbf{S}_{12}) \cdot \mathbf{S}_{23} = 0. \tag{9.43}$$

The simultaneous solution of these equations yields the desired fixed pivot $\mathbf{G}$. The moving pivot $\mathbf{W}^1$ is then calculated using (9.41).

Another approach is to use the quaternion equations of the triangles $\triangle S_{13}\mathbf{G}S_{23}$ and $\triangle S_{12}\mathbf{G}S_{23}$, given by

$$G(\frac{\beta_{12}}{2}) = S_{23}(\frac{\gamma_1}{2})S_{13}^*(\frac{\delta_1}{2}),$$

$$G(\frac{\beta_{13}}{2}) = S_{23}(\frac{\gamma_2}{2})S_{12}^*(\frac{\delta_2}{2}). \tag{9.44}$$

For the specified angles β_{12} and β_{13} these equations can be used to determine parameterized equations of the quadric cones $\mathscr{C}_{12}(\gamma_1)$ and $\mathscr{C}_{13}(\gamma_2)$. This is done by solving the scalar equations of the quaternions to determine the angles δ_1 and δ_2 in terms of γ_1 and γ_2 using (9.24). The parameterized version of the cone $\mathscr{C}_{12}(\gamma_1)$ can be substitute into the algebraic equation for $\mathscr{C}_{13}$ to obtain an equation for γ_1 that identifies the desired **G**.

9.4.5 Four Specified Orientations

In the previous section we found that for three task orientations every fixed axis is in one-to-one correspondence with a moving axis. Geometrically, this results from the fact that the perpendicular bisectors of the segments $\mathbf{W}^1\mathbf{W}^2$ and $\mathbf{W}^1\mathbf{W}^3$ always intersect in a line. In the case of four task orientations, the perpendicular bisector of $\mathbf{W}^1\mathbf{W}^4$ must also pass through this line. However, this cannot happen in general. This does not mean that there are no axes that have this property. In fact, there is a cubic cone of axes **G** that are centers for four moving axes $\mathbf{W}^1, i = 1,2,3,4$, that we can use to form RR chains that reach four orientations. This is the *center-axis cone*. The available moving axes form a cubic cone known as the *circling-axis cone*.

9.4.5.1 The Center-Axis Cone

Given four orientations M_i, $i = 1,2,3,4$, we have the relative rotations $[A(\phi_{1i}, \mathbf{S}_{1i})]$ and the design equations

$$\begin{bmatrix} A_2 & B_2 & C_2 \\ A_3 & B_3 & C_3 \\ A_4 & B_4 & C_4 \end{bmatrix} \begin{Bmatrix} \lambda \\ \mu \\ \nu \end{Bmatrix} = \begin{Bmatrix} 0 \\ 0 \\ 0 \end{Bmatrix}, \tag{9.45}$$

where A_i, B_i, and C_i are given by (9.28). In order to have a solution the 3×3 coefficient matrix $[M] = [A_i, B_i, C_i]$ of this equation must have the determinant $|M|$ equal to zero. This yields a cubic polynomial

$$\begin{aligned} \mathscr{R}(x,y,z): \ |M| = {}& a_{30}y^3 + (a_{21}x + a_{20}z)y^2 + (a_{12}x^2 + a_{11}xz + a_{10}z^2)y \\ & + a_{03}x^3 + a_{02}x^2z + a_{10}xz^2 + a_{00}z^3 = 0. \end{aligned} \tag{9.46}$$

This polynomial defines a cubic cone in the fixed frame, and any line on this cone may be chosen as a fixed axis **G** for the spherical RR chain. It is known as the *center-axis cone*.

The coefficients a_{ij} can be obtained by noting that each element of the coefficient matrix $[M]$ is linear in x, y, z, therefore $|M|$ has the form

$$|M| = \left| \mathbf{a}_1x + \mathbf{b}_1y + \mathbf{c}_1z, \ \mathbf{a}_2x + \mathbf{b}_2y + \mathbf{c}_2z, \ \mathbf{a}_3x + \mathbf{b}_3y + \mathbf{c}_3z \right| = 0, \tag{9.47}$$

where a_i, b_i, and c_i are constants defined by the task orientations. The linearity of the determinant allows an expansion that is identical to that presented in (5.37).

9.4.5.2 The Burmester–Roth Theorem

Roth [104] generalized Burmester's planar synthesis theory to spatial displacements in a way that included the design of spherical RR chains. He showed how to construct the spherical equivalent of Burmester's opposite-pole quadrilateral, that we call the *complementary-axis quadrilateral*, Figure 9.6. The fundamental result is that center axes must view opposite sides of the complementary-axis quadrilateral in angles that are equal or differ by π. This provides a way to generate the equation of the center-axis cone directly from the coordinates of the relative rotation axes.

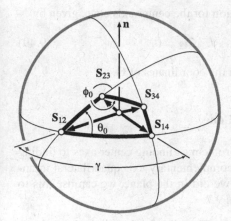

Fig. 9.6 The spherical complementary-axis quadrilateral $\mathcal{Q}$: $S_{12}S_{23}S_{34}S_{14}$.

Four task orientations define six relative rotation axes S_{ij}, $i < j = 1, 2, 3, 4$. Following Burmester's definition of the planar opposite-pole quadrilateral, Roth defines the three complementary pairs of relative rotation axes, $S_{12}S_{34}$, $S_{13}S_{24}$ and $S_{14}S_{23}$. The spherical *complementary-axis quadrilateral* is constructed from two sets of complementary pairs so that the pairs are opposite to each other along the diagonals of the quadrilateral. The fundamental result is the following:

Theorem 11 (The Burmester–Roth Theorem). *The center axis* G *of a spherical RR chain that can reach four specified orientations views opposite sides of a complementary-axis quadrilateral constructed from the relative rotation axes of the given orientations in angles that are equal, or differ by π.*

Proof. The definition of the complementary-axis quadrilateral ensures that opposite sides have the form $S_{ij}S_{ik}$ and $S_{mj}S_{mk}$. The center-axis theorem states that G views $S_{ij}S_{ik}$ in the dihedral angle $\beta_{jk}/2$ or $\beta_{jk}/2 + \pi$, where β_{jk} is the crank rotation of from position M_j to M_k. Similarly, it must view the $S_{mj}S_{mk}$ in $\beta_{jk}/2$ or $\beta_{jk}/2 + \pi$.

Consider the various combinations to see that **G** views these sides in angles that are equal, or differ by π. $\square$

This theorem provides a convenient derivation for the center-axis cone in terms of the coordinates of the relative rotation axes of the complementary-axis quadrilateral. Let the vertices of this quadrilateral be $\mathbf{S}_{12}\mathbf{S}_{23}\mathbf{S}_{34}\mathbf{S}_{14}$. Then the dihedral angle κ between the planes containing $\mathbf{GS}_{12}$ and $\mathbf{GS}_{23}$ must be the same as the angle between the planes $\mathbf{GS}_{14}$ and $\mathbf{GS}_{34}$, or differ by π. We compute $\tan\kappa = \tan(\kappa+\pi)$ to obtain

$$\tan\kappa = \frac{\mathbf{G}\times\mathbf{S}_{12}\cdot\mathbf{S}_{23}}{(\mathbf{G}\times\mathbf{S}_{12})\cdot(\mathbf{G}\times\mathbf{S}_{23})} = \frac{L_{12}}{C_{12}},$$

$$\tan\kappa = \frac{\mathbf{G}\times\mathbf{S}_{14}\cdot\mathbf{S}_{34}}{(\mathbf{G}\times\mathbf{S}_{14})\cdot(\mathbf{G}\times\mathbf{S}_{34})} = \frac{L_{34}}{C_{34}}. \tag{9.48}$$

Equate these expressions to obtain an equation for the center-axis cone, given by

$$\mathscr{R}(x,y,z): \; L_{12}C_{34} - L_{34}C_{12} = 0. \tag{9.49}$$

This is a homogeneous cubic polynomial in the coordinates of **G**.

9.4.5.3 The Parameterized Center-Axis Cone

The Burmester-Roth theorem reduces the problem of finding center axes to finding those axes that view opposite sides of the complementary-axis quadrilateral in angles that are equal, or differ by π. Just as we did in the plane, we can use this to obtain a construction for center axes, Figure 9.7.

Theorem 12 (Construction of Center Axes). *The axes that satisfy the Burmester–Roth theorem are obtained as follows:*

1. *Construct the complementary-axis quadrilateral $\mathscr{Q}$: $\mathbf{S}_{12}\mathbf{S}_{23}\mathbf{S}_{34}\mathbf{S}_{14}$ using the four task orientations.*
2. *Rotate the segment $\mathbf{S}_{12}\mathbf{S}_{23}$ by an angle θ about $\mathbf{S}_{12}$ and determine the new configuration $\mathscr{Q}'$ in order to obtain $\mathbf{S}'_{23}$ and $\mathbf{S}'_{34}$.*
3. *The axis $\mathbf{G}$ of the rotation of $\mathbf{S}'_{23}\mathbf{S}'_{34}$ from its original location $\mathbf{S}_{23}\mathbf{S}_{34}$ satisfies the Burmester–Roth theorem and is a center axis.*

Proof. Let **G** be the intersection of the perpendicular bisectors $\mathscr{V}_1 = (\mathbf{S}_{23}\mathbf{S}'_{23})^\perp$ and $\mathscr{V}_2 = (\mathbf{S}_{34}\mathbf{S}'_{34})^\perp$. Then **G** is the axis of rotation of the segment $\mathbf{S}_{23}\mathbf{S}_{34}$ by an angle κ to the position $\mathbf{S}'_{23}\mathbf{S}'_{34}$. The input crank formed by $\mathbf{S}_{12}\mathbf{S}_{23}$ has the dyad triangle $\triangle\mathbf{S}_{23}\mathbf{S}_{12}\mathbf{G}$ and **G** must view the $\mathbf{S}_{12}\mathbf{S}_{23}$ in the angle $\kappa/2$ or $\kappa/2+\pi$. Similarly, the geometry of the dyad triangle $\triangle\mathbf{S}_{34}\mathbf{S}_{14}\mathbf{G}$ requires that **G** view the segment $\mathbf{S}_{14}\mathbf{S}_{34}$ in either $\kappa/2$ or $\kappa/2+\pi$. Thus, **G** views the opposite sides $\mathbf{S}_{12}\mathbf{S}_{23}$ and $\mathbf{S}_{14}\mathbf{S}_{34}$ in angles that are equal, or differ by π. The same argument shows that **G** views the other two sides $\mathbf{S}_{23}\mathbf{S}_{34}$ and $\mathbf{S}_{12}\mathbf{S}_{14}$ in angles that are equal, or differ by π. Thus, **G** satisfies Burmester's theorem. $\square$

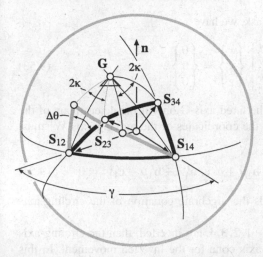

Fig. 9.7 Construction of the center-axis $\mathbf{G}$ using the spherical complementary-axis quadrilateral.

This construction yields a parameterized equation for the center-axis cone. The rotation of the coupler segment $\mathbf{S}_{23}\mathbf{S}_{34}$ from its initial configuration $\mathscr{Q}$ to another configuration $\mathscr{Q}'$ is the composite of a rotation by $\Delta\phi = \phi - \phi_0$ about $\mathbf{S}_{23}$, followed by a rotation of $\Delta\theta = \theta - \theta_0$ about $\mathbf{S}_{12}$. This composite rotation is given by

$$[G(\beta,\mathbf{G})] = [A(\Delta\theta,\mathbf{S}_{12})][A(\Delta\phi,\mathbf{S}_{23})]. \tag{9.50}$$

Rodrigues's formula yields the equation for the relative rotation axis $\mathbf{G}$ as

$$\tan\frac{\beta}{2}\mathbf{G} = \frac{\tan\frac{\Delta\theta}{2}\mathbf{S}_{12} + \tan\frac{\Delta\phi}{2}\mathbf{S}_{23} + \tan\frac{\Delta\theta}{2}\tan\frac{\Delta\phi}{2}\mathbf{S}_{12}\times\mathbf{S}_{23}}{1 - \tan\frac{\Delta\theta}{2}\tan\frac{\Delta\phi}{2}\mathbf{S}_{12}\cdot\mathbf{S}_{23}}. \tag{9.51}$$

The coupler angle $\phi(\theta)$ is determined from the driving crank rotation θ, using equation (7.36). We compute $\theta = \theta_0$ and $\phi = \phi_0$ in the initial configuration $\mathscr{Q}$ from the formulas

$$\tan\theta_0 = \frac{(\mathbf{S}_{12}\times\mathbf{S}_{14})\cdot\mathbf{S}_{23}}{(\mathbf{S}_{12}\times\mathbf{S}_{14})\cdot(\mathbf{S}_{12}\times\mathbf{S}_{23})}, \tan\phi_0 = \frac{(\mathbf{S}_{23}\times\mathbf{S}_{12})\cdot\mathbf{S}_{34}}{(\mathbf{S}_{23}\times\mathbf{S}_{12})\cdot(\mathbf{S}_{23}\times\mathbf{S}_{34})}. \tag{9.52}$$

The result is a formula (9.51) that generate the center-axis cone by varying the parameter θ.

9.4.5.4 The Circling-Axis Cone

Associated with each point on the center-axis cone is a solution of the design equations (9.45), which yields a moving axis $\mathbf{W}^1$. These axes form another cubic cone, called the *circling-axis cone*. We can compute this cone directly from the design equations (9.30). For a four position task, we have

$$\begin{bmatrix} A_2' & B_2' & C_2' \\ A_3' & B_3' & C_3' \\ A_4' & B_4' & C_4' \end{bmatrix} \begin{Bmatrix} x \\ y \\ z \end{Bmatrix} = \begin{Bmatrix} 0 \\ 0 \\ 0 \end{Bmatrix}. \tag{9.53}$$

These equations have a solution for the fixed axis $\mathbf{G}$ only if the determinant of the coefficient matrix $[M']$ is zero. Thus, the coordinates of the moving axis $\mathbf{W}^1$ must satisfy the condition

$$|M'| = \left| \mathbf{a}_1'\lambda + \mathbf{b}_1'\mu + \mathbf{c}_1'\nu, \; \mathbf{a}_2'\lambda + \mathbf{b}_2'\mu + \mathbf{c}_2'\nu, \; \mathbf{a}_3'\lambda + \mathbf{b}_3'\mu + \mathbf{c}_3'\nu \right| = 0. \tag{9.54}$$

An expansion similar to (5.37) yields the algebraic equation of the circling-axis cone.

If the four task orientations M_i, $i = 1, 2, 3, 4$, are inverted, then the circling-axis cone can be computed as the center-axis cone for the inverted movement. In this case, the cone is defined in the moving frame M. We then transform the coordinates of this curve to the fixed frame in the first orientation M_1 to obtain the set of moving pivots $\mathbf{W}^1$.

The complementary-axis quadrilateral for the inverted relative displacements in position M_1 is constructed from the image poles $\mathbf{S}_{12}^1$, $\mathbf{S}_{23}^1$, $\mathbf{S}_{34}^1$, and $\mathbf{S}_{14}^1$. Recall that $\mathbf{S}_{12}^1 = \mathbf{S}_{12}$ and $\mathbf{S}_{14}^1 = \mathbf{S}_{14}$, and further that $\mathbf{S}_{23}^1$ and $\mathbf{S}_{34}^1$ are the reflections of $\mathbf{S}_{23}$ and $\mathbf{S}_{34}$ through the planes containing $\mathbf{S}_{12}\mathbf{S}_{13}$ and $\mathbf{S}_{14}\mathbf{S}_{13}$, respectively. Thus, the image quadrilateral $\mathcal{Q}^\dagger : \mathbf{S}_{12}\mathbf{S}_{23}^1\mathbf{S}_{34}^1\mathbf{S}_{14}$ has the same dimensions as the complementary-axis quadrilateral $\mathcal{Q}$, and the same ground link $\mathbf{S}_{12}\mathbf{S}_{14}$, Figure 9.8.

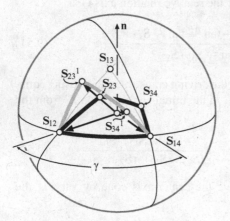

Fig. 9.8 The image complementary-axis quadrilateral $\mathcal{Q}^\dagger$ is obtained from $\mathcal{Q}$ by reflecting $\mathbf{S}_{23}$ through the plane $\mathbf{S}_{12}\mathbf{S}_{13}$ and $\mathbf{S}_{34}$ through $\mathbf{S}_{14}\mathbf{S}_{13}$.

The circling-axis cone can now be determined by applying the Burmester–Roth theorem to the image quadrilateral $\mathcal{Q}^\dagger$. The result is a parameterized equation for the moving axes $\mathbf{W}^1$.

9.4.6 Five Specified Orientations

Given five spatial orientations of a rigid body, we can determine as many as six spherical RR chains that can reach these positions. The solution procedure follows closely the solution of the planar design equations for five positions. We solve the design equations (9.25) using a two-step elimination procedure that yields a sixth degree polynomial. We can also solve this problem by finding the intersections of two center-axis cones.

9.4.6.1 Algebraic Elimination

Collect the coefficients of λ, μ, ν in the design equations to obtain

$$\begin{bmatrix} A_2 & B_2 & C_2 \\ A_3 & B_3 & C_3 \\ A_4 & B_4 & C_4 \\ A_5 & B_5 & C_5 \end{bmatrix} \begin{Bmatrix} \lambda \\ \mu \\ \nu \end{Bmatrix} = \begin{Bmatrix} 0 \\ 0 \\ 0 \end{Bmatrix}, \tag{9.55}$$

where A_i, B_i, and C_i are defined in (9.28). In order for this system of equations to have a solution the rank of the 4×3 coefficient matrix $[M]$ must be two.

Let $\mathcal{R}_j$ be the determinant of the 3×3 matrix formed from $[M]$ by removing row $5 - j$, so $\mathcal{R}_1$ is computed using the first three rows, $\mathcal{R}_2$ is obtained from the first two and last row, and so on. The result is four homogeneous cubic polynomials in x, y, and z, identical in structure to the center-axis cone (9.49),

$$\mathcal{R}_j(x,y,z): a_{30,j}y^3 + (a_{21,j}x + a_{20,j}z)y^2 + (a_{12,j}x^2 + a_{11,j}xz + a_{10,j}z^2)y$$
$$+ a_{03,j}x^3 + a_{02,j}x^2z + a_{01,j}xz^2 + a_{00,j}z^3 = 0, j = 1,2,3,4. \tag{9.56}$$

In the next step, we dehomogenize these equations by setting $z = 1$, and eliminate y to obtain a single polynomial in x.

The polynomials $\mathcal{R}_j$ are homogeneous in x, y, and z, which means that if $\mathbf{G} = (x,y,z)^T$ is a solution of these equations, then $\mathbf{G}' = k(x,y,z)^T$ is a solution as well. Therefore, we can set $z = 1$ and solve for the unique values of x and y, and simply recall that $\mathbf{G} = k(x,y,1)^T$ is also a solution.

In each of the polynomials $\mathcal{R}_j$ collect the coefficients of y so that it has the form

$$\mathcal{R}_j: d_{j0}y^3 + d_{j1}y^2 + d_{j2}y + d_{j3} = 0, j = 1,2,3,4. \tag{9.57}$$

Each coefficient d_{jk} is a polynomial in x of degree k.

Assemble the coefficients of these polynomials into a matrix that multiplies the vector $(y^3, y^2, y, 1)^T$, given by

$$\begin{bmatrix} d_{10} & d_{11} & d_{12} & d_{13} \\ \vdots & \vdots & \vdots & \vdots \\ d_{40} & d_{41} & d_{42} & d_{43} \end{bmatrix} \begin{Bmatrix} y^3 \\ y^2 \\ y \\ 1 \end{Bmatrix} = \begin{Bmatrix} 0 \\ 0 \\ 0 \\ 0 \end{Bmatrix}. \tag{9.58}$$

These four equations have a solution for the unknowns (y^3, y^2, y) only if the rank of the 4×4 coefficient matrix $[D] = [d_{j0}, d_{j1}, d_{j2}, d_{j3}]$ is three. Thus, this matrix has the determinant $|D| = 0$.

The determinant $|D|$ is a polynomial in the single variable x. The degree of this polynomial is the sum of the degrees of each of the columns of $[D]$, that is, $0 + 1 + 2 + 3 = 6$. Therefore, this step of the solution reduces the four cubic polynomials in x, y to a single sixth-degree polynomial

$$\mathscr{P}(x): \ |D| = \sum_{i=0}^{6} a_i x^i = 0. \tag{9.59}$$

This polynomial has six roots for which zero, two, four, or six may be real. Thus, there can be as many as six spherical RR chains that reach the five orientations.

To determine the spherical RR chains that reach five specified orientations, first formulate the polynomial $\mathscr{P}(x)$ and determine its roots $x_i, i = 1, \ldots, 6$. For each real root x_i, solve (9.58) to determine the coordinate y_i. This defines as many as six fixed axes $\mathbf{G}_i = (x_i, y_i, 1)^T$, called Burmester axes. To determine the associated moving axis $\mathbf{W}_i^1$, choose two of the constraint equations (9.25) and solve for λ, μ, and ν.

9.4.6.2 Intersecting Two Center-Axis Cones

An alternative solution for the fixed axes of spherical RR chains can be obtained using the parameterized form of the center-axis cone. The five task orientations define ten relative rotation axes $\mathbf{S}_{ij}, i < j = 1, \ldots, 5$. From these we can construct two complementary-axis quadrilaterals $\mathscr{Q}_{14}$: $\mathbf{S}_{12}\mathbf{S}_{23}\mathbf{S}_{34}\mathbf{S}_{14}$ and $\mathscr{Q}_{15}$: $\mathbf{S}_{12}\mathbf{S}_{23}\mathbf{S}_{35}\mathbf{S}_{15}$. A fixed axis compatible with five orientations must lie on the center-axis cone defined by $\mathscr{Q}_{14}$ and on the center-axis cone defined by $\mathscr{Q}_{15}$. Thus, the desired axes are the intersections of these two cones.

The complementary-axis quadrilaterals $\mathscr{Q}_{14}$ and $\mathscr{Q}_{15}$ share the same side $\mathbf{S}_{12}\mathbf{S}_{23}$. Therefore, the equations of the two center-axis cones are given by

$$\tan\frac{\kappa}{2}\mathbf{G} = \frac{\tan\frac{\Delta\theta_1}{2}\mathbf{S}_{12} + \tan\frac{\Delta\phi_1}{2}\mathbf{S}_{23} + \tan\frac{\Delta\theta_1}{2}\tan\frac{\Delta\phi_1}{2}\mathbf{S}_{12}\times\mathbf{S}_{23}}{1 - \tan\frac{\Delta\theta_1}{2}\tan\frac{\Delta\phi_1}{2}\mathbf{S}_{12}\cdot\mathbf{S}_{23}} \tag{9.60}$$

and

$$\tan\frac{\kappa}{2}\mathbf{G} = \frac{\tan\frac{\Delta\theta_2}{2}\mathbf{S}_{12} + \tan\frac{\Delta\phi_2}{2}\mathbf{S}_{23} + \tan\frac{\Delta\theta_2}{2}\tan\frac{\Delta\phi_2}{2}\mathbf{S}_{12}\times\mathbf{S}_{23}}{1 - \tan\frac{\Delta\theta_2}{2}\tan\frac{\Delta\phi_2}{2}\mathbf{S}_{12}\cdot\mathbf{S}_{23}}. \tag{9.61}$$

The angles $\Delta\phi_1$ and $\Delta\phi_2$ are functions of $\Delta\theta_1$ and $\Delta\theta_2$ defined by the dimensions of the two complementary-axis quadrilaterals.

The two equations (9.60) and (9.61) define the same axis $\mathbf{G}$ when

$$\Delta\theta_1 = \Delta\theta_2 \quad \text{and} \quad \Delta\phi_1 = \Delta\phi_2. \tag{9.62}$$

The first condition is satisfied by using the same parameter θ to drive $\mathbf{S}_{12}\mathbf{S}_{23}$ for both quadrilaterals. The second condition requires that the spherical triangle $\triangle\mathbf{S}_{23}\mathbf{S}_{34}\mathbf{S}_{35}$ have the same shape in each solution configuration. Thus, the fixed axis $\mathbf{G}$ is an axis of the relative rotation of the triangle $\triangle\mathbf{S}_{23}\mathbf{S}_{34}\mathbf{S}_{35}$ to each of the assemblies of the platform formed by the three RR chains $\mathbf{S}_{12}\mathbf{S}_{23}$, $\mathbf{S}_{14}\mathbf{S}_{34}$, and $\mathbf{S}_{15}\mathbf{S}_{35}$, Figure 9.9. This assembly of relative rotation axes is called the *spherical compatibility platform* and have the following theorem.

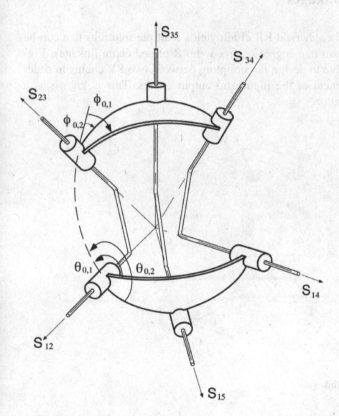

Fig. 9.9 The reference configuration of the spherical compatibility platform.

Theorem 13 (The Spherical Compatibility Platform). *The fixed axis of an RR chain that can reach with five specified orientations is a relative rotation axis of the spherical compatibility platform from its original configuration to one of its other assemblies.*

The analysis of this platform yields the two constraint equations

$$A_i \cos\phi + B_i \sin\phi = C_i, \; i = 1, 2, \tag{9.63}$$

one for each of the 4R chains $\mathscr{Q}_{14}$ and $\mathscr{Q}_{15}$. The solution of these equations shown in (A.11) yields an eighth degree polynomial. This means that the spherical 3RR platform can have as many as eight assemblies. One is the original configuration, so there are seven relative rotation axes for the displacement to the other seven assemblies. One of these axes is S_{13}, and the remaining six are the desired fixed axes.

9.5 Spherical 4R Linkages

In general, the design of a spherical RR chain yields multiple solutions that can be assembled in pairs to form one-degree-of-freedom 4R closed chain linkages, Figure 9.10. It is also possible to design the coupling between two RR chains in order to coordinate the movement of the input and output cranks. This is known as a *spherical function generator*.

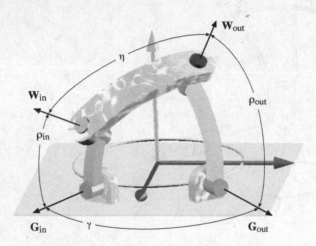

Fig. 9.10 The spherical 4R linkage.

Connecting the end-links of two RR chains constrains the range of movement of the individual chains. This can interfere with the smooth travel of the coupler between the task orientations. This is called the branching problem. A solution rectification strategy exists for spherical linkages that is analogous to the planar theory.

In the following sections we present the spherical version of solution rectification and then present discuss the design strategy for spherical function generators.

9.5.1 Solution Rectification

Bodduluri [80] shows that Filemon's construction generalizes to apply to spherical 4R linkage synthesis. Assume, as before, that the driven crank $\mathbf{G}_{out}\mathbf{W}^1_{out}$ has been determined. The output crank, when viewed from the coupler in each of the design positions, is seen to sweep two wedge shaped regions centered on $\mathbf{W}^1_{out}$. If the input moving axis $\mathbf{W}^1_{in}$ is chosen outside of these wedges, then the linkage will not jam between the design positions.

It is possible for this wedge to include the entire space. If this happens, then there are no solutions to Filemon's construction. A spherical version of Waldron's construction identifies output cranks that ensure that this does not occur. The result is a spherical 4R chain that moves smoothly through the task orientations.

9.5.1.1 Spherical Filemon's Construction

We focus our development on design for three task orientations. Given an output moving crank $\mathbf{G}_{out}\mathbf{W}^1_{out}$, the inverted positions of $\mathbf{G}_{out}$ are obtained using the inverse of the relative rotations

$$\mathbf{g}^i_{out} = [A^T_{1i}]\mathbf{G}. \qquad (9.64)$$

Let α_{12} and α_{23} be the angles measured around $\mathbf{W}^1_{out}$ from $\mathbf{g}^1_{out}$ to $\mathbf{g}^2_{out}$ and from $\mathbf{g}^2_{out}$ to $\mathbf{g}^3_{out}$, respectively. They are dihedral angles between the planes L_i defined by the axes $\mathbf{W}^1_{out}\mathbf{g}^i_{out}$, Figure 9.11. Assume that these angles are between π and $-\pi$. The angle τ of the wedge swept by movement of the output crank is the sum $\alpha_{12} + \alpha_{23}$, if these angles have the same sign. If these angles have different signs, then τ is the angle with the largest absolute value.

Select the moving pivot $\mathbf{W}^1_{in}$ outside of this wedge-shaped region, then the resulting spherical 4R linkage must pass through the design positions before it hits a singular configuration. This is the spherical version of Filemon's construction.

9.5.1.2 Spherical Waldron's Construction

Filemon's construction fails when the angle τ is greater than or equal to π, because in this case the wedges cover space and there are no input moving axes that avoid singular configurations. Waldron's planar solution to this problem generalizes to the spherical case. The goal is a condition on the design of the output crank $\mathbf{G}_{out}\mathbf{W}^1_{out}$ that ensures that there is a solution to Filemon's construction.

Given three task orientations, we have the pole triangle $\triangle \mathbf{S}_{12}\mathbf{S}_{23}\mathbf{S}_{13}$. Inverting the relative rotations to define the movement of the base frame relative to the coupler, we obtain the image pole triangle $\triangle \mathbf{S}_{12}\mathbf{S}^1_{23}\mathbf{S}_{13}$ in the first position M_1. The center-axis theorem requires the moving axis $\mathbf{W}^1_{out}$ to view the side $\mathbf{S}^1_{ij}\mathbf{S}^1_{jk}$ in the rotation angle $-\alpha_{ik}/2$ of the crank relative to the coupler. Thus, we have

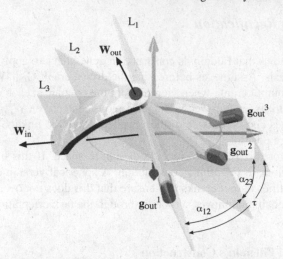

Fig. 9.11 The spherical version of Filemon's construction: The planes L_1 and L_3 bound the region excluded from selection for driving moving pivots $\mathbf{W}_{in}$.

$$\cos\left(\frac{\alpha_{ik}}{2}\right) = \frac{(\mathbf{W} \times \mathbf{S}_{ij}^1) \cdot (\mathbf{W} \times \mathbf{S}_{jk}^1)}{|\mathbf{W} \times \mathbf{S}_{ij}^1||\mathbf{W} \times \mathbf{S}_{jk}^1|}. \qquad (9.65)$$

If an output moving axis $\mathbf{W}_{out}^1$ has any one of the angles α_{ik} greater than or equal to π, then there is no solution to Filemon's construction. The points that have $\alpha_{ik}/2 = \pi/2$ lie on the quadric cones

$$\mathscr{C}_{ik} : (\mathbf{W} \times \mathbf{S}_{ij}^1) \cdot (\mathbf{W} \times \mathbf{S}_{jk}^1) = 0. \qquad (9.66)$$

The three cones $\mathscr{C}_{ik}$ pass through the sides $\mathbf{S}_{ij}^1\mathbf{S}_{jk}^1$ of the image pole triangle in a configuration analogous to Waldron's three-circle diagram. In this case the cones are general quadrics and are not circular.

These cones define regions within which an output axis $\mathbf{W}_{out}^1$ has coupler rotation angles α_{jk} all less than π. This guarantees that Filemon's construction yields a region from which an input axis $\mathbf{W}_{in}^1$ can be selected.

9.5.2 Function Generation

A spherical four-bar linkage can be designed to provide coordination between the input and output angular values θ_i, ψ_i, $i = 1, \ldots, n$, Figure 9.12. To do this, we arbitrarily select the fixed axes $\mathbf{O}$ and $\mathbf{C}$ for the input and output cranks of this linkage. Let $\gamma = \arccos(\mathbf{O} \cdot \mathbf{C})$ be the angle between these axes. Denote by $P_{\mathbf{O}}$ the plane through $\mathbf{O}$ that makes the dihedral angle θ_1 with the plane containing $\mathbf{OC}$.

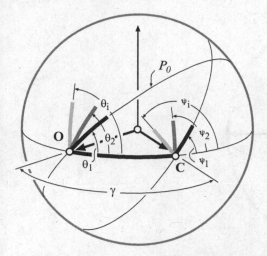

Fig. 9.12 Coordinated input and output angles for a spherical 4R function generator.

Following the approach we used for planar function generators, we define the exterior angles $\bar{\theta}_i = \pi - \theta_i$ around **O**. Introduce the coordinate frame F' attached the input crank so that its y-axis is perpendicular to the plane P_O. Then $\bar{\theta}_1$ is the angle from measured P_O to **OC**. The angles $\bar{\theta}_i$ and ψ_i are the joint angles of the spherical RR open chain formed by **OC** as it moves relative to F', Figure 9.13. The kinematics equations of this RR chain define the task orientations, given by

$$[A_i] = [Z(\bar{\theta}_i)][X(\gamma)][Z(\psi_i)], i = 1, \ldots, n. \tag{9.67}$$

Use these orientations $[A_i]$ to design a spherical RR chain **AB** to close the 4R chain.

The result is a spherical 4R chain that has the desired set of coordinated angles θ_i and ψ_i, $i = 1, \ldots, n$, between the input and output links. Notice that we can obtain a design for at most five coordinated values for these angles.

9.6 Summary

This chapter has presented the synthesis theory for spherical RR chains. The algebraic formulation of the design equations and their solution parallels the results presented for planar RR chains. Furthermore, the center-axis theorem leads to Roth's generalization of Burmester's planar constructions. And the quaternion equation for a spherical triangle generalizes the complex vector equation for a planar triangle. We also obtain spherical versions for Filemon and Waldron's constructions for solution rectification. The analogue between the planar and spherical RR design theories provides insight to spatial linkage analysis and design.

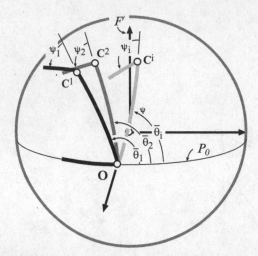

Fig. 9.13 Hold the input crank fixed and consider the output crank as the floating link of an RR chain to define task positions.

9.7 References

Dobrovolskii [26] is credited with the initial formalization of the synthesis theory for spherical linkages. The similarity of the geometry of planar and spherical linkages lead Meyer zur Capellen et al. [82] to present parallel formulations for their analysis. Suh and Radcliffe [133, 134] describe a constraint equation based formulation for spherical linkage synthesis. Roth [105, 106] generalized Burmester's planar constructions to space, which include the spherical results in this chapter as a special case. Dowler et al. [27] provide an example of spherical linkage design. The text by Chiang [12] is devoted to the analysis and design theory for spherical linkages. Computer-based design tools for these linkages are described Larochelle et al. [61], Ruth [110], and Furlong et al. [38].

Exercises

1. Consider the location of the continent of Africa 150 million years ago as M_1 and its location today as M_2. The longitude and latitude of points $\mathbf{P}^i$ and $\mathbf{Q}^i$ in the two locations are given in Table 9.1. Determine the relative rotation matrix $[A_{12}]$.
2. Design a 4R linkage to move the African continent between the two positions defined by Table 9.1.
3. Generate the center-point curve for the relative orientations $[A_{12}]$, $[A_{13}]$, and $[A_{14}]$ defined by the points $\mathbf{P}^i$ and $\mathbf{Q}^i$ in Table 8.1.
4. A spherical linkage support for a feeding device is to locate a spoon at the longitude, latitude, and roll coordinates $M_2 = (-22.5°, -5°, -30°)$ and $M_4 =$

Table 9.1 Longitude and latitude for two positions of points on the African continent

Point	M_1	M_2
P	$(35°, -13°)^T$	$(50°, 12°)^T$
Q	$(-22°, -41°)^T$	$(18°, -36°)^T$

$(25°, 20°, -10°)$ to pick up and deliver food, respectively. Select positions M_1 to provide a desired scooping movement and M_3 so that food is not lost in transit. Generate the center-point curve and design a spherical 4R linkage.

Chapter 10
Multiloop Spherical Linkages

In this chapter, the design procedure introduced for the synthesis of planar six- and eight-bar linkages is applied to the synthesis of spherical six- and eight-bar linkages. The process begins with the specification of a spherical 3R chain or a spherical 6R loop and five task orientations. The synthesis equations for RR chains are solved to constrain the spherical 3R serial chain to design a spherical six-bar linkage, and to constrain the 6R loop to design an eight-bar linkage. The result is one degree-of-freedom spherical linkages that move through the five task positions.

10.1 Synthesis of Spherical Six-bar Linkages

A spherical 3R serial chain consists of three revolute joints that pass through a fixed point in space, Figure 10.1. The graph of this chain is the same as in the planar case, consisting of four vertices 0, 1, 2, and 3. Each edge connecting two vertices represents a revolute joint C_i.

A spherical six-bar linkage consists of six links and seven joints, and has two topologically distinct configurations called the spherical Watt and Stephenson six-bar chains. See Figure 10.2 and Figure 10.3.

As in the plane, these two topologies yield the Watt I, Watt II, Stephenson I, Stephenson II, and Stephenson III linkages depending on which of the links is selected as ground. Furthermore, the movement of the linkage is defined by the seven lines through the origin that define the joint axes, and it is possible to connect these axes on either side of the origin, Figure 10.4. This yields $2^7 = 128$ ways to construct the same spherical six-bar linkage.

The ways in which two spherical RR chains can be added to the spherical 6R loop are the same in the planar case. See Figure 6.5 and 6.6, and Section 6.1.1.

J.M. McCarthy and G.S. Soh, *Geometric Design of Linkages*, Interdisciplinary Applied Mathematics 11, DOI 10.1007/978-1-4419-7892-9_10,
© Springer Science+Business Media, LLC 2011

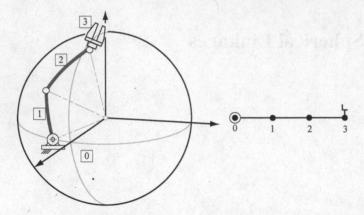

Fig. 10.1 The spherical 3R serial chain and its graph with a vertex for each link.

Fig. 10.2 The spherical Watt six-bar chain.

10.1.1 The Spherical RR Synthesis Equation

Our design process for a spherical six-bar linkage consists in sizing two spherical RR chains that constrain the spherical backbone chain to one degree of freedom. To do this, we formulate the synthesis equations for spherical RR constraint to accommodate the case when both links it connect change orientations.

Let $[B_{l,j}]$, $j = 1, \ldots, 5$, define five orientations of the lth moving link, and $[B_{k,j}]$, $j = 1, \ldots, 5$, define five orientations of the kth moving link measured in a world frame F. Let $\mathbf{g}$ be the coordinates of the R joint attached to the lth link measured in the link frame B_l. Similarly, let $\mathbf{w}$ be the coordinates of the other R joint measured in the link frame B_k. The five positions of these points as the two links move relative to each other are given by

$$\mathbf{G}^j = [B_{l,j}]\mathbf{g} \quad \text{and} \quad \mathbf{W}^j = [B_{k,j}]\mathbf{w}. \tag{10.1}$$

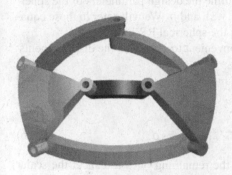

Fig. 10.3 The spherical Stephenson six-bar chain.

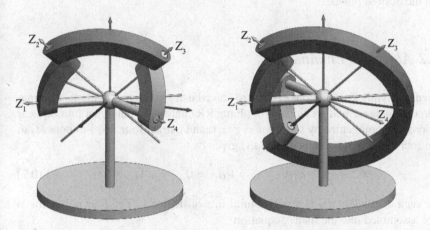

Fig. 10.4 This illustrates two of the sixteen ways that the same four-bar linkage can be assembled. In this case, the directions of axes Z_3 and Z_4 are reversed.

Now introduce the relative displacements

$$[R_{1j}] = [B_{l,j}][B_{l,1}]^{-1} \quad \text{and} \quad [S_{1j}] = [B_{k,j}][B_{k,1}]^{-1}, \qquad (10.2)$$

so we have

$$\mathbf{G}^j = [R_{1j}]\mathbf{G}^1 \quad \text{and} \quad \mathbf{W}^j = [S_{1j}]\mathbf{W}^1, \qquad (10.3)$$

where $[R_{11}] = [S_{11}] = [I]$ are the identity transformations.

The points $\mathbf{G}^j$ and $\mathbf{W}^j$ define the ends of a rigid link of angular length ρ; therefore we have the constraint equations

$$[R_{1j}]\mathbf{G}^1 \cdot [S_{1j}]\mathbf{W}^1 = \|\mathbf{G}^j\|\|\mathbf{W}^j\|\cos\rho. \qquad (10.4)$$

These five equations can be solved to determine the design parameters of the spherical RR constraint, $\mathbf{G}^1 = (u,v,w)$, $\mathbf{W}^1 = (x,y,z)$, and ρ. We will refer to these equations as the *general synthesis equations* for the spherical RR link.

To solve the synthesis equations, it is convenient to introduce the displacements

$$[D_{1j}] = [R_{1j}]^T[S_{1j}], \tag{10.5}$$

so these equations become

$$\mathbf{G}^1 \cdot [D_{1j}]\mathbf{W}^1 = \|\mathbf{G}^j\|\|\mathbf{W}^j\|\cos\rho. \tag{10.6}$$

Subtract the first of these equations from the remaining ones to cancel the scalar terms $\|\mathbf{G}^j\|\|\mathbf{W}^j\|\cos\rho$, and the square terms in the variables u, v, w and x, y, z. The resulting four bilinear equations can be solved algebraically by setting $w = z = 1$ to obtain the desired pivots.

10.1.2 Algebraic Elimination

The general synthesis equations (10.6) are solved using the same algebraic elimination procedure presented previously for planar RR chains. We construct four bilinear equations, dehomogenize by setting $w = z = 1$, and extract four 3×3 minors M_j to obtain four cubic polynomials in x and y, given by

$$\mathcal{R}_j : d_{j0}y^3 + d_{j1}y^2 + d_{j2}y + d_{j3} = 0, \quad j = 1,\dots,4, \tag{10.7}$$

where each coefficient d_{jk} is a polynomial in x of degree k. The four polynomials $\mathcal{R}_j$ are assembled into the matrix equation

$$\mathcal{R} = [D(x)]\mathbf{m} = \begin{bmatrix} d_{10}(x) & d_{11}(x) & d_{12}(x) & d_{13}(x) \\ \vdots & \vdots & \vdots & \vdots \\ d_{40}(x) & d_{41}(x) & d_{42}(x) & d_{43}(x) \end{bmatrix} \begin{Bmatrix} y^3 \\ y^2 \\ y \\ 1 \end{Bmatrix} = \begin{Bmatrix} 0 \\ 0 \\ 0 \\ 0 \end{Bmatrix}. \tag{10.8}$$

As we have seen, this equation can be solved for a non-zero $\mathbf{m} = (y^3, y^2, y, 1)$, only if the resultant matrix $[D(x)]$ had a determinant equal to zero.

This determinant can be expanded to obtain a sixth degree polynomial, or formed as a generalized eigenvalue problem (6.10) both of which yield at most six finite roots. Solve these either of these equations for x and y, and then solve bilinear form of the synthesis equations (10.6) to obtain u and v. The result is as as many as six solutions for $\mathbf{G}^1 = (u,v,1)$ and $\mathbf{W}^1 = (x,y,1)$.

10.1.3 The Number of Spherical Six-Bar Linkage Designs

The synthesis of a spherical RR constraint yields as many as six designs; therefore two constraints can yield as many as 36 designs. However, this process always yields one link of the 3R spherical chain, so there are at most 30 candidates.

The Stephenson I topology yields all 30 candidate designs. The Stephenson II has two ways to select the link that is the end effector, which yields 60 designs. The Stephenson III also has two ways to define the end effector link, one with 30 designs and the other with 15, for a total of 45 candidate linkages.

The Watt I chain has two sets of designs depending on the input crank, and both sets has an existing link of the 3R chain for both RR constraints, so there are 2×25 candidates. The result is as many as 185 different spherical six-bar linkage designs.

Combining this with the 128 ways to construct a particular spherical six-bar linkage, and we have as many as 23680 (185×128) designs.

10.2 Analysis of a Spherical Six-Bar Linkage

In order to animate the movement of a spherical six-bar linkages we determine its configuration angles for a series of input crank angles. For each input angle, we formulate the loop equations of component structures of the six-bar linkage.

For the cases of the Watt I, Stephenson I, and Stephenson IIIa, we obtain a sequence of spherical triangles that are analyzed to determine the configuration angles of the six-bar linkage. For the Stephenson II and Stephenson IIIb, we analyze a spherical pentad, Figure 10.5

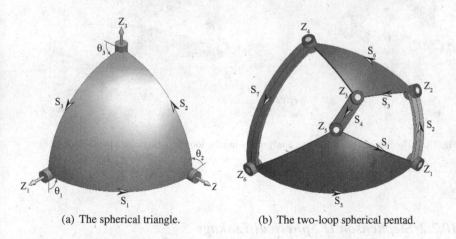

(a) The spherical triangle.   (b) The two-loop spherical pentad.

Fig. 10.5 Component structures used to analyze a multiloop spherical linkage.

10.2.1 Watt I Spherical Linkage

The analysis of a Watt I linkage decomposes into the analysis of two spherical triangles, Figure 10.6. Let the input crank be C_1C_2, which means the coordinates C_2 are known. The spherical triangle $C_2W_1G_1$ is analyzed to determine the joint axis W_1.

Given the joint angles of the first component triangle, we can determine the coordinates of C_3 and G_2. This yields a second component triangle $C_3W_2G_2$ that is analyzed to determine the joint axis W_2.

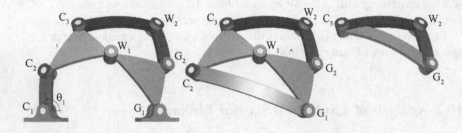

Fig. 10.6 The Watt I linkage is analyzed using the loop equations for two spherical triangle components.

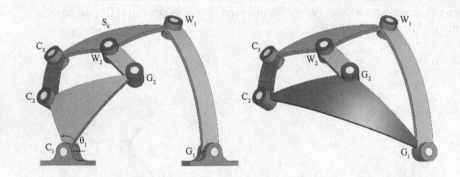

Fig. 10.7 The Stephenson II linkage analyzed using the loop equations for a pentad component.

10.2.2 Stephenson II Spherical Linkage

The spherical Stephenson II linkage with C_1C_2 as the input crank defines a spherical pentad for a given value of the input angle θ_1. This is because for a given value of

θ_1, the coordinates of $\mathbf{C}_2$ and $\mathbf{G}_2$ are known. The result is that this spherical six-bar reduces to the spherical pentad structure shown in Figure 10.7.

10.2.3 Component Loop Equations

The loop equations of the component structures of a spherical linkage provide the equations that we use to analyze multiloop spherical linkages.

Let $\mathbf{Z}_i, i = 1,\dots,k$ to be unit vectors that define the directions of each joints of a component, such as spherical triangle, pentad or three-loop structure. Let θ_i be joint angles that define rotation about the local z-axis at $\mathbf{Z}_i$. Introduce the angular length α_{ij} between any two pivots $\mathbf{Z}_i$ and $\mathbf{Z}_j$, and align the local y-axis is in the direction $\mathbf{Z}_i \times \mathbf{Z}_j$, so α_{ij} is the angle of rotation about this axis.

10.2.3.1 The Spherical Triangle

Using this notation, the component loop equations for a spherical triangle obtain

$$[Z(\theta_1)][Y(\alpha_{12})][Z(\theta_2)][Y(\alpha_{23})][Z(\theta_3)][Y(\alpha_{31})] = I. \qquad (10.9)$$

Multiply this equation on the right by $[Y(\alpha_{31})]^T$, and then select the $(3,3)$ components to eliminate both θ_1 and θ_3. This is done by pre- and post- multiplying by the unit vector $\vec{k}$, in order to and obtain the scalar equation

$$f(\theta_2) = \vec{k} \cdot ([Y(\alpha_{12})][Z(\theta_2)][Y(\alpha_{23})] - [Y(\alpha_{31})]^T)\vec{k} = 0. \qquad (10.10)$$

The result is a constraint equation that defines θ_2.

10.2.3.2 The Spherical Pentad

The loop equations for the pentad are obtained in the same way, but now we have two loops (i) $\mathbf{Z}_5\mathbf{Z}_1\mathbf{Z}_2\mathbf{Z}_3$, and (ii) $\mathbf{Z}_6\mathbf{Z}_1\mathbf{Z}_2\mathbf{Z}_4$, which yield the loop equations

$$[Z(\theta_5)][Y(\alpha_{51})][Z(\theta_1)][Y(\alpha_{12})][Z(\theta_2)][Y(\alpha_{23})][Z(\theta_3)][Y(\alpha_{35})] = I,$$
$$[Z(\theta_6)][Y(\alpha_{61})][Z(\theta_1 + \beta_1)][Y(\alpha_{12})][Z(\theta_2 + \beta_2)][Y(\alpha_{24})][Z(\theta_4)][Y(\alpha_{46})] = I.$$
$$(10.11)$$

The angle β_1 in the second of these equations is the offset between these loops measured around $\mathbf{Z}_1$ from $\mathbf{Z}_6$ to $\mathbf{Z}_5$. Similarly, β_2 is the offset around $\mathbf{Z}_2$ measured from $\mathbf{Z}_3$ to $\mathbf{Z}_4$.

These equations can be simplified in the same way as described above for the spherical triangle. Multiply the first equations on the right by $[Y(\alpha_{35})]^T$, and the second by $[Y(\alpha_{46})]^T$. Select the $(3,3)$ components to eliminate the variables θ_3 and

θ_5 from the first equation, and θ_4 and θ_6 from the second equation, to obtain

$$\mathscr{F}_1 : \vec{k} \cdot ([Y(\alpha_{51})][Z(\theta_1)][Y(\alpha_{12})][Z(\theta_2)][Y(\alpha_{23})] - [Y(\alpha_{35})]^T)\vec{k} = 0,$$
$$\mathscr{F}_2 : \vec{k} \cdot ([Y(\alpha_{61})][Z(\theta_1 + \beta_1)][Y(\alpha_{12})][Z(\theta_2 + \beta_2)][Y(\alpha_{24})] - [Y(\alpha_{46})]^T)\vec{k} = 0.$$
$$(10.12)$$

The result is two equations in the unknowns θ_1 and θ_2.

10.2.4 Eigenvalue-Based Elimination

The loop equations of the spherical pentad yield a system of constraint equations f_i, $i = 1,2$ that contain sines and cosines of the two joint angles, θ_i. Convert these constraint equations to polynomials by introducing the variables $x_i = \tan(\theta_i/2)$, such that

$$\cos\theta_i = \frac{1-x_i^2}{1+x_i^2} \quad \text{and} \quad \sin\theta_i = \frac{2x_i}{1+x_i^2}, \quad i = 1,2. \qquad (10.13)$$

Substitute these equations into the constraint equations to obtain two polynomials $(p_1, p_2) = 0$ in the two unknowns x_1 and x_2.

In order to solve the polynomial system (p_1, p_2), we introduce two additional polynomials $p_3 = x_1 p_1$ and $p_4 = x_1 p_2$. The monomials of the polynomial system (p_1, p_2, p_3, p_4) can be assembled into the vector $\mathbf{m} = (\mathbf{y}_0, \mathbf{y}_1, \mathbf{y}_2)^T$ such that $\mathbf{y}_0 = (1, x_1, x_1^2, x_1^3)$, $\mathbf{y}_1 = x_2 \mathbf{y}_0$, and $\mathbf{y}_2 = x_2^2 \mathbf{y}_0$. This polynomial system takes the form

$$[K]\mathbf{m} = 0, \qquad (10.14)$$

where $[K]$ is a 4×12 matrix having full row rank.

The polynomials (10.14) are solved as follows. Construct the two sets of 8 monomials $\mathbf{m}_1 = (\mathbf{y}_0, \mathbf{y}_1)$ and $\mathbf{m}_2 = (\mathbf{y}_1, \mathbf{y}_2)$. Notice that they satisfy the identities $x_2 \mathbf{m}_1 - \mathbf{m}_2 = 0$, so we obtain the system of 12 equations

$$\left\{ \begin{matrix} K\mathbf{m} \\ x_2\mathbf{m}_1 - \mathbf{m}_2 \end{matrix} \right\} = \begin{bmatrix} K_0 & K_1 & K_2 \\ x_2 I_4 & -I_4 & 0 \\ 0 & x_2 I_4 & -I_4 \end{bmatrix} \left\{ \begin{matrix} \mathbf{y}_0 \\ \mathbf{y}_1 \\ \mathbf{y}_2 \end{matrix} \right\} = 0, \qquad (10.15)$$

where $[I_4]$ is the 4×4 identity matrix and $[K_i]$ is the coefficient submatrix of $[K]$ associated with the monomial vector $\mathbf{y}_i$.

We now extract the subsystem of equations that forms a generalized eigenvalue problem with $\mathbf{m}_1 = (\mathbf{y}_0, \mathbf{y}_1)$ as its eigenvector. This is done by multiplying the equations (10.15) by a matrix that eliminates the coefficients of $\mathbf{y}_2$, that is

$$\begin{bmatrix} 0 & I_4 & 0 \\ I_4 & 0 & K_2 \end{bmatrix} \begin{bmatrix} K_0 & K_1 & K_2 \\ x_2 I_4 & -I_4 & 0 \\ 0 & x_2 I_4 & -I_4 \end{bmatrix} \begin{Bmatrix} \mathbf{y}_0 \\ \mathbf{y}_1 \\ \mathbf{y}_2 \end{Bmatrix} = \left[\begin{pmatrix} I_4 & 0 \\ 0 & K_2 \end{pmatrix} x_2 - \begin{pmatrix} 0 & I_4 \\ -K_0 & -K_1 \end{pmatrix} \right] \begin{Bmatrix} \mathbf{y}_0 \\ \mathbf{y}_1 \end{Bmatrix} = 0.$$

$$(10.16)$$

This reduces the system of 12 polynomial equations to the 8×8 generalized eigenvalue problem

$$\left[\begin{pmatrix} I_4 & 0 \\ 0 & K_2 \end{pmatrix} x_2 - \begin{pmatrix} 0 & I_4 \\ -K_0 & -K_1 \end{pmatrix} \right] \begin{Bmatrix} \mathbf{y}_0 \\ \mathbf{y}_1 \end{Bmatrix} = 0. \qquad (10.17)$$

For each eigenvalue x_2 we obtain an associated eigenvector $\mathbf{m}_1$. Notice that the scale factor μ of $\mathbf{m}_1$ can be determined from the value of the component 1 in $\mathbf{y}_0$. Divide by μ to obtain the value of the joint angle x_1.

10.2.5 Sorting Assemblies

As we analyze a spherical six-bar linkage for a sequence of input angles Θ_1^k, there are as many as eight sets of configuration angles $\vec{\Theta} = (\Theta_2, \Theta_3, \Theta_4, \Theta_5, \theta_6)_i$, $i = 1, \ldots, 8$ that define the assemblies of the linkage associated with each input angle. In order to sort the roots among the assemblies, we use the Jacobian of the loop equations.

In order to sort the roots among assemblies of the six-bar linkage, we compute the derivative of the complex loop equations (10.12) and assemble the equation

$$[\nabla \mathscr{F}(\vec{\Theta}_i^k)](\vec{\Psi} - \vec{\Theta}_i^k) = 0, \qquad (10.18)$$

where Ψ approximates the value $\vec{\Theta}_i^{k+1}$ associated with the input angle Θ_1^{k+1} and is near the assembly defined by $\vec{\Theta}_i^k$. It is then a matter of identifying which of the root $\vec{\Theta}_i^{k+1}$ is closest to Ψ on the ith circuit, in order to match the assemblies. This provides a rapid and exact method to determine a sequence of configuration angles for each assembly in order to animate the six-bar linkage.

10.3 Example: A Spherical Six-Bar Walker

As an example of the spherical six-bar linkage design process, we design a spherical Stephenson IIb linkage to guide the leg movement for a simple walking machine, Figure 10.8. The leg is to be attached to a ball joint at the hip, and the linkage guides the leg movement around this hip so that it moves along the forward direction.

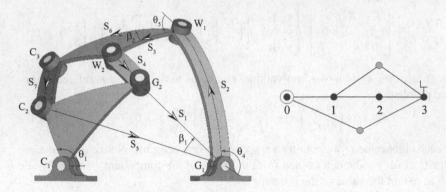

Fig. 10.8 A spherical 3R chain, $C_1C_2C_3$, constrained by two spherical RR chains, G_1W_1 and G_2W_2, to form a spherical Stephenson IIb linkage.

Table 10.1 Five task orientations for the end effector of the spherical 3R chain defined as longitude θ, latitude ϕ, and roll ψ.

Task	Orientation (θ, ϕ, ψ)
1	$(54°, 7°, -156°)$
2	$(53°, 25°, -155°)$
3	$(9°, 50°, -133°)$
4	$(-32°, 5°, -103°)$
5	$(24°, -11°, -146°)$

Step 1.

The task orientations that define the movement of the leg, $[T_i], i = 1, \ldots, 5$ are given in Table 10.1. These orientations guide the walker in a straight line while in contact with the ground and then lift and return to take another step.

The dimensions of the spherical 3R chain are chosen to be $\alpha_{12} = 29.1°$ and $\alpha_{23} = 10.5°$. The location of the base and moving pivots of the spherical 3R chain are specified by the orientation $[G]$, $(0°, 0°, 0°)$, and $[H]$, $(31.3°, 11.3°, 0°)$, respectively. Use this data to formulate the inverse kinematics equations of the spherical 3R chain and solve for the configuration angles $q_j = (\theta_1, \theta_2, \theta_3)_j$, $j = 1, \ldots, 5$, that reach the specified task orientations $[T_j]$, $j = 1, \ldots, 5$.

Notice that the inverse kinematics equations yield two sets of configuration angles corresponding to a spherical 3R chain with its elbow up and elbow down. In this example, we chose the configuration defined by the vectors $C_1 = (0, 0, 1)$, $C_2 = (0.48, -0.10, 0.87)$, and $C_3 = (0.48, -0.28, 0.83)$.

The five configurations of the spherical 3R chain provide the coordinate transformations for each link relative to the ground frame. Compute $[B_{1j}] = [G][Z(\theta_1)]$, which defines the jth orientation of the first link in F. The orientations of the second and third links are given by

$$[B_{2j}] = [G][Z(\theta_1)][X(\alpha_{12})][Z(\theta_2)],$$
$$\text{and } [B_{3j}] = [G][Z(\theta_1)][X(\alpha_{12})][Z(\theta_2)][X(\alpha_{23})][Z(\theta_3)], \quad j = 1, \dots, 5. \quad (10.19)$$

These transformations form the task requirements for the synthesis of the two RR constraints.

Step 2.

Use the five orientations $B_{3j}, j = 1, \dots, 5$ of link B_3 relative to ground to form the design equations for the first spherical RR constraint $G_1 W_1$, Figure 10.8. Solve the design equations (10.6) with $[D_{1i}] = [B_{3i}][B_{31}]^{-1}$, $i = 1, \dots, 5$ to obtain the values listed in Table 10.2. In this case, it can happen that the design equations may yield only complex solutions, which means that the original task positions or the dimensions of the spherical 3R chain must be adjusted.

Table 10.2 Step 1. Select the joint coordinates for the spherical 3R chain; Step 2. Solve for the first spherical RR chain; Step 3. Solve for the second spherical RR chain; The selected values in normalized form are highlighted in bold.

Step 1	C_1	C_2	C_3
	$(0,0,1)$	$(0.48, -0.10, 0.87)$	$(0.48, -0.28, 0.83)$

Step 2		G_1	W_1
1		$(-0.23, -0.97, 0.01)$	$(-0.55, -0.75, 0.37)$
2		$(-0.17, -0.76, 0.62)$	$(-0.56, 0.30, 0.77)$
3		$(0.10, -0.79, 0.60)$	$(-0.56, 0.73, 0.38)$
4		$(0.21, -0.17, 0.96)$	$(0.63, -0.35, 0.70)$
5		Complex	Complex
6		Complex	Complex

Step 3		G_2	W_2
1		Complex	Complex
2		Complex	Complex
3		$(0.48, -0.10, 0.87)$	$(0.48, -0.28, 0.83)$
4		Complex	Complex
5		Complex	Complex
6		$(0.54, -0.25, 0.80)$	$(0.39, 0.00, 0.92)$

Step 3.

For the Stephenson IIb, the second spherical RR constraint $G_2 W_2$ connects B_3 to B_1 and does not depend on the first constraint. Use the orientations obtained for the link B_1 and B_3 to assemble the design equations for $\mathbf{G}_2\mathbf{W}_2$. The solutions to these equations are shown in Table 10.2. The chain $\mathbf{C}_2\mathbf{C}_3$ appears among these solutions, and guarantees the existence of least one more real solution.

Analysis

In order to animate the movement of the walker linkage, we solve the component loop equations of the spherical pentad for a sequence of input angles. Let these input angles be θ_1^k, $k = 1 \ldots, N$, then, for each value, θ_1^k, solve the component loop equations (10.12) to determine the remaining configuration angles $\Theta^k = (\theta_4^k, \theta_5^k)^T$.

There can be as many as eight roots for each Θ^k, only one of which corresponds the assembly of the spherical six-bar specified in each of the task orientations. Identify the root Θ^k for the correct assembly by formulating and solving the approximation to the component loop equations (10.18), in order to match the assemblies. This calculation can be checked against the known configurations at each of the task orientations.

The results for our design are shown in Table 10.3. Figure 10.9 shows the image sequence of the chosen design passing through each of the specified task orientations and on a simple walking machine in the shape of a spider.

Table 10.3 Analysis results for the walker spherical six-bar linkage

	θ_1	θ_4	θ_5
1	30.54°	−93.05°	−114.23°
2	47.04°	−95.09°	−106.26°
3	63.54°	−96.95°	−99.50°
4	97.07°	−100.13°	−90.68°
5	147.37°	−102.01°	−93.12°
6	214.81°	−94.48°	−119.02°
7	282.24°	−83.79°	−141.45°
8	304.89°	−78.06°	−150.79°
9	327.54°	−80.12°	−148.43°
10	352.74°	−86.93°	−135.33°

10.4 Synthesis of a Spherical Eight-Bar Linkage

Our procedure for the synthesis of a spherical eight-bar linkage is identical to the procedure presented earlier for a planar eight-bar linkage. We start with a spherical 6R closed chain that is specified by the designer and a set of five task positions, and then solve for two RR constraints that guide the system through the given task.

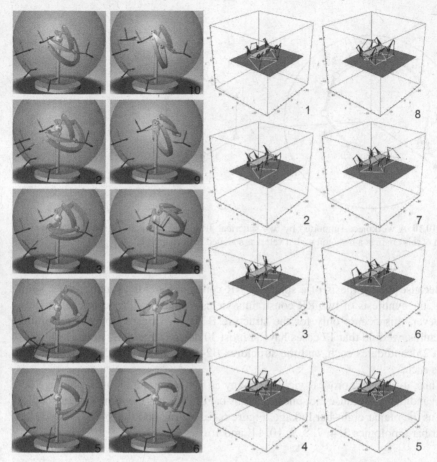

Fig. 10.9 The animation of the spherical Stephenson IIb linkage as it moves through task orientations, and the illustration of the use of this linkage for a simple walking machine.

The 6R loop is formed by two 3R chains that share the same end effector, Figure 10.10. The ways in which the two spherical constraints can be added to the spherical 6R loop are the same as that in the planar case. See Figure 6.15 and 6.14, and Section 6.4.1. As in the planar case, we limit our designs to two connections to the ground link B_0, so the designer specifies these base attachment points.

10.4.1 The Number of Spherical Eight-Bar Linkage Designs

The graph of a spherical eight-bar linkage is the same as that for a planar eight-bar linkage, which allows us to use the notation introduced previously to identify as BijBkl a spherical eight-bar linkage has the RR constraints B_iB_j and B_kB_l.

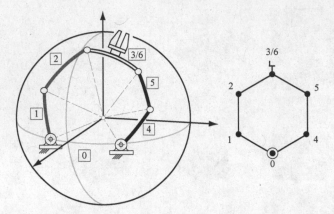

Fig. 10.10 A workpiece supported by two spherical 3R chains forms a 6R spherical loop. The graph of this chain is a hexagon with a vertex associated with each link.

Because the synthesis of a spherical RR constraint yields as many as six solutions, the synthesis of two RR constraints yields as many as 36 design candidates. However, this occurs only for the structure B15B24. Considering the remaining structures, we find that 17 cases have at most 30 design candidates each, 12 have at most 25 candidates each, and the structures B24B24, and B15B15 have 15 candidates each.

Thus, this design process yields as many as 876 spherical eight-bar linkage design candidates. Because the spherical eight-bar linkage has the same sixteen topologies as the planar eight-bar linkage, Figure 6.16, we see that this process yields the eight-bar topologies 3, 4, 7, 8, 9, 10, 11, and 16.

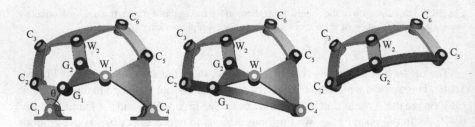

Fig. 10.11 The decomposition of a B14B37 linkage into a spherical triangle and pentad given the input angle θ_1.

10.5 Analysis of a Spherical Eight-bar Linkage

The analysis of a spherical eight-bar linkage follows the same process as described for the spherical six-bar linkage. Once the input crank is specified the coordinates of pivots attached to this crank are known and the linkage becomes a structure. The analysis of this structure involves identifying a sequence of component structures that are analyzed separately.

Consider the B14B37 eight-bar linkage shown in Figure 10.11. Let the input crank be C_1C_2, which means the coordinates of the points C_2 and G_1 are known. This allows us to isolate the spherical triangle $G_1W_1C_4$, which defines the joint angle around W_1.

Once W_1 is computed, we have G_2 and C_5 well. This isolates the spherical pentad shown in Figure 10.11, which is analyzed to determine the configuration of the eight-bar linkage.

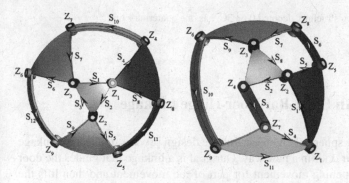

(a) The three-loop spherical struc- (b) The three-loop spherical structure, denoted type 3A   ture, denoted type 3B

Fig. 10.12 Two three-loop component structures constructed from three binary links and four ternary links.

The analysis of our spherical eight-bar linkages can be summarized as follows. Let the input link be C_1C_2, then the sequences of component structures necessary to analyze spherical eight-bar linkages are

1. A sequence of three spherical triangles for cases B14B27, B14B57, B15B47, B24B17, B14B15, B14B24, and B24B24;
2. A spherical triangle followed by a spherical pentad for the cases B14B37, B14B13, B14B25, and B14B34;
3. A spherical pentad followed by a spherical triangle for case B15B15;
4. A type 3B three-loop spherical chain for cases B15B27, B15B37, B13B57, B13B47, B25B17, B25B47, B24B37, B24B57, B34B17, B34B27, B15B24, B15B27, B15B34, B13B25, B13B24, and B25B34; and
5. A type 3C three-loop spherical chain for cases B15B13, B15B25, B13B34, B25B24, and B24B34.

Wampler [149] shows how to derive the loop equations for the three-loop component structures Type 3A, 3B and 3C, Figures 10.12, and 10.13 and how to generalize the eigenvalue-based elimination technique to analyze these spherical three-loop
structures.

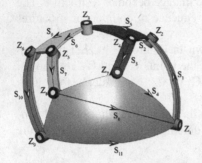

Fig. 10.13 The three-loop structure, denoted type 3C, has one quaternary link.

10.6 Example: An Eight-Bar Door Hinge Linkage

As an example of our spherical eight-bar linkage design process, consider a linkage
designed to open a car door in a new way. Our goal is a linkage that guides the door
through a standard opening movement for part of the movement and then lifts the
door and places it onto the top of the car.

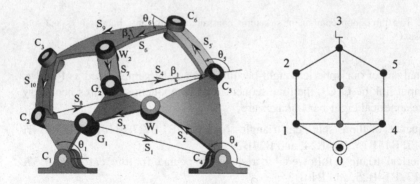

Fig. 10.14 The joint angle and link length parameters for the (8, 10) spherical eight-bar linkage.

Step 1.

For this design, we choose to design a B14B37 eight-bar linkage, Figure 10.14, using the five task orientations specified in Table 10.4. Let the 6R loop be defined by the coordinates of the pivots $C_1C_2C_3C_6C_5C_4$ listed in Table 10.5.

Table 10.4 Five task orientations for the end effector of the spherical 6R loop chain defined as longitude θ, latitude ϕ, and roll ψ.

Task Position (θ, ϕ, ψ)	
1	$(90°, -60°, 90°)$
2	$(70°, -7°, 77.7°)$
3	$(25.2°, -1°, 59.8°)$
4	$(-0.8°, -11°, 41.9°)$
5	$(-53.8°, -26°, 13.5°)$

The dimensions of the first spherical 3R chain are chosen to be $\alpha_{12} = \alpha_{23} = 50°$. The location of the base and moving pivots of this spherical 3R chain are specified by the orientation $[G]$, $(90°, -60°, 0°)$, and $[H]$, $(0°, 0°, 0°)$, respectively.

Similarly, the dimensions of the second spherical 3R chain are chosen to be $\alpha_{45} = \alpha_{56} = 60°$. The location of its base and moving pivots are specified by the orientation $[G]$, $(100°, -30°, 0°)$, and $[H]$, $(-50°, 0°, 0°)$, respectively.

The inverse kinematics equations of the 6R loop defined by these two 3R chains yield the configuration angles $\mathbf{q}_j = (\theta_1, \theta_2, \theta_3, \theta_4, \theta_5, \theta_6)$, $j = 1, \ldots, 5$, that reach the task orientations $[T_j]$, $j = 1, \ldots, 5$. Notice that there are four sets of solutions for $\mathbf{q}$ corresponding to elbow up and elbow down solutions for both spherical 3R chains.

The five configurations of the 6R loop provide the coordinate transformations for each link relative to the ground frame. Compute $[B_{1j}] = [G_1][Z(\theta_{1j})]$, which defines the five positions of the first link in F. The positions of the second, third, fourth, fifth, and sixth links are given by

$$
\begin{aligned}
[B_{2j}] &= [G_1][Z(\theta_{1j})][X(\alpha_{12})][Z(\theta_{2j})], \\
[B_{3j}] &= [G_1][Z(\theta_{1j})][X(\alpha_{12})][Z(\theta_{2j})][X(\alpha_{23})][Z(\theta_{3j})], \\
[B_{4j}] &= [G_2][Z(\theta_{4j})], \text{ and} \\
[B_{5j}] &= [G_2][Z(\theta_{4j})][X(\alpha_{45})][Z(\theta_{5j})].
\end{aligned} \tag{10.20}
$$

This provides the task orientation data used for formulating the synthesis equations for the two spherical RR constraints.

Step 2.

Use the orientations B_{1j}, for link B_1 and B_{4j} for link B_4 to assemble the design equations for a spherical RR chain $\mathbf{G}_1\mathbf{W}_1$, Figure 10.14. In this case, we obtain

Table 10.5 Step 1. Select the joint coordinates for the 6R loop; Step 2. Solve for the first spherical RR chain; Step 3. Solve for the second spherical RR chain; The selected values are highlighted in bold.

Step 1	C_1	C_2	C_3
	$(0.5,-0.87,0)$	$(0.32,-0.56,-0.77)$	$(0.5,-0.87,0)$

Step 1	C_4	C_5	C_6
	$(0.85,-0.5,-0.15)$	$(0.54,0.16,-0.83)$	$(0.99,-0.17,0)$

Step 2	G_1	W_1
1	$(0.82,-0.38,0.43)$	$(-0.34,0.93,0.17)$
2	$(0.5,-0.87,0)$	$(0.85,-0.50,-0.15)$
3	Complex	Complex
4	Complex	Complex
5	Complex	Complex
6	Complex	Complex

Step 3	G_2	W_2
1	Complex	Complex
2	Complex	Complex
3	$(-0.20,-0.68,0.70)$	$(-0.26,0.38,0.89)$
4	$(0.05,0.30,0.95)$	$(0.41,-0.59,0.70)$
5	Complex	Complex
6	Complex	Complex

(10.6) with $[D_{1j}] = ([B_{1,j}][B_{1,1}]^{-1})^T[B_{4,j}][B_{4,1}]^{-1}, i = 1,\ldots,5$. The solution to these equations are in Table 10.5.

The design equations yield two solutions for G_1W_1, one of which is the existing RR chain C_1C_4. This means we have only one choice for this RR constraint.

Step 3.

The RR chain G_1W_1 introduces a new link B_7, which takes the orientations $[B_{7j}], j = 1,\ldots,5$, when the end effector is in each of the specified task orientations.

Using the orientations of the end effector of the spherical 6R loop and the orientations of the link B_7, we assemble the design equations for a spherical RR constraint G_2W_2. The solution to these equations are listed in Table 10.5. Notice that it can happen that there are only complex solutions for these design equations. In this case, the task positions or 6R loop may have to be adjusted.

Analysis

In order to animate the movement of this B14B37 eight-bar linkage, we formulate the component loop equations of the spherical triangle (10.10) followed by the
spherical pentad (10.12). Let $\theta_1^k, k = 1 \ldots, N$ be the input crank angles, then we solve
these loop equations to determine the configuration angles $\Theta^k = (\theta_4^k, \theta_5^k, \theta_6^k)^T$.

Table 10.6 Analysis results for the eight-bar hinge linkage

	θ_1	θ_4	θ_5	θ_6
1	287.11°	103.86°	128.63°	−37.62°
2	312.29°	87.13°	71.87°	−27.31°
3	337.47°	74.90°	18.15°	−27.06°
4	7.31°	64.02°	−7.21°	−43.89°
5	37.14°	57.92°	−8.98°	−69.72°
6	53.10°	57.31°	−7.22°	−83.50°
7	69.06°	58.69°	−6.05°	−95.23°
8	99.84°	66.25°	−9.00°	−105.13°
9	130.63°	78.50°	−13.00°	−97.13°

There can be as many as $2 \times 8 = 16$ roots for each Θ^k, only one of which corresponds the assembly of the spherical eight-bar specified in each of the task orientations. Identify the root Θ^k for the correct assembly by formulating and solving
the approximation to the component loop equations in the same way as presented in
(10.18).

10.7 Summary

In this chapter we have extended the dimensional synthesis results for planar six-
and eight-bar linkages to the design of spherical linkages. Similar to planar linkages,
we can design spherical six-bar linkages by adding two RR chains to a spherical
3R serial chain. This yields as many as 185 six-bar design candidates. To design
spherical eight-bar linkages, we add two RR constraints to a spherical 6R loop and
obtain as many as 876 eight-bar design candidates. Example designs are presented
for a spherical six-bar linkage to guide the leg of a walking machine, and for a
spherical eight-bar linkage to guide a novel car door hinge.

We also presented the analysis of spherical six- and eight-bar linkages. This is
achieve by identifying a sequence of component structures that are analyzed to obtain a system of polynomial equations. An eigenvalue elimination technique is presented for the solution of these equations for the pentad structure.

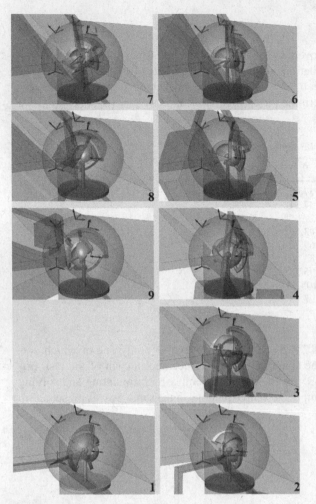

Fig. 10.15 Animation of the B14B37 spherical eight-bar linkage moving through each of the task orientations.

10.8 References

The dimensional synthesis for spherical multiloop mechanism can be found in Soh and McCarthy [121, 123, 122]. Hernadez et al. [47] design spherical Stephenson six-bar linkage for use as a robot wrist mechanism. Tsai [138] describes the use of graph theory in the design of linkages and enumerates the spherical six-bar linkage topologies.

Exercises

1. Use the task orientations in Table 10.4 to design a spherical Watt Ia six-bar door opening linkage. Choose an arbitrary spherical 3R chain. How many design candidates are obtained?
2. Analyze the the various design candidates for the above problem. How many of the design candidates pass smoothly through all of the five task orientations?
3. Design a spherical B34B17 eight-bar door opening linkage for the task orientations in Table 10.4, Figure 10.16. Choose an arbitrary spherical 6R chain. How many design candidates are obtained?

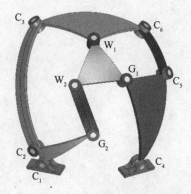

Fig. 10.16 The spherical B34B17 eight-bar linkage.

4. In how many ways can a particular design candidate for a spherical B34B17 eight-bar linkage be assembled? Determine the maximum number of linkages obtained for this particular structure.
5. Formulate the closed-form solution to the B34B17 linkage shown in Figure 10.16 using eigenvalue-based elimination [149]. Analyze the design candidates and determine how many pass through all five task orientations.

Chapter 11
Analysis of Spatial Chains

In this chapter we study spatial linkages. These systems have at least one link that moves through a general spatial displacement. We examine the TS and CC chains that are important to our design theory, as well as the TPS and TRS chains that appear in robotics. In addition, we study the 3R wrist which is actually a spherical linkage, however, it provides a convenient parameterization of the S-joint that is an important part of our spatial open chains. We determine the joint angles for these chains that position the end-effector in a desired location.

We then analyze the RSSR closed chain, which is closely related to both the planar and spherical 4R linkages, as well as the spatial 4R linkage, known as *Bennett's linkage*. We then examine the RSSP linkage, which is a spatial version of the slider-crank. Finally, we consider the spatial 4C closed chain. Remarkably, planar and spherical 4R linkages are also special cases of the 4C linkage.

11.1 The Kinematics Equations

The analysis of a linkage requires that coordinate frames be attached to each of the links in order to measure the joint parameters. While it is relatively easy to define these link frames for planar and spherical linkages, the assignment of frames for spatial linkages can be difficult. In addition, minor changes in the coordinate frame convention can yield different constraint equations. To standardize this process, we use the Denavit–Hartenberg convention to assign reference frames to the links in a spatial linkage.

11.1.1 Joint Axes

The general link of a spatial linkage is considered to be defined by two skew lines, which we denote here by S_1 and S_2. Let S_1 pass through the point $\mathbf{p}$ in the direction

J.M. McCarthy and G.S. Soh, *Geometric Design of Linkages*, Interdisciplinary Applied Mathematics 11, DOI 10.1007/978-1-4419-7892-9_11,
© Springer Science+Business Media, LLC 2011

S_1. Similarly, let S_2 pass through $\mathbf{q}$ in the direction S_2. Then these two lines are defined by the equations

$$S_1 : \mathbf{X}(t) = \mathbf{p} + t S_1 \quad \text{and} \quad S_2 : \mathbf{Y}(s) = \mathbf{q} + s S_2, \tag{11.1}$$

where t and s are arbitrary parameters. These lines are generally easy to identify because they are the axes of the joints that connect the link to the rest of the system.

We position the standard link frame B so that its z-axis is aligned with S_1, and its x-axis is along the common normal N directed from S_1 to S_2. This defines a unique position for the origin $\mathbf{c}$ of this frame along S_1.

In order to construct this coordinate frame we must be able to locate the common normal to two lines.

11.1.2 The Common Normal

The points of intersection $\mathbf{c}$ on S_1 and $\mathbf{r}$ on S_2 with the common normal N have the property that they minimize the distance $d(t,s) = |\mathbf{X}(t) - \mathbf{Y}(s)|$ for all points on these lines. The parameter values t', s' that define $\mathbf{c}$ and $\mathbf{r}$ can be computed by setting the partial derivatives of $d^2(s,t)$ to zero. In what follows, we obtain the same result using the fact that N is perpendicular to both S_1 and S_2.

Now introduce the parameters $t = t'$ and $s = s'$ that define the intersection points $\mathbf{c}$ and $\mathbf{r}$ on S_1 and S_2, so we have

$$\mathbf{c} = \mathbf{p} + t' S_1 \quad \text{and} \quad \mathbf{r} = \mathbf{q} + s' S_2. \tag{11.2}$$

The vector $\mathbf{r} - \mathbf{c}$ is given by

$$\mathbf{r} - \mathbf{c} = a\mathbf{N} = \mathbf{q} - \mathbf{p} + s' S_2 - t' S_1, \tag{11.3}$$

where $a = |\mathbf{r} - \mathbf{c}|$ and N is the unit vector along $S_1 \times S_2$ directed from $\mathbf{c}$ to $\mathbf{r}$.

Determine t' by computing the cross product of this equation with S_2 and then the dot product with N. Similarly, s' is obtained by computing the cross product with S_1, then the dot product with N. The results are

$$t' = \frac{(\mathbf{q} - \mathbf{p}) \times S_2 \cdot \mathbf{N}}{S_1 \times S_2 \cdot \mathbf{N}} \quad \text{and} \quad s' = \frac{(\mathbf{q} - \mathbf{p}) \times S_1 \cdot \mathbf{N}}{S_1 \times S_2 \cdot \mathbf{N}}. \tag{11.4}$$

Substitute t' and s' into (11.1) to define the points $\mathbf{c}$ and $\mathbf{r}$.

The angle α is measured from S_1 to S_2 around N, and is given by

$$\tan \alpha = \frac{S_1 \times S_2 \cdot \mathbf{N}}{S_1 \cdot S_2}. \tag{11.5}$$

In these calculations, we can set $\mathbf{N}$ to be the unit vector in the direction $\mathbf{S}_1 \times \mathbf{S}_2$ in order to determine these points. Then if needed, we change the sign of $\mathbf{N}$ so a is positive.

11.1.3 Coordinate Screw Displacements

To study the relative movement at each joint of a spatial linkage, we introduce three 4×4 matrices that we call *coordinate screw displacements*. Each of these matrices defines a translation along one coordinate axis combined with a rotation about that axis. This is the movement allowed by an RP open chain that has the axis of the revolute joint parallel to the guide of the slider. This assembly is called a cylindric joint, or C-joint, because trajectories traced by points in the moving body lie on cylinders about the joint axis.

Let $\mathbf{S}_1$ be the axis of a cylindric joint that connects a link $\mathbf{S}_1 \mathbf{S}_2$ to ground. Locate the fixed frame F so that its z-axis is along $\mathbf{S}_1$ and its origin is the point $\mathbf{p}$. Attach the link frame B so that its z-axis is along $\mathbf{S}_1$ and its x-axis is along the common normal $\mathbf{N}$ from $\mathbf{S}_1$ and $\mathbf{S}_2$. The displacement of B relative to F consists of a slide d and rotation θ along and around the z-axis of F. Combine the rotation matrix and translation vector for this displacement to form the 4×4 *homogeneous transform*, given by

$$\begin{Bmatrix} X \\ Y \\ Z \\ 1 \end{Bmatrix} = \begin{bmatrix} \cos\theta & -\sin\theta & 0 & 0 \\ \sin\theta & \cos\theta & 0 & 0 \\ 0 & 0 & 1 & d \\ 0 & 0 & 0 & 1 \end{bmatrix} \begin{Bmatrix} x \\ y \\ z \\ 1 \end{Bmatrix}, \tag{11.6}$$

or

$$\mathbf{X} = [Z(\theta,d)]\mathbf{x}. \tag{11.7}$$

This defines the transformation of coordinates $\mathbf{x}$ in B to $\mathbf{X}$ in F that represents the movement allowed by a cylindric joint. Notice that we do not distinguish between point coordinate vectors with and without the fourth component of 1. In what follows the difference should be clear from the context of our calculations.

The transform $[Z(\theta,d)]$ is the *coordinate screw displacement* about the z-axis. We can define similar screw displacements $[X(\cdot,\cdot)]$ and $[Y(\cdot,\cdot)]$ about the x- and y-axes,

$$[X(\theta,d)] = \begin{bmatrix} 1 & 0 & 0 & d \\ 0 & \cos\theta & -\sin\theta & 0 \\ 0 & \sin\theta & \cos\theta & 0 \\ 0 & 0 & 0 & 1 \end{bmatrix}, [Y(\theta,d)] = \begin{bmatrix} \cos\theta & 0 & \sin\theta & 0 \\ 0 & 1 & 0 & d \\ -\sin\theta & 0 & \cos\theta & 0 \\ 0 & 0 & 0 & 1 \end{bmatrix}. \tag{11.8}$$

We use these coordinate screw displacements to formulate the kinematics equations for spatial linkages.

It is useful to note that the inverse of a coordinate screw displacement can be obtained by negating its parameters. For example,

$$[Z(\theta,d)^{-1}] = [Z(-\theta,-d)] = \begin{bmatrix} \cos\theta & \sin\theta & 0 & 0 \\ -\sin\theta & \cos\theta & 0 & 0 \\ 0 & 0 & 1 & -d \\ 0 & 0 & 0 & 1 \end{bmatrix}. \qquad (11.9)$$

Notice that $[Z(\theta,d)^{-1}]$ is not the transpose of $[Z(\theta,d)]$.

11.1.4 The Denavit–Hartenberg Convention

A spatial open chain can be viewed as a sequence of joint axes S_i connected by common normal lines, Figure 11.1. Let A_{ij} be the common normal from joint axis S_i to S_j. The Denavit–Hartenberg convention attaches the link frame B_i such that its z-axis is directed along the axis S_i and its x-axis is directed along the common normal A_{ij}. This convention leaves undefined the initial and final coordinate frames F and M. These frames usually have their z-axes aligned with the first and last axes of the chain. However, their x-axes can be assigned any convenient direction.

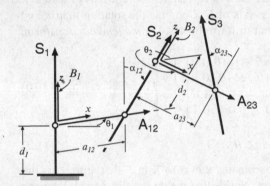

Fig. 11.1 Joint axes S_1, S_2, and S_3 and the link frames B_1 and B_2.

This assignment of standard frames B_i allows us to define the 4×4 transformation $[D]$ that locates the end-link of a spatial open chain as the sequence of transformations

$$[D] = [Z(\theta_1,d_1)][X(\alpha_{12},a_{12})][Z(\theta_2,d_2)] \cdots [X(\alpha_{n-1,n},a_{n-1,n})][Z(\theta_n,d_n)], \quad (11.10)$$

where α_{ij} and a_{ij} are the *twist angle* and *offset* between the axes S_i and S_j. This matrix equation defines the *kinematics equations* of the open chain.

The 4×4 transform $[T_j] = [X(\alpha_{ij},a_{ij})][Z(\theta_j,d_j)]$ is the transformation from frame B_i to B_j. Equation (11.10) is often written as

$$[D] = [T_1][T_2] \cdots [T_n]. \qquad (11.11)$$

Notice that $[T_1] = [Z(\theta_1, d_1)]$.

11.2 The Analysis of Spatial Open Chains

A *robot manipulator* is often designed as spatial open chain in which each joint is actuated. The kinematics equations of the open chain define the position of the end-effector for a given set of values for the joint parameters. It is also necessary to be able to compute the joint parameter values that provide a desired position for end-effector. This is known as the *inverse kinematics problem* in robotics.

11.2.1 The 3R Wrist (S-Joint)

The spherical 3R open chain that is designed so the second axis is perpendicular to both the first and third axes can reach every orientation in space. We use this 3R chain to parameterize an S-joint. This chain is also used as the wrist of a robot manipulator and is of sufficient importance that we formulate its kinematics equations separately. We then assemble it into RS and TS chains.

11.2.1.1 The Kinematics Equations

Introduce the frame F with its z-axis aligned with S_1 and its origin at the wrist center $\mathbf{a}$. Let A_{12} be the common normal to the axes S_1 and S_2. Introduce the link frame B_1 with its z-axis along S_1 as its its x-axis in the direction A_{12}. The angle ϕ_1 is measured from the x-axis of F to A_{12}. Thus, the transformation between these frames is defined by the coordinate screw displacement

$$[T_1] = [Z(\phi_1, 0)]. \tag{11.12}$$

Let A_{23} be the common normal to the axes S_2 and S_3. Introduce the link frame B_2 such that S_2 is its z-axis and A_{23} is its x-axis. The transformation $[X(\pi/2, 0)]$ rotates the frame B_1 around A_{12} to align its z-axis with S_2 and $[Z(\phi_2, 0)]$ rotates this frame into B_2. Thus, the transformation between these two frames is

$$[T_2] = [X(\frac{\pi}{2}, 0)][Z(\phi_2, 0)]. \tag{11.13}$$

The end-effector frame M has its z-axis aligned with S_3, and ϕ_3 is the angle measured from A_{23} to its x-axis. The transformation from B_2 to M consists of a rotation by $\pi/2$ about A_{23} to align S_2 with S_3 followed by a rotation about S_3 by the angle ϕ_3. Thus, we have

$$[T_3] = [X(\frac{\pi}{2},0)][Z(\phi_3,0)].$$ (11.14)

The kinematics equations of this chain are given by

$$[W] = [T_1][T_2][T_3],$$ (11.15)

or

$$\begin{bmatrix} a_{11} & a_{12} & a_{13} & 0 \\ a_{21} & a_{22} & a_{23} & 0 \\ a_{31} & a_{32} & a_{33} & 0 \\ 0 & 0 & 0 & 1 \end{bmatrix}$$

$$= \begin{bmatrix} c\phi_1c\phi_2c\phi_3+s\phi_1s\phi_3 & -c\phi_1c\phi_2s\phi_3+s\phi_1c\phi_3 & c\phi_1s\phi_2 & 0 \\ s\phi_1c\phi_2c\phi_3-c\phi_1s\phi_3 & -s\phi_1c\phi_2s\phi_3-c\phi_1c\phi_3 & s\phi_1s\phi_2 & 0 \\ s\phi_2c\phi_3 & -s\phi_2s\phi_3 & -c\phi_2 & 0 \\ 0 & 0 & 0 & 1 \end{bmatrix}.$$ (11.16)

Notice that this transformation defines a pure rotation parameterized by the three angles ϕ_i, $i = 1,2,3$.

11.2.1.2 Inverse Kinematics

If the orientation of M in a 3R wrist is known, then we have the rotation matrix $[A]$. The kinematics equations $[W] = [A,0]$ can be solved to determine the angles ϕ_i, $i = 1,2,3$. Equate the elements of the third column of (11.16) to obtain

$$\begin{Bmatrix} a_{13} \\ a_{23} \\ a_{33} \end{Bmatrix} = \begin{Bmatrix} c\phi_1s\phi_2 \\ s\phi_1s\phi_2 \\ -c\phi_2 \end{Bmatrix}.$$ (11.17)

This equation yields

$$\phi_1 = \arctan\frac{a_{23}}{a_{13}} \quad \text{and} \quad \phi_2 = \pm\arccos(-a_{33}).$$ (11.18)

The elements of the third row of (11.16) yield

$$\phi_3 = \arctan\frac{-a_{32}}{a_{31}}.$$ (11.19)

Notice that while we have unique solutions for ϕ_1 and ϕ_3, the angle ϕ_2 has two solutions. Both values $\pm\phi_2$ position the end-effector M in the desired orientation.

11.2.2 The RS Chain

The RS chain guides the center point $\mathbf{a}$ of the spherical joint in a circle around the axis R_1 of the revolute joint. In order to analyze this chain we replace the S-joint by the equivalent 3R wrist, such that the first axis of the wrist intersects R_1 in a right angle at $\mathbf{c}$. The distance $a = |\mathbf{a} - \mathbf{c}|$ is the length of the crank.

11.2.2.1 The Kinematics Equations

We position the base frame F so that its origin is at $\mathbf{c}$ and its z-axis is aligned with R_1. Let N be the common normal between R_1 and S_1 of the wrist. Because R_1 and S_1 intersect at $\mathbf{c}$, N passes through $\mathbf{c}$ perpendicular to these two lines. Introduce the link frame A_1 at $\mathbf{c}$ that has R_1 as its z-axis and its x axis aligned with N. The rotation θ_1 of the base revolute joint is measured from the x-axis of F to that of A_1, and we have

$$[R_1] = [Z(\theta_1, 0)]. \tag{11.20}$$

The transformation from A_1 to the link frame B_1 of the 3R wrist consists of a rotation by $\pi/2$ about N that brings R_1 into alignment with S_1, followed by a screw displacement along S_1 of distance a and angle ϕ_1. For convenience, we separate this into the product $[C][T_1]$ where

$$[C] = [X(\frac{\pi}{2}, 0)][Z(0, a)], \tag{11.21}$$

is a constant matrix. Then we have $[T_1] = [Z(\phi_1, 0)]$ as was defined above in (11.15) for the 3R wrist. The result is that the kinematics equations for the RS chain are given by

$$[D] = [R_1][C][T_1][T_2][T_3]. \tag{11.22}$$

11.2.2.2 Inverse Kinematics

If the position of the end-effector M is known, then the elements of 4×4 transform $[D] = [A, \mathbf{P}]$ are specified. Notice that because the wrist transformation $[W] = [T_1][T_2][T_3]$ is a pure rotation, it does not affect the displacement term in the product $[R_1][C]$, which is its fourth column. We equate these columns of $[D]$ and $[R_1][C]$ to obtain

$$\begin{Bmatrix} p_x \\ p_y \\ p_z \end{Bmatrix} = \begin{Bmatrix} a \sin \theta_1 \\ -a \cos \theta_1 \\ 0 \end{Bmatrix} \tag{11.23}$$

and

$$\theta_1 = \arctan \frac{p_x}{-p_y}. \tag{11.24}$$

This defines the transformation $[R_1]$. Because $[C]$ is known, we can compute

$$[W] = [C^{-1}][R_1^{-1}][D] \tag{11.25}$$

and use the inverse kinematic analysis for the 3R wrist to determine the angles ϕ_i, $i = 1,2,3$. The result is the set of joint parameter values needed to place the end-effector of the RS chain in the desired position M.

11.2.3 The TS Chain

We now consider the position analysis of the TS open chain. Figure 11.2. Crane and Duffy use the term T-joint for a pair of revolute joints with axes R_1 and R_2 that intersect at right angles. This joint is often called Hooke's joint and it is used as the gimbal mount for a gyroscope. Replace the S-joint by the equivalent 3R wrist such that the first axis of the wrist passes through the center c of the T-joint and is at a right angle to the second axis R_2. The distance $a = |\mathbf{a} - \mathbf{c}|$ is the length of this chain.

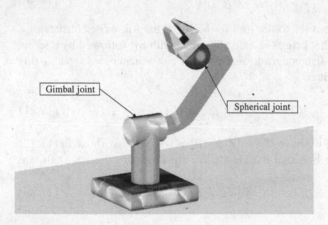

Gimbal joint

Spherical joint

Fig. 11.2 The TS open chain robot.

11.2.3.1 The Kinematics Equations

The RR chain that defines the T-joint is identical to the first two joints of the 3R wrist. As we have done previously, position F with its origin at c and its z-axis along R_1. Introduce the frame A_1 with its origin at c and its x-axis along the common normal N_{12} to the axes R_1 and R_2. The transformation from F to A_1 is $[R_1] = [Z(\theta_1, 0)]$ given above.

Let the N_{23} be the common normal between R_2 and S_1. Attach the frame A_2 with its x-axis along N_{23}. The transformation $[R_2]$ from A_1 to A_2 is

$$[R_2] = [X(\frac{\pi}{2},0)][Z(\theta_2,0)].\tag{11.26}$$

The transformation from A_2 to the link frame B_1 of the 3R wrist is achieved by the same transformation $[C][T_1]$ described above for the RS chain. Thus, the kinematics equations of the TS chain become

$$[D] = [R_1][R_2][C][T_1][T_2][T_3].\tag{11.27}$$

11.2.3.2 Inverse Kinematics

Given the position of the floating link M, we have the elements of the 4×4 matrix $[D] = [A, \mathbf{P}]$. Because the wrist transformation $[W] = [T_1][T_2][T_3]$ is a pure rotation, it does not affect the displacement term in the product $[R_1][R_2][C]$. We equate the fourth columns of $[D]$ and $[R_1][R_2][C]$ to obtain

$$\begin{Bmatrix} p_x \\ p_y \\ p_z \end{Bmatrix} = \begin{Bmatrix} a\cos\theta_1\sin\theta_2 \\ a\sin\theta_1\sin\theta_2 \\ -a\cos\theta_2 \end{Bmatrix}.\tag{11.28}$$

From this equation we can compute

$$\theta_1 = \arctan\frac{p_y}{p_x} \quad\text{and}\quad \theta_2 = \arctan\frac{p_y}{-p_z\sin\theta_1}.\tag{11.29}$$

Now the transformations $[R_1]$, $[R_2]$, and $[C]$ are known, so we have

$$[W] = [C^{-1}][R_2^{-1}][R_1^{-1}][D].\tag{11.30}$$

The inverse kinematic analysis for the 3R wrist determines the values of the angles ϕ_i, $i = 1, 2, 3$. This determines the joint parameter values needed to position the end-effector of the TS chain as required.

11.2.4 The TPS and TRS Chains

If the length a of the TS chain is allowed to vary, then we obtain a six-degree-of-freedom open chain that is often used as the structure for a robot arm. This variation can be introduced by a prismatic joint to form a TPS chain, or by a revolute joint to form a TRS chain.

The kinematics equations for the TS chain (11.27) can also be viewed as the kinematics equations for the TPS robot with the understanding that the length a is now a joint variable s. In order to solve the inverse kinematics for this chain we determine

$$s = \sqrt{p_x^2 + p_y^2 + p_z^2}.\tag{11.31}$$

The remaining joint parameters are obtained using the formulas above for the TS chain.

For the TRS chain, we have an elbow joint E inserted between the base point $\mathbf{c}$ and the wrist center $\mathbf{a}$. Let E be parallel to the joint R_2 and introduce $\mathbf{e}$ as the point on E closest to the base point $\mathbf{c}$. The distance $a_1 = |\mathbf{e} - \mathbf{c}|$ defines the length of the first link R_2E. Now let the axis S_1 of the first joint of the wrist intersect E at a right angle at $\mathbf{e}$. The distance $a_2 = |\mathbf{a} - \mathbf{e}|$ is the length of the second link.

The triangle $\triangle \mathbf{cea}$ is in the plane perpendicular to E, and the exterior angle θ_3 at $\mathbf{e}$ of the elbow joint controls the length $s = |\mathbf{a} - \mathbf{c}|$. The kinematics equations of the TS chain can be modified to include this elbow joint by redefining the matrix $[C]$, such that

$$[C] = [X(0,a_1)][Z(\theta_3,0)][X(\frac{\pi}{2},0)][Z(0,a_2)]. \tag{11.32}$$

The first two matrices define the transformation to the elbow joint E which is parallel to R_2. The second two matrices are the same as the original matrix $[C]$ but now adapted to the new joint axis.

The inverse kinematics solution for the TRS chain is essentially the same as that of the TPS chain. We compute the length s using (11.31). The joint angle θ_3 is obtained from the cosine law of the triangle $\triangle \mathbf{cea}$,

$$\theta_3 = \arccos \frac{s^2 - a_1^2 - a_2^2}{2a_1a_2}. \tag{11.33}$$

This yields two values $\pm\theta_3$ for the elbow joint. The formula for the joint angle θ_1 is the same as in (11.29). However, the equation for θ_2 must be adjusted to accommodate the angle $\psi = \angle \mathbf{eca}$, which is given by

$$\psi = \arctan \frac{a_2 \sin \theta_3}{a_1 + a_2 \cos \theta_3}, \tag{11.34}$$

The new formula for θ_2 is

$$\theta_2 = \arctan \frac{p_y}{-p_z \sin \theta_1} - \psi. \tag{11.35}$$

Once θ_1, θ_2, and θ_3 are known, we can use (11.30) to isolate the wrist transformation. Then the inverse kinematics solution for the 3R wrist completes the analysis.

11.2.5 The CC Chain

A CC open chain is formed by a link that is connected to ground by a fixed cylindric joint and to an end-effector a moving cylindric joint, Figure 11.3. Denote the axis of the base joint by O and the axis of the moving joint by A. The common normal N to these two axes identifies points $\mathbf{c}$ on O and $\mathbf{r}$ on A.

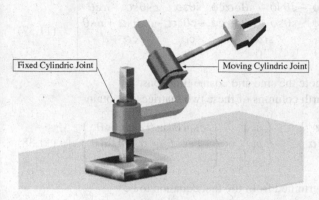

Fig. 11.3 The CC open chain robot.

Locate the fixed frame F at the base of this chain so that its z-axis lies on O. The displacement of this joint is measured by the distance d between **c** and the origin of F, and the rotation angle θ is measured from the x-axis of F to the common normal line N. Similarly, place the moving frame M in the end-effector so its z-axis lies on A. The displacement at this joint is measured by the distance c from **r** to the origin of M and the angle ϕ from N to the x-axis of M.

11.2.5.1 The Kinematics Equations

Now introduce the link frame B that has its z-axis aligned with O and its x-axis along N. Let $a = |\mathbf{r} - \mathbf{c}|$ be the length of the crank and let α be the twist angle measured around N from O to A. Then, the screw displacement $[X(\alpha, a)]$ rotates the axis O about N and aligns it with A. The result is the kinematics equations of the CC chain are given by

$$[D] = [Z(\theta, d)][X(\alpha, a)][Z(\phi, c)]. \tag{11.36}$$

If the slide distances d and c are constrained to be constant, then these kinematics equations become those of a spatial RR chain. The kinematics equations for the spatial RP, PR, and PP chains are obtained in the same way by constraining the appropriate rotation angle or slide distance of the fixed and moving C-joints to be a constant.

11.2.5.2 Inverse Kinematics

Let the desired position of the end-effector of a CC chain be specified by $[D] = [A, \mathbf{P}]$. We expand the right side of (11.36) to obtain equations that define the values of the joint parameters θ, d, ϕ and c, that is,

$$[A, \mathbf{P}] = \begin{bmatrix} c\theta c\phi - s\theta c\alpha s\phi & -c\theta s\phi - s\theta c\alpha c\phi & s\theta s\alpha & cs\theta s\alpha + ac\theta \\ s\theta c\phi + c\theta c\alpha s\phi & -s\theta s\phi + c\theta c\alpha c\phi & -c\theta s\alpha & -cc\theta s\alpha + as\theta \\ s\alpha s\phi & s\alpha c\phi & c\alpha & cc\alpha + d \\ 0 & 0 & 0 & 1 \end{bmatrix}. \quad (11.37)$$

In this equation s and c denote the sine and cosine functions.

Equate the third and fourth columns of these two matrices to obtain

$$\begin{Bmatrix} a_{13} \\ a_{23} \\ a_{33} \end{Bmatrix} = \begin{Bmatrix} \sin\theta\sin\alpha \\ -\cos\theta\sin\alpha \\ \cos\alpha \end{Bmatrix}, \quad \begin{Bmatrix} p_x \\ p_y \\ p_z \end{Bmatrix} = \begin{Bmatrix} c\sin\theta\sin\alpha + a\cos\theta \\ -c\cos\theta\sin\alpha + a\sin\theta \\ c\cos\alpha + d \end{Bmatrix}. \quad (11.38)$$

The crank rotation θ is determined from the first equation to be

$$\theta = \arctan\frac{a_{13}}{-a_{23}}. \quad (11.39)$$

Once θ is known, the parameters c and d are obtained from the second equation as

$$c = \frac{p_x - a\cos\theta}{\sin\theta\sin\alpha} \quad \text{and} \quad d = p_z - \frac{p_x - a\cos\theta}{\sin\theta\sin\alpha}\cos\alpha. \quad (11.40)$$

Finally, we determine the angle ϕ by equating the third rows of (11.36) and obtain

$$\phi = \arctan\frac{a_{31}}{a_{32}}. \quad (11.41)$$

These relations prescribe the configuration of the CC chain for a specified location of the end-effector.

11.3 Velocity Analysis of Spatial Open Chains

The kinematics equations of an open chain define the trajectory $\mathbf{X}(t) = [D(t)]\mathbf{x}$ of points in the end-link. The velocity of this trajectory is given by $\mathbf{V} = \dot{\mathbf{X}}(t) = [\dot{D}(t)]\mathbf{x}$. Substitute $\mathbf{x} = [D^{-1}]\mathbf{X}(t)$, so we have

$$\mathbf{V} = [\dot{D}][D^{-1}]\mathbf{X} = [S]\mathbf{X}. \quad (11.42)$$

Using the fact that the 4×4 homogeneous transform $[D(t)] = [A(t), \mathbf{d}(t)]$ consists of a rotation matrix $[A]$ and translation vector $\mathbf{d}$ we compute

$$\mathbf{V} = [\dot{A}][A^T]\mathbf{X} - [\dot{A}][A^T]\mathbf{d} + \dot{\mathbf{d}} = [\Omega](\mathbf{X} - \mathbf{d}) + \dot{\mathbf{d}}, \quad (11.43)$$

where $[\Omega]$ is the angular velocity matrix of the moving link. Introduce the angular velocity vector $\mathbf{w}$, so that this equation becomes the familiar definition of the velocity of a point in a moving body,

$$\mathbf{V} = \mathbf{w} \times (\mathbf{X} - \mathbf{d}) + \dot{\mathbf{d}}. \tag{11.44}$$

The 4×4 matrix $[S] = [\dot{D}][D^{-1}]$ is a generalization of the angular velocity matrix. It has the general form

$$[S] = \begin{bmatrix} \Omega & \mathbf{d} \times \mathbf{w} + \dot{\mathbf{d}} \\ 000 & 0 \end{bmatrix}. \tag{11.45}$$

The components of this matrix are assembled into the six-vector

$$\mathsf{T} = \left\{ \begin{matrix} \mathbf{w} \\ \mathbf{d} \times \mathbf{w} + \dot{\mathbf{d}} \end{matrix} \right\}, \tag{11.46}$$

called the *twist* of the moving body.

The axis of this twist $\mathsf{L} : \mathbf{P}(t) = \mathbf{c} + t\mathbf{s}$ is directed along the angular velocity vector $\mathbf{w} = |\mathbf{w}|\mathbf{s}$, and through the point

$$\mathbf{c} = \frac{\mathbf{w} \times (\mathbf{d} \times \mathbf{w} + \dot{\mathbf{d}}^*)}{\mathbf{w} \cdot \mathbf{w}}, \tag{11.47}$$

where $\dot{\mathbf{d}}^* = \dot{\mathbf{d}} - (\mathbf{s} \cdot \dot{\mathbf{d}})\mathbf{s}$. This line is the *instantaneous screw axis* of the motion. Substitute $\mathbf{P}(t)$ into (11.44) to see that the velocity of these points are directed along the line L.

The set of instantaneous screw axes generated as a body moves in space is called its *axode*. In the following section we compute the twist for a general spatial open chain.

11.3.1 Partial Twists of an Open Chain

Using the Denavit-Hartenberg convention, the kinematics equations for six-degree-of-freedom open chain can be written in the form

$$[D] = [Z(\theta_1, d_1)][X(\alpha_{12}, a_{12})][Z(\theta_2, d_2)] \cdots [X(\alpha_{56}, a_{56})][Z(\theta_6, d_6)]. \tag{11.48}$$

Only six of the twelve joint parameters θ_i, d_i, $i = 1, \ldots, 6$, are variable. The parameters α_{ij} and a_{ij} define the angle of twist and length of each link.

Let the joint angles θ_i, $i = 1, \ldots, 6$ be variable and the joint slides constant, that is, $\dot{d}_i = 0$. The analysis is the same if any of the slides are variable and the joint angles are constant. Compute the partial derivative matrices

$$[S_i] = \left[\frac{\partial D}{\partial \theta_i} \right] [D^{-1}], \tag{11.49}$$

so we have

$$[S] = \dot{\theta}_1 [S_1] + \dot{\theta}_2 [S_2] + \cdots + \dot{\theta}_6 [S_6]. \tag{11.50}$$

Each matrix $[S_i]$ has the form

$$[S_i] = [D_i][K][D_i^{-1}], \tag{11.51}$$

where
$$[D_i] = [Z(\theta_1, d_1)][X(\alpha_{12}, a_{12})] \cdots [X(\alpha_{i-1,i}, a_{i-1,i})] \tag{11.52}$$

is the transformation to the ith joint of the chain. The 4×4 matrix $[K]$ in each of these terms is

$$[K] = \begin{bmatrix} 0 & -1 & 0 & 0 \\ 1 & 0 & 0 & 0 \\ 0 & 0 & 0 & 0 \\ 0 & 0 & 0 & 0 \end{bmatrix}. \tag{11.53}$$

Let the transformation $[D_i]$ consist of the rotation matrix $[A_i]$ and translation vector $\mathbf{d}_i$, and compute

$$[S_i] = \begin{bmatrix} A_i & \mathbf{d}_i \\ 000 & 1 \end{bmatrix} \begin{bmatrix} K & 0 \\ 000 & 0 \end{bmatrix} \begin{bmatrix} A_i^T & -[A_i^T]\mathbf{d}_i \\ 000 & 1 \end{bmatrix} = \begin{bmatrix} A_i K A_i^T & -A_i K A_i^T \mathbf{d}_i \\ 000 & 0 \end{bmatrix}. \tag{11.54}$$

Here $[K]$ is the upper left 3×3 submatrix of (11.53) that performs the cross product by $\vec{k}$, that is, $[K]\mathbf{y} = \vec{k} \times \mathbf{y}$. Thus, the elements of each $[S_i]$ can be assembled into the *partial twist* $\mathsf{S}_i = (\mathbf{S}_i, \mathbf{d}_i \times \mathbf{S}_i)^T$, where $\mathbf{S}_i = [A_i]\vec{k}$. The result is that (11.50) can be written as

$$\mathsf{T} = \begin{Bmatrix} \mathbf{w} \\ \mathbf{d} \times \mathbf{w} + \dot{\mathbf{d}} \end{Bmatrix} = \begin{bmatrix} \mathbf{S}_1 & \mathbf{S}_2 & \cdots & \mathbf{S}_6 \\ \mathbf{d}_1 \times \mathbf{S}_1 & \mathbf{d}_2 \times \mathbf{S}_2 & \cdots & \mathbf{d}_6 \times \mathbf{S}_6 \end{bmatrix} \begin{Bmatrix} \dot{\theta}_1 \\ \vdots \\ \dot{\theta}_6 \end{Bmatrix}. \tag{11.55}$$

The 6×6 matrix in this equation is closely related to the Jacobian defined for a robot manipulator.

11.3.2 The Jacobian of a Spatial Open Chain

The Jacobian of a six-degree-of-freedom open chain relates the joint rates $\dot{\theta}_i$ to the velocity of its end-effector M. This velocity is usually defined as the six-vector $\mathsf{V} = (\mathbf{d}, \mathbf{w})^T$, where $\mathbf{d}$ locates the origin of M and $\mathbf{w}$ is its angular velocity.

Notice that V can be obtained from the screw T in (11.46) by subtracting $\mathbf{d} \times \mathbf{w}$ from the second 3-vector component and then interchanging the two sets of vectors, that is,

$$\mathsf{V} = \begin{Bmatrix} \mathbf{d} \\ \mathbf{w} \end{Bmatrix} = \begin{bmatrix} 0 & I \\ I & 0 \end{bmatrix} \left(\begin{Bmatrix} \mathbf{w} \\ \mathbf{d} \times \mathbf{w} + \dot{\mathbf{d}} \end{Bmatrix} - \begin{Bmatrix} 0 \\ \mathbf{d} \times \mathbf{w} \end{Bmatrix} \right). \tag{11.56}$$

Using the identity $\mathbf{w} = \dot{\theta}_1 \mathbf{S}_1 + \cdots + \dot{\theta}_6 \mathbf{S}_6$ we obtain

$$\begin{Bmatrix} \mathbf{d} \\ \mathbf{w} \end{Bmatrix} = \begin{bmatrix} (\mathbf{d}_1 - \mathbf{d}) \times \mathbf{S}_1 & (\mathbf{d}_2 - \mathbf{d}) \times \mathbf{S}_2 & \cdots & 0 \\ \mathbf{S}_1 & \mathbf{S}_2 & \cdots & \mathbf{S}_6 \end{bmatrix} \begin{Bmatrix} \dot{\theta}_1 \\ \vdots \\ \dot{\theta}_6 \end{Bmatrix}, \qquad (11.57)$$

where $\mathbf{d}_i - \mathbf{d}$ is the vector from the origin of M to the ith joint axis. Notice that $\mathbf{d}_6 = \mathbf{d}$. Therefore, this term cancels. This is the *Jacobian* of the spatial open chain.

11.4 The RSSR Linkage

The spatial RSSR four-bar linkage can be constructed by rigidly connecting the end-links of two RS chains. The resulting system has two degrees of freedom, because the sum of the freedom at each joint is eight. One of these degrees of freedom is the rotation of the coupler about the axis joining the two S-joints. This freedom is independent of the configuration of the input and output cranks, and can be arbitrarily specified. In what follows, we focus on the constraint equations that relate the crank rotations of the chain.

Let the lines O and C be the axes of the two revolute joints and let N be their common normal. We assume that the two S-joints rotate in planes perpendicular to the axes O and C. If $\mathbf{c}_1$ and $\mathbf{c}_2$ are the points of intersection of N with the two axes, then we locate these cranks at the distances p and q from $\mathbf{c}_1$ and $\mathbf{c}_2$, respectively. The length of the ground link is $g = |\mathbf{c}_2 - \mathbf{c}_1|$ and its twist angle γ is measured from O to C around N.

We locate the fixed frame F so that its origin is $\mathbf{c}_1$, its z-axis is along $\mathbf{O}$, and its x-axis is along N. Let the radius of the input crank be a, then the coordinates $\mathbf{a}$ of the center of the input S-joint are given by

$$\mathbf{a} = [Z(\theta, p)](a\vec{\imath}), \qquad (11.58)$$

where θ is the rotation angle of the input crank.

To determine the coordinates $\mathbf{b}$ of the center of the output S-joint, we introduce the frame F' with its origin at $\mathbf{c}_2$, its z-axis aligned with C, and its x-axis along N. In this frame we have $\mathbf{b}' = [Z(\psi, q)](b\vec{\imath})$ where b is the radius of the crank and ψ is the output rotation angle. It is now easy to see that the coordinates of $\mathbf{b}$ in F are given by

$$\mathbf{b} = [X(\gamma, g)][Z(\psi, q)](b\vec{\imath}). \qquad (11.59)$$

Thus, the coordinates of the S-joints are given by

$$\mathbf{a} = \begin{Bmatrix} a\cos\theta \\ a\sin\theta \\ p \end{Bmatrix} \quad \text{and} \quad \mathbf{b} = \begin{Bmatrix} b\cos\psi + g \\ b\cos\gamma\sin\psi - q\sin\gamma \\ b\sin\gamma\sin\psi + q\cos\gamma \end{Bmatrix}. \qquad (11.60)$$

11.4.1 The Output Angle

The input and output crank of the RSSR linkage must move in a way that maintains a constant distance h between the centers $\mathbf{a}$ and $\mathbf{b}$ of the two S-joints. This yields the constraint equation for the chain as

$$(\mathbf{b} - \mathbf{a}) \cdot (\mathbf{b} - \mathbf{a}) = h^2. \tag{11.61}$$

Substitute (11.60) into this equation to obtain

$$A(\theta)\cos\phi + B(\theta)\sin\phi = C(\theta), \tag{11.62}$$

where

$$
\begin{aligned}
A(\theta) &= 2ba\cos\theta - 2bg, \\
B(\theta) &= 2ba\cos\gamma\sin\theta + 2bp\sin\gamma, \\
C(\theta) &= g^2 + q^2 + a^2 + b^2 + p^2 - h^2 - 2ag\cos\theta + 2qa\sin\gamma\sin\theta - 2qp\cos\gamma.
\end{aligned}
\tag{11.63}
$$

This equation is solved, as shown in (A.1), to yield

$$\psi(\theta) = \arctan\frac{B}{A} \pm \arccos\frac{C}{\sqrt{A^2 + B^2}}. \tag{11.64}$$

Note that there are two output crank angles ψ for each input θ.

We now show that the constraint equations for the planar and spherical 4R chains can be obtained as special cases of (11.62).

11.4.1.1 Planar 4R Linkage

A planar 4R linkage can be viewed as an RSSR linkage that has parallel revolute axes O and C, which means that $\gamma = 0$. In order to have the cranks in the same plane we set $p = q$. The result is the coefficients of (11.62) become

$$
\begin{aligned}
A(\theta) &= 2ba\cos\theta - 2bg, \\
B(\theta) &= 2ba\sin\theta, \\
C(\theta) &= g^2 + a^2 + b^2 - h^2 - 2ag\cos\theta.
\end{aligned}
\tag{11.65}
$$

Compare these coefficients with (2.50) to see that this is the constraint equation of a planar 4R linkage.

11.4.1.2 Spherical 4R Linkage

A spherical 4R linkage can be considered to be an RSSR linkage in which the axes O and C intersect, which means that $g = 0$. Set the location of the revolute joints at the distance $p = q = 1$ from the point of intersection $\mathbf{c}$ of the two axes. This allows us to define the angular dimensions α and β of the input and output links such that

$$a = \tan\alpha \quad \text{and} \quad b = \tan\beta. \tag{11.66}$$

The coupler angle η is defined by the triangle $\triangle\mathbf{acb}$, with sides of lengths $|\mathbf{a} - \mathbf{c}| = 1/\cos\alpha$ and $|\mathbf{b} - \mathbf{c}| = 1/\cos\beta$, respectively. The cosine law yields the relation

$$h^2 = \frac{1}{\cos^2\alpha} + \frac{1}{\cos^2\beta} - \frac{2\cos\eta}{\cos\alpha\cos\beta}. \tag{11.67}$$

Substitute these formulas for a, b, and h and the values for g, p, and q into the equations (11.63). The term $2/\cos\alpha\cos\beta$ cancels in all three coefficients to yield

$$
\begin{aligned}
A(\theta) &= \sin\beta\sin\alpha\cos\theta, \\
B(\theta) &= \sin\beta\sin\alpha\cos\gamma\sin\theta + \sin\beta\cos\alpha\sin\gamma, \\
C(\theta) &= \cos\eta + \sin\alpha\cos\beta'\sin\gamma\sin\theta - \cos\gamma\cos\alpha\cos\beta.
\end{aligned}
\tag{11.68}
$$

Compare these coefficients to those derived in (7.27). The difference between the two sets is due to different conventions for measuring the input and output angles. Denote by θ' and ψ' the input and output angles used in (7.27). Then substitute $\theta = \theta' - \pi/2$ and $\psi = \psi' - \pi/2$ into (11.68) and notice that $\sin\theta$ becomes $-\cos\theta'$ and $\cos\theta$ becomes $\sin\theta'$. This also changes A into B' and B into $-A'$. The result is that these coefficients are transformed into those in (7.27).

11.4.1.3 Spatial 4R Linkage (Bennett's Linkage)

An interesting special case of the RSSR is the spatial 4R chain, called *Bennett's linkage*. For this linkage, the base revolute joints must be located along the common normal line N of their axes, which means that $p = q = 0$. Furthermore, the opposite sides of the linkage must be equal, so $g = h$ and $a = b$.

We replace the S-joints with revolute joints that have axes A and B. These axes are positioned at $\mathbf{a}$ and $\mathbf{b}$ so that the input crank OA and the output crank CB have the same twist angle α. Furthermore, the twist angle along the floating link AB must be γ, which is the twist angle of the ground link OC. The result is an assembly of four joints that have consecutive common normals that intersect, and opposite sides that have the same dimensions.

Even with these constraints, the linkage will not move unless the dimensions satisfy the additional relationship

$$\frac{a}{\sin \alpha} = \frac{g}{\sin \gamma} \tag{11.69}$$

known as Bennett's condition.

Substitute $p = q = 0$, $g = h$, and $a = b$ into (11.62) to obtain the equation

$$(a\cos \theta - g)\cos \psi + (a\cos \gamma \sin \theta)\sin \psi = a - g\cos \theta. \tag{11.70}$$

Divide this equation by a and substitute $g/a = \sin \gamma / \sin \alpha$ in order to obtain the constraint equation for Bennett's linkage,

$$A(\theta)\cos \psi + B(\theta)\sin \psi = C(\theta), \tag{11.71}$$

where

$$A(\theta) = \sin \alpha \cos \theta - \sin \gamma,$$
$$B(\theta) = \sin \alpha \cos \gamma \sin \theta,$$
$$C(\theta) = \sin \alpha - \sin \gamma \cos \theta. \tag{11.72}$$

The solution of this equation yields ψ as a function of θ.

It is useful to examine another approach to the solution of (11.71). Introduce the parameters $u = \tan(\theta/2)$ and $v = \tan(\psi/2)$ so we have

$$\cos \theta = \frac{1-u^2}{1+u^2}, \sin \theta = \frac{2u}{1+u^2}, \cos \psi = \frac{1-v^2}{1+v^2}, \sin \psi = \frac{2uv}{1+v^2}. \tag{11.73}$$

Substitute these formulas into (11.71) to obtain

$$u^2(\sin \alpha + \sin \gamma) - 2\sin \alpha \cos \gamma uv + v^2(\sin \alpha - \sin \gamma) = 0. \tag{11.74}$$

The solution of this quadratic equation for the ratio v/u yields

$$\frac{\tan \frac{\psi}{2}}{\tan \frac{\theta}{2}} = \frac{\sin \alpha \cos \gamma \pm \cos \alpha \sin \gamma}{\sin \alpha - \sin \gamma}, \tag{11.75}$$

which shows that this ratio remains constant as the linkage moves.

11.5 The RSSP Linkage

The RSSP linkage is the spatial version of a slider-crank linkage. Let O be the axis of the R-joint and let **a** be the center of the S-joint for the input RS crank. The output link is the slider of the PS chain. Let **b** be the center of the output S-joint **b**. Then its path is a line C that we consider to be the guide of the P-joint. The common normal N between O and C defines the points c_1 and c_2, respectively, on these lines.

Locate the base frame F with its origin at c_1, its z-axis along O, and its x-axis along N. In this frame the coordinates of a are the same as are given in (11.60). The coordinates of b are given by

$$\mathbf{b} = [Z(\gamma, g)](s\vec{k}) = \left\{ \begin{array}{c} g \\ -s\sin\gamma \\ s\cos\gamma \end{array} \right\}. \tag{11.76}$$

The fact that the distance between the two pivots a and b is the constant length h yields the constraint equation for the RSSP chain as

$$s^2 + 2(a\sin\gamma\sin\theta - p\cos\gamma)s + (g^2 + a^2 + p^2 - h^2 - 2ag\cos\theta) = 0. \tag{11.77}$$

For each value of the input crank θ we solve this quadratic equation to obtain two values for the location s of the output slider.

There are several interesting special cases for this linkage.

11.5.0.4 The Planar RRRP

The planar slider-crank occurs when the axes O and C are at right angles and the revolute joint is located in the xy-plane. This means that $\gamma = \pi/2$ and $p = 0$. In this case, equation (11.77) becomes

$$s^2 + 2(a\sin\theta)s + (g^2 + a^2 - h^2 - 2ag\cos\theta) = 0. \tag{11.78}$$

Compare this to the constraint equation of the planar slider-crank (2.23). Notice that because $r = a$, $L = h$, and $e = g$, these equations are the same.

11.5.0.5 The Symmetric RSSP

If the axis O of the input RS crank intersects the guide C of the slider, then slider moves in the yz-plane as the input crank rotates about the z-axis. In this case $g = 0$ and the constraint equation takes the form

$$s^2 + 2(a\sin\gamma\sin\theta - p\cos\gamma)s + (a^2 + p^2 - h^2) = 0. \tag{11.79}$$

Notice that because $\sin\theta = \sin(\pi - \theta)$, this equation is the same for positions of the input crank that are symmetric relative to the yz-plane. The result is a movement of the output link that is a symmetric function of the input crank angle.

11.5.0.6 The Sinusoidal RSSP

Choose the dimensions of the symmetric RSSP so that the constant term is zero, that is,

$$h^2 = a^2 + p^2. \tag{11.80}$$

Then (11.79) has one solution $s = 0$ and a second solution

$$s = 2p\cos\gamma - 2a\sin\gamma\sin\theta, \tag{11.81}$$

which is a sinusoidal function of the input angle θ of amplitude $A = 2a\sin\gamma$.

11.6 The 4C Linkage

A 4C closed chain is formed by connecting the end-links of two CC chains. Each of the four cylindric joints has two degrees of freedom for a total of eight joint freedoms, which means that the chain has degree of freedom of two. The two independent parameters are the slide d and rotation θ of the input crank its fixed axis. See Figure 11.4.

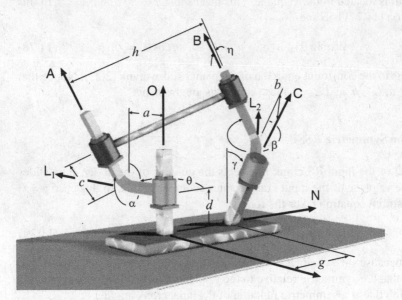

Fig. 11.4 The 4C linkage.

11.6.0.7 The Link Dimensions

Let the fixed and moving axes of the input crank be O and A, respectively. Denote the common normal between these lines by L_1 and its points of intersection by $\mathbf{a}_1$ on O and $\mathbf{r}_1$ on A. This crank has length $a = |\mathbf{r}_1 - \mathbf{a}_1|$ and the twist angle angle α measured about L_1 from O to A. For convenience assemble these parameters into the ordered pair $\hat{\alpha} = (\alpha, a)$.

Let C and B be the fixed and moving axes of the output crank. In this case, let L_2 be the common normal and $\mathbf{a}_2$ and $\mathbf{r}_2$ its points of intersection with the axes C and B. The length of this crank is $b = |\mathbf{r}_2 - \mathbf{a}_2|$ and its twist angle is β measured around L_2. Collect these parameters into the ordered pair $\hat{\beta} = (\beta, b)$.

The floating link is formed by the common normal M between the moving axes A and B, which defines points $\mathbf{b}_1$ and $\mathbf{b}_2$. The distance $h = |\mathbf{b}_2 - \mathbf{b}_1|$ and angle η from A to B define its dimensions, denoted by $\hat{\eta} = (\eta, h)$. The ground link of the chain is defined by the common normal N between the base joints O and C and $\mathbf{c}_1$ and $\mathbf{c}_2$ be the points of intersection with these axes. Its dimensions are given by $\hat{\gamma} = (\gamma, g)$.

11.6.1 The Kinematics Equations

Position the fixed frame F with its origin at $\mathbf{c}_1$, its z-axis along O and its x-axis along N. Introduce the link frame T_1 with its origin at $\mathbf{a}_1$ and its x-axis aligned with L_1. The displacement of the input crank is given by the screw displacement of T_1 along O of distance $d = |\mathbf{a}_1 - \mathbf{c}_1|$ and angle θ is measured from the x-axis of F to L_1. We assemble these joint parameters into the pair $\hat{\theta} = (\theta, d)$.

The position of the coupler relative to the input crank is defined by attaching the frame M with its origin at $\mathbf{b}_1$, its z-axis along A, and its x-axis aligned with M. The distance $c = |\mathbf{b}_1 - \mathbf{r}_1|$ is the slide of the moving joint, and ϕ is its rotation angle measured from L_1 to M. Assemble the joint parameters into the pair $\hat{\phi} = (\phi, c)$. These definitions yield the kinematics equations of the input crank OA as

$$[D_{\text{in}}] = [Z(\theta, d)][X(\alpha, a)][Z(\phi, c)]. \tag{11.82}$$

Let the frame F' be located on the output crank CB so that its origin is $\mathbf{c}_2$, its z-axis is C, and its x-axis is L_2. We can attach link frames T_2 and M' in exactly the same way to obtain the kinematics equations for this crank as

$$[D_{\text{out}}] = [Z(\psi, e)][X(\beta, b)][Z(\zeta, f)], \tag{11.83}$$

where $\hat{\psi} = (\psi, e)$ and $\hat{\zeta} = (\zeta, f)$ define the rotation angle and sliding distance at the fixed and moving joints, respectively.

To define the kinematics equations of the 4C closed chain, we introduce transformation $[X(\gamma, g)]$ from F to F' and the transformation $[X(\eta, h)]$ from M to M'. The result is that the coordinate transformation $[T]$ that locates the frame M' in F is defined in two ways, that is,

$$[T] = [D_{in}][X(\eta,h)] = [X(\gamma,g)][D_{out}]. \tag{11.84}$$

Substitute (11.82) and (11.83) into this equation to obtain the kinematics equations of the 4C chain.

11.6.2 The Spherical Image

Associated with a 4C linkage is the spherical linkage with joint axes formed by the direction vectors **O**, **A**, **B**, and **C** of the lines O, A, B, and C. This linkage is known as the *spherical image* of the 4C chain. We follow a slightly different derivation and obtain the same constraint equation for this spherical 4R chain.

11.6.2.1 The Output Angle

The direction vectors **A** and **B** of the moving axes of the spherical image can be obtained from the third column of the transformations $[D_{in}]$ and $[X(\gamma,g)][D_{out}]$, respectively. These computations yield

$$\mathbf{A} = \begin{Bmatrix} \sin\theta\sin\alpha \\ -\cos\theta\sin\alpha \\ \cos\alpha \end{Bmatrix}, \quad \mathbf{B} = \begin{Bmatrix} \sin\psi\sin\beta \\ -\cos\gamma\cos\psi\sin\beta - \sin\gamma\cos\beta \\ -\sin\gamma\cos\psi\sin\beta + \cos\gamma\cos\beta \end{Bmatrix}. \tag{11.85}$$

The fact that $\mathbf{A}\cdot\mathbf{B} = \cos\eta$ for all positions of the linkage yields the constraint equation

$$A(\theta)\cos\psi + B(\theta)\sin\psi = C(\theta), \tag{11.86}$$

where

$$A(\theta) = \cos\theta\sin\alpha\cos\gamma\sin\beta - \cos\alpha\sin\gamma\sin\beta,$$
$$B(\theta) = \sin\theta\sin\alpha\sin\beta,$$
$$C(\theta) = \cos\eta - \cos\theta\sin\alpha\sin\gamma\cos\beta - \cos\alpha\cos\gamma\cos\beta. \tag{11.87}$$

This constraint equation is the same as was derived previously as (7.27). Solve this equation to determine the output angle ψ as a function of the crank angle θ.

11.6.2.2 The Coupler Angle

The kinematics equations provide two ways to define the direction **B**. It is the third column of $[X(\gamma,g)][D_{out}]$ and the third column of $[D_{in}][X(\eta,h)]$. Focusing only on the coordinate rotations, we can write this as

$$[Z(\theta)][X(\alpha)][Z(\phi)][X(\eta)]\vec{k} = \mathbf{B}, \qquad (11.88)$$

where $\mathbf{B}$ is defined above in (11.85), To simplify this equation, introduce the notation $\mathbf{Y} = [X(\alpha)][Z(\phi)][X(\eta)]\vec{k}$, so we have

$$[Z(\theta)]\mathbf{Y} = \mathbf{B}, \quad \text{or} \quad \mathbf{Y} = [Z(\theta)]^T \mathbf{B}, \qquad (11.89)$$

which yields the relations

$$\left\{ \begin{array}{c} s\phi s\eta \\ -c\alpha c\phi s\eta - s\alpha c\eta \\ -s\alpha c\phi s\eta + c\alpha c\eta \end{array} \right\} = \left\{ \begin{array}{c} c\theta(s\psi s\beta) - s\theta(c\gamma c\psi s\beta + s\gamma c\beta) \\ -s\theta(s\psi s\beta) - c\theta(c\gamma c\psi s\beta + s\gamma c\beta) \\ -s\gamma c\psi s\beta + c\gamma c\beta \end{array} \right\}. \qquad (11.90)$$

Solving for ϕ from the first and second components of these vectors, we obtain

$$\sin\phi = \frac{c\theta(s\psi s\beta) - s\theta(c\gamma c\psi s\beta + s\gamma c\beta)}{\sin\eta},$$

$$\cos\phi = \frac{s\theta(s\psi s\beta) + c\theta(c\gamma c\psi s\beta + s\gamma c\beta) + s\alpha c\eta}{c\alpha s\eta}. \qquad (11.91)$$

These equations yield the appropriate coupler angle ϕ for either solution selected for the output angle ψ.

11.6.3 The Vector Loop Equation

We now formulate equations that define the output slides e at C, f at B, and the coupler slide c at A in terms of the rotation angle θ and sliding distance d of the input crank OA. To do this, we equate the fourth columns of the kinematics equations and obtain the *vector loop equations* of the 4C chain.

Introduce the notation $\mathbf{P}_{in}$ and $\mathbf{P}_{out}$ for the fourth columns of the left and right sides of the kinematics equations, that is, of the matrices $[D_{in}][X(\eta, h)]$ and $[X(\gamma, g)][D_{out}]$, respectively. Expansion of these equations yields

$$\mathbf{P}_{in} = \left\{ \begin{array}{c} ac\theta + h(c\theta c\phi - s\theta s\phi c\alpha) \\ as\theta + h(s\theta\phi + c\theta c\phi c\alpha) \\ d + hs\alpha s\phi \end{array} \right\} + c \left\{ \begin{array}{c} s\theta s\alpha \\ -c\theta s\alpha \\ c\alpha \end{array} \right\} \qquad (11.92)$$

and

$$\mathbf{P}_{out} = \left\{ \begin{array}{c} g + bc\psi \\ bs\psi c\gamma \\ bs\psi s\gamma \end{array} \right\} + e \left\{ \begin{array}{c} 0 \\ -s\gamma \\ c\gamma \end{array} \right\} + f \left\{ \begin{array}{c} s\psi s\beta \\ -c\psi s\beta c\gamma - c\beta s\gamma \\ -c\psi s\beta s\gamma + c\beta c\gamma \end{array} \right\}. \qquad (11.93)$$

Because $\mathbf{P}_{in} = \mathbf{P}_{out}$, we can construct the matrix equation

$$\begin{bmatrix} -s\theta s\alpha & 0 & s\psi s\beta \\ c\theta s\alpha & -s\gamma & -c\psi s\beta c\gamma - c\beta s\gamma \\ -c\alpha & c\gamma & -c\psi s\beta s\gamma + c\beta c\gamma \end{bmatrix} \begin{Bmatrix} c \\ e \\ f \end{Bmatrix}$$

$$= \begin{Bmatrix} ac\theta + h(c\theta c\phi - s\theta s\phi c\alpha) - g - bc\psi \\ as\theta + h(s\theta\phi + c\theta c\phi c\alpha) - bs\psi c\gamma \\ d + hs\alpha s\phi - bs\psi s\gamma \end{Bmatrix}. \qquad (11.94)$$

If the angles ϕ and ψ have been determined, then this equation can be solved to obtain a unique set of joint sliding distances c, e, and f.

11.6.3.1 The Planar 4R

It is useful to see how the vector loop equations (11.94) become the loop equations for the planar 4R when all four joint axes are parallel. Let $\alpha = \beta = \gamma = \eta = 0$. Then (11.94) becomes

$$\begin{bmatrix} 0 & 0 & 0 \\ 0 & 0 & 0 \\ -1 & 1 & 1 \end{bmatrix} \begin{Bmatrix} c \\ e \\ f \end{Bmatrix} = \begin{Bmatrix} ac\theta + h(c\theta c\phi - s\theta s\phi) - g - bc\psi \\ as\theta + h(s\theta\phi + c\theta c\phi) - bs\psi \\ d \end{Bmatrix}. \qquad (11.95)$$

These equations can be written in the form

$$a\cos\theta + h\cos(\theta+\phi) = g + b\cos\psi,$$
$$a\sin\theta + h\sin(\theta+\phi) = b\sin\psi,$$
$$e + f = c + d. \qquad (11.96)$$

Compare with (2.55) to see that the equations are the same. Notice that movement of this linkage in the plane perpendicular to the joint axes is independent of sliding movement along these axes.

11.7 The 5TS Spatial Linkage

The platform linkage constructed from five parallel TS chains is a one-degree-of-freedom linkage, Figure 11.5. The workspace of this system is the intersection of the workspaces of the individual TS chains. In order to define this workspace we use an *implicit* formulation based on the geometric constraints imposed by each of the TS chains.

For the five TS chains, let $\mathbf{B}_i$ be the center of the ith T-joint in F and let $\mathbf{p}_i$ be the center of the ith S-joint measured in M, so $\mathbf{P}_i = [D]\mathbf{p}_i$ locates this point in F. A displacement $[D]$ of the platform is in the workspace of this linkage if it maintains constant lengths R_i for the five chains. This means that $[D]$ must satisfy the set of constraint equations

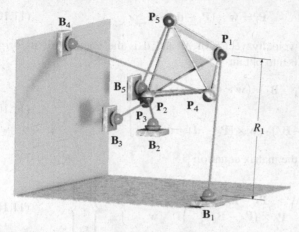

Fig. 11.5 The 5TS platform linkage.

$$\mathbf{F} = \left\{ \begin{array}{c} ([D]\mathbf{p}_1 - \mathbf{B}_1) \cdot ([D]\mathbf{p}_1 - \mathbf{B}_1) - R_1^2 \\ \vdots \\ ([D]\mathbf{P}_5 - \mathbf{B}_5) \cdot ([D]\mathbf{p}_5 - \mathbf{B}_5) - R_5^2 \end{array} \right\} = 0. \tag{11.97}$$

The analytical solution of these equations, known as solving the *direct kinematics* of a platform, is beyond the scope this text.

A numerical solution can be obtained by introducing an extra TPS chain to actuate the linkage. This chain does not impose any constraints on the platform but does push it along the path allowed by the five supporting TS chains. Suppose a displacement $[D] = [A, \mathbf{d}]$ satisfies the equations (11.97), then we can compute the slide s of the actuating leg from the equation

$$([D]\mathbf{p}_6 - \mathbf{B}_6) \cdot ([D]\mathbf{p}_6 - \mathbf{B}_6) - s^2 = 0. \tag{11.98}$$

We now formulate the solution as a root-finding problem. The time-derivative of the ith constraint equation can be written in the form

$$([D]\mathbf{p}_i - \mathbf{B}_i) \cdot ([\dot{D}]\mathbf{p}_i) - R_i \dot{R}_i = 0, \tag{11.99}$$

where $\dot{R}_i = 0$ for $i = 1, \ldots, 5$ and $\dot{R}_6 = \dot{s}$. Substitute $\mathbf{p}_i = [D^{-1}]\mathbf{P}_i$ to obtain

$$\dot{\mathbf{F}} = \left\{ \begin{array}{c} (\mathbf{P}_1 - \mathbf{B}_1) \cdot ([S]\mathbf{P}_1) \\ \vdots \\ (\mathbf{P}_6 - \mathbf{B}_6) \cdot ([S]\mathbf{P}_6) - s\dot{s} \end{array} \right\} = 0, \tag{11.100}$$

where $[S] = [\dot{D}D^{-1}]$. The operation $[S]\mathbf{P}_i$ computes the velocity of the point $\mathbf{P}_i$ as the platform M moves. Recall that this is equivalent to the formula

$$\dot{P}_i = \mathbf{w} \times (\mathbf{P}_i - \mathbf{d}) + \dot{\mathbf{d}}, \tag{11.101}$$

where $\mathbf{w}$ is the angular velocity vector of M and $\dot{\mathbf{d}}$ is the velocity of its origin. Substitute this into (11.100) to obtain

$$\dot{\mathbf{F}} = \left\{ \begin{array}{c} (\mathbf{P}_1 - \mathbf{B}_1) \cdot \left(\mathbf{w} \times (\mathbf{P}_1 - \mathbf{d}) + \dot{\mathbf{d}} \right) \\ \vdots \\ (\mathbf{P}_6 - \mathbf{B}_6) \cdot \left(\mathbf{w} \times (\mathbf{P}_6 - \mathbf{d}) + \dot{\mathbf{d}} \right) - s\dot{s} \end{array} \right\} = 0. \tag{11.102}$$

This can be rewritten as the matrix equation

$$\begin{bmatrix} \mathbf{P}_1 - \mathbf{B}_1 & \cdots & \mathbf{P}_6 - \mathbf{B}_6 \\ \mathbf{P}_1 \times (\mathbf{P}_1 - \mathbf{B}_1) & \cdots & \mathbf{P}_6 \times (\mathbf{P}_6 - \mathbf{B}_6) \end{bmatrix}^T \left\{ \begin{array}{c} \dot{\mathbf{d}} + \mathbf{d} \times \mathbf{w} \\ \mathbf{w} \end{array} \right\} = \left\{ \begin{array}{c} 0 \\ \vdots \\ s\dot{s} \end{array} \right\}. \tag{11.103}$$

The six-vector $\mathsf{P}_i = (\mathbf{P}_i - \mathbf{B}_i, \mathbf{P}_i \times (\mathbf{P}_i - \mathbf{B}_i))^T$ is known as the *Plücker vector* of the line along the ith leg. The matrix $[\Gamma]$ formed by these Plücker vectors is the *Jacobian* of this platform linkage.

For a given position $[D_0]$ of the platform and actuator rate $\dot{s}$, we solve (11.103) to determine the angular velocity $\mathbf{w}$ and velocity $\dot{\mathbf{d}}$ of the platform. This provides an approximation to the solution $[D(t)] = (I + [S]t)[D_0]$. The result is an algorithm to trace the trajectory of the floating link.

When the matrix $[\Gamma]$ loses rank, the linkage is said to be in a *singular configuration*. This is discussed in more detail in the last chapter.

11.8 Summary

This chapter has presented the direct and inverse kinematic analysis of the 3R wrist, the RS, TS, and CC open chains, as well as the TPS and TRS robots. The velocity analysis of these systems yields the *Jacobian* of robotic theory. The analysis of the RSSR linkage yields the results for planar and spherical 4R linkages as special cases. It also provides the constraint equation for the spatial 4R closed chain, known as Bennett's linkage. We also analyzed the RSSP, which has several special cases including the planar slider-crank. The analysis of the 4C linkage was presented which also specializes to the spherical and planar 4R linkages. Finally, we outlined a numerical solution for the analysis of the 5TS platform. An analytical solution exists for this system (Su et al. [127]) but it is beyond the scope of our work.

11.9 References

The analysis of spatial open chains is found in robotics texts by Craig [15], Crane and Duffy [16], and Tsai [136]. Also see Crane and Duffy's analysis of the CCC manipulator. The analysis of the RSSR and RSSP linkages is taken from Hunt [50]. The analysis of a 4C linkage presented here follows Suh and Radcliffe's [134] analysis of the RCCC linkage. A classification of the platform manipulator systems can be found in Faugere and Lazard [31], and Husty [51] presents a general algorithm for the analysis of these systems.

Exercises

1. Let a spatial RR open chain have length $a = 10$ and angle of twist $\alpha = 90°$. Determine the kinematics equations of the chain. Let $\mathbf{P} = (0,0,-10)^T$ be a point in the end-link of this chain, and determine its coordinates in F when $\theta = 90°$ and $\phi = 180°$ (Mabie and Reinholtz [70]).

2. Consider the TS chain with length $a = 4$. Set the joint angles to the values $\theta_1 = 60°$, $\theta_2 = 120°$, and $\phi_1 = 135°$, $\phi_2 = -60°$, $\phi_3 = 45°$, and determine the 4×4 transform that locates the end-link (Kinzel and Waldron 1999).

3. Derive the Jacobian for a TS chain.

4. Consider an RSSR linkage that has the fixed axes in the directions $\mathbf{O} = \vec{k}$ and $\mathbf{C} = \vec{\imath}$. Let $\mathbf{a}_0 = (0,0,0)^T$ and $\mathbf{a}_1 = (0,1,0)^T$ be the endpoints of the input crank, and $\mathbf{b}_0 = (2,0,0)^T$ and $\mathbf{b}_1 = (2,0,1)^T$ the end points of the output crank. Determine the input-output equation for this linkage (Suh and Radcliffe [134]).

5. An RSSP linkage has $\vec{\jmath}$ as the direction of its input fixed axis, $\mathbf{O}$. Let $\mathbf{a}_0 = (0,2,4)^T$ and $\mathbf{a}_1 = (0,1,4)^T$ be the endpoints of the input crank in a reference position, and let $h = 10$ be the length of the coupler. Finally, let the output moving pivot follow the line $\mathbf{L} = t\vec{\jmath}$. Analyze this linkage to determine the output slide s as a function of the input crank rotation θ (Sandor and Erdman [112]).

6. Analyze the Bennett linkage with twist angles $\alpha = 30°$ and $\gamma = 60°$. Consider the same linkage with $\alpha = -30°$. Show that the output functions of these two linkages are different (Hunt [50]).

7. Consider the 4C linkage with dimensions $\hat{\alpha} = (30°,2)$, $\hat{\eta} = (55°,4)$, $\hat{\beta} = (45°,3)$, and $\hat{\gamma} = (60°,5)$. Determine the joint angles and offsets of this linkage for the input $\hat{\theta} = (\theta,0)$ (Suh and Radcliffe [134]).

Chapter 12
Spatial Kinematics

In this chapter we develop the geometry of spatial displacements defined by coordinate transformations consisting of spatial rotations and translations. We consider the invariants of these transformations and find that there are no invariant points. Instead there is an invariant line, called the screw axis. Thus, the geometry of lines becomes important to our study of spatial kinematics. We find that a configuration of three lines, called a spatial triangle, generalizes our results for planar and spherical triangles to three-dimensional space.

A convenient set of coordinates for lines, known as Plücker coordinates, are introduced, then generalized to yield screws. Dual vector algebra manipulates these coordinates using the same rules as the usual vector algebra. This yields a screw form of Rodrigues's formula that defines the screw axis of a composite displacement in terms of the screw axes of the two factor displacements.

12.1 Spatial Displacements

A spatial displacement is the composition of a spatial rotation followed by a spatial translation. This transformation takes the coordinates $\mathbf{x} = (x, y, z)^T$ of a point in the moving frame M and computes its coordinates $\mathbf{X} = (X, Y, Z)^T$ in the fixed frame F, by the formula

$$\mathbf{X} = T(\mathbf{x}) = [A]\mathbf{x} + \mathbf{d}, \tag{12.1}$$

where $[A]$ is a 3×3 *rotation matrix* and $\mathbf{d}$ is a 3×1 *translation vector*. A spatial displacement preserves the distance between points measured in both M and F.

J.M. McCarthy and G.S. Soh, *Geometric Design of Linkages*, Interdisciplinary Applied
Mathematics 11, DOI 10.1007/978-1-4419-7892-9_12,
© Springer Science+Business Media, LLC 2011

12.1.1 Homogeneous Transforms

The transformation that defines a spatial displacement is not a linear operation. To
see this compute $T(\mathbf{x}+\mathbf{y})$. The result does not equal to $T(\mathbf{x})+T(\mathbf{y})$. This can be
attributed to the inhomogeneous translation term in (12.1). A standard strategy to
adjust for this inhomogeneity is to add a fourth component to our position vectors
that will always equal 1. Then we have the 4×4 *homogeneous transform*

$$\begin{Bmatrix} \mathbf{X} \\ 1 \end{Bmatrix} = \begin{bmatrix} A & \mathbf{d} \\ 000 & 1 \end{bmatrix} \begin{Bmatrix} \mathbf{x} \\ 1 \end{Bmatrix}, \tag{12.2}$$

which we write as

$$\mathbf{X} = [T]\mathbf{x}. \tag{12.3}$$

Notice that we have not distinguished between the point coordinates that have a 1
as their fourth component. In general, these vectors will have three components.
Please assume the addition of the fourth component, when it is appropriate for the
use of these 4×4 transforms. We use $[T] = [A, \mathbf{d}]$ to denote the 4×4 homogeneous
transform with rotation matrix $[A]$ and translation vector $\mathbf{d}$.

12.1.2 Composition of Displacements

The set of matrices that have the structure shown in (12.2) form a matrix group,
denoted by SE(3), with matrix multiplication as its operation. The matrix product of
$[T_1] = [A_1, \mathbf{d}_1]$ and $[T_2] = [A_2, \mathbf{d}_2]$ yields

$$[T_3] = [A_1, \mathbf{d}_1][A_2, \mathbf{d}_2] = [A_1 A_2, \mathbf{d}_1 + A_1 \mathbf{d}_2]. \tag{12.4}$$

It is easy to see that the 4×4 transform $[T_3]$ has the same structure as (12.2) with
$A_3 = A_1 A_2$ as its rotation matrix and $\mathbf{d}_3 = \mathbf{d}_1 + [A_1]\mathbf{d}_2$ as its translation vector.

The composition of the displacements $[T_1] = [A_1, \mathbf{d}_1]$ and $[T_2] = [A_2, \mathbf{d}_2]$ can be
interpreted as follows. Let $[T_1]$ define the position of a frame M' relative to F such
that $\mathbf{X} = [A_1]\mathbf{y} + \mathbf{d}_1$. Then the position of M relative to M' is defined by $[T_2]$ such
that $\mathbf{y} = [A_2]\mathbf{x} + \mathbf{d}_2$. Thus, the position of M relative to F is given by

$$\mathbf{X} = [A_1 A_2]\mathbf{x} + \mathbf{d}_1 + A_1 \mathbf{d}_2. \tag{12.5}$$

Compare this equation to (12.4) to see that the product of two homogeneous transforms defines this composition of displacements.

Similarly, the matrix inverse $[T]^{-1} = [A, \mathbf{d}]^{-1}$ defines the inverse displacement

$$[T^{-1}] = [A, \mathbf{d}]^{-1} = [A^T, -A^T \mathbf{d}]. \tag{12.6}$$

It is easy to use (12.4) to check that $[A, \mathbf{d}][A^T, -A^T \mathbf{d}] = [I]$.

12.1.2.1 Changing Coordinates of a Displacement

Consider the displacement $\mathbf{X} = [T]\mathbf{x}$ that defines the position of M relative to F. We now consider the transformation $[T']$ between the frames M' and F' that are displaced by the same amount from both M and F. In particular, let $[R] = [B, \mathbf{c}]$ be the displacement that transforms the coordinates between the primed and unprimed frames, that is, $\mathbf{Y} = [R]\mathbf{X}$ and $\mathbf{y} = [R]\mathbf{x}$ are the coordinates in F' and M', respectively. Then, from $\mathbf{X} = [T]\mathbf{x}$ we can compute

$$\mathbf{Y} = [R][T][R^{-1}]\mathbf{y}. \tag{12.7}$$

Thus, the original matrix $[T]$ is transformed by the change of coordinates into $[T'] = [R][T][R^{-1}]$.

12.1.3 Relative Displacements

For a set of displacements $[T_i] = [A_i, \mathbf{d}_i]$, $i = 1, \ldots, n$, the relative displacement between any two is given by

$$[T_{ij}] = [T_j][T_i^{-1}]. \tag{12.8}$$

If $\mathbf{X}^i = [T_i]\mathbf{x}$ denotes the coordinates in F for points in position M_i, then we have

$$\mathbf{X}^j = [T_{ij}]\mathbf{X}^i. \tag{12.9}$$

Notice that both $\mathbf{X}^i$ and $\mathbf{X}^j$ are measured in the fixed reference frame F; they are the coordinates of corresponding points of M in positions M_i and M_j.

12.1.3.1 Relative Inverse Displacements

The relative inverse displacement $[T_{ik}^{\dagger}]$ between two inverse positions F_i and F_k is given by

$$[T_{ik}^{\dagger}] = [T_k^{-1}][T_i]. \tag{12.10}$$

Notice that this is not the inverse of the relative displacement $[T_{ik}]$, which would be $[T_{ik}^{-1}] = [T_i][T_k^{-1}]$.

The relative inverse displacement $[T_{ik}^{\dagger}]$ is defined from the point of view of the moving frame M. However, we can choose a specific position M_j and transform this displacement by $[T_j]$, to obtain

$$[T_{ik}^{j}] = [T_j][T_{ik}^{\dagger}][T_j^{-1}]. \tag{12.11}$$

This is known as the *image* of the relative inverse transformation for position M_j in F.

Notice that if M_j is one of the frames used in computing the relative inverse displacement, for example $j = i$, then we have

$$[T_{ik}^i] = [T_i]([T_k^{-1}][T_i])[T_i^{-1}] = [T_i][T_k^{-1}] = [T_{ik}^{-1}]. \tag{12.12}$$

This same result is obtained when $j = k$. Thus, for $j = i$ or $j = k$ the image of the relative inverse displacement $[T_{ik}^j]$ is the inverse of the relative displacement.

12.1.4 Screw Displacements

We now consider the invariants of spatial displacements. If a point C has the same coordinates before and after a spatial displacement $[T]$, then it satisfies the equation

$$C = [T]C, \quad \text{or} \quad [I - T]C = 0, \tag{12.13}$$

which simplifies to

$$[I - A]C = d. \tag{12.14}$$

Recall that all spatial rotations have 1 as an eigenvalue. Therefore, the 3×3 matrix $[I - A]$ is singular. Thus, a spatial displacement has no fixed points.

While there are no fixed points, there is a line, called the *screw axis*, that remains fixed during a spatial displacement. To determine this line, we decompose the translation component of the displacement $[T] = [A, d]$ into vectors parallel and perpendicular to the rotation axis S of $[A]$, that is,

$$d = d^* + kS, \quad \text{where} \quad k = d \cdot S. \tag{12.15}$$

The displacement $[T]$ can now be written as the composition of the rotational displacement $[R] = [A, d^*]$ and the translation $[S] = [I, dS]$,

$$[T] = [S][R] = [I, kS][A, d^*] = [A, d^* + kS] = [A, d]. \tag{12.16}$$

Notice that all spatial displacements can be decomposed in this way.

We have already seen in (8.81) that a rotational displacement $[R] = [A, d^*]$ has a fixed point is given by

$$C = \frac{b \times (d^* - b \times d^*)}{2b \cdot b}, \tag{12.17}$$

where $b = \tan(\phi/2)S$ is Rodrigues's vector of the rotation $[A]$. Now consider the line S through this point in the direction of the rotation axis of $[A]$, defined by

$$S : P(t) = C + tS. \tag{12.18}$$

Points on this line remain fixed during the rotational displacement $[R] = [A, d^*]$. Furthermore, the translation $[S] = [I, kS]$ slides points along this line the distance k. This line remains fixed during the displacement. Thus, a general spatial displacement consists of a rotation by ϕ about this line and the sliding distance k along it. This is called a *screw displacement* and the line S is called the *screw axis*.

12.1.5 The Screw Matrix

It is often convenient to define a spatial displacement in terms of its screw axis S and the angle ϕ and slide k around and along it. We have already determined a formula for a rotation matrix $[A(\phi, S)]$ in terms of its rotation axis and angle. From (12.16) we see that the translation vector is given by

$$d = [I - A]C + kS. \qquad (12.19)$$

Now use the notation $\hat{\phi} = (\phi, k)$ for the rotation and slide of the screw displacement and define the *screw matrix*

$$[T(\hat{\phi}, S)] = [A(\phi, S), [I - A]C + kS]. \qquad (12.20)$$

This is the 4×4 homogeneous transform with elements defined in terms of the screw parameters of the displacement.

This form of a spatial displacement allows us to write the transformation of x in M to X in F as

$$X - C = [A](x - C) + kS, \qquad (12.21)$$

which shows directly that the displacement consists of a rotation about C followed by a translation along the screw axis S.

12.2 Lines and Screws

The geometry of the screw axis of a spatial displacement is best studied using Plücker's coordinates that define the line directly. Plücker coordinates for a line are six-vectors assembled from the direction of the line and its moment about the origin of the reference frame. The generalization of these coordinates, called a *screw*, is familiar from the study of elementary statics and dynamics where it appears as a the pair formed by the resultant force and moment on a body.

12.2.1 Plücker Coordinates of a Line

Consider the line S through two points C and Q in space, given by the parameterized equation

$$S : P(t) = C + tS, \qquad (12.22)$$

where $\mathbf{C}$ has been selected as a reference point on the line, and S is the unit vector along $\mathbf{Q} - \mathbf{C}$. To eliminate the free parameter t in the definition of S, we introduce the *Plücker coordinates* of the line

$$\mathsf{S} = \left\{ \begin{array}{c} \mathbf{S} \\ \mathbf{C} \times \mathbf{S} \end{array} \right\}. \tag{12.23}$$

The vector $\mathbf{C} \times \mathbf{S}$ is the moment of the line about the origin of the reference frame. Notice that these coordinates do not depend on the choice of the reference point $\mathbf{C}$, because any other point $\mathbf{C}' = \mathbf{C} + k\mathbf{S}$ yields the same moment $\mathbf{C}' \times \mathbf{S} = \mathbf{C} \times \mathbf{S}$.

A general pair of vectors $\mathsf{W} = (\mathbf{W}, \mathbf{V})^T$ can be the Plücker coordinates of a line only if $\mathbf{W} \cdot \mathbf{V} = 0$. This is equivalent to saying that there must be a vector $\mathbf{C}$ such that

$$\mathbf{C} \times \mathbf{W} = \mathbf{V}. \tag{12.24}$$

Solve this equation by computing the vector product of both sides by $\mathbf{W}$ to obtain

$$\mathbf{C} = \frac{\mathbf{W} \times \mathbf{V}}{\mathbf{W} \cdot \mathbf{W}}. \tag{12.25}$$

This formula defines the coordinates for the reference point directly in terms of the Plücker coordinates of the line.

Plücker coordinates are homogeneous, which means that $\mathbf{W} = w\mathbf{S}$ defines the same line as the unit vector $\mathbf{S}$. For convenience, we normalize our the Plücker coordinates so $\mathsf{S} = (\mathbf{S}, \mathbf{C} \times \mathbf{S})$, where $|\mathbf{S}| = 1$.

12.2.2 Screws

A general pair of vectors $\mathsf{W} = (\mathbf{W}, \mathbf{V})^T$ for which $\mathbf{W} \cdot \mathbf{V} \neq 0$ and $|\mathbf{W}| = w \neq 1$ is called a *screw*. We can associate with any screw W a line S, called the *axis* of the screw. To do this, decompose the second vector $\mathbf{V}$ into components parallel and perpendicular to $\mathbf{W}$, so we have $\mathbf{V} = p_w \mathbf{W} + \mathbf{V}^*$. Since $\mathbf{W} \cdot \mathbf{V}^* = 0$, we can determine a point $\mathbf{C}$ such that $\mathbf{C} \times \mathbf{W} = \mathbf{V}^*$. This is given by

$$\mathbf{C} = \frac{\mathbf{W} \times \mathbf{V}^*}{\mathbf{W} \cdot \mathbf{W}} = \frac{\mathbf{W} \times \mathbf{V}}{\mathbf{W} \cdot \mathbf{W}}. \tag{12.26}$$

Notice that the vector product with $\mathbf{W}$ automatically eliminates the component of $\mathbf{V}$ in the direction $\mathbf{W}$.

The line $\mathsf{S} = (\mathbf{W}, \mathbf{C} \times \mathbf{W})^T$ is the axis of the screw W. Let $\mathbf{W} = w\mathbf{S}$. Then the components of this screw can be written in the form

$$\mathsf{W} = \left\{ \begin{array}{c} w\mathbf{S} \\ w\mathbf{C} \times \mathbf{S} + wp_w\mathbf{S} \end{array} \right\}. \tag{12.27}$$

The parameter $w = |\mathbf{W}|$ is called the *magnitude* of the screw, and

$$p_w = \frac{\mathbf{W} \cdot \mathbf{V}}{\mathbf{W} \cdot \mathbf{W}} \qquad (12.28)$$

is its *pitch*. Lines are often called *zero-pitch screws*.

12.2.3 Dual Vector Algebra

We now introduce *dual vector algebra*, which allows us to manipulate the pairs of vectors that define lines and screws using the same operations as vector algebra.

12.2.3.1 The Dual Magnitude of a Screw

A multiplication operation can be defined so that a general screw W, given by (12.27), can be obtained as the product of the pair of scalars $\hat{w} = (w, wp_w)$ with the pair of vectors $\mathsf{S} = (\mathbf{S}, \mathbf{C} \times \mathbf{S})$. This operation is formulated by introducing the dual unit ε that has all the properties of a real scalar with the additional feature that $\varepsilon^2 = 0$. Using this symbol, we define the dual number

$$\hat{w} = (w, wp_w) = w + \varepsilon w p_w. \qquad (12.29)$$

Notice that we do not distinguish symbolically between the dual number written as a pair of numbers or written using the dual unit ε. Similarly, we can define the dual vector

$$\mathsf{S} = (\mathbf{S}, \mathbf{C} \times \mathbf{S})^T = \mathbf{S} + \varepsilon \mathbf{C} \times \mathbf{S}. \qquad (12.30)$$

Again, we do not distinguish between the screw written as a pair of vectors or a dual vector.

Now multiply the dual scalar $\hat{w}$ and the components of the dual vector S and impose the rule $\varepsilon^2 = 0$ to obtain

$$\hat{w}\mathsf{S} = (w + \varepsilon w p_w)(\mathbf{S} + \varepsilon \mathbf{C} \times \mathbf{S}) = w\mathbf{S} + \varepsilon(w\mathbf{C} \times \mathbf{S} + w p_w \mathbf{S}). \qquad (12.31)$$

Compare this to (12.27) to see that this equation defines a general screw W. The dual number $\hat{w} = w + \varepsilon w p_w$ is the *dual magnitude* of the screw W.

12.2.3.2 Dual Numbers

The set of dual numbers $\hat{a} = a + \varepsilon a°$, where a and $a°$ are real numbers and $\varepsilon^2 = 0$, has all the properties of complex numbers. Addition and subtraction are obtained componentwise, and multiplication is performed as though these numbers were polynomials in ε. Division is defined so that for $\hat{a} = a + \varepsilon a°$ and $\hat{b} = b + \varepsilon b°$, we have

$$\frac{\hat{b}}{\hat{a}} = \frac{(b+\varepsilon b^\circ)(a-\varepsilon a^\circ)}{(a+\varepsilon a^\circ)(a-\varepsilon a^\circ)} = \frac{b}{a} + \varepsilon \frac{b^\circ a - ba^\circ}{a^2}. \tag{12.32}$$

Notice that if the dual number has a zero real part then this division operation is undefined. Such numbers are known as *pure* dual numbers. See Appendix D for a summary of the properties of dual numbers.

12.2.3.3 The Dual Scalar Product

The linearity of the scalar product of vectors allows us to define the dual scalar product as

$$\mathsf{W} \cdot \mathsf{V} = (\mathbf{W} + \varepsilon \mathbf{W}^\circ) \cdot (\mathbf{V} + \varepsilon \mathbf{V}^\circ) = \mathbf{W} \cdot \mathbf{V} + \varepsilon (\mathbf{W} \cdot \mathbf{V}^\circ + \mathbf{W}^\circ \cdot \mathbf{V}). \tag{12.33}$$

This equation can be written in matrix form by introducing the 6×6 matrix $[\varPi]$ defined by

$$[\varPi]\mathsf{V} = \begin{bmatrix} 0 & I \\ I & 0 \end{bmatrix} \left\{ \begin{matrix} \mathbf{V} \\ \mathbf{V}^\circ \end{matrix} \right\} = \left\{ \begin{matrix} \mathbf{V}^\circ \\ \mathbf{V} \end{matrix} \right\}. \tag{12.34}$$

The second component of the dual scalar product (12.33) is $\mathsf{W}^T[\varPi]\mathsf{V}$, and we have

$$\mathsf{W} \cdot \mathsf{V} = (\mathbf{W}^T \mathbf{V}, \mathsf{W}^T[\varPi]\mathsf{V}) \tag{12.35}$$

as the matrix form of the dual scalar product.

Notice that if a screw is given by $\mathsf{W} = \hat{w}\mathsf{S}$, where S is the Plücker coordinates of its axis, then we can compute

$$\mathsf{W} \cdot \mathsf{W} = \hat{w}^2(\mathsf{S} \cdot \mathsf{S}) = \hat{w}^2. \tag{12.36}$$

This is because $\mathsf{S} \cdot \mathsf{S} = 1$ for normalized Plücker coordinates. For this reason Plücker vectors are often called *unit screws*.

The dual magnitude of a screw $\mathsf{W} = (\mathbf{W}, \mathbf{W}^\circ)^T$ can be computed using the dual scalar product to obtain

$$|\mathsf{W}| = (\mathsf{W} \cdot \mathsf{W})^{1/2}. \tag{12.37}$$

Furthermore, the axis of a screw W can be found by dividing by its dual magnitude, that is,

$$\mathsf{S} = \frac{\mathsf{W}}{|\mathsf{W}|} = \frac{\mathbf{W} + \varepsilon \mathbf{W}^\circ}{w + \varepsilon w p_w} = \frac{1}{w}\mathbf{W} + \varepsilon \frac{1}{w}(\mathbf{W}^\circ - p_w \mathbf{W}). \tag{12.38}$$

This yields the same screw axis as was defined above.

12.2.3.4 The Dual Vector Product

The linearity of the vector vector product allows its extension to dual vectors as well. Consider the two screws $\mathsf{W} = (\mathbf{W}, \mathbf{W}^\circ)^T$ and $\mathsf{V} = (\mathbf{V}, \mathbf{V}^\circ)^T$, and compute

$$W \times V = (\mathbf{W} + \varepsilon \mathbf{W}^\circ) \times (\mathbf{V} + \varepsilon \mathbf{V}^\circ) = \mathbf{W} \times \mathbf{V} + \varepsilon (\mathbf{W} \times \mathbf{V}^\circ + \mathbf{W}^\circ \times \mathbf{V}). \quad (12.39)$$

In what follows we show that this screw has as its axis the common normal to the axes of the screws W and V.

12.2.4 Orthogonal Components of a Line

Let the Plücker coordinates of the x, y, and z axes of the fixed frame F be $\mathsf{I} = (\vec{i}, \mathbf{o} \times \vec{i})^T$, $\mathsf{J} = (\vec{j}, \mathbf{o} \times \vec{j})^T$, and $\mathsf{K} = (\vec{k}, \mathbf{o} \times \vec{k})^T$, where $\mathbf{o}$ is the origin of F. We now determine the orthogonal components of a general line measured against these coordinate lines.

Consider the line S that intersects the z-axis K in a right angle at a distance d from $\mathbf{o}$, such that it lies at an angle θ measured from the x-axis I. The direction of S is $\mathbf{S} = \cos \theta \vec{i} + \sin \theta \vec{j}$, and its moment term is $(\mathbf{o} + d\vec{k}) \times \mathbf{w}$. Therefore, we have

$$S = \left\{ \begin{array}{c} \cos \theta \vec{i} + \sin \theta \vec{j} \\ (\mathbf{o} + d\vec{k}) \times (\cos \theta \vec{i} + \sin \theta \vec{j}) \end{array} \right\}$$

$$= \left\{ \begin{array}{c} \cos \theta \vec{i} \\ \cos \theta \mathbf{o} \times \vec{i} - d \sin \theta \vec{i} \end{array} \right\} + \left\{ \begin{array}{c} \sin \theta \vec{j} \\ \sin \theta \mathbf{o} \times \vec{j} + d \cos \theta \vec{j} \end{array} \right\}. \quad (12.40)$$

Thus, the coordinates of S can be written as the sum of two screws. The first screw has the dual magnitude $(\cos \theta, -d \sin \theta)$ and the line I as its axis. The second screw has the dual magnitude $(\sin \theta, d \cos \theta)$ and J as its axis.

12.2.4.1 The Dual Angle

We now introduce the *dual angle* $\hat{\theta} = \theta + \varepsilon d$, which measures the angle θ and distance d around and along the axis K from the x-axis I to the line S, Figure 12.1. The cosine and sine functions of this dual angle are defined such that

$$\cos \hat{\theta} = \cos \theta - \varepsilon d \sin \theta, \quad \sin \hat{\theta} = \sin \theta + \varepsilon d \cos \theta. \quad (12.41)$$

Equation (12.40) can now be written in the form

$$S = \cos \hat{\theta} \mathsf{I} + \sin \hat{\theta} \mathsf{J}, \quad (12.42)$$

where the screws $\cos \hat{\theta} \mathsf{I}$ and $\sin \hat{\theta} \mathsf{J}$ are the *orthogonal components* of S in the frame F.

The dual scalar product can be used to compute the component of S along the line I,

$$S \cdot \mathsf{I} = \mathbf{S} \cdot \vec{i} + \varepsilon (\mathbf{S} \cdot \mathbf{o} \times \vec{i} + \mathbf{C} \times \mathbf{S} \cdot \vec{i}) = \cos \hat{\theta}. \quad (12.43)$$

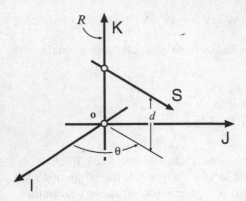

Fig. 12.1 The dual angle $\hat{\theta}$ to a line S in the reference frame R formed by the mutually orthogonal lines I, J, and K.

This calculation uses the fact that $\mathbf{S} \times \vec{\imath} = -\sin\theta \vec{k}$ and that the component of $\mathbf{C} - \mathbf{o}$ along $\vec{k}$ is d. Thus, the dual scalar product allows us to calculate the dual angle between any two lines about their common normal. In fact, for general screws W and V, we have

$$\mathbf{W} \cdot \mathbf{V} = |\mathbf{W}||\mathbf{V}| \cos\hat{\theta}, \tag{12.44}$$

where $|\mathbf{W}|$ and $|\mathbf{V}|$ are the dual magnitudes of these screws and $\hat{\theta}$ is the dual angle between their axes.

12.2.4.2 The Intersection of Two Lines

We now consider the meaning of a zero value for the dual scalar product between two screws, that is, $\mathbf{W} \cdot \mathbf{V} = 0$. Let the dual magnitudes of these screws be $|\mathbf{W}| = w(1 + \varepsilon p_w)$ and $|\mathbf{V}| = v(1 + \varepsilon p_v)$, so we have

$$\mathbf{W} \cdot \mathbf{V} = wv(1 + \varepsilon p_w)(1 + \varepsilon p_v)(\cos\theta - \varepsilon d \sin\theta). \tag{12.45}$$

This shows that $\mathbf{W} \cdot \mathbf{V} = 0$ implies that $\cos\hat{\theta} = 0$, which means that $\hat{\theta} = (\pi/2, 0)$. Thus, the axes of the two screws must intersect in right angles.

Two screws that satisfy the weaker condition $\mathbf{W}^T [\Pi] \mathbf{V} = 0$ are said to be *reciprocal*. Notice that this is equivalent to the requirement that $\mathbf{W} \cdot \mathbf{V} = k$, where k is a real constant. Expand this relation to obtain

$$\mathbf{W}^T [\Pi] \mathbf{V} = \mathbf{w} \cdot \mathbf{v}^\circ + \mathbf{w}^\circ \mathbf{v} = wv\big((p_w + p_v)\cos\theta - d \sin\theta\big) = 0. \tag{12.46}$$

Thus, the condition that two screws are reciprocal is

$$d \tan\theta = p_w + p_v. \tag{12.47}$$

Applying this condition to two lines for which $p_w = p_v = 0$, we see that to be reciprocal the lines must be parallel ($\theta = 0$), or they must intersect ($d = 0$).

12.2.4.3 The Common Normal

The dual vector product between two lines defines a screw that has the common normal between the lines as its axis. To see this, we first consider the dual vector product between the coordinate axes I and J. By direct computation we obtain

$$\mathsf{I} \times \mathsf{J} = \vec{\imath} \times \vec{\jmath} + \varepsilon\left(\vec{\imath} \times (\mathbf{o} \times \vec{\jmath}) + (\mathbf{o} \times \vec{\imath}) \times \vec{\jmath}\right)$$
$$= \vec{\imath} \times \vec{\jmath} + \varepsilon \mathbf{o} \times (\vec{\imath} \times \vec{\jmath}) = \mathsf{K}. \tag{12.48}$$

The simplification of the dual component in this equation uses the identity for triple vector products

$$\mathbf{a} \times (\mathbf{b} \times \mathbf{c}) + \mathbf{b} \times (\mathbf{c} \times \mathbf{a}) + \mathbf{c} \times (\mathbf{a} \times \mathbf{b}) = 0, \tag{12.49}$$

which in our case yields the relation

$$-\mathbf{o} \times (\vec{\jmath} \times \vec{\imath}) = \vec{\imath} \times (\mathbf{o} \times \vec{\jmath}) + \vec{\jmath} \times (\vec{\imath} \times \mathbf{o}). \tag{12.50}$$

Similar calculations show that $\mathsf{J} \times \mathsf{K} = \mathsf{I}$ and $\mathsf{K} \times \mathsf{I} = \mathsf{J}$.

Consider a general pair of lines $\mathsf{S} = (\mathbf{S}, \mathbf{C} \times \mathbf{S})^T$ and $\mathsf{L} = (\mathbf{W}, \mathbf{Q} \times \mathbf{W})^T$. Let N be the common normal between these lines, and let $\mathbf{p}$ and $\mathbf{r}$ be its points of intersection with S and L, respectively, Figure 12.2. Now use these two points to define the moment terms in the Plücker coordinates for these lines, so we have $\mathsf{S} = (\mathbf{S}, \mathbf{p} \times \mathbf{S})^T$ and $\mathsf{L} = (\mathbf{W}, \mathbf{r} \times \mathbf{W})^T$. We can now compute the vector product

$$\mathsf{S} \times \mathsf{L} = (\mathbf{S} + \varepsilon \mathbf{p} \times \mathbf{S}) \times (\mathbf{W} + \varepsilon \mathbf{r} \times \mathbf{W}),$$
$$= \mathbf{S} \times \mathbf{W} + \varepsilon\left(\mathbf{S} \times (\mathbf{r} \times \mathbf{W}) + (\mathbf{p} \times \mathbf{S}) \times \mathbf{W}\right). \tag{12.51}$$

Notice that $\mathbf{r} = \mathbf{p} + d\mathbf{N}$, where N is the direction of the common normal. Substitute this into the equation above, and use the vector identity (12.49) to obtain

$$\mathsf{S} \times \mathsf{L} = \sin\theta \mathbf{N} + \varepsilon(\cos\theta \mathbf{N} + \sin\theta \mathbf{o} \times \mathbf{N}) = \sin\hat{\theta} \mathbf{N}. \tag{12.52}$$

Thus, the dual vector product is an operation that computes the common normal to two given lines. Furthermore, for general screws W and V we have

$$\mathsf{W} \times \mathsf{V} = |\mathsf{W}||\mathsf{V}| \sin\hat{\theta} \mathbf{N}, \tag{12.53}$$

where $|\mathsf{W}|$ and $|\mathsf{V}|$ are the dual magnitudes of W and V.

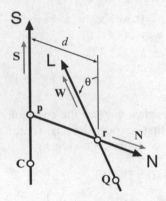

Fig. 12.2 Two general lines S and L define a common normal line N.

12.2.5 The Spatial Displacement of Screws

A spatial displacement $[T] = [A, \mathbf{d}]$ transforms the coordinates of points that form a line. By applying this transformation to two points on the line we obtain a 6×6 transformation $[\hat{T}]$ for Plücker coordinates. This transformation applies to general screws as well.

Consider the line $\mathsf{x} = (\mathbf{x}, \mathbf{p} \times \mathbf{x})^T$ for which every point is displaced by the 4×4 homogeneous transform $[T] = [A, \mathbf{d}]$ to define a new line $\mathsf{X} = (\mathbf{X}, \mathbf{P} \times \mathbf{X})^T$. We now determine the associated transformation $[\hat{T}]$ that acts directly on Plücker coordinates such that

$$\mathsf{X} = [\hat{T}]\mathsf{x}. \tag{12.54}$$

Let $\mathbf{q}$ be a point on the line x a unit distance from $\mathbf{p}$, so $\mathbf{x} = \mathbf{q} - \mathbf{p}$. Then, we can compute the new coordinates $\mathbf{P}$ and $\mathbf{Q}$ to define the line X

$$\left\{\begin{matrix} \mathbf{X} \\ \mathbf{P} \times \mathbf{X} \end{matrix}\right\} = \left\{\begin{matrix} \mathbf{Q} - \mathbf{P} \\ \mathbf{P} \times (\mathbf{Q} - \mathbf{P}) \end{matrix}\right\} = \left\{\begin{matrix} [A]\mathbf{x} \\ [D][A]\mathbf{x} + [A](\mathbf{p} \times \mathbf{x}) \end{matrix}\right\}. \tag{12.55}$$

This calculation uses the skew-symmetric matrix $[D]$ defined by $[D]\mathbf{y} = \mathbf{d} \times \mathbf{y}$ for any $\mathbf{y}$. Thus, we obtain $[\hat{T}]$ as the 6×6 matrix

$$[\hat{T}] = \begin{bmatrix} A & 0 \\ DA & A \end{bmatrix}. \tag{12.56}$$

The inverse of this transformation is easily obtained as

$$[\hat{T}^{-1}] = \begin{bmatrix} A^T & 0 \\ A^T D^T & A^T \end{bmatrix}. \tag{12.57}$$

Note that because $[D]$ is skew-symmetric, we have $[D + D^T] = 0$.

12.2.5.1 The Transformation of Screws

The transformation $[\hat{T}]$ defined by (12.56) applies to general screws as well. To see this, consider the screw $w = (\mathbf{w}, \mathbf{v})^T$ and compute $[\hat{T}]w$ to obtain

$$\begin{Bmatrix} \mathbf{W} \\ \mathbf{V} \end{Bmatrix} = \begin{Bmatrix} [A]\mathbf{w} \\ [D][A]\mathbf{w} + [A]\mathbf{v} \end{Bmatrix}. \tag{12.58}$$

Clearly, the transformation of the direction $\mathbf{w} = k\mathbf{s}$ of the screw is the same as for lines, that is,

$$\mathbf{W} = w\mathbf{S} = w[A]\mathbf{s} = [A]\mathbf{w}. \tag{12.59}$$

Therefore, we focus attention on the term $\mathbf{V} = [D][A]\mathbf{w} + [A]\mathbf{v}$.

Let w be written in terms of its axis $\mathbf{s} = (\mathbf{s}, \mathbf{p} \times \mathbf{s})^T$, so we have $w = (w\mathbf{s}, w\mathbf{p} \times \mathbf{s} + w p_w \mathbf{s})^T$, where w is the magnitude and p_w the pitch of w. Clearly, we have

$$[D][A](w\mathbf{s}) = w\mathbf{d} \times \mathbf{S}. \tag{12.60}$$

We can now compute

$$[A]\mathbf{v} = [A](w\mathbf{p} \times \mathbf{s} + w p_w \mathbf{s}) = w([A]\mathbf{p}) \times \mathbf{S} + w p_w \mathbf{S}. \tag{12.61}$$

Combining these results we have

$$\mathbf{V} = [D][A]\mathbf{w} + [A]\mathbf{v} = \mathbf{P} \times \mathbf{W} + p_w \mathbf{W}, \tag{12.62}$$

where $\mathbf{P} = [A]\mathbf{p} + \mathbf{d}$. Thus, for a general screw w in M we obtain

$$\mathbf{W} = [\hat{T}]w. \tag{12.63}$$

The transformation $[\hat{T}]$ preserves the magnitude and pitch of screws.

The matrix form of the dual scalar product makes it easy to show that the transformation $[\hat{T}]$ preserves the dual magnitude of a screw. In particular, we show that $\mathbf{W} \cdot \mathbf{W} = \mathbf{w} \cdot \mathbf{w}$ by the calculation

$$\mathbf{W} \cdot \mathbf{W} = ([\hat{T}]w) \cdot ([\hat{T}]w) = \left(\mathbf{w}^T [A]^T [A]\mathbf{w}, \mathbf{w}^T [\hat{T}]^T [\Pi][\hat{T}]w \right)$$
$$= \left(\mathbf{w}^T \mathbf{w}, \mathbf{w}^T [\Pi]w \right) = \mathbf{w} \cdot \mathbf{w}. \tag{12.64}$$

This computation uses the identities $[A]^T[A] = [I]$ and $[\hat{T}]^T[\Pi][\hat{T}] = [\Pi]$.

12.3 The Geometry of Screw Axes

12.3.1 The Screw Axis of a Displacement

We have seen that for every spatial displacement there is a fixed line, called its screw axis. Here we show that the Plücker coordinates $S = (\mathbf{S}, \mathbf{V})$ of this screw axis satisfy the condition

$$S = [\hat{T}]S. \tag{12.65}$$

This shows that the screw axis is an invariant of the 6×6 transformation matrix $[\hat{T}]$.

We rewrite (12.65) as the equation

$$[I - \hat{T}]S = 0, \tag{12.66}$$

and seek solutions other than $S = 0$. This is easily done if we separate it into the pair of 3×1 vector equations

$$[I - A]\mathbf{S} = 0 \quad \text{and} \quad [I - A]\mathbf{V} - [DA]\mathbf{S} = 0. \tag{12.67}$$

We already know how to determine the vector $\mathbf{S} = (s_x, s_y, s_z)^T$, which is the rotation axis of the rotation matrix $[A]$.

Notice that because $[A]\mathbf{S} = \mathbf{S}$, the second equation of (12.67) can be written as

$$[I - A]\mathbf{V} = [D]\mathbf{S}. \tag{12.68}$$

Now $[D]\mathbf{S} = \mathbf{d} \times \mathbf{S}$ must be orthogonal to $\mathbf{S}$. Therefore, it does not have a component in the direction of the null space of $[I - A]$, which is $\mathbf{S}$. This means that we can solve this equation for $\mathbf{V}$.

Substitute Cayley's formula for $[A]$, and simplify to obtain

$$[B]\mathbf{V} = \frac{1}{2}[I - B][S]\mathbf{d}; \tag{12.69}$$

for convenience we have introduced $[D]\mathbf{S} = -[S]\mathbf{d}$. Using the fact that $[B] = \tan(\phi/2)[S]$, we can write this equation as

$$[S]\mathbf{V} = \frac{1}{2\tan\frac{\phi}{2}}[S][I - B]\mathbf{d} = \mathbf{C}. \tag{12.70}$$

Finally, multiply both sides by $[S]$ and simplify to obtain

$$\mathbf{V} = \frac{-1}{2\tan\frac{\phi}{2}}[S^2][I - B]\mathbf{d}. \tag{12.71}$$

Because $\mathbf{S} \cdot \mathbf{V} = 0$, we see that $S = (\mathbf{S}, \mathbf{V})^T$ are the Plücker coordinates of a line.

The reference point $\mathbf{C}$ for S is determined by $\mathbf{C} = \mathbf{S} \times \mathbf{V}$, which is given in (12.70) above. The line $\mathsf{S} = (\mathbf{S}, \mathbf{C} \times \mathbf{S})^T$ is exactly the screw axis that was formulated earlier for the a spatial displacement $[T] = [A, \mathbf{d}]$.

12.3.2 Perpendicular Bisectors and the Screw Axis

A spatial displacement preserves the distances and angles between all points in the moving body. Therefore, it preserves the dual angles between lines in the body. In particular, the dual angle $\hat{\alpha}$ between the screw axis S of a displacement and the axis of a general screw must be the same before and after the displacement, Figure 12.3.

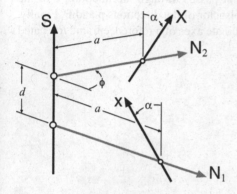

Fig. 12.3 Corresponding positions of lines x and X lie at the same dual angle $\hat{\alpha}$ relative to the screw axis S of a spatial displacement.

This is seen by letting x be the coordinates of a screw in the initial position, so we have $\mathsf{X} = [\hat{T}]\mathsf{x}$ as its coordinates after the displacement. From the fact that $\mathsf{S} = [\hat{T}]\mathsf{S}$, we can compute

$$\mathsf{S} \cdot \mathsf{X} = ([\hat{T}]\mathsf{S}) \cdot ([\hat{T}]\mathsf{x}) = \mathsf{S} \cdot \mathsf{x}. \tag{12.72}$$

Thus, the screw axis S forms the same dual angle $\hat{\alpha}$ with the axes of both x and its corresponding screw X.

This allows us to compute

$$\mathsf{S} \cdot (\mathsf{X} - \mathsf{x}) = 0, \tag{12.73}$$

which shows that the axis of the difference of any two corresponding screws $\mathsf{X} - \mathsf{x}$ must intersect S in a right angle.

12.3.2.1 The Screw Perpendicular Bisector

We now examine equation (12.73) in detail. To do this focus on the pair of corresponding screws $x = p$ and $X = P$ and consider all the screws Y that satisfy the equation

$$Y \cdot (P - p) = 0. \tag{12.74}$$

For example, the screw $(P + p)/2$ is a member of this set, as can be seen from the calculation

$$\frac{P+p}{2} \cdot (P - p) = \frac{P \cdot P - p \cdot p}{2} = 0. \tag{12.75}$$

Recall from (12.64) that $|P| = |p|$.

Let D be the common normal to the lines p and P with points of intersection c_1 on p and c_2 on P. Introduce the line V that passes through the midpoint c of the segment $c_2 - c_1$ and is directed along the bisector of the directions $\mathbf{p}$ and $\mathbf{P}$. Finally, let $N = D \times V$, so D, V, and N are the coordinate axes of a reference frame R located at c. See Figure 12.4.

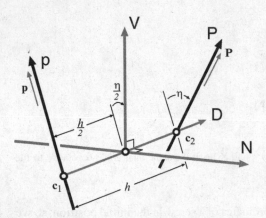

Fig. 12.4 The screws $P + p$ and $P - p$ have V and N as their respective axes.

If we denote the dual angle between p and P by $\hat{\eta} = (\eta, h)$, then we can write the components of these screws as

$$p = |p| \left(\cos \frac{\hat{\eta}}{2} V - \sin \frac{\hat{\eta}}{2} N \right) \quad \text{and} \quad P = |p| \left(\cos \frac{\hat{\eta}}{2} V + \sin \frac{\hat{\eta}}{2} N \right). \tag{12.76}$$

This allows us to compute

$$P + p = 2|p| \cos \frac{\hat{\eta}}{2} V. \tag{12.77}$$

The line V is the axis of the midpoint screw. Furthermore, from

$$P - p = 2|p| \sin \frac{\hat{\eta}}{2} N \tag{12.78}$$

we see that N is the axis of the screw difference $P - p$.

Thus, the set of screws Y that satisfy (12.74) must have axes that intersect N in right angles. This defines a two parameter family of screws that we call a *screw perpendicular bisector*.

12.3.2.2 Constructing the Screw Axis

The equation (12.73) shows that the screw axis S must lie on the screw perpendicular bisector for all segments $X - x$ in the moving body. This provides a convenient way to construct the screw axis of a displacement. Consider two specific segments $P - p$ and $Q - q$ formed by the two screws p, q and their corresponding screws $P = [\hat{T}]p$ and $Q = [\hat{T}]q$. This defines two screw perpendicular bisectors

$$Y \cdot (P - p) = 0 \quad \text{and} \quad Y \cdot (Q - q) = 0. \tag{12.79}$$

Let N_1 be the axis of $P - p$, and let N_2 be the axis of $Q - q$. Then the screw axis of the displacement S must intersect both of these axes in right angles. Thus, S must be the common normal to the axes N_1 and N_2.

The algebra of dual vectors allows us to compute S from the dual vector product of the screws $P - p$ and $Q - q$, that is,

$$S = \frac{(P - p) \times (Q - q)}{|(P - p) \times (Q - q)|}. \tag{12.80}$$

This provides a direct way to compute the screw axis of a spatial displacement from data that define the positions of two screws.

12.3.2.3 The Dual Displacement Angle

We can determine the dual angle $\hat{\phi}$ of a spatial displacement using any screw p and its corresponding displaced screw $P = [\hat{T}]p$. This is done by computing the dual scalar and vector products

$$\sin \hat{\phi} = \frac{(S \times p) \times (S \times P) \cdot S}{|(S \times p) \cdot (S \times P)|}, \quad \cos \hat{\phi} = \frac{(S \times p) \cdot (S \times P)}{|(S \times p) \cdot (S \times P)|}. \tag{12.81}$$

Thus, we have

$$\tan \hat{\phi} = \frac{(S \times p) \cdot P}{(S \times p) \cdot (S \times P)}. \tag{12.82}$$

The simplification in the numerator is obtained using dual vector identities that are identical to those of vector algebra.

12.3.3 Rodrigues's Equation for Screws

We now examine in more detail the geometric relationship between the screw axis S of a displacement $[\hat{T}]$ and the initial and final positions of a general screw. Because $X = [\hat{T}]x$, we have $X \cdot X - x \cdot x = 0$, which can also be written as

$$(X - x) \cdot (X + x) = 0. \qquad (12.83)$$

This can be interpreted as stating that the axes of the diagonals of a screw rhombus must intersect at right angles. In what follows, we will determine the components of these diagonals and obtain a screw version of Rodrigues's equation.

Let the common normals between the screw axis S and the axes of the corresponding screws x and X be the lines $N_1 = (\mathbf{N}_1, \mathbf{r}_1 \times \mathbf{N}_1)^T$ and $N_2 = (\mathbf{N}_2, \mathbf{r}_2 \times \mathbf{N}_2)^T$, where $\mathbf{r}_i$ are the respective points of intersection with S, Figure 12.5. Also introduce the lines V_1 and V_2, given by $V_i = S \times N_i$. Then x and X can be expanded into components

$$x = |x|(\cos \hat{\alpha} S - \sin \hat{\alpha} V_1), \quad X = |x|(\cos \hat{\alpha} S - \sin \hat{\alpha} V_2). \qquad (12.84)$$

From these equations we obtain the screws $X - x$ and $X + x$ as

$$\begin{aligned} X - x &= -|x| \sin \hat{\alpha}(V_2 - V_1), \\ X + x &= |x|\left(2 \cos \hat{\alpha} S - \sin \hat{\alpha}(V_2 + V_1)\right). \end{aligned} \qquad (12.85)$$

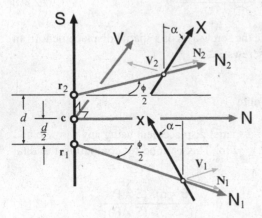

Fig. 12.5 The components of the screws $X + x$ and $X - x$ can be determined in the S, N, V frame.

To simplify (12.85), we introduce the line N through the midpoint **c** of the segment $\mathbf{r}_2 - \mathbf{r}_2$ along S. Choose the direction of N so that it bisects the rotation angle ϕ, that is, so **N** is aligned with the vector $(\mathbf{N}_1 + \mathbf{N}_2)/2$. We complete the frame at **c** by introducing V, given by $V = S \times N$. In the S, N, V frame the lines N_1 and N_2

become

$$N_1 = \cos\frac{\hat{\phi}}{2}N - \sin\frac{\hat{\phi}}{2}V, \quad N_2 = \cos\frac{\hat{\phi}}{2}N + \sin\frac{\hat{\phi}}{2}V. \tag{12.86}$$

Notice that $V_2 - V_1 = S \times (N_2 - N_1) = 2\sin(\hat{\phi}/2)S \times V$. Therefore,

$$X - x = 2|x|\sin\hat{\alpha}\sin\frac{\hat{\phi}}{2}N. \tag{12.87}$$

From the fact that $V_1 + V_2 = S \times (N_1 + N_2) = 2\cos(\hat{\phi}/2)V$, we have

$$X + x = 2|x|\left(\cos\hat{\alpha}S - \sin\hat{\alpha}\cos\frac{\hat{\phi}}{2}V\right). \tag{12.88}$$

From equations (12.87) and (12.88) we obtain the screw form of Rodrigues's equation as

$$X - x = \tan\frac{\hat{\phi}}{2}S \times (X + x). \tag{12.89}$$

The screw $B = \tan(\hat{\phi}/2)S$ is known as *Rodrigues's screw*.

12.4 The Spatial Screw Triangle

12.4.1 The Screw Axis of a Composite Displacement

Rodrigues's equation can be used to derive a formula that defines the screw axis of a composite displacement in terms of the screw axes of the two individual displacements.

Let $[T(\hat{\alpha}, A)]$ be the displacement with screw axis A and rotation angle and slide distance $\hat{\alpha} = (\alpha, a)$. Given another displacement $[T(\hat{\beta}, B)]$, we can compute the composite displacement by matrix multiplication

$$[T(\hat{\gamma}, C)] = [T(\hat{\beta}, B)][T(\hat{\alpha}, A)]. \tag{12.90}$$

Our goal is to obtain a formula for the screw axis C and the dual angle $\hat{\gamma}$ in terms of $\hat{\beta}$, B and $\hat{\alpha}$, A.

The displacement $[T(\hat{\alpha}, A)]$ has the associated 6×6 transformation $[\hat{T}(\hat{\alpha}, A)]$ that transforms screws x in M to y in M'. The displacement $[T(\hat{\beta}, B)]$ also has an associated 6×6 transformation $[\hat{T}(\hat{\beta}, B)]$ that transforms screws y in M' to X in F. This sequence of displacements can be written using Rodrigues's equation (12.89) to yield

$$y - x = \tan\frac{\alpha}{2}A \times (y + x),$$

$$X - y = \tan\frac{\beta}{2}B \times (X + y). \tag{12.91}$$

We eliminate the screw y in these equations in order to obtain a formula for the screw axis C and dual angle $\hat{\gamma}$.

The following calculations use dual vector algebra and follow exactly the derivation for the spherical version of Rodrigues's formula. The first step is to introduce X in the first equation and x in the second, so we have

$$y - x = \tan\frac{\hat{\alpha}}{2}A \times (X + x - (X - y)),$$

$$X - y = \tan\frac{\hat{\beta}}{2}B \times (X + x + (y - x)). \tag{12.92}$$

Add these equations and use (12.91) to obtain

$$X - x = (\tan\frac{\hat{\beta}}{2}B + \tan\frac{\hat{\alpha}}{2}A) \times (X + x) - \tan\frac{\hat{\alpha}}{2}A \times \left(\tan\frac{\hat{\beta}}{2}B \times (X + y)\right)$$

$$+ \tan\frac{\hat{\beta}}{2}B\left(\tan\frac{\hat{\alpha}}{2}A \times (y + x)\right). \tag{12.93}$$

Triple product identities for dual vectors that are identical to those for vectors simplify this equation to yield

$$X - x = \tan\frac{\hat{\gamma}}{2}C \times (X + x), \tag{12.94}$$

where

$$\tan\frac{\hat{\gamma}}{2}C = \frac{\tan\frac{\hat{\beta}}{2}B + \tan\frac{\hat{\alpha}}{2}A + \tan\frac{\hat{\beta}}{2}\tan\frac{\hat{\alpha}}{2}B \times A}{1 - \tan\frac{\hat{\beta}}{2}\tan\frac{\hat{\alpha}}{2}B \cdot A}. \tag{12.95}$$

This is *Rodrigues's formula* for screws.

12.4.2 The Spatial Triangle

We now show that (12.95) is the equation of an assembly of three lines A, B, and C known as a *spatial triangle*. Introduce the common normal N directed from the line B to A. Now introduce the common normals N_A and N_B directed from A and B to C, respectively. The lines A, B, and C form the *vertices* of the spatial triangle, and the common normals N, N_A, and N_B form its *sides*, Figure 12.6.

Let the interior dual angle between the sides N_A and N be $\hat{\alpha}/2$. Similarly, let the interior dual angle between the sides N and N_B be $\hat{\beta}/2$. We now show that for

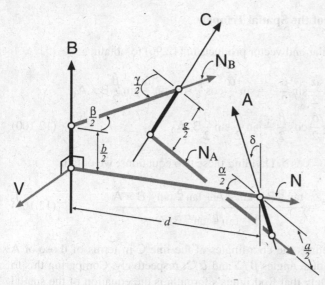

Fig. 12.6 The spatial triangle formed from the lines A, B, and C and their common normals N, N_a, and N_b.

this configuration the line C and exterior dual angle $\hat{\gamma}/2$ are defined by Rodrigues's formula (12.95).

From these definitions we see that

$$N_A \times N_B = \sin\frac{\hat{\gamma}}{2}C \quad \text{and} \quad N_A \cdot N_B = \cos\frac{\hat{\gamma}}{2}. \tag{12.96}$$

A formula for C is easily obtained by determining N_A and N_B explicitly in terms of A and B.

Let $V = N \times B$ complete the reference frame formed by N, B, and V. The line N_B intersects B and lies parallel to the NV plane at the angle $\hat{\beta}/2$ relative to N. Therefore,

$$N_B = \cos\frac{\hat{\beta}}{2}N - \sin\frac{\hat{\beta}}{2}V. \tag{12.97}$$

Note that N and V are computed from the given lines A and B.

Introduce the line $T = N \times A$. Then a computation similar to (12.97) yields the coordinates of T as

$$T = -\sin\hat{\delta}B + \cos\hat{\delta}V. \tag{12.98}$$

The line N_A lies parallel to the TN plane such that the dual angle measured from N_A to N is $\hat{\alpha}/2$. Thus, N_A is given by

$$N_A = \sin\frac{\hat{\alpha}}{2}T + \cos\frac{\hat{\alpha}}{2}N. \tag{12.99}$$

12.4.2.1 The Equation of the Spatial Triangle

We now compute the scalar and vector products in (12.96) to obtain

$$\sin\frac{\hat{\gamma}}{2}\mathsf{C} = \cos\frac{\hat{\alpha}}{2}\sin\frac{\hat{\beta}}{2}\mathsf{B} + \sin\frac{\hat{\alpha}}{2}\cos\frac{\hat{\beta}}{2}\mathsf{A} + \sin\frac{\hat{\alpha}}{2}\sin\frac{\hat{\beta}}{2}\mathsf{B}\times\mathsf{A},$$

$$\cos\frac{\hat{\gamma}}{2} = \cos\frac{\hat{\alpha}}{2}\cos\frac{\hat{\beta}}{2} - \sin\frac{\hat{\alpha}}{2}\sin\frac{\hat{\beta}}{2}\mathsf{B}\cdot\mathsf{A}. \tag{12.100}$$

Notice that $\mathsf{B}\cdot\mathsf{A} = \mathsf{T}\cdot\mathsf{V} = \cos\hat{\delta}$. Dividing these two equations, we obtain

$$\tan\frac{\hat{\gamma}}{2}\mathsf{C} = \frac{\tan\frac{\hat{\beta}}{2}\mathsf{B} + \tan\frac{\hat{\alpha}}{2}\mathsf{A} + \tan\frac{\hat{\beta}}{2}\tan\frac{\hat{\alpha}}{2}\mathsf{B}\times\mathsf{A}}{1 - \tan\frac{\hat{\beta}}{2}\tan\frac{\hat{\alpha}}{2}\mathsf{B}\cdot\mathsf{A}}. \tag{12.101}$$

Equation (12.101) defines the coordinates of the line C in terms of those of A and B and their interior dual angles $\hat{\beta}/2$ and $\hat{\alpha}/2$, respectively. Comparing this to (12.95) we see immediately that Rodrigues's formula is the equation of the spatial triangle formed by the three screw axes A, B, and C.

12.4.3 The Composite Screw Axis Theorem

The screw form of Rodrigues's equation is separated into two parts by the dual unit ε. The real part is simply Rodrigues's equation for the composition of rotations. We have seen that this defines a spherical triangle $\triangle\mathbf{ABC}$, which we call the *spherical image* of the spatial triangle $\triangle\mathsf{ABC}$.

We have already seen that there are two forms of the spherical image $\triangle\mathbf{ABC}$ depending on the magnitude of the rotation angles α, β, and γ.

1. If $\sin(\gamma/2) > 0$, that is, $\gamma < 2\pi$, then the vertex $\mathbf{C}$ has a positive component along $\mathbf{B}\times\mathbf{A}$. In this case $\alpha/2$ and $\beta/2$ are the interior angles of $\triangle\mathbf{ABC}$ at the vertices $\mathbf{A}$ and $\mathbf{B}$. The angle $\gamma/2$ is the exterior angle at $\mathbf{C}$.
2. If $\sin(\gamma/2) < 0$, that is, $\gamma > 2\pi$, then the vertex $\mathbf{C}$ is directed opposite to the vector $\mathbf{B}\times\mathbf{A}$. The angles $\alpha/2$ and $\beta/2$ are the exterior angles of $\triangle\mathbf{ABC}$ at $\mathbf{A}$ and $\mathbf{B}$, respectively. If the angle κ is the interior angle at $\mathbf{C}$, then $\gamma/2 = \kappa + \pi$.

The dual part of Rodrigues's formula (12.95) is linear in the slide parameters of the dual angles. This means that the spatial configuration of lines can be adjusted by changing the slide parameters a and b without changing the directions of any of the lines A, B, or C. Thus, we have the following theorem:

Theorem 14 (The Composite Screw Axis Theorem). *The axis* C *of a composite displacement* $[T_C] = [T_B][T_A]$ *forms a spatial triangle with the screw axes* B *and* A *of the displacements* $[T_B]$ *and* $[T_A]$. *If* $\sin(\gamma/2) > 0$, *then the interior dual angles of this triangle at* A *and* B *are* $\hat{\alpha}/2$ *and* $\hat{\beta}/2$, *respectively. If* $\sin(\gamma/2) < 0$, *then* $\hat{\alpha}/2$ *and*

$\hat{\beta}/2$ are the exterior dual angles at these vertices. In this case, if $\hat{\kappa}$ is the interior dual angle at C, then $\hat{\gamma}/2 = \hat{\kappa} + \pi$.

12.4.4 Dual Quaternions and the Spatial Triangle

W. K. Clifford [13] generalized Hamilton's quaternions to obtain hypercomplex numbers known as *dual quaternions* (Yang and Freudenstein [154]). A dual quaternion $\hat{P}$ is the formal sum of a dual number $\hat{p}_0 = (p, p^\circ)$ and a screw P = (**p**, **a**), written as $\hat{P} = \hat{p}_0 + \text{P}$. Dual quaternions can be added together componentwise, and multiplied by a scalar like eight-dimensional vectors. They can also be multiplied by dual scalars like four-dimensional vectors of dual numbers.

Furthermore, Clifford extended Hamilton's product for quaternions to a product for dual quaternions, given by the formula

$$\hat{R} = \hat{P}\hat{Q} = (\hat{p}_0 + \text{P})(\hat{q}_0 + \text{Q})$$
$$= (\hat{p}_0\hat{q}_0 - \text{P} \cdot \text{Q}) + (\hat{q}_0\text{P} + \hat{p}_0\text{Q} + \text{P} \times \text{Q}). \quad (12.102)$$

The scalar and vector products are operations between dual vectors.

The *conjugate* of a dual quaternion $\hat{Q} = \hat{q}_0 + \text{Q}$ is $\hat{Q}^* = \hat{q}_0 - \text{Q}$, and the product $\hat{Q}\hat{Q}^*$ is the dual number

$$\hat{Q}\hat{Q}^* = (\hat{q}_0 + \text{Q})(\hat{q}_0 - \text{Q}) = \hat{q}_0^2 + \text{Q} \cdot \text{Q} = |\hat{Q}|^2. \quad (12.103)$$

The dual number $|\hat{Q}|$ is called the *norm* of the dual quaternion.

We are interested in dual quaternions $\hat{Q}$ of unit norm, which means that $|\hat{Q}| = 1$. These unit dual quaternions can be written in the form

$$\hat{Q} = \cos\frac{\hat{\phi}}{2} + \sin\frac{\hat{\phi}}{2}\text{S}, \quad (12.104)$$

where S = $(\textbf{S}, \textbf{C} \times \textbf{S})^T$ is the Plücker coordinate vector of a line.

Now consider the product of the two unit dual quaternions $\hat{A} = \cos(\hat{\alpha}/2) + \sin(\hat{\alpha}/2)\text{A}$ and $\hat{B} = \cos(\hat{\beta}/2) + \sin(\hat{\beta}/2)\text{B}$, that is,

$$\hat{C} = \cos\frac{\hat{\gamma}}{2} + \sin\frac{\hat{\gamma}}{2}\text{C} = \left(\cos\frac{\hat{\beta}}{2} + \sin\frac{\hat{\beta}}{2}\text{B}\right)\left(\cos\frac{\hat{\alpha}}{2} + \sin\frac{\hat{\alpha}}{2}\text{A}\right). \quad (12.105)$$

Expanding this product, we obtain

$$\cos\hat{\gamma} + \sin\hat{\gamma}\text{C} = \left(\cos\frac{\hat{\beta}}{2}\cos\frac{\hat{\alpha}}{2} - \sin\frac{\hat{\beta}}{2}\sin\frac{\hat{\alpha}}{2}\text{B} \cdot \text{A}\right)$$
$$+ \left(\sin\frac{\hat{\beta}}{2}\cos\frac{\hat{\alpha}}{2}\text{B} + \sin\frac{\hat{\alpha}}{2}\cos\frac{\hat{\beta}}{2}\text{A} + \sin\frac{\hat{\beta}}{2}\sin\frac{\hat{\alpha}}{2}\text{B} \times \text{A}\right). \quad (12.106)$$

Compare this to (12.100) to see that dual quaternion multiplication computes the Plücker coordinates of one vertex of a spatial triangle and its associated exterior dual angle from the Plücker coordinates of the other two vertices and their interior dual angles. Thus, the algebra of dual quaternions provides a useful tool for exploiting the geometry of the spatial triangle.

12.4.5 The Triangle of Relative Screw Axes

Given three positions M_i, M_j, and M_k for a moving body, we have the three relative transformations $[\hat{T}_{ij}] = [\hat{T}_j][\hat{T}_i^{-1}]$, $[\hat{T}_{jk}] = [\hat{T}_k][\hat{T}_j^{-1}]$, and $[\hat{T}_{ik}] = [\hat{T}_k][\hat{T}_i^{-1}]$. From the fact that

$$[\hat{T}_{ik}] = [\hat{T}_{jk}][\hat{T}_{ij}],\tag{12.107}$$

we can use Rodrigues's formula for screws to obtain

$$\tan\frac{\hat{\phi}_{ik}}{2}S_{ik} = \frac{\tan\frac{\hat{\phi}_{jk}}{2}S_{jk} + \tan\frac{\hat{\phi}_{ij}}{2}S_{ij} + \tan\frac{\hat{\phi}_{jk}}{2}\tan\frac{\hat{\phi}_{ij}}{2}S_{jk} \times S_{ij}}{1 - \tan\frac{\hat{\phi}_{jk}}{2}\tan\frac{\hat{\phi}_{ij}}{2}S_{jk} \cdot S_{ij}}.\tag{12.108}$$

The spatial triangle formed by the three relative screw axes S_{ij}, S_{ik}, and S_{jk} is analogous to the pole triangle for three planar displacements and is called the *screw triangle*, Figure 12.7.

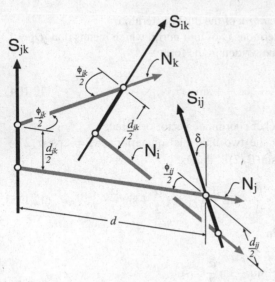

Fig. 12.7 The screw triangle formed by the screw axes S_{ij}, S_{jk}, and S_{ik}.

12.5 Summary

This chapter has developed the basic geometric properties of spatial rigid displacements. The central role played by the screw axis of a displacement lead to the introduction of line geometry, screws, and dual vector algebra. We obtain the screw form of Rodrigues's formula and see that it is the equation of a spatial triangle formed by three lines and their common normals. Its properties generalize results for the planar and spherical pole triangles.

12.6 References

The kinematics of spatial displacements is developed in detail in Roth [104, 105] and Bottema and Roth [5]. Dimentberg [25] introduced the algebra of dual vectors and Yang [155] applied it to the analysis of spatial linkages. Also see Woo and Freudenstein [152]. Yang and Freudenstein [154] formulated dual quaternion algebra for use in spatial kinematic theory. Pennock and Yang [93] use dual-number matrices to solve the inverse kinematics of robots. Fischer [34] presents the kinematic, static, and dynamic analysis of spatial linkages using dual vectors and matrices.

Exercises

1. A spatial displacement has as its axis the line through the origin in the direction $\mathbf{S} = (0, \cos(45°), \sin(45°))^T$. Let the rotation and slide around and along this axis be $(45°, \sqrt{2})$. Determine the 4×4 homogeneous transform.
2. Determine the spatial displacement $[D] = [T_1][T_2][T_2]$ defined by a sequence of transformations: (i) $[T_1]$, a translation by $(5, 4, 1)^T$; (ii) $[T_2]$, a rotation by $30°$ about the x-axis; and (iii) $[T_3]$, a rotation by $60°$ about the unit vector through the point $(2, 0, 2)^T$ (Crane and Duffy [16]).
3. Determine the 4×4 homogeneous transform $[T_{12}]$ from the initial and final positions of four points by constructing the matrix equation using homogeneous coordinates, $[\mathbf{A}^2, \mathbf{B}^2, \mathbf{C}^2, \mathbf{D}^2] = [T_{12}][\mathbf{A}^1, \mathbf{B}^1, \mathbf{C}^1, \mathbf{D}^1]$. Solve this equation for $[T_{12}]$ using the coordinates in Table 12.1 (Sandor and Erdman [112]).

Table 12.1 Point coordinates defining two spatial positions

Point	M_1	M_2
$\mathbf{A}^i$	$(0, 3, 7)^T$	$(1.90, 11.23, 7.19)^T$
$\mathbf{B}^i$	$(2, 7, 10)^T$	$(3.29, 14.44, 11.29)^T$
$\mathbf{C}^i$	$(0, 5, 10)^T$	$(4.26, 13.41, 8.84)^T$
$\mathbf{D}^i$	$(-2, 5, 7)^T$	$(3.92, 9.83, 8.59)^T$

4. Use the coordinates in Table 12.1 to determine the screw axis S and rotation and slide around and along this axis for the displacement $[T_{12}]$.
5. Given the screw axis S and rotation and slide ϕ and d let N_1 and N_2 be two lines that intersect S at right angles that are separated by the dual angle $\hat{\phi}/2 = (\phi/2, d/2)$. Show that this displacement is equivalent to the sequence of reflections through N_1 and then N_2 (Bottema and Roth [5]).
6. Show that Rodrigues's screw B can be computed from the initial and final positions of two lines p, P and q, Q to obtain $B = (P-p) \times (Q-q)/(P-p) \cdot (Q+q)$.
7. Show that three positions $\mathbf{P}^i$, $i = 1,2,3$, of a point can be obtained from the reflection of a cardinal point $\mathbf{P}^*$ through the three sides of the screw triangle defined by three specified spatial positions.
8. Obtain three positions L^i, $i = 1,2,3$, of a line by the reflection of a cardinal line L^* through the three sides of the screw triangle.

Chapter 13
Algebraic Synthesis of Spatial Chains

In this chapter we consider the design of spatial TS, CC, and RR chains. Our approach is the same as that used to design planar and spherical linkages. For each chain we determine the geometric constraints that characterize the chain, and formulate *design equations* that are solved for a given set of task positions.

The maximum number of task positions that can be prescribed decreases with the degree of freedom of the chain. In particular, the five-degree-of-freedom TS chain can reach as many as seven specified task positions, the four-degree-of-freedom CC chain can reach five positions, and the two-degree-of-freedom spatial RR can only reach three positions.

Our focus is primarily on solving the design equations for the maximum number of task positions. However, in practice, the use of fewer task positions can increase the dimension of the set of solutions providing flexibility to address other aspects of the design. This feature of spatial linkage design has yet to be exploited in a systematic way.

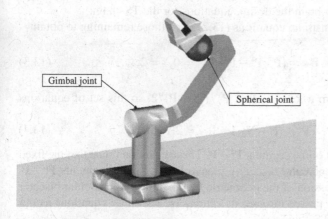

Gimbal joint

Spherical joint

Fig. 13.1 The TS spatial open chain.

J.M. McCarthy and G.S. Soh, *Geometric Design of Linkages*, Interdisciplinary Applied
Mathematics 11, DOI 10.1007/978-1-4419-7892-9_13,
© Springer Science+Business Media, LLC 2011

13.1 The Geometry of a TS Spatial Chain

A TS chain is a link connected to ground by a T-joint, also known as a gimbal mount, and to an end-link by a spherical joint, Figure 13.1. The revolute joints that make up the T-joint are often oriented to provide a slew rotation about a vertical axis combined with an elevation rotation about a horizontal axis. The S-joint can be constructed as a ball-in-socket or as a 3R wrist. In either case it is assumed to provide full orientation freedom of the floating link about its center.

Let the center of the T-joint be the fixed point $\mathbf{B}$, and let the center of the S-joint coincide with a point $\mathbf{p}$ in M that has coordinates $\mathbf{P} = [T]\mathbf{p}$ in F. Then the TS chain constrains the floating link to move so that the point $\mathbf{P}$ lies on a sphere about $\mathbf{B}$, that is,

$$(\mathbf{P} - \mathbf{B}) \cdot (\mathbf{P} - \mathbf{B}) = R^2, \tag{13.1}$$

where R is the length of the crank. This constraint characterizes the geometry of the TS chain.

13.1.1 Perpendicular Bisectors

Let the task for this chain be defined by n spatial positions M_i, $i = 1, \ldots, n$, of the end-link, which means that we have the 4×4 transforms $[T_i]$, $i = 1, \ldots, n$. The constraint equation (13.1) must be satisfied by the coordinates $\mathbf{P}^i$, $i = 1, \ldots, n$, of the moving pivot in each position. Therefore, we have

$$(\mathbf{P}^i - \mathbf{B}) \cdot (\mathbf{P}^i - \mathbf{B}) = R^2, i = 1, \ldots, n. \tag{13.2}$$

Notice that this equation remains correct even for point coordinate vectors $\mathbf{P}^i$ and $\mathbf{B}$ that are homogeneous with the fourth component normalized to 1. We now manipulate these constraints to obtain the design equations for the TS chain.

Subtract the first of constraint equations (13.2) from those remaining to obtain

$$(\mathbf{P}^i - \mathbf{P}^1) \cdot \mathbf{B} - \frac{1}{2}(\mathbf{P}^i \cdot \mathbf{P}^i - \mathbf{P}^1 \cdot \mathbf{P}^1) = 0, i = 2, \ldots, n. \tag{13.3}$$

We factor the second term to obtain $(\mathbf{P}^i - \mathbf{P}) \cdot (\mathbf{P}^i + \mathbf{P})/2$, so this set of equations becomes

$$(\mathbf{P}^i - \mathbf{P}^1) \cdot (\mathbf{B} - \mathbf{V}_{1i}) = 0, i = 2, \ldots, n, \tag{13.4}$$

where $\mathbf{V}_{1i}$ is the midpoint of the segment $\mathbf{P}^i - \mathbf{P}$. These equations state that the fixed pivot $\mathbf{B}$ lies on the perpendicular bisecting plane of each of the segments $\mathbf{P}^i - \mathbf{P}$. This is an algebraic expression of the geometric fact that the perpendicular bisector of any chord of a sphere passes through its center. We use these equations as our *design equations* for the TS chain.

13.1.2 The Design Equations

As we have done previously, choose the first position M_1 of the task as a reference position, and determine the $n-1$ relative displacements $[T_{1i}] = [T_i][T_1^{-1}]$, $i = 2, \ldots, n$. This allows us to define the positions $\mathbf{P}^i$ of the moving pivot in terms of $\mathbf{P}^1$, so we have

$$\mathscr{D}_{1i}: \ ([T_{1i} - I]\mathbf{P}^1) \cdot \left(\mathbf{B} - \frac{1}{2}[T_{1i} + I]\mathbf{P}^1\right) = 0, \, i = 2, \ldots, n. \tag{13.5}$$

Notice that this equation assumes that $\mathbf{P}^1$ and $\mathbf{B}$ are normalized homogeneous coordinates.

Introduce the relative rotation matrix $[A_{1i}]$ and relative translation vector $\mathbf{d}_{1i}$, so that $[T_{1i}] = [A_{1i}, \mathbf{d}_{1i}]$. Then, given the relative screw axis S_{1i}, we have the identity

$$\mathbf{d}_{1i} = [I - A_{1i}]\mathbf{C}_{1i} + d_{1i}\mathsf{S}_{1i}, \tag{13.6}$$

where $\mathbf{C}_{1i}$ is a point on the screw axis and $d_{1i}\mathsf{S}_{1i}$ is the slide along this axis. This allows us to compute

$$[T_{1i} - I]\mathbf{P}^1 = [A_{1i} - I](\mathbf{P}^1 - \mathbf{C}_{1i}) + d_{1i}\mathsf{S}_{1i}$$

and

$$[T_{1i} + I]\mathbf{P}^1 = [A_{1i} + I](\mathbf{P}^1 - \mathbf{C}_{1i}) + d_{1i}\mathsf{S}_{1i} + 2\mathbf{C}_{1i}. \tag{13.7}$$

Substitute these equations into (13.5) to obtain

$$([A_{1i} - I](\mathbf{P}^1 - \mathbf{C}_{1i}) + d_{1i}\mathsf{S}_{1i})$$
$$\cdot \left((\mathbf{B} - \mathbf{C}_{1i}) - \frac{1}{2}[A_{1i} + I](\mathbf{P}^1 - \mathbf{C}_{1i}) - \frac{1}{2}d_{1i}\mathsf{S}_{1i}\right) = 0. \tag{13.8}$$

Expanding this expression, we obtain after some cancellation

$$\mathscr{D}_{1i}: \ (\mathbf{B} - \mathbf{C}_{1i}) \cdot [A_{1i} - I](\mathbf{P}^1 - \mathbf{C}_{1i}) + d_{1i}\mathsf{S}_{1i} \cdot (\mathbf{B} - \mathbf{P}^1) - \frac{d_{1i}^2}{2} = 0,$$
$$i = 2, \ldots, n. \tag{13.9}$$

This form of the design equations shows that they are bilinear in the unknown components of the vectors $\mathbf{B}$ and $\mathbf{P}^1$.

Let the components of the fixed and moving pivots be denoted by $\mathbf{B} = (x, y, z)^T$ and $\mathbf{P}^1 = (u, v, w)^T$. Then seven spatial positions yield six design equations in these six unknown components. Collect these equations into the matrix equation

$$\begin{bmatrix} A_2(x,y,z) & B_2(x,y,z) & C_2(x,y,z) & D_2(x,y,z) \\ \vdots & \vdots & \vdots & \vdots \\ A_7(x,y,z) & B_7(x,y,z) & C_7(x,y,z) & D_7(x,y,z) \end{bmatrix} \begin{Bmatrix} u \\ v \\ w \\ 1 \end{Bmatrix} = \begin{Bmatrix} 0 \\ \vdots \\ 0 \end{Bmatrix}. \tag{13.10}$$

We solve these equations in the sections that follow.

13.1.3 Four Specified Spatial Positions

Before determining the solution to the general case, it is interesting to consider the design of TS chains to reach four task positions. In this case we have three bilinear equations in the six components of the fixed point $\mathbf{B}$ and the moving point $\mathbf{P}^1$. We can choose to select either of these points and solve the resulting linear equations for the other.

Suppose, for example, that we select the moving point $\mathbf{P}^1 = (\lambda, \mu, \nu)^T$. Then we can gather the coefficients of the design equations (13.5) into the matrix equation

$$\begin{bmatrix} A_2' & B_2' & C_2' & D_2' \\ A_3' & B_3' & C_3' & D_3' \\ A_4' & B_4' & C_4' & D_4' \end{bmatrix} \begin{Bmatrix} x \\ y \\ z \\ 1 \end{Bmatrix} = \begin{Bmatrix} 0 \\ 0 \\ 0 \end{Bmatrix}. \tag{13.11}$$

This equation has a unique solution $\mathbf{B} = (x,y,z)^T$, which is the center of the sphere that passes through the four points $\mathbf{P}^i$, $i = 1,2,3,4$.

13.1.3.1 The RS Chain

Notice that if the rank of the coefficient matrix in (13.11) is two, not three, then the four points $\mathbf{P}^i$, $i = 1,2,3,4$, lie on a circle and do not define a sphere. This can be viewed as the condition for the design of an RS chain that reaches the four positions. Each of the four 3×3 minors yields a cubic polynomial in λ, μ, ν that defines a cubic surface in F. These four surfaces intersect in a finite number of points that are the moving pivots $\mathbf{P}^1$ with four positions on a circle. Solve (13.11) to obtain the associated fixed pivot.

13.1.4 Seven Specified Spatial Positions

The solution of the TS design equations for seven positions follows the same procedure as that used to solve the planar RR design equations. Considering (13.10) as six linear equations in three unknowns, we see that there is a solution only if each of the fifteen 4×4 minors of the 6×4 matrix $[M] = [A_i, B_i, C_i, D_i]$ is zero. Computing each of these determinants we obtain 15 quartic polynomials in x, y, z,

$$\mathcal{R}_j : \sum_{i=1}^{35} a_{ji}x^l y^m z^n = 0,\ l,m,n = 0,\ldots,4,\ l+m+n \le 4,\ j = 1,\ldots,15. \quad (13.12)$$

Each $\mathcal{R}_j$ has 35 terms.

13.1.4.1 Eliminate x and y

Rewrite these polynomials so that z is absorbed into the coefficients and only x and y appear explicitly, that is,

$$\mathcal{R}_j : d_{j1}x^4 + d_{j2}x^3 y + d_{j3}x^2 y^2 + d_{j4}xy^3 + d_{j5}y^4 + d_{j6}x^3 + d_{j7}x^2 y$$
$$+ d_{j8}xy^2 + d_{j9}y^3 + d_{j10}x^2 + d_{j11}xy + d_{j12}y^2 + d_{j13}x + b_{j14}y$$
$$+ d_{j15} = 0,\ j = 1,\ldots,15. \quad (13.13)$$

Note that the total degree of any term is at most 4. Assemble these these polynomials $\mathcal{R}_i$, $i = 1,\ldots,15$, into the matrix equation

$$[M_{15\times 15}]\mathbf{V} = 0, \quad (13.14)$$

where $\mathbf{V}$ is the vector of 15 power products

$$\mathbf{V} = (x^4, x^3 y, x^2 y^2, xy^3, y^4, x^3, x^2 y, xy^2, y^3, x^2, xy, y^2, x, y, 1)^T. \quad (13.15)$$

This equation has a solution only if the determinant of the coefficient matrix is zero, that is,

$$|M_{15\times 15}| = 0. \quad (13.16)$$

This determinant is a polynomial of degree 20, because it has five columns of constant terms, four columns of first-degree terms, three of degree two, two of degree three, and one of degree four, which yields the degree $4 + 6 + 6 + 4 = 20$.

13.1.4.2 Reduction of the Constant Terms

The first five columns of constant terms in the determinant of $[M_{15\times 15}]$ can be row reduced by Gaussian elimination to yield

$$|M_{15 \times 15}| = \begin{vmatrix} c_{1,1} & c_{1,2} & c_{1,3} & c_{1,4} & c_{1,5} & c_{1,6} & \cdots & c_{1,15} \\ 0 & c_{2,2} & c_{2,3} & c_{2,4} & c_{2,5} & c_{2,6} & \cdots & c_{2,15} \\ 0 & 0 & c_{3,3} & c_{3,4} & c_{3,5} & c_{3,6} & \cdots & c_{3,15} \\ 0 & 0 & 0 & c_{4,4} & c_{4,5} & c_{4,6} & \cdots & c_{4,15} \\ 0 & 0 & 0 & 0 & c_{5,5} & c_{5,6} & \cdots & c_{5,15} \\ 0 & 0 & 0 & 0 & 0 & c_{6,6} & \cdots & c_{6,15} \\ \vdots & \vdots & \vdots & \vdots & \vdots & \vdots & & \vdots \\ 0 & 0 & 0 & 0 & 0 & c_{15,6} & \cdots & c_{15,15} \end{vmatrix} = 0. \tag{13.17}$$

In general, the five constants, $c_{1,1}, c_{2,2}, c_{3,3}, c_{4,4}, c_{5,5}$, must be nonzero, for otherwise the determinant of $[M_{15 \times 15}]$ is always zero. Thus, the polynomial defining the TS chains is given by the 10×10 determinant

$$|M_{10 \times 10}| = \begin{vmatrix} c_{6,6} & c_{6,7} & \cdots & c_{6,15} \\ c_{7,6} & c_{7,7} & \cdots & c_{7,15} \\ \vdots & \vdots & & \vdots \\ c_{15,6} & c_{15,7} & \cdots & c_{15,15} \end{vmatrix} = 0. \tag{13.18}$$

This 20th degree polynomial can be generated and solved using algebraic manipulation software on a personal computer.

The real roots of the polynomial defined by (13.18) yields as many as 20 values for z. For each real root $z_j = B_{z,j}$, we solve (13.14) to obtain $\mathbf{V}_j$, which in turn yields $x_j = B_{x,j}$ and $y_j = B_{y,j}$. The result is as many as 20 points $\mathbf{B}_j$ that are the fixed pivots of the TS chains.

For a given fixed pivot $\mathbf{B}_j$ use (13.10) to compute the associated moving pivot $\mathbf{P}^1$. The length of the link joining these two pivots is R_j, which is obtained using (13.1).

13.2 The Geometry of a CC Chain

A spatial CC chain has a fixed C-joint connected by a link to a moving C-joint and allows the end-link four degrees of freedom, Figure 13.2. Let G be the axis of the base joint of the chain and W the axis of the moving joint. Then the link connecting these axes maintains a constant distance r and a constant twist angle ρ along the common normal between the fixed axis G and every location W^i of the moving axis. This means that the dual angle $\hat{\rho} = (\rho, r)$ remains constant as the chain moves.

The spatial positions of the floating link are defined by the 4×4 homogeneous transforms $[T] = [A, \mathbf{d}]$. Associated with this transform is the 6×6 transformation $[\hat{T}]$ for screws. If w is the Plücker coordinate vector of the moving axis in M, then in F we have the corresponding locations of the moving axis, given by

$$W^i = [\hat{T}_i]w, \tag{13.19}$$

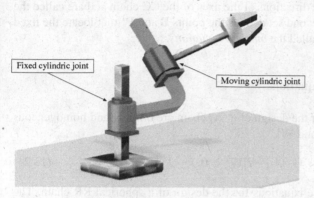

Fig. 13.2 The CC open chain.

for each position $[\hat{T}_i]$ of M.

In order to maintain a constant dual angle $\hat{\rho}$ the coordinates G and W^i must satisfy the constraint equations

$$G \cdot W^i = |G||W^i| \cos \hat{\rho}, \, i = 1, \dots, n. \tag{13.20}$$

Because $|w| = |W^i|$ for each of the positions of w, we see that the right side of these equations are identical. Subtract the first equation from those remaining to obtain

$$G \cdot (W^i - W^1) = 0, \, i = 2, \dots, n. \tag{13.21}$$

Compare these equations to (12.73) to see that G lies on the screw perpendicular bisector of each segment $W^i - W^1$.

Now introduce the relative transformations $[\hat{T}_{1i}] = [\hat{T}_i][\hat{T}_1^{-1}]$ such that $W^i = [\hat{T}_{1i}]W^1$. Then we obtain

$$G \cdot [\hat{T}_{1i} - I]W^1 = 0, \, i = 2, \dots, n. \tag{13.22}$$

These are the *design equations* for a CC chain.

13.2.1 Direction and Moment Equations

Let the Plücker vectors of the fixed and moving axes be $G = (G, B \times G)^T$ and $W = (W^i, P^i \times W^i)^T$, where **B** is a reference point on the fixed axis and P^i is the reference point for the moving axis. We now separate the dual scalar product of the design equations into the two sets of equations

$$G \cdot [A_{1i} - I]W^1 = 0, \quad G^T[\Pi][T_{1i} - I]W^1 = 0, \, i = 2, \dots, n. \tag{13.23}$$

The first set determines the directions of the axes of the CC chain and are called the *direction equations*. The second set defines the points **B** and $\mathbf{P}^1$ that locate the fixed and moving axes and are called the *moment equations*.

13.2.1.1 The Direction Equations

The direction equations for the design of a CC chain are bilinear and homogeneous in the direction vectors **G** and $\mathbf{W}^1$,

$$\mathscr{P}_{1i}: \ \mathbf{G} \cdot [A_{1i} - I]\mathbf{W}^1 = 0, i = 2, \ldots, n. \tag{13.24}$$

In fact, they are exactly the equations for the design of a spherical RR chain. The solution to these equations presented in chapter 8 yields a finite number of the directions for the axes of a CC chain that reaches five task positions.

13.2.1.2 The Moment Equations

For convenience let the moment terms of the fixed and moving axes be denoted by $\mathbf{R} = \mathbf{B} \times \mathbf{G}$ and $\mathbf{V}^1 = \mathbf{P}^1 \times \mathbf{W}^1$. Then the moment equations can be written as

$$\mathscr{M}_{1i}: \ \begin{Bmatrix} \mathbf{G} \\ \mathbf{R} \end{Bmatrix}^T \begin{bmatrix} D_{1i}A_{1i} \ A_{1i} - I \\ A_{1i} - I & 0 \end{bmatrix} \begin{Bmatrix} \mathbf{W}^1 \\ \mathbf{V}^1 \end{Bmatrix} = 0, i = 2, \ldots, n. \tag{13.25}$$

The 6×6 matrix in this equation is $[\Pi][\hat{T}_{1i} - I]$.

The vectors **G** and $\mathbf{W}^1$ are known from the direction equations (13.24). Therefore, the moment equations are linear in the unknown components of **R** and $\mathbf{V}^1$, that is,

$$\mathbf{R} \cdot [A_{1i} - I]\mathbf{W}^i + \mathbf{G} \cdot [A_{1i} - I]\mathbf{V}^1 + \mathbf{G} \cdot [D_{1i}A_{1i}]\mathbf{W}^1 = 0, \tag{13.26}$$

or

$$\mathbf{L}_i \cdot \mathbf{R} + \mathbf{M}_i \cdot \mathbf{V}^1 + N_i = 0, i = 2, \ldots, n, \tag{13.27}$$

where

$$\mathbf{L}_i = [A_{1i} - I]\mathbf{W}^i, \quad \mathbf{M}_i = [A_{1i} - I]^T \mathbf{G}, \quad N_i = \mathbf{G} \cdot [D_{1i}A_{1i}]\mathbf{W}^1. \tag{13.28}$$

We must add to these equations the requirements that $\mathbf{G} \cdot \mathbf{R} = 0$ and $\mathbf{W}^1 \cdot \mathbf{V}^1 = 0$, so $G = (\mathbf{G}, \mathbf{R})^T$ and $W^1 = (\mathbf{W}^1, \mathbf{V}^1)^T$ are Plücker coordinates of lines. This yields the matrix equation

$$\begin{bmatrix} \mathbf{G}^T & 0 \\ 0 & \mathbf{W}^{1T} \\ \mathbf{L}_2^T & \mathbf{M}_2^T \\ \vdots & \vdots \\ \mathbf{L}_n^T & \mathbf{M}_n^T \end{bmatrix} \begin{Bmatrix} \mathbf{R} \\ \mathbf{V}^1 \end{Bmatrix} = \begin{Bmatrix} 0 \\ 0 \\ -N_2 \\ \vdots \\ -N_n \end{Bmatrix}. \tag{13.29}$$

Solve these equations for the moments $\mathbf{R}$ and $\mathbf{V}^1$. Then the reference points $\mathbf{B}$ and $\mathbf{P}^1$ on G and W^1 are given by

$$\mathbf{B} = \frac{\mathbf{G} \times \mathbf{R}}{\mathbf{G} \cdot \mathbf{G}} \quad \text{and} \quad \mathbf{P}^1 = \frac{\mathbf{W}^1 \times \mathbf{V}^1}{\mathbf{W}^1 \cdot \mathbf{W}^1}. \tag{13.30}$$

13.2.2 Five Specified Spatial Positions

Given five spatial positions of the moving body, we know that the direction equations (13.24) can be solved to determine up to six pairs of directions $\mathbf{G}$ and $\mathbf{W}^1$. For $n = 5$, the moment equations (13.29) become

$$\begin{bmatrix} \mathbf{G}^T & 0 \\ 0 & \mathbf{W}^{1T} \\ \mathbf{L}_2^T & \mathbf{M}_2^T \\ \vdots & \vdots \\ \mathbf{L}_5^T & \mathbf{M}_5^T \end{bmatrix} \left\{ \begin{array}{c} \mathbf{R} \\ \mathbf{V}^1 \end{array} \right\} = \left\{ \begin{array}{c} 0 \\ 0 \\ -N_2 \\ \vdots \\ -N_5 \end{array} \right\}. \tag{13.31}$$

Solve these six linear equations to define the unique set of moments $\mathbf{R}$, $\mathbf{V}$ for each of the directions obtained for $\mathbf{G}$ and $\mathbf{W}^1$. The six fixed axes G are known as *Burmester lines* and are the spatial analogue to the Burmester points in the plane. Thus, as many as six CC chains are obtained that reach five spatial task positions.

13.2.3 The Transfer Principle

We have seen that algebra of dual vectors has all of the properties of vector algebra. In fact, every calculation using vectors has an analogous calculation using dual vectors. This is easily verified, or it can be viewed as the result of the differentiability of the rational operations and trigonometric functions that make up vector algebra, Appendix D. The relationship between these two algebras is called the *transfer principle*.

What is interesting about the transfer principle is the connection that it provides between the geometry of points and the geometry of lines. Geometric facts in point geometry that are defined using vector equations become geometric facts in line geometry obtained from the dual-vector versions of the same equations. We have already seen that Rodrigues's formula for screws (12.95) is the dual-vector form of Rodrigues's formula for rotation axes (8.84). Similarly, the the screw perpendicular bisector (13.21) is the equation of a perpendicular bisecting plane (9.11) written using dual vectors. In fact, the substitution of dual vectors transforms the entire design theory for spherical RR chains into the design theory for CC chains.

In what follows we focus on geometric results for CC chains that arise from the spatial dyad triangle and the central-axis theorem. The relationship between these results and the identical results for the spherical and planar RR design theories provides a unifying framework for spatial design theory.

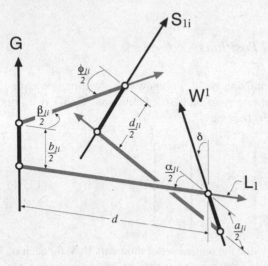

Fig. 13.3 The spatial dyad triangle for the CC chain.

13.2.4 The Spatial Dyad Triangle

The displacement of the end-link of a CC chain from position M_i to M_j can be viewed as the result of a screw displacement by $\hat{\alpha}_{ij}$ about the moving axis W^i followed by a screw displacement by $\hat{\beta}_{ij}$ about the fixed axis G. This results in a relative displacement from M_i to M_j by the amount $\hat{\phi}_{ij}$ about the screw axis S_{ij} defined by the transformation equation

$$[T(\hat{\phi}_{ij}, S_{ij})] = [T(\hat{\beta}_{ij}, G)][T(\hat{\alpha}_{ij}, W^i)]. \tag{13.32}$$

Rodrigues's formula for screws yields the equation of the spatial triangle formed by S_{ij}, G, and W^i,

$$\tan\frac{\hat{\phi}_{ij}}{2}S_{ij} = \frac{\tan\frac{\hat{\beta}_{ij}}{2}G + \tan\frac{\hat{\alpha}_{ij}}{2}W^i + \tan\frac{\hat{\beta}_{ij}}{2}\tan\frac{\hat{\alpha}_{ij}}{2}G \times W^i}{1 - \tan\frac{\hat{\beta}_{ij}}{2}\tan\frac{\hat{\alpha}_{ij}}{2}G \cdot W^i}. \tag{13.33}$$

Thus, the internal angles at W^i and G are $\hat{\alpha}_{ij}/2$ and $\hat{\beta}_{ij}/2$, respectively, and the external angle at S_{ij} is $\hat{\phi}_{ij}/2$, Figure 13.3.

13.2.5 *The Central-Axis Theorem*

Consider the CC chain that guides a body through positions M_i, M_j, and M_k, which means that we have locations W^i, W^j, and W^k of the moving axis, and

$$G \cdot (W^j - W^i) = 0 \quad \text{and} \quad G \cdot (W^k - W^i) = 0. \tag{13.34}$$

This states that G lies on the screw perpendicular bisectors of these segments.

Let L_i and L_j be the common normal lines between G and the two lines W^i and W^j. The dual angle $\hat{\beta}_{ij}$ between these common normals is the dual crank rotation angle about the fixed pivot. Introduce the midpoint screw $M_{ij} = (W^i + W^j)/2$ that has the axis V_{ij}. Notice that V_{ij} bisects the dual angle $\hat{\beta}_{ij}$.

Similarly, we consider the position W^k and construct its common normal L_k to G. The dual crank angle $\hat{\beta}_{jk}$ is measured from the line L_j to L_k. Let V_{jk} be the axis of the midpoint screw $M_{jk} = (W^j + W^k)/2$. The dual angle between the lines V_{ij} and V_{jk} is easily seen to be

$$\frac{\hat{\beta}_{ij}}{2} + \frac{\hat{\beta}_{jk}}{2} = \frac{\hat{\beta}_{ik}}{2}. \tag{13.35}$$

The relative screw axis S_{ij} of the displacement from position M_i to M_j must also lie on the screw perpendicular bisector. Therefore,

$$S_{ij} \cdot (W^j - W^i) = 0. \tag{13.36}$$

This means that S_{ij} and S_{jk} have the lines V_{ij} and V_{jk}, respectively, as common normals with the fixed axis G.

We can now compute the dual angle $\hat{\beta}_{ik}$ in terms of the relative screw axes S_{ij} and S_{jk} and the fixed axis using the formula

$$\tan \frac{\hat{\beta}_{ik}}{2} = \frac{G \times S_{ij} \cdot S_{jk}}{(G \times S_{ij}) \cdot (G \times S_{jk})}. \tag{13.37}$$

Thus, a central axis G views the relative screw axes S_{ij} and S_{jk} in the dual angle $\hat{\beta}_{ik}/2$ or $\hat{\beta}_{ik}/2 + \pi$. This latter possibility arises because the tangent function of both angles are equal. We restate this as a theorem.

Theorem 15 (The Central-Axis Theorem). *The central axis G of a CC chain that reaches the spatial positions M_i, M_j, and M_k views the relative screw axes S_{ij} and S_{jk} in the dual angle $\hat{\beta}_{ik}/2$ or $\hat{\beta}_{ik}/2 + \pi$, where $\hat{\beta}_{ik}$ is the crank rotation angle from position M_i to M_k.*

13.2.6 Roth's Theorem

Given four positions of a body M_i, $i = 1,2,3,4$, we can determine six relative rotation axes S_{ij}, $i < j = 1,2,3,4$. Collect the axes into pairs of *complementary axes* for which no subscript is repeated, that is, $S_{12}S_{34}$, $S_{13}S_{24}$, and $S_{14}S_{23}$. Two sets of complementary axes are used to construct a spatial quadrilateral with the four lines as vertices and their common normals as sides. The normal lines connecting complementary pairs of axes are the diagonals of this *complementary-screw quadrilateral*. The six relative screw axes define three complementary-screw quadrilaterals, $S_{12}S_{13}S_{34}S_{24}$, $S_{12}S_{14}S_{34}S_{23}$, and $S_{13}S_{14}S_{24}S_{23}$.

Theorem 16 (Roth's Theorem). *The central axis G of a spatial CC chain that can reach four spatial positions views opposite sides of a complementary-screw quadrilateral constructed from the relative screw axes of the given positions in dual angles that are equal, or differ by π.*

Proof. The definition of the complementary-screw quadrilateral ensures that opposite sides have the form $S_{ij}S_{ik}$ and $S_{mj}S_{mk}$. The central-axis theorem states that G views $S_{ij}S_{ik}$ in the dual angle $\hat{\beta}_{jk}/2$ or $\hat{\beta}_{jk}/2 + \pi$, where $\hat{\beta}_{jk}$ is crank rotation angle and slide from position M_j to M_k. Similarly, it must view the $S_{mj}S_{mk}$ in $\hat{\beta}_{jk}/2$ or $\hat{\beta}_{jk}/2 + \pi$. The result is that G views the two sides in dual angles that are equal, or differ by π. $\square$

13.2.6.1 The Parameterized Central-Axis Congruence

Roth's theorem reduces the problem of finding fixed axes, or *central axes*, to finding those axes that view opposite sides of complementary-screw quadrilateral in equal dual angles. The following construction generates these points.

Theorem 17 (Construction of Central Axes). *The axes that satisfy Roth's theorem are obtained as follows:*

1. *Use the four spatial task positions to construct the complementary-screw quadrilateral $\mathcal{Q}$: $S_{12}S_{23}S_{34}S_{14}$ and consider it to form a 4C linkage.*
2. *Rotate and slide the side $S_{12}S_{23}$ about S_{12} by the dual angle $\hat{\theta}$ and determine the new configuration $\mathcal{Q}'$. This yields a new location $S'_{23}S'_{34}$ for the coupler of this chain.*
3. *The screw axis G of the displacement of $S'_{23}S'_{34}$ relative to its original location $S_{23}S_{34}$ satisfies Roth's theorem and is a central axis.*

Proof. Let G be the intersection of the screw perpendicular bisectors $\mathcal{V}_1 = (S_{23}S'_{23})^\perp$ and $\mathcal{V}_2 = (S_{34}S'_{34})^\perp$. Then G is the screw axis of the displacement of the segment $S_{23}S_{34}$ by the dual angle $\hat{\kappa}$ to the position $S'_{23}S'_{34}$. The input CC chain formed by $S_{12}S_{23}$ has the spatial dyad triangle $\triangle S_{23}S_{12}G$ and G must view the $S_{12}S_{23}$ in the dual angle $\hat{\kappa}/2$ or $\hat{\kappa}/2 + \pi$. Similarly, the geometry of the spatial dyad triangle

$\triangle S_{34}S_{14}G$ requires that G view the segment $S_{14}S_{34}$ in either $\hat{\kappa}/2$ or $\hat{\kappa}/2+\pi$. Thus, G views the opposite sides $S_{12}S_{23}$ and $S_{14}S_{34}$ in angles that are equal, or differ by π. The same argument shows that G views the other two sides $S_{23}S_{34}$ and $S_{12}S_{14}$ in angles that are equal, or differ by π. Thus, G satisfies Roth's theorem. $\square$

The movement of the coupler $S_{23}S_{34}$ of the complementary-screw quadrilateral from its position in $\mathcal{Q}$ to another position specified by $\hat{\theta}$ can be obtained as a composite displacement. First displace this link about the axis S_{23} by the coupler angle $\Delta\hat{\phi} = (\phi_i - \phi_0, c_i - c_0)$. Follow this with the displacement about the axis S_{12} by the drive angle $\Delta\hat{\theta} = (\theta_i - \theta_0, d_i - d_0)$. The result is the transformation equation

$$[T(\kappa, G)] = [T(\Delta\hat{\theta}, S_{12})][T(\Delta\hat{\phi}, S_{23})]. \tag{13.38}$$

Rodrigues's formula yields the equation for the central axis G as

$$\tan\frac{\hat{\kappa}}{2}G = \frac{\tan\frac{\Delta\hat{\theta}}{2}S_{12} + \tan\frac{\Delta\hat{\phi}}{2}S_{23} + \tan\frac{\Delta\hat{\theta}}{2}\tan\frac{\Delta\hat{\phi}}{2}S_{12} \times S_{23}}{1 - \tan\frac{\Delta\hat{\theta}}{2}\tan\frac{\Delta\hat{\phi}}{2}S_{12} \cdot S_{23}}. \tag{13.39}$$

The coupler angle $\hat{\phi}$ is obtained from the analysis of the 4C linkage for each value of the crank parameter $\hat{\theta}$.

The initial dual angle $\hat{\theta}_0$ of this linkage is determined directly from the screws $S_{12} \times S_{14}$, and $S_{12} \times S_{23}$. The result is

$$\tan\hat{\theta}_0 = \frac{(S_{12} \times S_{14}) \cdot S_{23}}{(S_{12} \times S_{14}) \cdot (S_{12} \times S_{23})}. \tag{13.40}$$

A similar equation defines the initial dual angle $\hat{\phi}_0$ obtained using the screws $S_{12} \times S_{23}$ and $S_{23} \times S_{34}$.

Equation (13.39) defines the set of lines G known as the *central-axis congruence*, which is parameterized by the dual angle $\hat{\theta} = (\theta, d)$. For each value θ we obtain a single direction for G. Values for d displace this line to form a plane of parallel lines. The result is a two-dimensional set of lines in space assembled as planes of parallel lines.

13.2.7 The Spatial Compatibility Platform

Five spatial positions of a body determine ten relative screw axes $S_{ij}, i < j = 1, \ldots, 5$. Consider the two complementary-screw quadrilaterals $\mathcal{Q}_{14} : S_{12}S_{23}S_{34}S_{14}$ and $\mathcal{Q}_{15} : S_{12}S_{23}S_{35}S_{15}$. A fixed axis compatible with five positions must lie on the central-axis congruence generated by $\mathcal{Q}_{14}$ and on the central-axis congruence generated by $\mathcal{Q}_{15}$. Thus, these lines are the intersection of these two congruences.

Both complementary-screw quadrilaterals $\mathcal{Q}_{14}$ and $\mathcal{Q}_{15}$ are driven by the crank $S_{12}S_{23}$, therefore we have

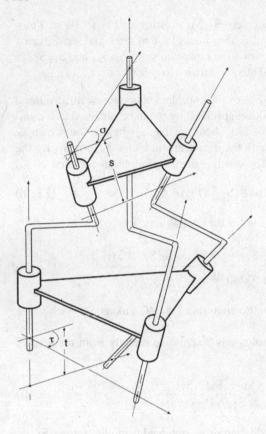

Fig. 13.4 The spatial platform.

$$\tan\frac{\hat{\kappa}}{2}G = \frac{\tan\frac{\Delta\hat{\theta}_1}{2}S_{12} + \tan\frac{\Delta\hat{\phi}_1}{2}S_{23} + \tan\frac{\Delta\hat{\theta}_1}{2}\tan\frac{\Delta\hat{\phi}_1}{2}S_{12}\times S_{23}}{1 - \tan\frac{\Delta\hat{\theta}_1}{2}\tan\frac{\Delta\hat{\phi}_1}{2}S_{12}\cdot S_{23}}. \tag{13.41}$$

and

$$\tan\frac{\hat{\kappa}}{2}G = \frac{\tan\frac{\Delta\hat{\theta}_2}{2}S_{12} + \tan\frac{\Delta\hat{\phi}_2}{2}S_{23} + \tan\frac{\Delta\hat{\theta}_2}{2}\tan\frac{\Delta\hat{\phi}_2}{2}S_{12}\times S_{23}}{1 - \tan\frac{\Delta\hat{\theta}_2}{2}\tan\frac{\Delta\hat{\phi}_2}{2}S_{12}\cdot S_{23}}. \tag{13.42}$$

The angles $\Delta\hat{\phi}_1$ and $\Delta\hat{\phi}_2$ are functions of $\Delta\hat{\theta}_1$ and $\Delta\hat{\theta}_2$ defined by the dimensions of the two complementary-screw quadrilaterals.

It is easy to see that equations (13.41) and (13.42) define the same screw axis G when the two quadrilaterals $\mathcal{Q}_{14} : S_{12}S_{23}S_{34}S_{14}$ and $\mathcal{Q}_{15} : S_{12}S_{23}S_{35}S_{15}$ are displaced such that

$$\Delta\hat{\theta}_1 = \Delta\hat{\theta}_2 \quad \text{and} \quad \Delta\hat{\phi}_1 = \Delta\hat{\phi}_2. \tag{13.43}$$

The first equation is satisfied by using the same dual angle parameter $\hat{\theta}$ to drive $S_{12}S_{23}$ for both screw quadrilaterals. The second equation requires that the spatial triangle $\triangle S_{34}S_{23}S_{35}$ have the same shape in each solution configuration. Thus, the fixed axis G is the screw axis of the relative displacement of this spatial triangle from its initial position to each of the assemblies of the 3CC platform formed by the chains $S_{12}S_{23}$, $S_{14}S_{34}$, and $S_{15}S_{35}$, Figure 13.4. This assembly of relative screw axes is called the *spatial compatibility platform* and we have the following theorem:

Theorem 18 (Murray's Compatibility Platform Theorem). *The fixed axis of a CC chain that can reach five spatial task positions is a screw axis of the displacement of the spatial compatibility platform from its initial configuration to one of its other assemblies.*

Analysis of the spatial compatibility platform separates into the analysis of its spherical image, which is identical to the spherical 3RR platform analyzed in chapter 8, and the solution of two sets of 4C vector loop equations discussed in chapter 9. The zero, two, four, or six axes obtained from the spherical image combine with the linear solution of the loop equations to yield zero, two, four, or six CC chains that reach five spatial task positions.

13.3 The Geometry of Spatial RR Chains

A spatial RR chain connects a floating link to ground using two revolute joints, Figure 13.5. Notice that the axes of these joints must be skew to each other in space, because parallel axes define a planar RR chain, and intersecting axes define a spherical RR chain.

Let G be the fixed axis and W the moving axis. Because a revolute joint allows only pure rotation, the common normal L between G and W intersects the fixed axis in the same point **B** for all positions of the chain. The same is true for the moving axis where the point **P** on the common normal to W traces a circle around **B**.

In order to define a spatial RR chain we need the directions **G** and **W** of the two lines and the coordinates of the points **B** and **P** on their common normal. Thus, there are ten design parameters that define this chain. We now determine the design equations.

13.3.1 The Constraint Equations

The RR chain combines the constraints of both the CC chain and the TS chain. It requires that the axes $G = (\mathbf{G}, \mathbf{B} \times \mathbf{G})^T$ and $W^i = (\mathbf{W}^i, \mathbf{P}^i \times \mathbf{W}^i)^T$ maintain a constant dual angle $\hat{\rho} = (\rho, R)$ in all positions of the chain, while at the same time the distances between **B** and the points $\mathbf{P}^i$ remain constant. In addition, we have the

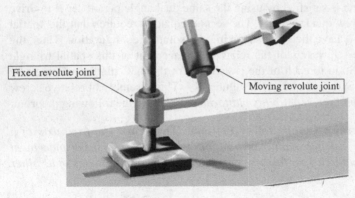

Fig. 13.5 A spatial RR open chain.

constraint that the relative vector $\mathbf{P}^i - \mathbf{B}$ must lie on the common normal between G and $\mathbf{W}^i$ in each position.

For n task positions, the geometry of the CC chain provides direction and moment constraint equations:

$$\mathscr{P}_{1i} : \mathbf{G} \cdot [A_{1i} - I]\mathbf{W}^1 = 0,$$

$$\mathscr{M}_{1i} : \mathbf{R} \cdot [A_{1i} - I]\mathbf{W}^i + \mathbf{G} \cdot [A_{1i} - I]\mathbf{V}^1 + \mathbf{G} \cdot [D_{1i}A_{1i}]\mathbf{W}^1 = 0,$$

$$i = 2,\ldots,n. \qquad (13.44)$$

The geometry of the TS chain provides the distance constraint

$$\mathscr{D}_{1i} : (\mathbf{B} - \mathbf{C}_{1i}) \cdot [A_{1i} - I](\mathbf{P}^1 - \mathbf{C}_{1i}) + d_{1i}\mathbf{S}_{1i} \cdot (\mathbf{B} - \mathbf{P}^1) - \frac{d_{1i}^2}{2} = 0,$$

$$i = 2,\ldots,n. \qquad (13.45)$$

And finally, we have the equations

$$\mathscr{N}_i : \quad \mathbf{G} \cdot (\mathbf{P}^i - \mathbf{B}) = 0, \quad \mathbf{W}^i \cdot (\mathbf{P}^i - \mathbf{B}) = 0, i = 1,\ldots,n. \qquad (13.46)$$

that ensure that $\mathbf{B}$ and $\mathbf{P}^i$ are on the common normal between G and $\mathbf{W}^i$ in each position.

13.3.1.1 Redundancy of the Moment Constraints

We now show that the direction, distance, and common normal constraint equations combine to satisfy the moment constraints. To do this we begin with (13.21) and compute

$$\mathbf{G} \cdot (\mathbf{W}^i - \mathbf{W}^1) = 0,$$
$$\mathbf{G} \cdot (\mathbf{P}^i \times \mathbf{W}^i - \mathbf{P}^1 \times \mathbf{W}^1) + \mathbf{B} \times \mathbf{G} \cdot (\mathbf{W}^i - \mathbf{W}^1) = 0. \qquad (13.47)$$

The first equation yields the direction constraints $\mathscr{P}_{1i}$ that the angle ρ between these axes must be constant. The second equation is the moment constraint $\mathscr{M}_{1i}$ and we combine terms to obtain

$$(\mathbf{P}^i - \mathbf{B}) \cdot \mathbf{G} \times \mathbf{W}^i - (\mathbf{P}^i - \mathbf{B}) \cdot \mathbf{G} \times \mathbf{W}^1 = 0. \qquad (13.48)$$

Notice that $\mathbf{G} \times \mathbf{W}^i = \sin\rho \mathbf{L}_i$, where $\mathbf{L}_i$ is the direction of the common normal to $\mathbf{G}$ and $\mathbf{W}^i$. The normal constraints $\mathscr{N}_i$ require that $\mathbf{P}^i - \mathbf{B} = R\mathbf{L}_i$. Therefore, we have

$$(R\mathbf{L}_i) \cdot (\sin\rho \mathbf{L}_i) - (R\mathbf{L}_1) \cdot (\sin\rho \mathbf{L}_1) = 0. \qquad (13.49)$$

Thus, the moment constraint equations are identically satisfied.

For n task positions we have $2(n-1)$ angle and distance constraints and $2n$ common normal constraints. Thus, for three specified positions, $n = 3$, there are ten equations in the ten unknown design parameters for the RR chain. We formulate a solution to these equations below.

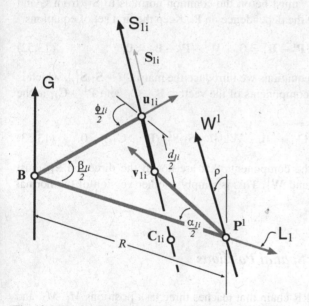

Fig. 13.6 The spatial dyad triangle for the spatial RR chain.

13.3.1.2 The Spatial Dyad Triangle Constraints

The constraint equations for the RR chain can be simplified by examining the geometry of the dyad triangles associated with each specified position, see Figure 13.6.

Let $\mathbf{u}_{1i}$ be the point of intersection of the common normal from the fixed axis G to each screw axis $\mathbf{S}_{1i}$. Similarly, let $\mathbf{v}_{1i}$ be the intersection of each screw axis with its the common normal to W^1. The geometry of the dyad triangle yields the relation $|\mathbf{u}_{1i} - \mathbf{v}_{1i}| = d_{1i}/2$, where d_{1i} is the slide associated with the displacement from position M_1 to M_i. The component of $\mathbf{B} - \mathbf{P}^1$ in the direction $\mathbf{S}_{1i}$ clearly equals $\mathbf{u}_{1i} - \mathbf{v}_{1i}$. The result is a simplified version of the distance constraint

$$\mathscr{D}_{1i} : \mathbf{S}_{1i} \cdot (\mathbf{B} - \mathbf{P}^1) - \frac{d_{1i}}{2} = 0, i = 2, \ldots, n. \tag{13.50}$$

This can also be obtained by showing that

$$(\mathbf{B} - \mathbf{C}_{1i}) \cdot [A_{1i} - I](\mathbf{P}^1 - \mathbf{C}_{1i}) = 0 \tag{13.51}$$

for RR chains, which simplifies (13.45) to yield the new constraint.

We now simplify the common normal constraints. The geometry of the RR dyad triangle shows that $\mathbf{B}$ and $\mathbf{P}^1$ must be on the common normals to $\mathbf{S}_{1i}$ from G and W^1. We use this to eliminate the dependence on $\mathbf{P}^i$. Keep the first set of equations

$$\mathscr{N}_1 : \mathbf{G} \cdot (\mathbf{P}^1 - \mathbf{B}) = 0, \quad \mathbf{W}^1 \cdot (\mathbf{P}^1 - \mathbf{B}) = 0. \tag{13.52}$$

However, for the remaining equations we introduce the matrix $[I - \mathbf{S}_{1i}\mathbf{S}_{1i}^T]$, which is an operator that cancels the components of the vectors $\mathbf{B} - \mathbf{C}_{1i}$ and $\mathbf{P}^1 - \mathbf{C}_{1i}$ in the direction $\mathbf{S}_{1i}$. The result is

$$\mathscr{N}_i : \mathbf{G} \cdot [I - \mathbf{S}_{1i}\mathbf{S}_{1i}^T](\mathbf{B} - \mathbf{C}_{1i}) = 0, \quad \mathbf{W}^1 \cdot [I - \mathbf{S}_{1i}\mathbf{S}_{1i}^T](\mathbf{P}^1 - \mathbf{C}_{1i}) = 0. \tag{13.53}$$

These equations state that the components that are not in the direction $\mathbf{S}_{1i}$ must be perpendicular to both $\mathbf{G}$ and $\mathbf{W}^1$. This is simply another version of the normal constraint.

13.3.2 Three Specified Spatial Positions

In order to design a spatial RR chain that reaches three task positions M_1, M_2, and M_3, we must solve the ten design equations

$$\mathscr{P}_{12} : \mathbf{G} \cdot [A_{12} - I] \mathbf{W}^1 = 0,$$

$$\mathscr{P}_{13} : \mathbf{G} \cdot [A_{13} - I] \mathbf{W}^1 = 0,$$

$$\mathscr{D}_{12} : \mathbf{S}_{12} \cdot (\mathbf{B} - \mathbf{P}^1) - \frac{d_{12}}{2} = 0,$$

$$\mathscr{D}_{13} : \mathbf{S}_{13} \cdot (\mathbf{B} - \mathbf{P}^1) - \frac{d_{13}}{2} = 0,$$

$$\mathscr{N}_1 : \mathbf{G} \cdot (\mathbf{P}^1 - \mathbf{B}) = 0, \quad \mathbf{W}^1 \cdot (\mathbf{P}^1 - \mathbf{B}) = 0,$$

$$\mathscr{N}_2 : \mathbf{G} \cdot [I - \mathbf{S}_{12} \mathbf{S}_{12}^T](\mathbf{B} - \mathbf{C}_{12}) = 0, \quad \mathbf{W}^1 \cdot [I - \mathbf{S}_{12} \mathbf{S}_{12}^T](\mathbf{P}^1 - \mathbf{C}_{12}) = 0,$$

$$\mathscr{N}_3 : \mathbf{G} \cdot [I - \mathbf{S}_{13} \mathbf{S}_{13}^T](\mathbf{B} - \mathbf{C}_{13}) = 0, \quad \mathbf{W}^1 \cdot [I - \mathbf{S}_{13} \mathbf{S}_{13}^T](\mathbf{P}^1 - \mathbf{C}_{13}) = 0. \quad (13.54)$$

The relative screw axes $\mathsf{S}_{12} = (\mathbf{S}_{12}, \mathbf{C}_{12} \times \mathbf{S}_{12})$ and $\mathsf{S}_{13} = (\mathbf{S}_{13}, \mathbf{C}_{13} \times \mathbf{S}_{13})$ are known from the task positions. The unknowns are the six point coordinates $\mathbf{B} = (x, y, z)^T$ and $\mathbf{P}^1 = (u, v, w)^T$ and the four parameters that define the directions of $\mathbf{G}$ and $\mathbf{W}^1$. Tsai and Roth [135] show that these equations can be solved to obtain two RR chains that form a Bennett linkage.

To solve these equations we introduce a special coordinate system associated with the two relative displacement screws S_{12} and S_{13}. These two screw axes lie on the two-system of relative screw axes generated by the movement of a Bennett linkage. In the principal frame of this two-system we find that six of the ten equations are identically satisfied and the number of variables is reduced to four.

13.3.2.1 The Cylindroid

The set of relative screw displacements $\mathsf{S}(\beta, \alpha)$ generated by an RR chain is obtained from the product of the two dual quaternions $\hat{G}(\beta/2) = \cos(\beta/2) + \sin(\beta/2)\mathsf{G}$ and $\hat{W}^1 = \cos(\alpha/2) + \sin(\alpha/2)\mathsf{W}^1$, which yields

$$\sin \frac{\hat{\phi}}{2} \mathsf{S} = \sin \frac{\beta}{2} \cos \frac{\alpha}{2} \mathsf{G} + \sin \frac{\alpha}{2} \cos \frac{\beta}{2} \mathsf{W}^1 + \sin \frac{\beta}{2} \sin \frac{\alpha}{2} \mathsf{G} \times \mathsf{W}^1,$$

$$\cos \frac{\hat{\phi}}{2} = \cos \frac{\beta}{2} \cos \frac{\alpha}{2} - \sin \frac{\beta}{2} \sin \frac{\alpha}{2} \mathsf{G} \cdot \mathsf{W}^1. \quad (13.55)$$

Notice that in these equations the angles α and β are real, not dual, angles.

Huang [49] reports the interesting result that if the angles β and α of the RR chain are constrained such that

$$\frac{\tan \frac{\alpha}{2}}{\tan \frac{\beta}{2}} = k, \quad (13.56)$$

where k is an arbitrary constant, then the axes $\mathsf{S}(\beta, \alpha)$ in (13.55) trace a *cylindroid*. Recall from (11.75) that this is a characteristic of the movement of Bennett's linkage.

A cylindroid is a ruled surface traced by the axes of the real linear combination of two independent screws, known as a *two-system*. The geometry of the cylindroid is well known, see Hunt [50]. In particular, it has a natural set of *principal axes* that simplify its description.

13.3.2.2 The Two-System

We now construct the two-system defined by the screws

$$S_a = \sin \frac{\hat{\phi}_{12}}{2} S_{12}, \quad S_b = \sin \frac{\hat{\phi}_{13}}{2} S_{13}, \tag{13.57}$$

which are obtained from the three specified positions. The RR chain that we are seeking must generate this two-system.

A general screw in this two-system is given by

$$F = a S_a + b S_b, \tag{13.58}$$

where a and b are real constants. The screws S_a and S_b can be written in the form

$$S_a = \sin \frac{\phi_{12}}{2}(1,P_a) S_{12}, \quad S_b = \sin \frac{\phi_{13}}{2}(1,P_b) S_{13}, \tag{13.59}$$

where

$$P_a = \frac{d_{12}}{2 \tan \frac{\phi_{12}}{2}} \quad \text{and} \quad P_b = \frac{d_{13}}{2 \tan \frac{\phi_{13}}{2}} \tag{13.60}$$

are the *pitches* of the two screws. We can absorb the scalar magnitudes of these screws into the constants a and b, so the equation of the two-system becomes

$$F = a(1,P_a) S_{12} + b(1,P_b) S_{13}. \tag{13.61}$$

Now introduce the common normal K between the axes S_{12} and S_{13}. Let $S_{12} = I$ and $J = K \times I$, so the line S_{13} is given by

$$S_{13} = \cos \hat{\delta} I + \sin \hat{\delta} J. \tag{13.62}$$

The dual angle $\hat{\delta} = (\delta,d)$ locates this axis in the frame I, J, and K. Substitute this into equation (13.61) to obtain

$$F = a(1,P_a) I + b(1,P_b)(\cos \hat{\delta} I + \sin \hat{\delta} J)$$
$$= \big(a + b\cos\delta, aP_a + b(P_b\cos\delta - d\sin\delta)\big) I + \big(b\sin\delta, b(P_b\sin\delta + d\cos\delta)\big) J. \tag{13.63}$$

The axis of each screw F intersects K and is parallel to the plane defined by I and J. Let $\hat{\phi} = (\phi,z)$ denote the dual angle from I to each axis. Then we have

$$F = f(1,P)(\cos \hat{\phi} I + \sin \hat{\phi} J)$$
$$= f(\cos\phi, -z\sin\phi + P\cos\phi) I + f(\sin\phi, z\cos\phi + P\sin\phi), \tag{13.64}$$

where f is the magnitude of F. Equate the real and dual components of equations (13.63) and (13.64) to obtain

$$f \begin{Bmatrix} \cos\phi \\ \sin\phi \end{Bmatrix} = \begin{bmatrix} 1 & \cos\delta \\ 0 & \sin\delta \end{bmatrix} \begin{Bmatrix} a \\ b \end{Bmatrix}. \tag{13.65}$$

and

$$f \begin{bmatrix} -\sin\phi & \cos\phi \\ \cos\phi & \sin\phi \end{bmatrix} \begin{Bmatrix} z \\ P \end{Bmatrix} = \begin{bmatrix} P_a & P_b\cos\delta - d\sin\delta \\ 0 & P_b\sin\delta + d\cos\delta \end{bmatrix} \begin{Bmatrix} a \\ b \end{Bmatrix}. \tag{13.66}$$

Solve the first of these matrix equations for $(a,b)^T$, and substitute into the second equation in order to obtain

$$\begin{Bmatrix} z(\phi) \\ P(\phi) \end{Bmatrix} = \begin{bmatrix} -\sin\phi & \cos\phi \\ \cos\phi & \sin\phi \end{bmatrix} \begin{bmatrix} P_a & (P_b - P_a)\cot\delta - d \\ 0 & P_b + d\cot\delta \end{bmatrix} \begin{Bmatrix} \cos\phi \\ \sin\phi \end{Bmatrix}. \tag{13.67}$$

This result defines the pitch P and the offset z of each screw in the two-system as a function of the angle ϕ measured around K from $I = S_{12}$.

13.3.2.3 The Principal Axes

The principal axes of the two-system are the axes of the screws that have the maximum and minimum values for the pitch $P(\phi)$. We determine these axes by computing the derivative of (13.67), which is

$$\frac{d}{d\phi} \begin{Bmatrix} z \\ P \end{Bmatrix} = \begin{bmatrix} -(P_b - P_a)\cot\delta + d & (P_b - P_a) + d\cot\delta \\ (P_b - P_a) + d\cot\delta & (P_b - P_a)\cot\delta - d \end{bmatrix} \begin{Bmatrix} \sin 2\phi \\ \cos 2\phi \end{Bmatrix}. \tag{13.68}$$

Set the second equation equal to zero to determine the angle $\phi = \sigma$ that locates the screws with extreme values of pitch, that is,

$$\tan 2\sigma = \frac{-(P_b - P_a)\cot\delta + d}{(P_b - P_a) + d\cot\delta}. \tag{13.69}$$

Notice that this equation defines two values, $\phi = \sigma$ and $\phi = \sigma + \pi/2$, which identify the *principal screws* of the two-system. Their axes X and Y are the *principal axes*.

Determine the offset $z(\sigma)$ from (13.67) and assemble the dual angle $\hat{\sigma} = (\sigma, z(\sigma))$ to locate the principal screws of the two-system. The result is

$$X = \cos\hat{\sigma} I + \sin\hat{\sigma} J \quad \text{and} \quad Y = -\sin\hat{\sigma} I + \cos\hat{\sigma} J. \tag{13.70}$$

We use the coordinate system X, Y, K to reformulate our design equations for the RR chain.

The angle $\phi = \tau$ to the screw with extreme values for the offset z from S_{12} is obtained from the first equation of (13.68) as

$$\tan 2\tau = \frac{(P_b - P_a) + d\cot\delta}{(P_b - P_a)\cot\delta - d}.\qquad(13.71)$$

This defines two angles $\phi = \tau$ and $\phi = \tau + \pi/2$. Notice that $\tan 2\sigma \tan 2\tau = -1$, therefore the angles σ and τ differ by $45°$. This means that the screws with the maximum and minimum offset along K are directed at $45°$ to the principal axes.

Thus, with the specification of the three task positions comes a coordinate frame F' aligned with the principal axes of the two-system constructed from $\sin\hat{\phi}_{12}S_{12}$ and $\sin\hat{\phi}_{13}S_{13}$. Transform the coordinates of the RR design problem so that the three positions are defined relative to F'. If $[R]$ is the 4×4 transform that defines the position of F', then the new positions M_1', M_2', M_3' are defined by the transformations $[T_i'] = [R][T_i][R^{-1}]$.

13.3.3 Bennett Linkage Coordinates

We now introduce a set of coordinates originally used to capture the symmetry inherent in a Bennett linkage (Yu [156]). These coordinates are adapted to the tetrahedron formed by the joints of a Bennett linkage, which is centered on the principal axes of our cylindroid, Figure 13.7. These coordinates reduce the number of design parameters from ten to four and at the same time identically satisfy six of the constraint equations. The result is four nonlinear equations in four unknowns, which we solve numerically.

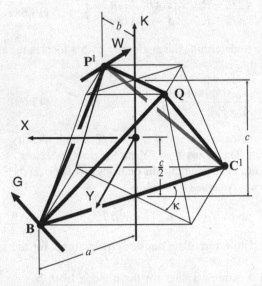

Fig. 13.7 The coordinate frame adapted to a Bennett linkage.

Let the points at the four vertices of the tetrahedron be denoted by $\mathbf{B}$, $\mathbf{P}^1$, $\mathbf{Q}$, and $\mathbf{C}^1$, and denote the edges defined by $\mathbf{B} - \mathbf{C}^1$ and $\mathbf{P}^1 - \mathbf{Q}$ by E_1 and E_2. The tetrahedron is oriented so that the common normal to E_1 and E_2 is aligned with the axis K of the cylindroid. The dimensions of the Bennett linkage ensure that in this configuration K bisects both segments $\mathbf{B} - \mathbf{C}^1$ and $\mathbf{P}^1 - \mathbf{Q}$. Let the lengths of these edges be $2a = |\mathbf{B} - \mathbf{C}^1|$ and $2b = |\mathbf{P}^1 - \mathbf{Q}|$ and let c and κ be the distance and angle between E_1 and E_2 measured along and around K. Finally, the tetrahedron is located so that the principal axis X is midway between E_1 and E_2 and it bisects the angle κ. These four parameters completely specify the tetrahedron, that is, we have

$$\mathbf{B} = \left\{ \begin{array}{c} a\cos\frac{\kappa}{2} \\ a\sin\frac{\kappa}{2} \\ -\frac{c}{2} \end{array} \right\}, \quad \mathbf{P}^1 = \left\{ \begin{array}{c} b\cos\frac{\kappa}{2} \\ -b\sin\frac{\kappa}{2} \\ \frac{c}{2} \end{array} \right\}$$

and

$$\mathbf{Q} = \left\{ \begin{array}{c} -b\cos\frac{\kappa}{2} \\ b\sin\frac{\kappa}{2} \\ \frac{c}{2} \end{array} \right\}, \quad \mathbf{C}^1 = \left\{ \begin{array}{c} -a\cos\frac{\kappa}{2} \\ -a\sin\frac{\kappa}{2} \\ -\frac{c}{2} \end{array} \right\}. \tag{13.72}$$

The axes of the Bennett linkage are perpendicular to each of the four faces of this tetrahedron. Thus, we have

$$\mathbf{G} = k_g(\mathbf{Q} - \mathbf{B}) \times (\mathbf{P}^1 - \mathbf{B}) = k_g \left\{ \begin{array}{c} 2bc\sin\frac{\kappa}{2} \\ 2bc\cos\frac{\kappa}{2} \\ 4ab\cos\frac{\kappa}{2}\sin\frac{\kappa}{2} \end{array} \right\}$$

and

$$\mathbf{W}^1 = k_w(\mathbf{B} - \mathbf{P}^1) \times (\mathbf{C}^1 - \mathbf{P}^1) = k_w \left\{ \begin{array}{c} -2ac\sin\frac{\kappa}{2} \\ 2ac\cos\frac{\kappa}{2} \\ 4ab\cos\frac{\kappa}{2}\sin\frac{\kappa}{2} \end{array} \right\}. \tag{13.73}$$

The constants k_g and k_w normalize these vectors. Notice that the screws $\mathbf{G} = (\mathbf{G}, \mathbf{B} \times \mathbf{G})^T$ and $\mathbf{W}^1 = (\mathbf{W}^1, \mathbf{P}^1 \times \mathbf{W}^1)$ depend only on the four parameters a, b, c, and κ.

Using these coordinates, the six design equations that define the common normal constraints $\mathcal{N}_i$, $i = 1, 2, 3$, are identically satisfied. This leaves the four design equations

$$\mathscr{P}_{12} : \mathbf{G} \cdot [A'_{12} - I]\mathbf{W}^1 = 0, \quad \mathscr{D}_{12} : \mathbf{S}'_{12} \cdot (\mathbf{B} - \mathbf{P}^1) - \frac{d'_{12}}{2} = 0,$$

$$\mathscr{P}_{13} : \mathbf{G} \cdot [A'_{13} - I]\mathbf{W}^1 = 0, \quad \mathscr{D}_{13} : \mathbf{S}'_{13} \cdot (\mathbf{B} - \mathbf{P}^1) - \frac{d'_{13}}{2} = 0. \tag{13.74}$$

These equations can be expanded using algebraic manipulation software and solved numerically to determine the values of a, b, c, and κ. Notice that by determining these parameters we actually define a Bennett linkage that guides the coupler through the three specified positions.

13.4 Platform Linkages

The design theory associated with each of the TS, CC, and RR spatial chains yields multiple solutions for a finite set of task positions. Thus, it is possible to connect the end-links of various solutions together and obtain a platform linkage with a reduced number of degrees of freedom. We have already seen that the 5TS and 4C linkages have one and two degrees of freedom, respectively. These chains can be also be combined to define the 3TS-CC and a TSRR linkages, both with one degree of freedom. As in the cases of planar and spherical linkages, the combination of chains introduces constraints on the range of movement of the individual cranks that can affect the smooth movement of the coupler or platform. The analysis of this problem for general spatial platforms is the focus of much research. Currently, there is little rectification theory for spatial linkage design.

13.4.1 The 4C Spatial Linkage

Our solutions for CC chains can yield as many as six of these chains for a five position task. Thus, we can form up to 15 two-degree-of-freedom 4C linkages to accomplish this task. While the design theory for this linkage is the generalization of the planar and spherical 4R design theories, its use depends on the development of computer-aided design and simulation systems that allow the designer to visualize and plan its movement in space.

13.4.2 The 5TS Spatial Linkage

The solution of the constraint equations for a TS chain that can reach seven specified goal positions may yield many as 20 of these chains. Thus, there can be as many as 15504 5TS linkages that can reach seven specified positions. The analysis of these linkages to determine their smooth movement through the goal positions is required to complete the design.

13.4.3 Function Generation

Spatial open chains can be connected in a way that coordinates the input and output parameters to form *spatial function generators*. Here we show that RSSR and RSSP function generators can be designed using the theory presented in the previous sections.

Suppose that we have a table of n input angular values θ_i and output angles ψ_i for the RSSR, and output slides s_i for the RSSP linkage. Following the strategy used for both the plane and the sphere, invert this problem to consider the movement of the output link relative to the input link and design the TS chain that reaches the specified positions.

For both the RSSR and the RSSP linkages, we select two skew lines O and C to be the axes of the fixed joints. Let $\hat{\gamma} = (\gamma, g)$ be the dual angle between O and C. Denote the common normal between O and C by N, and have it intersect O at the point **c**.

13.4.3.1 The RSSR linkage

For the RSSR linkage, the lines O and C are the axes of the fixed revolute joints. Let θ_i and ψ_i, $i = 1, \ldots, n$, be the desired table of values. Now convert the input angles θ_i to exterior angles $\bar{\theta}_i = \pi - \theta_i$. Let P_O be the line through **c** orthogonal to O and oriented such that $\bar{\theta}_1 = \pi - \theta$ is the angle measured from P_O around O to N. Introduce a new fixed frame F' that has its origin at **c**, its z-axis is along O and its x-axis along P_O.

The angles $\bar{\theta}_i$ and ψ_i can now be viewed as the joint angles of the spatial RR open chain formed by OC in F'. The kinematics equations of this RR chain define task positions in F', given by

$$[D_i] = [Z(\bar{\theta}_i, 0)][X(\gamma, g)][Z(\psi_i, 0)], \quad i = 1, \ldots, n. \tag{13.75}$$

Our design results yield TS chains that reach as many as seven positions $[D_i]$. The TS chain design combines with the selected RR chain to form an RTSR linkage, or equivalently an RSSR linkage, that has the desired set of coordinated angles θ_i and ψ_i.

13.4.3.2 The RSSP linkage

For the RSSP linkage, the line O is the axis of the revolute joint and C is the guide for the prismatic joint. Let θ_i and s_i, $i = 1, \ldots, n$, be the desired table of values. As we did above, invert the design problem and determine the exterior angles $\bar{\theta}_i = \pi - \theta_i$. Let P_O be the line through **c** orthogonal to O, and oriented such that $\bar{\theta}_1 = \pi - \theta$ is the angle measured from P_O around O to N. Locate the frame F' with its origin at **c**, its z-axis is along O, and its x-axis is along P_O. In this frame the kinematic equations of the spatial RP chain are given by

$$[D_i] = [Z(\bar{\theta}_i, 0)][X(\gamma, g)][Z(0, s_i)], i = 1, \dots, n. \qquad (13.76)$$

The positions $[D_i]$ are used to design a TS chain that connects the input crank and output slider. The result is an RTSP linkage, or equivalently an RSSP linkage, that provides the desired coordination between the crank angles θ_i and the slider translations s_i.

13.5 Summary

This chapter has developed the design theory for TS, CC and spatial RR open chains. These chains can be combined to form the 5TS, 4C and spatial 4R spatial linkages, as well as RSSR and RSSP function generators. The focus on these chains reflects the attention they have received in the design literature. There are other chains available for spatial linkages, for example the CS and spatial RPR chains, for which the design theory is not as well-developed.

13.6 References

Suh [132] and Chen and Roth [10] introduced the geometric design of spatial linkages as the solution of sets of constraint equations. This approach is presented in detail in the text Suh and Radcliffe [134]. Our results for the design of TS chains draw on the work by Innocenti [52] and Liao [66]. For CC chains, we use [77, 78] and Murray [85]. See Ahlers [79] for the design of a CC robot that approximate a specified trajectory. Tsai and Roth [135] solved the design equations for spatial RR chains. Our formulation follows Perez [97], who uses the principal frame of the cylindroid to simplify these equations.

Exercises

1. Determine the two spatial RR chains that reach the three positions specified in Table 13.1 (Tsai and Roth [135]).
2. Determine the four CC chains that reach the five positions specified in Table 13.2 (Murray and McCarthy [85]).
3. Determine the 20 TS chains that reach the seven positions listed in Table 13.3 (Innocenti [52]).

Table 13.1 Three spatial positions

M_i	$\mathbf{S}_i$	$\mathbf{C}_i$	(θ_i, d_i)
1	$(0,0,1)^T$	$(0,0,0)^T$	$(0,0)$
2	$(0,0,1)^T$	$(0,0,0)^T$	$(40°, 0.80)$
3	$(\sin 30°, 0, \cos 30°)^T$	$(0,1,0)^T$	$(70°, 0.60)$

Table 13.2 Five spatial positions

M_i	$\mathbf{S}_i$	$\mathbf{C}_i$	(θ_i, d_i)
1	$(0,0,1)^T$	$(0,0,0)^T$	$(0,0)$
2	$(0.04, 0.10, 0.90)^T$	$(0.94, 0.52, -0.41)^T$	$(133°, 1.38)$
3	$(0.15, -0.04, 0.10)^T$	$(-0.61, 0.84, 0.12)^T$	$(70.6°, 1.90)$
4	$(0.43, -0.25, 0.87)^T$	$(0.73, 0.69, -0.16)^T$	$(87.9°, -1.12)$
5	$(0.40, -0.02, -0.92)^T$	$(5.90, 0.10, 2.60)^T$	$(25.2°, 2.56)$

Table 13.3 Seven spatial positions

M_i	$\mathbf{d}_i$	(long., lat., roll)
1	$(0,0,0)^T$	$(0°, 0°, 0°)$
2	$(1, -0.74, -0.13)^T$	$(6.18°, 4.28°, -97.93°)$
3	$(0.32, -0.51, -0.80)^T$	$(-83.26°, -18.23°, 73.61°)$
4	$(-0.18, -1.78, -1.04)^T$	$(-170.03°, 39.54°, -50.94°)$
5	$(-1.26, 0.84, -1.50)^T$	$(-84.74°, -29.18°, 150.3°)$
6	$(-3.59, 2.73, -2.03)^T$	$(-8.30°, 5.04°, -68.25°)$
7	$(-0.05, 0.57, -1.48)^T$	$(118.3°, -33.80°, 139.0°)$

Chapter 14
Synthesis of Spatial Chains with Reachable Surfaces

In this chapter, we consider the design of spatial serial chains that guide a body such that a point in the body moves on a specific algebraic surface. The problem originates with Schoenflies [113], who sought points that remained in a given configuration for a given set of spatial displacements. Burmester [7] applied this idea to planar mechanism design by seeking the points in a planar moving body that remain on a circle. Chen and Roth [10] generalized this problem to find points and lines in a moving body that take positions on surfaces associated with the articulated chains used to build robot manipulators.

Our focus is on serial chains that support a spherical wrist. The center of this wrist traces a surface that is said to be "reachable" by the chain. Considering the various ways of assembling these articulated chains, we obtain seven reachable algebraic surfaces, and the problem reduces to computing the dimensions of these chains from a set of polynomial equations.

14.1 Spatial Serial Chains

In this chapter, we focus on five degree-of-freedom spatial serial chains that include a spherical wrist. The surface traced by a point $\mathbf{P}$ at the wrist center under the movement of two remaining joints is called its *reachable surface*. Considering only revolute and prismatic joints, we can enumerate the seven possibilities:

1. the PPS chain, for which the wrist center, $\mathbf{P}$, lies on a plane—notice that the angle between the slide can be any angle α except zero; similarly, the distance ρ between the slides can be any value because a prismatic joint guides all points in the body in the same direction;
2. the TS chain that has $\mathbf{P}$ on a sphere—recall that the T joint is constructed from two perpendicular intersecting revolute joints, that is, with link angle $\alpha = \pi/2$ and length $\rho = 0$;

J.M. McCarthy and G.S. Soh, *Geometric Design of Linkages*, Interdisciplinary Applied Mathematics 11, DOI 10.1007/978-1-4419-7892-9_14,
© Springer Science+Business Media, LLC 2011

Table 14.1 The basic serial chains and their associated reachable surfaces.

Case	Chain	Angle	Length	Surface
1	PPS	$\alpha \neq 0$	ρ	plane
2	TS	$\alpha = \pi/2$	$\rho = 0$	sphere
3	CS	$\alpha = 0$	ρ	circular cylinder
4	RPS	$\alpha \neq 0$	ρ	circular hyperboloid
5	PRS	$\alpha \neq 0$	ρ	elliptic cylinder
6	right RRS	$\alpha = \pi/2$	ρ	circular torus
7	RRS	α	ρ	general torus

3. the CS chain for which **P** lies on a cylinder—the C joint is constructed from a PR chain for which the direction of the prismatic slide is parallel to the axis of the revolute joint, that is, $\alpha = 0$; note that ρ can be any value;
4. the RPS chain that guides **P** on the surface of a right circular hyperboloid—the link angle α can be any value except zero;
5. the PRS chain in which the angle between the prismatic slide and the axis of the revolute is not zero guides **P** on an elliptic cylinder—the link angle α can be any value except zero;
6. the "right" RRS chain in which the revolute joints are perpendicular but do not intersect has **P** trace a right circular torus—the linkage angle $\alpha = \phi/2$; and
7. the general RRS chain in which the revolute joint axes are neither perpendicular nor intersecting guides the wrist center on a general circular torus—the linkage angle cannot be $\alpha = 0, \pi/2$.

The result is seven articulated chains and the associated algebraic surfaces that are reachable by their wrist centers, Table 14.1. The algebraic equations of these surfaces are used to formulate the synthesis equations for these seven spatial serial chains. In what follows, we determine the number of free parameters for each chain, the associated number of task positions that define these parameters, and assemble the synthesis equations. These equations can be solved using polynomial continuation.

14.2 Linear Product Decomposition

The synthesis equations for the seven spatial serial chains described above result in polynomial systems of very high degree. Bézout's theorem states that the number of solutions to a polynomial system is less than or equal to the degree of the polynomial system, which is obtained by multiplying the degrees of each of the polynomials in the system. In what follows, we find that the synthesis equations of these serial chains have so much internal structure that the total degree overestimates the number of solutions by two orders of magnitude.

In order to efficiently use polynomial continuation techniques to find all of the solutions to our synthesis equations, it is useful to have a better estimate for the number of solutions than the total degree. Here we present the linear product decomposition of a polynomial system and then use it to determine a bound on the number of solutions for each of our systems of synthesis equations. The linear product decomposition also serves as a convenient start system for polynomial continuation algorithms.

Morgan et al. [84] show that a "generic" system of polynomials that includes every monomial of a specified system of polynomials will have at least as many solutions as the specified polynomial system. The *linear product decomposition* of a specified system of polynomials is a way of constructing this generic polynomial system that includes all of the monomials of the specified system, so that it allows convenient computation of the number of roots. Each polynomial in the linear product decomposition consists of polynomials formed by the products of linear combinations of the variables and all of the monomials of the corresponding original polynomial.

Let $\langle x, y, 1 \rangle$ represent the set of linear combinations of parameters x, y, and 1, which means that a typical term is $\alpha x + \beta y + \gamma \in \langle x, y, 1 \rangle$, where α, β, and γ are arbitrary constants. Using this notation, we define the product $\langle x, y, 1 \rangle \langle u, v, 1 \rangle$ as the set of linear combinations of the product of the elements of the two sets, that is,

$$\langle x, y, 1 \rangle \langle u, v, 1 \rangle = \langle xu, xv, yu, yv, x, y, u, v, 1 \rangle. \tag{14.1}$$

This product commutes, which means that $\langle x \rangle \langle y \rangle = \langle y \rangle \langle x \rangle$, and it distributes over unions, such that $\langle x \rangle \langle y \rangle \cup \langle x \rangle \langle z \rangle = \langle x \rangle (\langle y \rangle \cup \langle z \rangle) = \langle x \rangle \langle y, z \rangle$. Furthermore, we represent repeated factors using exponents, so $\langle x, y, 1 \rangle \langle x, y, 1 \rangle = \langle x, y, 1 \rangle^2$.

In order to illustrate the construction of the linear product decomposition consider the synthesis equations of the TS chain presented in the previous chapter, given by

$$(\mathbf{P}^i - \mathbf{B}) \cdot (\mathbf{P}^i - \mathbf{B}) = R^2, \quad i = 1, \dots, 7, \tag{14.2}$$

where the dot denotes the vector dot product. Now subtract the first equation from the rest in order to eliminate R^2. This reduces the problem to six equations in the unknowns $\mathbf{z} = (x, y, z, u, v, w)$, given by

$$\mathscr{S}_j(\mathbf{z}) = (\mathbf{P}^{j+1} \cdot \mathbf{P}^{j+1} - \mathbf{P}^1 \cdot \mathbf{P}^1) - 2\mathbf{B} \cdot (\mathbf{P}^{j+1} - \mathbf{P}^1) = 0, \quad j = 1, \dots, 6. \tag{14.3}$$

We now focus attention on the monomials formed by the unknowns.

Recall that $\mathbf{P}^i = [A_i]\mathbf{p} + \mathbf{d}_i$, where $[A_i]$ and $\mathbf{d}_i$ are known, so it is easy to see that

$$2\mathbf{B} \cdot (\mathbf{P}^{j+1} - \mathbf{P}^1) \in \langle u, v, w \rangle \langle x, y, z, 1 \rangle. \tag{14.4}$$

It is also possible to compute

$$\mathbf{P}^{j+1} \cdot \mathbf{P}^{j+1} - \mathbf{P}^1 \cdot \mathbf{P}^1 = 2\mathbf{d}_{j+1} \cdot [A_{j+1}]\mathbf{p} - 2\mathbf{d}_1 \cdot [A_1]\mathbf{p} + \mathbf{d}_{j+1}^2 - \mathbf{d}_1^2 \in \langle x, y, z, 1 \rangle. \tag{14.5}$$

Thus, we find that each of the equations in (14.3) has the monomial structure given by

$$\langle x,y,z,1\rangle \cup \langle u,v,w\rangle \langle x,y,z,1\rangle \subset \langle x,y,z,1\rangle \langle u,v,w,1\rangle. \tag{14.6}$$

This allows us to construct a generic set of polynomials as a product of linear factors that contains our synthesis equations as a special case, that is,

$$Q(\mathbf{z}) = \left\{ \begin{array}{c} (a_1 x + b_1 y + c_1 z + d_1)(e_1 u + f_1 v + g_1 w + h_1) \\ \vdots \\ (a_6 x + b_6 y + c_6 z + d_6)(e_6 u + f_6 v + g_6 w + h_6) \end{array} \right\} = 0, \tag{14.7}$$

where the coefficients are known constants. This is the *linear product decomposition* of the synthesis equations for the TS chain.

This linear product decomposition provides a convenient way to determine a bound on the number of solutions for the original polynomial system. This is done by assembling all combinations of the linear factors, one from each equation, that can be set to zero and solved for the unknown parameters. The number of combinations that yield solutions is the LPD bound for the original polynomial system.

In the example above, select three factors $a_i x + b_i y + c_i z + d_i = 0$ from the six equations, and combine these with the three factors $e_i u + f_i v + g_i w + h_i = 0$ in the remaining equations. A solution of this set of six linear equations is a root of (14.7). Thus, we find that this system has $\binom{6}{3} = 20$ solutions, which matches the known result for (14.3).

In the following sections, we determine the synthesis equations for each of the seven spatial serial chains with a reachable surface. We evaluate its total degree, compute its linear product decomposition bound, and then numerically solve a generic problem to find the number of articulated chains that reach a specified set of displacements.

14.3 The Plane

The PPS serial chain has the property that the wrist center $\mathbf{P}$ is constrained to lie on a plane (Figure 14.1). A point $\mathbf{P} = (X,Y,Z)$ lies on a plane with the surface normal $\mathbf{G} = (a,b,c)$ if it satisfies the equation

$$aX + bY + cZ - d = \mathbf{G} \cdot \mathbf{P} - d = 0. \tag{14.8}$$

The parameter d is the product of the magnitude $|G|$ and the signed normal distance to the plane.

Given a set of spatial displacements, $\hat{Q}_i$, $i = 1,\ldots,n$, we have the images $\mathbf{P}^i = [T(\hat{Q}_i)]\mathbf{p}$ of a single point $\mathbf{p}$ in the moving frame M. We seek both the plane P : $(\mathbf{G},d)$ and the point $\mathbf{p} = (x,y,z)$, so each $\mathbf{P}^i$ lies on this plane.

There are seven parameters in this problem. However, the components of $\mathbf{G}$ are not independent because only its direction is important to defining the plane, not

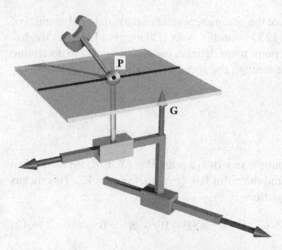

Fig. 14.1 A plane as traced by a point at the wrist center of a PPS serial chain.

its magnitude. A convenient way to constrain this magnitude is to choose a vector **m** and a scalar e, and require that $\mathbf{m} \cdot \mathbf{G} = e$. Add to this six equations obtained by evaluating (14.8) on six arbitrary displacements, given by

$$\mathbf{G} \cdot \mathbf{P}^i - d = 0, \quad i = 1,\ldots,6. \tag{14.9}$$

Subtract the first of these equations from the remaining to eliminate d, and the result is the polynomial system

$$\begin{aligned}
\mathscr{P}_j : \quad & \mathbf{G} \cdot (\mathbf{P}^{j+1} - \mathbf{P}^1) = 0, \quad j = 1,\ldots,5, \\
\mathscr{C} : \quad & \mathbf{m} \cdot \mathbf{G} - 1 = 0.
\end{aligned} \tag{14.10}$$

This is a set of five quadratic equations and one linear equation in the six unknowns $\mathbf{z} = (a,b,c,x,y,z)$. The total degree of this system of polynomials is $2^5 = 32$, which means that for six arbitrary displacements there are most 32 points in the moving body that have six positions on a plane.

The linear combinations of monomials that contain the plane equations (14.10) are given by

$$\begin{aligned}
\mathscr{P}_j \in \langle a,b,c \rangle \langle x,y,z,1 \rangle |_j = 0, \quad j = 1,\ldots,5, \\
\mathscr{C} \in \langle a,b,c,1 \rangle = 0.
\end{aligned} \tag{14.11}$$

This is the linear product decomposition (LPD) of the polynomial system. The root count for this linear product decomposition is given by the combinations of linear factors that can be set to zero and solved for the unknown parameters. In this case, we have $\binom{5}{2} = 10$ roots, which means that there are at most 10 points in the moving body that have six positions on a plane.

In this case, direct elimination of the parameters yields a univariate polynomial of degree 10, which shows that this LPD bound is exact (DiGregorio [23] and Raghavan [102]). Once the plane P and point $\mathbf{p}$ are defined, then it is possible to determine a PPS chain that guides this point through the specified positions.

14.4 The Sphere

We now return to our opening example in which a point $\mathbf{P} = (X, Y, Z)$ is constrained to lie on a sphere of radius R around the point $\mathbf{B} = (u, v, w)$, Figure 14.2. This means that its coordinates satisfy the equation

$$(X - u)^2 + (Y - v)^2 + (Z - w)^2 - R^2 = (\mathbf{P} - \mathbf{B})^2 - R^2 = 0. \qquad (14.12)$$

We now consider $\mathbf{P}^i$ to be the image of a point $\mathbf{p} = (x, y, z)$ in a moving frame M that takes positions in space defined by the displacements $\hat{\mathbf{Q}}_i$, $i = 1, \ldots, n$.

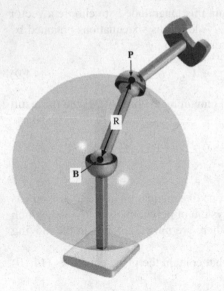

Fig. 14.2 A sphere traced by a point at the wrist center of a TS serial chain.

This problem has seven parameters, and therefore we can evaluate (14.12) on $n = 7$ displacements. We reduce these equations to the set of six quadratic polynomials,

$$\mathscr{S}_j : (\mathbf{P}^{j+1^2} - \mathbf{P}^{1^2}) - 2\mathbf{B} \cdot (\mathbf{P}^{j+1} - \mathbf{P}^1) = 0, \quad j = 1, \ldots, 6. \qquad (14.13)$$

This system has total degree of $2^6 = 64$.

We have already seen that the system (14.13) has the linear product decomposition

$$\mathscr{S}_j \in \langle x, y, z, 1 \rangle \langle u, v, w, 1 \rangle|_j = 0, \quad j = 1, \ldots, 6. \tag{14.14}$$

From this we can compute the LPD bound $\binom{6}{3} = 20$. Parameter elimination yields a univariate polynomial of degree 20, so we see that this bound is exact. Innocenti [52] presents an example that results in 20 real roots. Also see Liao and McCarthy [66] and Raghavan [103].

The conclusion is that given seven arbitrary spatial positions there can be as many as 20 points in the moving body that have positions lying on a sphere. For each real point, it is possible to determine an associated TS chain.

14.5 The Circular Cylinder

In order to define the equation of a circular cylinder, let the line $L(t) = \mathbf{B} + t\mathbf{G}$ be its axis. A general point $\mathbf{P}$ on the cylinder lies on a circle about the point $\mathbf{Q}$ closest to it on the axis $L(t)$. See Figure 14.3. Introduce the unit vectors $\mathbf{u}$ and $\mathbf{v}$ along $\mathbf{G}$ and the radius R of the cylinder, respectively, so we have

$$\mathbf{P} - \mathbf{B} = d\mathbf{u} + R\mathbf{v}, \tag{14.15}$$

where d is the distance from $\mathbf{B}$ to $\mathbf{Q}$. Compute the cross product of this equation with $\mathbf{G}$, in order to cancel d before squaring both sides. The result is

$$((\mathbf{P} - \mathbf{B}) \times \mathbf{G})^2 = R^2 \mathbf{G}^2. \tag{14.16}$$

In this calculation we use the fact that $(\mathbf{v} \times \mathbf{G})^2 = \mathbf{G}^2$.

Another version of the equation of the cylinder is obtained by substituting $d = (\mathbf{P} - \mathbf{B}) \cdot \mathbf{u}$ into (14.15) and squaring both sides to obtain

$$(\mathbf{P} - \mathbf{B})^2 - ((\mathbf{P} - \mathbf{B}) \cdot \mathbf{G})^2 \frac{1}{\mathbf{G} \cdot \mathbf{G}} - R^2 = 0. \tag{14.17}$$

Notice that we allow $\mathbf{G}$ to have an arbitrary magnitude. This form of the cylinder is related to the equation of the circular hyperboloid, which is discussed in the next section.

Equation (14.16) has 10 parameters: the radius R and three each in the vectors $\mathbf{P} = (X, Y, Z)$, $\mathbf{B} = (u, v, w)$, and $\mathbf{G} = (a, b, c)$. However, because only the direction of $\mathbf{G}$ is important to the definition of the cylinder, its three components are not independent. We set the magnitude of $\mathbf{G}$ as we did above for the equation of the plane. Choose an arbitrary vector $\mathbf{m}$ and scalar e and require the components of $\mathbf{G}$ to satisfy the constraint,

$$\mathscr{C}_1: \quad \mathbf{G} \cdot \mathbf{m} - e = 0. \tag{14.18}$$

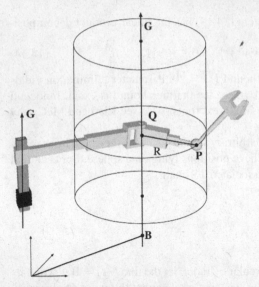

Fig. 14.3 The circular cylinder reachable by a CS serial chain.

The components of the point **B** are also not independent, but for a different reason. It is because any point on the line $L(t)$ can be selected as the reference point **B**. We identify this point by requiring **B** to lie on a specific plane $U : (\mathbf{n}, f)$, that is,

$$\mathscr{C}_2 : \quad \mathbf{B} \cdot \mathbf{n} - f = 0, \tag{14.19}$$

where $\mathbf{n}$ and f are chosen arbitrarily to avoid the possibility that the line $L(t)$ may lie on U.

Eight more polynomials are obtained by evaluating (14.17) with **P** specified as the image of $\mathbf{p} = (x, y, z)$ for eight spatial displacements, that is, $\mathbf{P}^i = [T(\hat{\mathbf{Q}}_i)]\mathbf{p}$, $i = 1, \ldots, 8$. The result is

$$((\mathbf{P}^i - \mathbf{B}) \times \mathbf{G})^2 - R^2 \mathbf{G}^2 = 0, \quad i = 1, \ldots, 8. \tag{14.20}$$

Subtract the first equation from the remaining to eliminate R:

$$((\mathbf{P}^{j+1} - \mathbf{B}) \times \mathbf{G})^2 - ((\mathbf{P}^1 - \mathbf{B}) \times \mathbf{G})^2 = 0, \quad j = 1, \ldots, 7. \tag{14.21}$$

Expand the terms in this equation to obtain the system of polynomials

$$\mathscr{P}_i : \quad (\mathbf{P}^{j+1} \times \mathbf{G})^2 - (\mathbf{P}^1 \times \mathbf{G})^2 - 2((\mathbf{P}^{j+1} - \mathbf{P}^1) \times \mathbf{G}) \cdot (\mathbf{B} \times \mathbf{G}) = 0,$$
$$j = 1, \ldots, 7,$$

$$\mathscr{C}_1 : \quad \mathbf{G} \cdot \mathbf{m} - e = 0,$$
$$\mathscr{C}_2 : \quad \mathbf{B} \cdot \mathbf{n} - f = 0. \tag{14.22}$$

This is a set of seven polynomials of degree four and two linear equations. The total degree is $4^7 = 16384$. See Nielsen and Roth [89] and Su et al. [129] for additional details about this problem.

We now consider the monomial structure of polynomial system (14.22). The polynomials $\mathscr{P}_i$ are linear combinations of monomials in the set generated by

$$(\langle x,y,z,1\rangle\langle a,b,c\rangle)^2 \cup \langle x,y,z,1\rangle\langle a,b,c\rangle\langle u,v,w\rangle\langle a,b,c\rangle. \tag{14.23}$$

The products commute, $\langle a\rangle\langle b\rangle = \langle b\rangle\langle a\rangle$, and they distribute over unions, $\langle a\rangle\langle b\rangle \cup \langle a\rangle\langle c\rangle = \langle a\rangle(\langle b\rangle \cup \langle c\rangle) = \langle a\rangle\langle b,c\rangle$—therefore (14.23) becomes

$$\langle a,b,c\rangle^2(\langle x,y,z,1\rangle^2 \cup \langle x,y,z,1\rangle\langle u,v,w\rangle), \tag{14.24}$$

which can be written as

$$\langle a,b,c\rangle^2\langle x,y,z,1\rangle\langle x,y,z,u,v,w,1\rangle. \tag{14.25}$$

This shows that the polynomial system (14.22) has the monomial structure

$$
\begin{aligned}
\mathscr{P}_j &\in \langle a,b,c\rangle^2\langle x,y,z,1\rangle\langle x,y,z,u,v,w,1\rangle|_j = 0, \quad j = 1,\ldots,7, \\
\mathscr{C}_1 &\in \langle u,v,w,1\rangle = 0, \\
\mathscr{C}_2 &\in \langle a,b,c,1\rangle = 0.
\end{aligned} \tag{14.26}
$$

In order to estimate the number of roots we see that to specify $\mathbf{G} = (a,b,c)$ we must combine $\mathscr{C}_2$ with two terms $\langle a,b,c\rangle$ from the seven polynomials $\mathscr{P}_i$. Because this term is squared, the number of choices is increased by a factor of $2^2 = 4$. For the remainder of the parameters, we can choose from zero to three of the terms $\langle x,y,z,1\rangle$ from the remaining five polynomials to define $\mathbf{p} = (x,y,z)$. The third term in what is left and $\mathscr{C}_1$ define the remaining parameters. This yields the LPD bound of

$$2^2\binom{7}{2}\sum_{i=0}^{3}\binom{5}{i} = 2\,184, \tag{14.27}$$

which is much reduced from the total degree of $16\,384$.

For polynomial systems with a large number of roots, elimination is not attractive, but we may find all solutions using polynomial continuation. We used the software PHC (Verschelde [140]) and POLSYS-GLP (Su et al. [130]) to compute the roots for random test cases and determined that the exact root count for this problem is 804. Clearly, there is more structure in this system of polynomials than what is shown in the linear product decomposition.

Thus, we find that for eight arbitrary spatial positions we can find as many as 804 points in the moving body each of which has all eight positions on a circular cylinder. For each of these points, we can determine an associated CS chain.

14.6 The Circular Hyperboloid

A circular hyperboloid is generated by rotating one line around another, so that every point on the moving line traces a circle around the fixed line, which is the axis of the hyperboloid, Figure 14.4. Of all of these circles there is one with the smallest radius, R, and its center $\mathbf{B} = (u, v, w)$ is the center of the hyperboloid. Let $\mathbf{G} = (a, b, c)$ be the direction of the axis, and denote its Plücker coordinates $G = (\mathbf{G}, \mathbf{B} \times \mathbf{G})$. A unit vector $\mathbf{N}$ perpendicular to $\mathbf{G}$ though $\mathbf{B}$ is the common normal between the axis G and one of the generated lines, H. The generator is located at the distance R along $\mathbf{N}$, and lies at an angle α around $\mathbf{N}$ relative to the axis G.

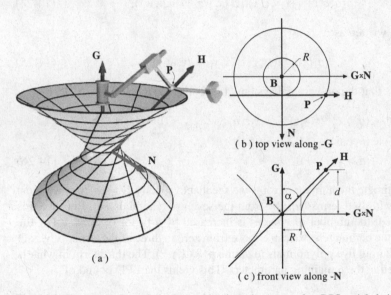

Fig. 14.4 The circular hyperboloid traced by the wrist center of an RPS serial chain.

If point $\mathbf{P}$ is a point on the generator H, then its d measured along the axis G from G is given by

$$d = \frac{(\mathbf{P} - \mathbf{B}) \cdot \mathbf{G}}{\sqrt{\mathbf{G} \cdot \mathbf{G}}}. \tag{14.28}$$

Notice that we are not assuming that $\mathbf{G}$ is a unit vector. The magnitude of $\mathbf{P} - \mathbf{B}$ is now computed to be

$$(\mathbf{P} - \mathbf{B})^2 = R^2 + d^2 + (d \tan \alpha)^2. \tag{14.29}$$

Substitute d into this equation to obtain the equation of a circular hyperboloid

$$(\mathbf{P} - \mathbf{B})^2 - ((\mathbf{P} - \mathbf{B}) \cdot \mathbf{G})^2 \left(\frac{1 + \tan^2 \alpha}{\mathbf{G} \cdot \mathbf{G}} \right) - R^2 = 0. \tag{14.30}$$

When $\alpha = 0$, this equation becomes the equation of a cylinder presented in the previous section.

Figure 14.4(a) shows the RPS chain associated with the circular hyperboloid. The R joint axis is G, and its P joint axis is in the direction α measured around the common normal. The point **P** is the center of the S joint, and lies at the distance R in the direction **N** of the common normal.

Expand equation (14.30) and collect terms to obtain

$$k_0 \mathbf{P} \cdot \mathbf{P} + 2\mathbf{K} \cdot \mathbf{P} - (\mathbf{P} \cdot \mathbf{G})^2 - \zeta = 0, \qquad (14.31)$$

where we have introduce the parameters k_0, $\mathbf{K} = (k_1, k_2, k_3)$, and ζ defined by

$$k_0 = \frac{\mathbf{G} \cdot \mathbf{G}}{1 + \tan^2 \alpha}, \quad \mathbf{K} = (\mathbf{B} \cdot \mathbf{G})\mathbf{G} - k_0 \mathbf{B}, \quad \zeta = (\mathbf{B} \cdot \mathbf{G})^2 - k_0 \mathbf{B} \cdot \mathbf{B} + k_0 R^2. \quad (14.32)$$

Given values for ζ, k_0, **K**, and **G**, we can compute **B** by solving the linear equations

$$\begin{Bmatrix} k_1 \\ k_2 \\ k_3 \end{Bmatrix} = \begin{bmatrix} a^2 - k_0 & ab & ac \\ ab & b^2 - k_0 & ac \\ ac & bc & c^2 - k_0 \end{bmatrix} \begin{Bmatrix} u \\ v \\ w \end{Bmatrix}. \qquad (14.33)$$

Then the length and twist parameters, R and α, are obtained from the formulas

$$\alpha = \arccos\left(\sqrt{\frac{k_0}{\mathbf{G} \cdot \mathbf{G}}}\right), \quad R = \sqrt{\frac{\zeta - (\mathbf{B} \cdot \mathbf{G})^2 + k_0 \mathbf{B} \cdot \mathbf{B}}{k_0}}. \qquad (14.34)$$

Thus, the 11 dimensional parameters ζ, k_0, **K**, **G**, and **P** define a circular hyperboloid.

As we have done previously, we set the length of **G** by choosing an arbitrary vector **m** and scalar e to define the constraint

$$\mathscr{C}: \quad \mathbf{G} \cdot \mathbf{m} - e = 0. \qquad (14.35)$$

This means that given 10 arbitrary displacements $[T_i] = [A_i, \mathbf{d}_i]$, we can map a point $\mathbf{p} = (x, y, z)$ to its displaced positions $\mathbf{P}^i = [T_i]\mathbf{p}$, $i = 1, \ldots, 10$. Evaluating the equation of the hyperboloid on these 10 points, we obtain

$$k_0 \mathbf{P}^i \cdot \mathbf{P}^i + 2\mathbf{K} \cdot \mathbf{P}^i - (\mathbf{P}^i \cdot \mathbf{G})^2 - \zeta = 0, \quad i = 1, \ldots, 10. \qquad (14.36)$$

Subtract the first of these equations from the remaining in order to eliminate ζ and define the system of polynomials

$$\mathscr{H}_j: \quad k_0(\mathbf{P}^{j+1^2} - \mathbf{P}^{1^2}) + 2\mathbf{K} \cdot (\mathbf{P}^{j+1} - \mathbf{P}^1) - (\mathbf{P}^{j+1} \cdot \mathbf{G})^2 + (\mathbf{P}^1 \cdot \mathbf{G})^2 = 0,$$

$$j = 1, \ldots, 9,$$

$$\mathscr{C}: \quad \mathbf{G} \cdot \mathbf{m} - e = 0. \qquad (14.37)$$

This is a system of nine fourth–degree polynomials $\mathcal{H}_j$ and one linear equation $\mathcal{C}$, which has a total degree of $4^9 = 262\,144$. See Nielsen and Roth [89] and Kim and Tsai [55] for other formulations of this problem.

A better bound on the number of solutions can be obtained by considering the monomial structure of these equations. Recall that the term $\mathbf{P}^{j+1\,2} - \mathbf{P}^{1\,2}$ is linear in x, y, and z, because the quadratic terms cancel; see (14.5). This means that the polynomials $\mathcal{H}_j$ have the monomial structure

$$\mathcal{H}_j \in \langle k_0 \rangle \langle x, y, z, 1 \rangle \cup \langle k_1, k_2, k_3 \rangle \langle x, y, z, 1 \rangle \cup (\langle x, y, z, 1 \rangle \langle a, b, c \rangle)^2. \tag{14.38}$$

This simplifies to yield the linear product decomposition for the system (14.37) as

$$\mathcal{H}_j \in \langle a, b, c \rangle^2 \langle x, y, z, 1 \rangle \langle x, y, z, k_0, k_1, k_2, k_3, 1 \rangle |_j, \quad j = 1, \ldots, 9,$$
$$\mathcal{C} \in \langle a, b, c, 1 \rangle. \tag{14.39}$$

This structure allows us to count the number of roots from the number of admissible sets of linear equations that yield solutions for the unknown parameters. In this case we obtain

$$\mathrm{LPD} = 2^2 \binom{9}{2} \sum_{j=0}^{3} \binom{7}{j} = 9216. \tag{14.40}$$

The result is that for ten arbitrary spatial positions we can find as many as $9\,216$ points that have all 10 positions on a circular hyperboloid. For each of these points we can find an associated RPS chain.

14.7 The Elliptic Cylinder

An elliptic cylinder is generated by a circle that has its center swept along a line $L(t) = \mathbf{B} + t\mathbf{S}_1$ such that the vector through the center normal to the plane of the circle maintains a constant direction $\mathbf{S}_2$ at an angle α relative to the direction $\mathbf{S}_1$ of $L(t)$; see Figure 14.5. The major axis of the elliptic cross-section is the radius R of the circle and the minor axis is $R\cos\alpha$. This surface is generated by the wrist center a PRS chain that has its P joint aligned with the axis $L(t)$ and its R joint positions so that its axis is along $\mathbf{S}_2$.

Consider a general point on the cylinder $\mathbf{P}$, and let $\mathbf{Q}$ be the center of the circle. The point $\mathbf{Q}$ moves along the axis $L(t)$, which has the Plücker coordinates $\mathbf{S}_1 = (\mathbf{S}_1, \mathbf{B} \times \mathbf{S}_1)$. The distance from the reference point $\mathbf{B}$ to $\mathbf{Q}$ is denoted by d. These definitions allow us to express the location of $\mathbf{P}$ relative to $\mathbf{B}$ as

$$\mathbf{P} - \mathbf{B} = d\mathbf{S}_1 + R\mathbf{u}, \tag{14.41}$$

where $\mathbf{u}$ is a unit vector in the direction $\mathbf{S}_1 \times \mathbf{S}_2$. Compute the cross product with $\mathbf{S}_1$ to eliminate d, and the cross product with $\mathbf{S}_2$ to obtain

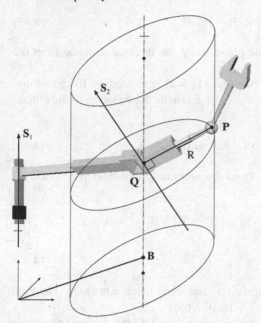

Fig. 14.5 The elliptic cylinder reachable by a PRS serial chain.

$$\mathbf{S}_2 \times ((\mathbf{P} - \mathbf{B}) \times \mathbf{S}_1) = R(\mathbf{S}_2 \cdot \mathbf{S}_1)\mathbf{u}. \tag{14.42}$$

The magnitude of this vector identity yields our equation of the elliptic cylinder

$$\left(\mathbf{S}_2 \times ((\mathbf{P} - \mathbf{B}) \times \mathbf{S}_1)\right)^2 = R^2(\mathbf{S}_1 \cdot \mathbf{S}_2)^2. \tag{14.43}$$

This equation has 13 dimensional parameters: the radius R, three each for the directions $\mathbf{S}_1$, $\mathbf{S}_2$, and the points $\mathbf{P}$ and $\mathbf{B}$. Notice that if $\mathbf{S}_1 = \mathbf{S}_2 = \mathbf{G}$ this simplifies to the equation of a circular cylinder.

There are actually only 10 independent parameters in (14.43), and we can determine three additional linear constraints. First, note that it is the directions of $\mathbf{S}_1$ and $\mathbf{S}_2$ that matter, not their magnitudes. We specify these magnitudes by introducing two arbitrary planes $V_k : (\mathbf{m}_k, e_k)$, $k = 1, 2$. In general, the lines through the origin parallel to S_i must intersect these planes, respectively. We select these points of intersection to be $\mathbf{S}_i$; that is, we require

$$\mathscr{C}_k : \quad \mathbf{m}_k \cdot \mathbf{S}_k - e_k = 0, \quad k = 1, 2. \tag{14.44}$$

Next, we note that any point along the line S_1 can serve as the reference point $\mathbf{B}$ for the axis of the cylinder. We determine $\mathbf{B}$ by specifying an arbitrary plane $U : (\mathbf{n}, f)$. In general, the line G must intersect this plane, and we select this point at $\mathbf{B}$. Thus, $\mathbf{B}$ satisfies the linear equation

$$\mathscr{C}_3: \quad \mathbf{n} \cdot \mathbf{B} - f = 0. \tag{14.45}$$

Notice that $\mathbf{n}$ is the unit normal to the plane and f the directed distance from the origin to the plane.

We now consider the images of a point $\mathbf{p} = (x, y, z)$ generated by 10 spatial displacements, that is $\mathbf{P}^i = [T(\hat{\mathbf{Q}}_i)]\mathbf{p}$, $i = 1, \ldots, 10$. Evaluate the equation of the elliptic cylinder on these 10 points to obtain

$$\left(\mathbf{S}_2 \times ((\mathbf{P}^i - \mathbf{B}) \times \mathbf{S}_1)\right)^2 - R^2 (\mathbf{S}_1 \cdot \mathbf{S}_2)^2 = 0, \quad i = 1, \ldots, 10. \tag{14.46}$$

Subtract the first of these equations from the remaining to obtain the system of polynomials

$$\begin{aligned}
\mathscr{E}_j: & \quad \left(\mathbf{S}_2 \times ((\mathbf{P}^{j+1} - \mathbf{B}) \times \mathbf{S}_1)\right)^2 - \left(\mathbf{S}_2 \times ((\mathbf{P}^1 - \mathbf{B}) \times \mathbf{S}_1)\right)^2 = 0, \quad j = 1, \ldots, 9, \\
\mathscr{C}_k: & \quad \mathbf{m}_k \cdot \mathbf{S}_k - e_k = 0, \quad k = 1, 2, \\
\mathscr{C}_3: & \quad \mathbf{n} \cdot \mathbf{B} - f = 0.
\end{aligned} \tag{14.47}$$

The result is nine polynomials of degree six, and three linear equations. The total degree of this polynomial system is $6^9 = 10{,}077{,}696$.

The total degree of this system can be reduced as follows. Expand the triple product

$$\begin{aligned}
\mathbf{S}_2 \times ((\mathbf{P} - \mathbf{B}) \times \mathbf{S}_1) &= (\mathbf{S}_1 \cdot \mathbf{S}_2)(\mathbf{P} - \mathbf{B}) - ((\mathbf{P} - \mathbf{B}) \cdot \mathbf{S}_2)\mathbf{S}_1 \\
&= (\mathbf{S}_1 \cdot \mathbf{S}_2)(\mathbf{P} - (\mathbf{P} \cdot \mathbf{K})\mathbf{S}_1 + \mathbf{Q}),
\end{aligned} \tag{14.48}$$

where

$$\mathbf{K} = \frac{\mathbf{S}_2}{\mathbf{S}_1 \cdot \mathbf{S}_2} \quad \text{and} \quad \mathbf{Q} = (\mathbf{B} \cdot \mathbf{K})\mathbf{S}_1 - \mathbf{B}. \tag{14.49}$$

Add to this the constraints

$$\mathbf{S}_1 \cdot \mathbf{S}_1 = 1, \quad \mathbf{K} \cdot \mathbf{S}_1 = 1, \quad \mathbf{Q} \cdot \mathbf{K} = 0. \tag{14.50}$$

This combines with the other constraints to reduce the degree of these polynomials to four, so we have

$$\begin{aligned}
& (\mathbf{P} - (\mathbf{P} \cdot \mathbf{K})\mathbf{S}_1 + \mathbf{Q})^2 \\
&= \mathbf{P}^2 + (\mathbf{P} \cdot \mathbf{K})^2 + \mathbf{Q}^2 - 2(\mathbf{P} \cdot \mathbf{S}_1)(\mathbf{P} \cdot \mathbf{K}) + 2\mathbf{P} \cdot \mathbf{Q} - 2(\mathbf{P} \cdot \mathbf{K})(\mathbf{Q} \cdot \mathbf{S}_1).
\end{aligned} \tag{14.51}$$

The result is a new version of the polynomial system

$$\mathcal{E}_j' : (\mathbf{P}^{j+1} - (\mathbf{P}^{j+1} \cdot \mathbf{K})\mathbf{S}_1 + \mathbf{Q})^2 - (\mathbf{P}^1 - (\mathbf{P}^1 \cdot \mathbf{K})\mathbf{S}_1 + \mathbf{Q})^2 = 0,$$

$$j = 1, \dots, 9,$$

$$\mathcal{C}_1' : \mathbf{S}_1 \cdot \mathbf{S}_1 - 1 = 0,$$
$$\mathcal{C}_2' : \mathbf{K} \cdot \mathbf{S}_1 - 1 = 0,$$
$$\mathcal{C}_3' : \mathbf{Q} \cdot \mathbf{K} = 0, \tag{14.52}$$

which has total degree $(2^3)(4^9) = 2\,097\,152$.

As we have done previously, we examine the monomial structure of the equations $\mathcal{E}_j'$. Let $\mathbf{S}_1 = (a, b, c)$, $\mathbf{K} = (k_1, k_2, k_3)$, and $\mathbf{Q} = (q_1, q_2, q_3)$, and recall that the quadratic terms in $\mathbf{P}^{j+1^2} - \mathbf{P}^{1^2}$ cancel, as does the term $\mathbf{Q}^2$. Thus, the polynomials $\mathcal{E}_j'$ have the monomial structure

$$\langle x, y, z, 1 \rangle \cup \langle x, y, z, 1 \rangle^2 \langle k_1, k_2, k_3 \rangle^2 \cup \langle x, y, z, 1 \rangle^2 \langle k_1, k_2, k_3 \rangle \langle a, b, c \rangle$$
$$\cup \langle x, y, z, 1 \rangle \langle q_1, q_2, q_3 \rangle \cup \langle x, y, z, 1 \rangle \langle k_1, k_2, k_3 \rangle \langle a, b, c \rangle \langle q_1, q_2, q_3 \rangle. \tag{14.53}$$

This leads to the linear product decomposition for this polynomial system

$$\mathcal{E}_j' \in \langle x, y, z, 1 \rangle \langle x, y, z, q_1, q_2, q_3, 1 \rangle \langle k_1, k_2, k_3, 1 \rangle \langle k_1, k_2, k_3, a, b, c, 1 \rangle |_j = 0,$$

$$j = 1, \dots, 9,$$

$$\mathcal{C}_1' \in \langle a, b, c, 1 \rangle^2 = 0,$$
$$\mathcal{C}_2' \in \langle k_1, k_2, k_3, 1 \rangle \langle a, b, c, 1 \rangle = 0,$$
$$\mathcal{C}_3' \in \langle k_1, k_2, k_3, 1 \rangle \langle q_1, q_2, q_3, 1 \rangle = 0. \tag{14.54}$$

The LPD bound for this system is $247\,968$, and it was solved using POLSYS_GLP on 128 nodes of the Blue Horizon supercomputer at the San Diego Supercomputer Center. The result was $18\,120$ real and complex solutions, each which defines the RPS chains.

14.8 The Circular Torus

A circular torus is generated by sweeping a circle around an axis so that its center traces a second circle. Let the axis be $L(t) = \mathbf{B} + t\mathbf{G}$, with Plücker coordinates $G = (\mathbf{G}, \mathbf{B} \times \mathbf{G})$. See Figure 14.6. Introduce a unit vector $\mathbf{v}$ perpendicular to this axis so that the center of the generating circle is given by $\mathbf{Q} - \mathbf{B} = \rho\mathbf{v}$. Now define $\mathbf{u}$ to be the unit vector in the direction $\mathbf{G}$—then a point $\mathbf{P}$ on the torus is defined by the vector equation

$$\mathbf{P} - \mathbf{B} = \rho\mathbf{v} + R(\cos\phi\mathbf{v} + \sin\phi\mathbf{u}), \tag{14.55}$$

where ϕ is the angle measured from $\mathbf{v}$ to the radius vector of the generating circle.

An algebraic equation of the torus is obtained from (14.55) by first computing the magnitude

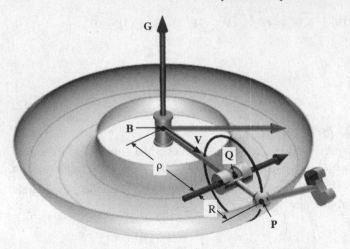

Fig. 14.6 The circular torus traced by the wrist center of a right RRS serial chain.

$$(\mathbf{P}-\mathbf{B})^2 = \rho^2 + R^2 + 2\rho R \cos\phi. \tag{14.56}$$

Next compute the dot product with $\mathbf{u}$, to obtain

$$(\mathbf{P}-\mathbf{B})\cdot\mathbf{u} = R\sin\phi. \tag{14.57}$$

Finally, eliminate $\cos\phi$ and $\sin\phi$ from these equations, and the result is

$$\mathbf{G}^2((\mathbf{P}-\mathbf{B})^2 - \rho^2 - R^2)^2 + 4\rho^2((\mathbf{P}-\mathbf{B})\cdot\mathbf{G})^2 = 4\rho^2\mathbf{G}^2 R^2. \tag{14.58}$$

This is the equation of a circular torus. It has 11 parameters, the scalars ρ and R, and the three vectors $\mathbf{G}$, $\mathbf{P}$, and $\mathbf{B}$.

In contrast to what we have done previously, here we set the magnitude of $\mathbf{G}$ to a constant, in order to simplify the polynomial (14.58),

$$\mathscr{G} : \mathbf{G}\cdot\mathbf{G} = 1. \tag{14.59}$$

Unfortunately, this doubles the number of solutions, since $-\mathbf{G}$ and $\mathbf{G}$ define the same torus—however, it reduces this polynomial from degree six to degree four.

Let $[T_i] = [A_i, \mathbf{d}_i]$ be a specified set of displacements, so we have the 10 positions $\mathbf{P}^i = [T_i]\mathbf{p}$ of a point $\mathbf{p} = (x, y, z)$ that is fixed in the moving frame M. Evaluating (14.58) on these points, we obtain the polynomial system

$$\begin{aligned}
\mathscr{T}_i: \quad & ((\mathbf{P}^i - \mathbf{B})^2 - \rho^2 - R^2)^2 + 4\rho^2((\mathbf{P}^i - \mathbf{B})\cdot\mathbf{G})^2 - 4\rho^2 R^2 = 0, \quad i = 1,\dots,10, \\
\mathscr{G}: \quad & \mathbf{G}\cdot\mathbf{G} - 1 = 0.
\end{aligned} \tag{14.60}$$

The total degree of this system is $2(4^{10}) = 2\,097\,152$.

In order to simplify the polynomials $\mathscr{T}_i$ we introduce the parameters

$$\mathbf{H} = 2\rho\mathbf{G} \quad \text{and} \quad k_1 = \mathbf{B}^2 - \rho^2 - R^2, \tag{14.61}$$

which yields the identity

$$4\rho^2 R^2 = \mathbf{H}^2 \left(\mathbf{B}^2 - \frac{\mathbf{H}^2}{4} - k_1 \right). \tag{14.62}$$

Substitute these relations into $\mathscr{T}_i$ to obtain

$$\mathscr{T}_i': \quad ((\mathbf{P}^i)^2 - 2\mathbf{P}^i \cdot \mathbf{B} + k_1)^2 + ((\mathbf{P}^i - \mathbf{B}) \cdot \mathbf{H})^2 - \mathbf{H}^2 \left(\mathbf{B}^2 - \frac{\mathbf{H}^2}{4} - k_1 \right) = 0.$$

$$i = 1, \ldots, 10, \quad (14.63)$$

It is difficult to find a simplified formulation for these equations, even if we subtract the first equation from the remaining in order to cancel terms.

Expanding the polynomial $\mathscr{T}_i'$ and examining each of the terms, we can identify the linear product decomposition

$$\mathscr{T}_i' \in \langle x, y, z, h_1, h_2, h_3, 1 \rangle^2 \langle x, y, z, h_1, h_2, h_3, u, v, w, k_1, 1 \rangle^2. \tag{14.64}$$

This allows us to compute the LPD bound on the number of roots as

$$\text{LPD} = 2^{10} \sum_{j=0}^{6} \binom{10}{j} = 868\,352. \tag{14.65}$$

Our POLSYS_GLP algorithm obtained 94,622 real and complex solutions for a random set of specified displacements. However, this problem needs further study to provide an efficient way to evaluate and sort the large number of right RRS chains.

14.9 The General Torus

A general torus is defined by sweeping a circle that has a general orientation in space about an arbitrary axis. See Figure 14.7. Let $S_1 = (\mathbf{S}_1, \mathbf{B} \times \mathbf{S}_1)$ be the Plücker coordinates of the line that forms the axis of the torus, and let $S_2 = (\mathbf{S}_2, \mathbf{Q} \times \mathbf{S}_2)$ define the line through the center of the circle that is perpendicular to the plane of the circle. These two lines define a common normal N, and we choose its intersection with S_1 and S_2 to be the reference points $\mathbf{B}$ and $\mathbf{Q}$, respectively. The normal angle and distance between these lines around and along their common normal are denoted by α and ρ. Finally, we identify the center of the circle as lying a distance d along S_2 measured from $\mathbf{Q}$.

In this derivation, we constrain $\mathbf{S}_1$ and $\mathbf{S}_2$ to be unit vectors, in order to reduce the degree of the resulting equation. This allows us to define the unit vector in the common normal direction as $\mathbf{n} = (\mathbf{S}_1 \times \mathbf{S}_2)/\sin\alpha$, so we obtain a general point $\mathbf{P}$ on the torus from the vector equation

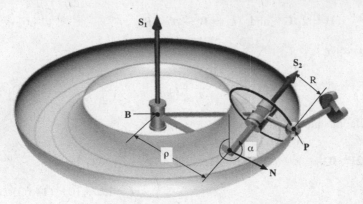

Fig. 14.7 The general torus reachable by the wrist center of an RRS serial chain.

$$\mathbf{P} - \mathbf{B} = \rho \mathbf{n} + d\mathbf{S}_2 + R(\cos\phi\,\mathbf{n} + \sin\phi(\mathbf{S}_2 \times \mathbf{n})). \tag{14.66}$$

The algebraic equation for the torus is obtained by first computing

$$(\mathbf{P} - \mathbf{B})^2 = \rho^2 + d^2 + R^2 + 2\rho R\cos\phi \tag{14.67}$$

and

$$(\mathbf{P} - \mathbf{B}) \cdot (\mathbf{v} \times \mathbf{n}) = R\sin\phi. \tag{14.68}$$

Notice that $\mathbf{S}_2 \times \mathbf{n}$ is

$$\mathbf{S}_2 \times \frac{\mathbf{S}_1 \times \mathbf{S}_2}{\sin\alpha} = \frac{1}{\sin\alpha}(\mathbf{S}_1 - \cos\alpha\,\mathbf{S}_2). \tag{14.69}$$

Now eliminate ϕ between these two equations to obtain

$$((\mathbf{P} - \mathbf{B})^2 - \rho^2 - d^2 - R^2)^2 + \frac{4\rho^2}{\sin^2\alpha}((\mathbf{P} - \mathbf{B}) \cdot \mathbf{S}_1 - d\cos\alpha)^2 - 4\rho^2 R^2 = 0. \tag{14.70}$$

This equation has the four scalar parameters ρ, α, d, and R, and the three vector parameters $\mathbf{P}$, $\mathbf{B}$, and $\mathbf{S}_1$. These 13 parameters combine with the constraint that $|\mathbf{S}_1| = 1$ to yield 12 independent parameters.

In order to simplify the use of equation (14.70), we introduce the parameters

$$k_1 = \mathbf{B} \cdot \mathbf{B} - \rho^2 - R^2 - d^2,$$

$$k_2 = (\mathbf{B} \cdot \mathbf{S}_1 + d\cos\alpha)\frac{2\rho}{\sin\alpha},$$

$$k_3 = 4\rho^2 R^2,$$

$$\mathbf{H} = \frac{2\rho}{\sin\alpha}\mathbf{S}_1. \tag{14.71}$$

These parameters allow us to write (14.70) in the form

$$(\mathbf{P}^2 - 2\mathbf{P}\cdot\mathbf{B} + k_1)^2 + (\mathbf{P}\cdot\mathbf{H} - k_2)^2 - k_3 = 0. \tag{14.72}$$

This is a quartic polynomial in the three scalars k_1, $i = 1,2,3$ and three vectors $\mathbf{P}$, $\mathbf{B}$, and $\mathbf{H}$.

Given a set of displacements $\hat{\mathbf{Q}}_i$, $i = 1,\dots,12$, we evaluate (14.72) on the points $\mathbf{P}^i = [T(\hat{\mathbf{Q}}_i)]\mathbf{p}, i = 1,\dots,12$. Subtract the first of these equations from the remaining to cancel k_3 and obtain

$$\mathscr{GT}_j:\ (\mathbf{P}^{j+1^2} - 2\mathbf{P}^{j+1}\cdot\mathbf{B} + k_1)^2 - (\mathbf{P}^{1^2} - 2\mathbf{P}^1\cdot\mathbf{B} + k_1)^2$$
$$+ (\mathbf{P}^{j+1}\cdot\mathbf{H} - k_2)^2 + (\mathbf{P}^1\cdot\mathbf{H} - k_2)^2 = 0,\quad j = 1,\dots,11. \tag{14.73}$$

The total degree of this system of polynomials is $4^{11} = 4\,194\,304$.

We can refine the estimate of the number of roots of this polynomial system by using the linear product decomposition. Expanding the polynomial $\mathscr{GT}_j$, we obtain the terms

$$\mathbf{P}^{j+1^4} - \mathbf{P}^{1^4} \in \langle x,y,z,1\rangle^3,$$
$$(2\mathbf{P}^{j+1}\cdot\mathbf{B})^2 - (2\mathbf{P}^1\cdot\mathbf{B})^2 \in \langle x,y,z,1\rangle^2\langle u,v,w\rangle^2,$$
$$-4\mathbf{P}^{j+1^2}(\mathbf{P}^{j+1}\cdot\mathbf{B}) + 4\mathbf{P}^{1^2}(\mathbf{P}^1\cdot\mathbf{B}) \in \langle x,y,z,1\rangle^3\langle u,v,w\rangle,$$
$$2k_1(\mathbf{P}^{j+1^2} - \mathbf{P}^{1^2} - 2\mathbf{P}^{j+1}\cdot\mathbf{B} + 2\mathbf{P}^1\cdot\mathbf{B}) \in \langle x,y,z,1\rangle\langle u,v,w,1\rangle\langle k_1\rangle,$$
$$(\mathbf{P}^{j+1}\cdot\mathbf{H}))^2 - (\mathbf{P}^1\cdot\mathbf{H})^2 \in \langle x,y,z,1\rangle^2\langle h_1,h_2,h_3\rangle^2$$
$$-2k_2(\mathbf{P}^{j+1}\cdot\mathbf{H} - \mathbf{P}^1\cdot\mathbf{H}) \in \langle x,y,z,1\rangle\langle h_1,h_2,h_3\rangle\langle k_2\rangle. \tag{14.74}$$

Notice that the quartic terms in the first expression cancel. We combine these monomials into the linear product decomposition:

$$\mathscr{GT}_j:\ \langle x,y,z,1\rangle^2\langle u,v,w,h_1,h_2,h_3,1\rangle\langle x,y,z,u,v,w,h_1,h_2,h_3,k_1,k_2,1\rangle|_j,$$
$$j = 1,\dots,11. \tag{14.75}$$

This allows us to compute the LPD bound of $448\,702$.

The parallel POLSYS_GLP homotopy solver computed $42\,615$ solutions, and each real solution can be used to design an RRS chain to reach the specified displacements. The distribution and utility of these solutions requires further study.

14.10 Polynomial Continuation Algorithms

Our concern is finding all of the solutions of the n polynomial equations in n unknowns that form the synthesis equations for a spatial serial chain. As we have seen, the synthesis equations for the PPS and TS chains can be solved by direct elimination of the unknown parameters to obtain a univariate polynomial. However, the

Table 14.2 Total degree, LPD bound, and number of solutions to the design equations

Case	Surface	Total Degree	LPD Bound	Number of Roots
1	plane	32	10	10
2	sphere	64	20	20
3	circ. cylinder	16,384	2,184	804
4	circ. hyperboloid	262,144	9,216	1,024
5	elliptic cylinder	2,097,152	247,968	18,120
6	right torus	2,097,152	868,352	94,622
7	general torus	4,194,304	448,702	42,615

remaining chains yield systems of polynomials that are too complicated to solve by direct parameter elimination; therefore we use polynomial continuation to solve these equations.

Polynomial continuation solves a polynomial system $P(\mathbf{z}) = 0$ by starting with a polynomial system $Q(\mathbf{z}) = 0$ that has the same structure as $P(\mathbf{z})$ and a known set of solutions. A good example of a start system $Q(\mathbf{z})$ is obtained from the linear project decomposition of $P(\mathbf{z})$. The system $Q(\mathbf{z})$ is continuously transformed into $P(\mathbf{z})$ such that the solutions of $Q(\mathbf{z}) = 0$ move to become the solutions of $P(\mathbf{z}) = 0$. This can be viewed as the numerical solution of a set of ordinary differential equations where the solutions of $Q(\mathbf{z})$ are the initial conditions.

To see how this is done, consider the array of polynomials $P(\mathbf{z})$ that form any of our synthesis equations,

$$P(\mathbf{z}) = \begin{Bmatrix} S_1(\mathbf{z}) \\ S_2(\mathbf{z}) \\ \vdots \\ S_n(\mathbf{z}) \end{Bmatrix} = 0, \tag{14.76}$$

where $\mathbf{z} = (z_1, z_2, \ldots, z_n)$ is the design parameter vector. Now construct the convex combination homotopy map

$$H(\lambda, \mathbf{z}) = (1 - \lambda)Q(\mathbf{z}) + \lambda P(\mathbf{z}), \tag{14.77}$$

where $\lambda \in [0, 1)$ is the real-valued homotopy parameter. The coefficients of our polynomial system $P(\mathbf{z}) = 0$ are real; however, its solutions $\mathbf{z}$ need not be. Therefore, the homotopy $H(\lambda, \mathbf{z})$ must be viewed as an array of n complex functions in n complex variables $\mathbf{z}$ together with a single real variable λ.

For each root of the start system $Q(\mathbf{z}) = 0$, denoted by $\mathbf{z} = \mathbf{a}_j$, $j = 1, \ldots, N$, the homotopy equation $H(\lambda, \mathbf{z}) = 0$ has an associated zero curve γ_a, which is the connected component of $H^{-1}(0)$ containing the start point $(0, \mathbf{a}_j)$. The zero curve leads either to a point $(1, \mathbf{z}_a)$ where $P(\mathbf{z}_a) = 0$, or diverges to a root "at infinity."

Each zero curve can be parameterized by its arc length s, so γ_a has the form $(\lambda(s), \mathbf{z}(s))$. Tracking this curve involves numerical computation of points $\mathbf{y}_i \approx (\lambda(s_i), \mathbf{z}(s_i))$, where $\{s_i\}$ is an increasing sequence of arc lengths. Along the zero curve γ_a, we have the identity $H(\gamma(s), \mathbf{z}(s)) = 0$; therefore we can compute

$$\frac{d}{ds}H(\lambda,\mathbf{z}) = \begin{bmatrix} H_\lambda & H_\mathbf{z} \end{bmatrix} \begin{Bmatrix} d\lambda/ds \\ d\mathbf{z}/ds \end{Bmatrix} = 0, \qquad (14.78)$$

where $[J_H] = [H_\lambda, H_\mathbf{z}]$ is the $n \times (n+1)$ matrix of partial derivatives of the homotopy $H(\lambda,\mathbf{z})$. Notice that the vector $\mathbf{v} = (d\lambda/ds, d\mathbf{z}/ds)^T$ is tangent to the zero curve γ_a and is in the null space of the Jacobian matrix $[J_H]$.

The unit vector $\mathbf{v}_i$ in the direction of increasing arc length at a point $\mathbf{y}_i$ on γ_a is used to predict a value for the next point $\mathbf{y}_{i+1}^0$, that is

$$\mathbf{y}_{i+1}^0 = \mathbf{y}_i + (s_{i+1} - s_i)\mathbf{v}_i, \qquad (14.79)$$

where $s_{i+1} - s_i$ is the chosen arc-length step. The predicted value of $\mathbf{y}_{i+1}^0$ is corrected using the Taylor series expansion of the homotopy given by

$$H(\mathbf{y}_{i+1}^0) + [J_H(\mathbf{y}_{i+1}^0)](\mathbf{y}_{i+1}^1 - \mathbf{y}_{i+1}^0) \approx 0, \qquad (14.80)$$

which yields the correction formula

$$\mathbf{y}_{i+1}^1 = \mathbf{y}_{i+1}^0 - [J_H(\mathbf{y}_{i+1}^0)]^\dagger H(\mathbf{y}_{i+1}^0). \qquad (14.81)$$

The dagger denotes the Moore-Penrose pseudoinverse of the $n \times (n+1)$ Jacobian matrix. Geometrically, this iteration of the correction formula moves $\mathbf{y}_{i+1}^k$ toward the zero curve γ_a along a normal direction.

The predictor can be improved by interpolation at previously computed points along the zero curve, and a projective transformation can be used to bound the arc length of all of the paths so that none diverge to infinity. Finally, an "end-game" strategy can improve the calculation of $\mathbf{y}$ near $\gamma = 1$. See Wise et al.[151] and Sommese and Wampler [115] for details.

14.11 Summary

In this chapter, we have formulated the synthesis equations for a set of seven spatial serial chains that can position a spherical wrist center on an algebraic surface. We show how to generate a linear product decomposition and obtain the linear product bound for the number of solutions to these equations. Finally, we present the basic formulation of polynomial continuation algorithms that can be used to solve these equations. Table 14.2 lists the number of solutions for the polynomial systems for each of the seven spatial serial chains.

14.12 References

This chapter is based of the work of Su et al. [128], which expands the work by Chen and Roth [10]. It was Schoenflies [113] who introduced the problem of seeking points and lines in a moving body that take positions on surfaces associated with the spatial serial chains and Burmester [7] who used it for linkage design—also see Suh and Radcliffe [134]. Nielsen and Roth [89] and Husty [51] apply polynomial elimination methods to robotic systems. Our discussion of numerical polynomial homotopy is based on Wise et al. [151] and Tsai and Morgan [137]. Bernshtein [3] and Morgan et al. [84] analyze the bounds on the number of solutions of systems of polynomials, also Verschelde and Haegemans [139]. The POLSYS_GLP algorithm is described in Su et al. [130]. See Sommese and Wampler [115] for the theory and applications of the Bertini polynomial continuation software to a range of robotics problems.

Chapter 15
Clifford Algebra Synthesis of Serial Chains

In this chapter we formulate design equations for a spatial serial chain using the matrix exponential form of its kinematics equations. These equations define the position and orientation of the end effector in terms of rotations about the joint axes of the chain. Because the coordinates of these axes appear explicitly, we can specify a set of task positions, and solve these equations to determine the location of the joints. At the same time we are free to specify joint parameters or certain dimensions to ensure that the resulting robotic system has certain features. The structure of these design equations can be simplified by using the even Clifford algebra $C^+(P^3)$, known as dual quaternions.

15.1 The Product-of-Exponentials Form of the Kinematics Equations

The synthesis equations for a spatial serial chain are obtained from the matrix exponential form of its kinematics equations. This form of the kinematics equations replaces the Denavit-Hartenberg parameters with the coordinates of the n joint axes $S_i, i = 1, \ldots, n$. It is the coordinates of these axes that are the unknowns of the design problem.

Consider a displacement defined such that the moving body rotates the angle ϕ and slides the distance k around and along the screw axis $\mathsf{S} = (\mathbf{S}, \mathbf{C} \times \mathbf{S})$. Let $\mu = k/\phi$, then we can introduce the screw $\mathsf{J} = (\mathbf{S}, \mathbf{V}) = (\mathbf{S}, \mathbf{C} \times \mathbf{S} + \mu \mathbf{S})$, where μ is called the *pitch* of the screw. The components of J define the 4×4 twist matrix

$$J = \begin{bmatrix} 0 & -s_z & s_y & v_x \\ s_z & 0 & -s_x & v_y \\ -s_y & s_x & 0 & v_z \\ 0 & 0 & 0 & 0 \end{bmatrix}, \tag{15.1}$$

J.M. McCarthy and G.S. Soh, *Geometric Design of Linkages*, Interdisciplinary Applied Mathematics 11, DOI 10.1007/978-1-4419-7892-9_15,
© Springer Science+Business Media, LLC 2011

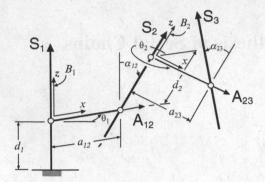

Fig. 15.1 Local frames for a serial chain.

and we find that the 4×4 homogeneous transform representing a rotation ϕ and a translation k about and along an axis S, $[T(\phi, k, S)]$, is defined as the matrix exponential

$$[T(\phi, k, S)] = e^{\phi J}. \tag{15.2}$$

The matrix exponential takes a simple form for the matrices $[Z(\theta_i, d_i)]$ and $[X(\alpha_{i,i+1}, a_{i,i+1})]$. The screws defined for these two transformations are $K = (\vec{k}, v\vec{k})$ and $I = (\vec{\imath}, \lambda\vec{\imath})$, where $v = d_i/\theta_i$ and $\lambda = a_{i,i+1}/\alpha_{i,i+1}$ are their respective pitches. Thus, we have

$$[Z(\theta_i, d_i)] = e^{\theta_i K} \quad \text{and} \quad [X(\alpha_{i,i+1}, a_{i,i+1})] = e^{\alpha_{i,i+1} I}, \tag{15.3}$$

and the kinematics equations become

$$[D] = [G] e^{\theta_1 K} e^{\alpha_{12} I} e^{\theta_2 K} \cdots e^{\alpha_{n-1,n} I} e^{\theta_n K} [H]. \tag{15.4}$$

This is one way to write the product of exponentials form of the kinematics equations. In the next section, we modify this slightly for use as our design equations.

15.1.1 Relative Displacements

If we choose a reference position for the end-effector, denoted by $[D_0]$, then the associated joint angle vector $\vec{\theta}_0$ can be determined, as well as the world frame coordinates of each of the joint axes. The transformation $[D_0]$ is often selected to be the configuration in which the joint parameters are zero and is called the *zero reference position* by Gupta [41].

The displacement of the serial chain relative to this reference configuration is defined by $[D(\Delta\vec{\theta})] = [D(\vec{\theta})][D(\vec{\theta}_0)]^{-1}$ and yields a convenient formulation for the kinematics equations. Assume that $[D_0]$ is a general position of the end-effector defined by joint parameters $\vec{\theta}_0$, so $\Delta\vec{\theta} = \vec{\theta} - \vec{\theta}_0$. Then, using the usual kinematics equations, we have

$$[D(\Delta\vec{\theta})] = ([G][Z(\theta_1,d_1)]\cdots[Z(\theta_n,d_n)][H])$$
$$([G][Z(\theta_{10},d_{10})]\cdots[Z(\theta_{n0},d_{n0})][H]))^{-1}. \quad (15.5)$$

In order to expand this equation, we introduce the partial displacements

$$[A_{i0}] = [G][Z(\theta_{10},d_{10})][X(\alpha_{12},a_{12})]\cdots[X(\alpha_{i-1,i},a_{i-1,i})], \quad (15.6)$$

where, for example,

$$[A_{10}] = [G], \quad \text{and} \quad [A_{20}] = [G][Z(\theta_{10},d_{10})][X(\alpha_{12},a_{12})].$$

Now insert the identity $[Z(\theta_{i,0})]^{-1}[A_{i0}]^{-1}[A_{i0}][Z(\theta_{i0})] = [I]$ after the first $n-1$ joint transforms $[Z(\theta_i,d_i)]$ in equation (15.5), in order to obtain the sequence of terms

$$[T(\Delta\theta_i,S_i)] = [A_{i0}][Z(\theta_i,d_i)][Z(\theta_{i,0})]^{-1}[A_{i0}]^{-1} = [A_{i0}][Z(\Delta\theta_i,\Delta d_i)][A_{i0}]^{-1}. \quad (15.7)$$

The result is the relative transformation that takes the form

$$[D(\Delta\vec{\theta})] = [T(\Delta\theta_1,S_1)][T(\Delta\theta_2,S_2)]\cdots[T(\Delta\theta_n,S_n)], \quad (15.8)$$

where S_i are the Plücker coordinates of each joint axis obtained by transforming the joint screw K to the world frame by the coordinate transformations defined in (15.7).

Using the exponential form the transformations $[T(\Delta\theta_i,S_i)]$, we write the relative kinematics equations (15.8) as

$$[D(\Delta\vec{\theta})] = e^{\Delta\theta_1 S_1} e^{\Delta\theta_2 S_2} \ldots e^{\Delta\theta_n S_n}, \quad (15.9)$$

where the matrices S_i are defined as

$$S_i = A_{i0} K A_{i0}^{-1}. \quad (15.10)$$

The product-of-exponentials form of the kinematics equations (15.4) is now obtained as

$$[D] = [D(\Delta\vec{\theta})][D_0] = e^{\Delta\theta_1 S_1} e^{\Delta\theta_2 S_2} \ldots e^{\Delta\theta_n S_n}[D_0]. \quad (15.11)$$

The difference between this equation and (15.4) is that here the coordinates of the joint axes of the serial chain are defined in the world frame.

15.2 The Even Clifford Algebra $C^+(P^3)$

The Clifford algebra of the projective three-space P^3 is a sixteen-dimensional vector space with a product operation that is defined in terms of a scalar product, McCarthy

[76]. The elements of even rank form an eight-dimensional subalgebra $C^+(P^3)$ that can be identified with the set of 4×4 homogeneous transforms.

The typical element of $C^+(P^3)$ can be written as the eight-dimensional vector given by

$$\hat{A} = a_0 + a_1 i + a_2 j + a_3 k + a_4 \varepsilon + a_5 i \varepsilon + a_6 j \varepsilon + a_7 k \varepsilon, \qquad (15.12)$$

where the basis elements i, j, and k are the well-known quaternion units, and ε is called the dual unit. The quaternion units satisfy the multiplication relations

$$i^2 = j^2 = k^2 = -1, \quad ij = k, \, jk = i, \, ki = j, \quad \text{and} \quad ijk = -1. \qquad (15.13)$$

The dual number ε commutes with i, j, and k, and multiplies by the rule $\varepsilon^2 = 0$.

In our calculations, it is convenient to consider the linear combination of quaternion units to be a vector in three dimenions, so we use the notation $\mathbf{A} = a_1 i + a_2 j + a_3 k$ and $\mathbf{A}^\circ = a_5 i + a_6 j + a_7 k$; the small circle superscript is often used to distinguish coefficients of the dual unit. This allows us to write the Clifford algebra element (15.12) as

$$\hat{A} = a_0 + \mathbf{A} + a_4 \varepsilon + \mathbf{A}^\circ \varepsilon. \qquad (15.14)$$

Now collect the scalar and vector terms so that this element takes the form

$$\hat{A} = (a_0 + a_4 \varepsilon) + (\mathbf{A} + \mathbf{A}^\circ \varepsilon) = \hat{a} + \mathsf{A}. \qquad (15.15)$$

The dual vector $\mathsf{A} = \mathbf{A} + \mathbf{A}^\circ \varepsilon$ can be identified with the pairs of vectors that define lines and screws.

Using this notation. the Clifford algebra product of elements $\hat{A} = \hat{a} + \mathsf{A}$ and $\hat{B} = \hat{b} + \mathsf{B}$ takes the form

$$\hat{C} = (\hat{b} + \mathsf{B})(\hat{a} + \mathsf{A}) = (\hat{b}\hat{a} - \mathsf{B} \cdot \mathsf{A}) + (\hat{a}\mathsf{B} + \hat{b}\mathsf{A} + \mathsf{B} \times \mathsf{A}), \qquad (15.16)$$

where the usual vector dot and cross products are extended linearly to dual vectors.

15.2.1 Exponential of a Vector

The product operation in the Clifford algebra allows us to compute the exponential of a vector $\theta \mathsf{S}$, where $|\mathsf{S}| = 1$, as

$$e^{\theta \mathsf{S}} = 1 + \theta \mathsf{S} + \frac{\theta^2}{2} \mathsf{S}^2 + \frac{\theta^3}{3!} \mathsf{S}^3 + \cdots. \qquad (15.17)$$

Using (15.16) we can write $\mathsf{S} = 0 + \mathsf{S}$ and compute

$$\mathsf{S}^2 = (0 + \mathsf{S})(0 + \mathsf{S}) = -1, \quad \mathsf{S}^3 = -\mathsf{S}, \quad \mathsf{S}^4 = 1, \quad \text{and} \quad \mathsf{S}^5 = \mathsf{S}, \qquad (15.18)$$

which means that we have

$$e^{\theta S} = \left(1 - \frac{\theta^2}{2} + \frac{\theta^4}{4!} + \cdots\right) + \left(\theta - \frac{\theta^3}{3!} + \frac{\theta^5}{5!} + \cdots\right) S = \cos\theta + \sin\theta S. \quad (15.19)$$

This is the well-known *unit quaternion* that represents a rotation around the axis S by the angle $\phi = 2\theta$. The rotation angle ϕ is double that given in the quaternion, because the Clifford algebra form of a rotation requires multiplication by both $Q = \cos\theta + \sin\theta S$ and its conjugate $Q^* = \cos\theta - \sin\theta S$. In particular, if x and X are the coordinates of a point before and after the rotation, then we have the quaternion coordinate transformation equation

$$X = QxQ^*. \quad (15.20)$$

For this reason the quaternion is often written in terms of one-half the rotation angle, that is, $Q = \cos\frac{\phi}{2} + \sin\frac{\phi}{2}S$

15.2.2 Exponential of a Screw

The Plücker coordinates $S = (S, C \times S)$ of a line can be identified with the Clifford algebra element $S = S + \varepsilon C \times S$. Similarly, the screw $J = (S, V) = (S, C \times S + \mu S)$ becomes the element $J = S + \varepsilon V = (1 + \mu\varepsilon)S$. Using the Clifford product we can compute the exponential of the screw θJ,

$$e^{\theta J} = 1 + J + \frac{\theta^2}{2}J^2 + \frac{\theta^3}{3!}J^3 + \cdots. \quad (15.21)$$

Notice that $S^2 = -1$; therefore

$$J^2 = -(1+\mu\varepsilon)^2 = -(1+2\mu\varepsilon), \quad J^3 = -(1+3\mu\varepsilon)S,$$
$$J^4 = 1 + 4\mu\varepsilon, \quad \text{and} \quad J^5 = (1+5\mu\varepsilon)S, \quad (15.22)$$

and we obtain

$$e^{\theta J} = \left(1 - \frac{\theta^2}{2} + \frac{\theta^4}{4!} + \cdots\right) + \left(\theta - \frac{\theta^3}{3!} + \frac{\theta^5}{5!} + \cdots\right)S \quad (15.23)$$

$$- \theta\mu\varepsilon\left(\theta - \frac{\theta^3}{3!} + \cdots\right) + \theta\mu\varepsilon\left(1 - \frac{\theta^2}{2} + \cdots\right)S$$

$$= (\cos\theta - d\sin\theta\varepsilon) + (\sin\theta + d\cos\theta\varepsilon)S. \quad (15.24)$$

Let $d = \theta\mu$ be the slide along the screw axis associated with the angle θ. At this point it is convenient to introduce the dual angle $\hat{\theta} = \theta + d\varepsilon$, so we have the identities

$$\sin\hat{\theta} = \sin\theta + d\cos\theta\varepsilon \quad \text{and} \quad \cos\hat{\theta} = \cos\theta - d\sin\theta\varepsilon, \quad (15.25)$$

which are derived using the series expansions of sine and cosine.

Equation (15.24) introduces the *unit dual quaterion*, which is identified with spatial displacements. To see the relationship we factor out the rotation term to obtain

$$\hat{Q} = \cos\hat{\theta} + \sin\hat{\theta}S = (1 + \mathbf{t}\varepsilon)(\cos\theta + \sin\theta S), \tag{15.26}$$

where

$$\mathbf{t} = d\mathbf{S} + \sin\theta\cos\theta\,\mathbf{C} \times \mathbf{S} - \sin^2\theta(\mathbf{C} \times \mathbf{S}) \times \mathbf{S}. \tag{15.27}$$

This vector is one-half the translation $\mathbf{d} = 2\mathbf{t}$ of the spatial displacement associated with this dual quaternion in the same way that we saw that the rotation angle is $\phi = 2\theta$. This is because the transformation of line coordinates x to X by the rotation ϕ around an axis S with the translation $\mathbf{d}$ involves multiplication by both the Clifford algebra element $\hat{Q} = \cos\hat{\theta} + \sin\hat{\theta}S$ and its conjugate $\hat{Q}^* = \cos\hat{\theta} - \sin\hat{\theta}S$, given by

$$X = \hat{Q}x\hat{Q}^*. \tag{15.28}$$

For this reason the unit dual quaternion is usually written in terms of the half rotation angle and half displacement vector,

$$\hat{Q} = \cos\frac{\hat{\phi}}{2} + \sin\frac{\hat{\phi}}{2}S = \left(1 + \frac{1}{2}\mathbf{d}\varepsilon\right)\left(\cos\frac{\phi}{2} + \sin\frac{\phi}{2}S\right), \tag{15.29}$$

where

$$\mathbf{d} = 2\left(\frac{k}{2}\mathbf{S} + \sin\frac{\phi}{2}\cos\frac{\phi}{2}\mathbf{C} \times \mathbf{S} - \sin^2\frac{\phi}{2}(\mathbf{C} \times \mathbf{S}) \times \mathbf{S}\right). \tag{15.30}$$

Notice that we introduced the slide along S given by $k = \phi\mu$, so we have the dual angle $\hat{\phi} = \phi + k\varepsilon$.

15.2.3 Clifford Algebra Kinematics Equations

The exponential of a screw defines a relative displacement from an initial position to a final position in terms of a rotation around and slide along an axis. This means that the composition of Clifford algebra elements defines the relative kinematics equations for a serial chain that are equivalent to (15.9).

Consider the nC serial chain in which each joint can rotate through an angle θ_i around, and slide the distance d_i along, the axis S_i, for $i = 1, \ldots, n$. Let $\vec{\theta}_0$ and $\vec{d}_0$ be the joint parameters of this chain when in the reference configuration, so we have

$$\Delta\hat{\vec{\theta}} = (\vec{\theta} + \vec{d}\varepsilon) - (\vec{\theta}_0 + \vec{d}_0\varepsilon) = (\Delta\hat{\theta}_1, \Delta\hat{\theta}_2, \ldots, \Delta\hat{\theta}_n). \tag{15.31}$$

Then, the movement from this reference configuration is defined by the kinematics equations

$$\hat{D}(\Delta\hat{\theta}) = e^{\frac{\Delta\hat{\theta}_1}{2}S_1} e^{\frac{\Delta\hat{\theta}_2}{2}S_2} \cdots e^{\frac{\Delta\hat{\theta}_n}{2}S_n},$$

$$= \left(c\frac{\Delta\hat{\theta}_1}{2} + s\frac{\Delta\hat{\theta}_1}{2}S_1 \right) \left(c\frac{\Delta\hat{\theta}_2}{2} + s\frac{\Delta\hat{\theta}_2}{2}S_2 \right) \cdots \left(c\frac{\Delta\hat{\theta}_n}{2} + s\frac{\Delta\hat{\theta}_n}{2}S_n \right).$$

$$(15.32)$$

Note that s and c denote the sine and cosine functions, respectively.

15.3 Design Equations for a Serial Chain

The goal of our design problem is to determine the dimensions of a spatial serial chain that can position a tool held by its end-effector in a given set of task positions. The location of the base of the robot, the position of the tool frame, as well as the link dimensions and joint angles are considered to be design variables.

15.3.1 Specified Task Positions

Identify a set of task positions $[P_j]$, $j = 1,\ldots,m$. Then, the physical dimensions of the chain are defined by the requirement that for each position $[P_j]$ there be a joint parameter vector $\vec{\theta}_j$ such that the kinematics equations of the chain satisfy the relations

$$[P_j] = [D(\vec{\theta}_j)], \quad i = 1,\ldots,m. \quad (15.33)$$

Now choose $[P_1]$ as the reference position and compute the relative displacements $[P_j][P_1]^{-1} = [P_{1j}]$, $j = 2,\ldots,m$.

For each of these relative displacements $[P_{1j}]$ we can determine the dual unit quaternion $\hat{P}_{1j} = \cos\frac{\Delta\hat{\phi}_{1j}}{2} + \sin\frac{\Delta\hat{\phi}_{1j}}{2}P_{1j}$, $j = 2,\ldots,m$. The dual angle $\Delta\hat{\phi}_{1j}$ defines the rotation about and slide along the axis P_{1j} that defines the displacement from the first to the jth position. Now writing equation (15.32) for the $m - 1$ relative displacements, we obtain

$$\hat{P}_{1j} = e^{\frac{\Delta\hat{\theta}_{1j}}{2}S_1} e^{\frac{\Delta\hat{\theta}_{2j}}{2}S_2} \cdots e^{\frac{\Delta\hat{\theta}_{nj}}{2}S_n}, \quad j = 2,\ldots,m. \quad (15.34)$$

The result is $8(m-1)$ design equations. The unknowns are the n joint axes S_i, $i = 1,\ldots,n$, and the $n(m-1)$ pairs of joint parameters $\Delta\hat{\theta}_{ij} = \Delta\theta_{ij} + \Delta d_{ij}\varepsilon$.

15.3.2 The Independent Synthesis Equations

The eight components of the unit Clifford algebra kinematics equations (15.34) are
not independent. It is easy to see that a dual unit quaternion satisfies the identity

$$\hat{Q}\hat{Q}^* = e^{\frac{\Delta\hat{\phi}}{2}S}e^{-\frac{\Delta\hat{\phi}}{2}S} = 1, \tag{15.35}$$

which imposes two constraints. Thus, only six of the eight synthesis equations obtained for each relative task position are independent, which means that there are
only $6(n-1)$ independent synthesis equations for an n position task. Furthermore,
the axis S has unit magnitude, which means that only four of its six components are
independent.

In order to count the number of independent equations and unknowns in the
Clifford algebra synthesis equations, it is useful to identify the relationship between
the constraints on a dual unit quaternion and the constraints on the dual unit vector
that generates it.

Property 1 (Normality Condition). The dual quaternion arising from the product of
dual quaternions has unit magnitude if and only if each factor is the exponential of
a dual unit vector.

Proof. For the screw displacement $\hat{Q} = e^{\frac{\Delta\hat{\phi}}{2}S}$ the unit condition yields

$$\hat{Q}\hat{Q}^* = \left(c\frac{\Delta\hat{\phi}}{2} + s\frac{\Delta\hat{\phi}}{2}S\right)\left(c\frac{\Delta\hat{\phi}}{2} - s\frac{\Delta\hat{\phi}}{2}S\right) = c\frac{\Delta\hat{\phi}}{2}c\frac{\Delta\hat{\phi}}{2} + s\frac{\Delta\hat{\phi}}{2}s\frac{\Delta\hat{\phi}}{2}S\cdot S. \tag{15.36}$$

Notice that if $S\cdot S = 1$, then

$$\hat{Q}\hat{Q}^* = c\frac{\Delta\hat{\phi}}{2}c\frac{\Delta\hat{\phi}}{2} + s\frac{\Delta\hat{\phi}}{2}s\frac{\Delta\hat{\phi}}{2} = c\frac{\Delta\phi}{2}^2 + s\frac{\Delta\phi}{2}^2 = 1. \tag{15.37}$$

Now, for a dual quaternion obtained as the composition of transformations about n
joint axes, we have

$$\hat{Q}\hat{Q}^* = \left(e^{\frac{\Delta\phi_1}{2}S_1}\cdots e^{\frac{\Delta\phi_n}{2}S_n}\right)\left(e^{\frac{\Delta\phi_1}{2}S_1}\cdots e^{\frac{\Delta\phi_n}{2}S_n}\right)^*. \tag{15.38}$$

Expand this product and use the associative property of the Clifford algebra to obtain

$$\hat{Q}\hat{Q}^* = e^{\frac{\Delta\phi_1}{2}S_1}\cdots(e^{\frac{\Delta\phi_n}{2}S_n}e^{\frac{-\Delta\phi_n}{2}S_n})\cdots e^{\frac{-\Delta\phi_1}{2}S_1}, \tag{15.39}$$

such that the terms $e^{\frac{\Delta\phi_n}{2}S_n}e^{\frac{-\Delta\phi_n}{2}S_n}$ equal 1 when $S_n\cdot S_n = 1$. The result is

$$\hat{Q}\hat{Q}^* = 1 \quad \Longleftrightarrow \quad S_i\cdot S_i = 1, \quad i = 1,\ldots,n. \tag{15.40}$$

□

This condition shows that six of the eight components of the dual quaternion kinematics equations combine with the normal conditions on the Plücker coordinates of the joint axes to define the minimum set of independent synthesis equations for the serial chain problem.

15.3.3 Counting the Equations and Unknowns

Consider a spatial serial chain that consists of r revolute joints and p prismatic joints. A purely prismatic joint is defined by the unit vector S that defines the slide direction, so it has two independent parameters. The revolute joint axis is defined by Plücker coordinate vectors $S_i = S + C \times S\varepsilon$ that have four independent components due to the normal conditions

$$|S| = 1 \quad \text{and} \quad S \cdot (C \times S) = 0. \tag{15.41}$$

Thus, the joint axes that define this chain have $K = 6r + 3p$ components, minus $2r + p$ Plücker constraints, which yields $4r + 2p$ independent unknowns.

Revolute and prismatic joints each have a single joint parameter, either a rotation angle or slide distance, which means that our chain has $(r + p)(m - 1)$ unknown joint parameters that define the m relative positions.

Subtracting the number of equations from the number of unknowns, we obtain

$$E = 4r + 2p + (r + p)(m - 1) - 6(m - 1)$$
$$= (3r + p + 6) + (r + p - 6)m, \tag{15.42}$$

where E is the excess of unknowns over equations. This excess can be made to equal zero for chains with degree of freedom dof $= r + p \leq 5$, in which case we specify

$$m = \frac{3r + p + 6 - c}{6 - (r + p)}, \tag{15.43}$$

task positions. If fewer than this number of task positions are defined, or if the chain has six or more degrees of freedom, then we are free to select values for the excess design parameters. In (15.43) we have added c to denote any extra constraint that may be imposed on the axes. Table 15.1 presents the maximum number of positions that can be defined for some chains with five degrees of freedom.

It is interesting to notice that because the composition of displacements has structure of semidirect product, rotations are obtained only from operating rotations only. This means that a counting scheme can be generated specifically for the rotations by considering only the rotation component of the dual quaternion only. The result is the maximum number of task rotations is given by

$$m_R = \frac{3 + r}{3 - r}. \tag{15.44}$$

In some cases with $r = 1$ or 2, the rotation part of the design equations can be used to determine the directions of these axes independently. Perez and McCarthy [97] call these chains "orientation limited."

Table 15.1 The number of task positions that determine the structural parameters for five-degree-of-freedom serial chains.

Chain	K	Task positions	Total Equations
PRPRP	21	15	91
RPRPR	24	17	104
RRRRP	27	19	117
RRRRR	30	21	130

15.3.4 Special Cases: T, S and PP Joints

The counting formula in (15.43) is used for revolute and prismatic joints assembled in serial chains. The RR and RRR chains can be further specialized by introducing geometric constraints between their joint axes to define the universal joint and spherical joint, respectively. Also two consecutive prismatic joints span the group of displacements T_P, planar translations on plane P, and they form a special type of joint that we call PP. We are going to show how in some of these cases, the number of design parameters is less than those considering directly the geometric constraints on the axes.

15.3.4.1 The T Joint

Consider the RR chain formed by axes S_i and S_{i+1}. If we require these axes to intersect in a right angle, then we obtain Hooke's joint, also called a universal joint, which we denote by a T following Crane and Duffy [16]. This geometric constraint is defined by the dual vector equation

$$T: \quad S_i \cdot S_{i+1} = 0, \tag{15.45}$$

which expands to define the two constraints

$$T: \quad \mathbf{S}_i \cdot \mathbf{S}_{i+1} = 0 \quad \text{and} \quad \mathbf{S}_i \cdot \mathbf{S}_{i+1}^\circ + \mathbf{S}_i^\circ \cdot \mathbf{S}_{i+1} = 0. \tag{15.46}$$

The design equations for the RRR chain, for instance, are easily transformed into design equations for the TR chain by including these two constraint equations with the appropriate indices.

15.3.4.2 The S Joint

In the same way, a sequence of three revolute joints, and RRR chain, can be constrained such that they intersect in a point, and the pairs in sequence are perpendicular. This is a common construction for an active spherical joint, denoted by S, which allows full orientation freedom around the intersection point. However, for synthesis applications it can be shown that any three axes create the same spherical joint.

Label three axes S_i, S_{i+1}, and S_{i+2}. Then the equations that define this joint consist of the dual vector constraints

$$S: \quad S_i \cdot S_{i+1} = 0, \quad S_i \cdot S_{i+2} = 0 \quad \text{and} \quad S_{i+1} \cdot S_{i+2} = 0, \tag{15.47}$$

If we write the spherical joint as the dual quaternion product of these individual axes,

$$\hat{S}(\theta_1, \theta_2, \theta_3) = \hat{S}_1(\theta_1)\hat{S}_2(\theta_2)\hat{S}_3(\theta_3), \tag{15.48}$$

when expanded, we obtain

$$\hat{S}(\theta_1, \theta_2, \theta_3) = \alpha_4 + \alpha_1 S_1 + \alpha_2 S_2 + \alpha_3 S_3, \tag{15.49}$$

where each α_i appears as combinations of the joint variables,

$$\begin{aligned}
\alpha_1 &= \sin\frac{\theta_1}{2}\cos\frac{\theta_2}{2}\cos\frac{\theta_3}{2} + \cos\frac{\theta_1}{2}\sin\frac{\theta_2}{2}\sin\frac{\theta_3}{2}, \\
\alpha_2 &= \cos\frac{\theta_1}{2}\sin\frac{\theta_2}{2}\cos\frac{\theta_3}{2} - \sin\frac{\theta_1}{2}\cos\frac{\theta_2}{2}\sin\frac{\theta_3}{2}, \\
\alpha_3 &= \sin\frac{\theta_1}{2}\sin\frac{\theta_2}{2}\cos\frac{\theta_3}{2} + \cos\frac{\theta_1}{2}\cos\frac{\theta_2}{2}\sin\frac{\theta_3}{2}, \\
\alpha_4 &= \cos\frac{\theta_1}{2}\cos\frac{\theta_2}{2}\cos\frac{\theta_3}{2} - \sin\frac{\theta_1}{2}\sin\frac{\theta_2}{2}\sin\frac{\theta_3}{2}.
\end{aligned} \tag{15.50}$$

Now we show any directions S_1, S_2, S_3 can be used to define the spherical joint. Equate (15.48) to a goal displacement $\hat{P} = (p_w + \varepsilon p_w^0) + (\mathbf{P} + \varepsilon \mathbf{P}^0)$,

$$\hat{S}(\theta_1, \theta_2, \theta_3) = \hat{P}, \tag{15.51}$$

and solve linearly for the combinations of joint variables in the α_i factors using the real part of the dual quaternion equation,

$$\begin{bmatrix} S_1 & S_2 & S_3 & \vec{0} \\ 0 & 0 & 0 & 1 \end{bmatrix} \begin{Bmatrix} \alpha_1 \\ \alpha_2 \\ \alpha_3 \\ \alpha_4 \end{Bmatrix} = \begin{Bmatrix} \mathbf{P} \\ p_w \end{Bmatrix}, \tag{15.52}$$

where we write the scalar term as the fourth row. The values obtained for the joint angles,

$$\alpha_1 = \mathbf{S}_1 \cdot \mathbf{P}, \quad \alpha_2 = \mathbf{S}_2 \cdot \mathbf{P}, \quad \alpha_3 = \mathbf{S}_3 \cdot \mathbf{P}, \quad \alpha_4 = p_w, \qquad (15.53)$$

are related by the following expression,

$$\mathscr{R}: \quad (\mathbf{S}_1 \cdot \mathbf{P})^2 + (\mathbf{S}_2 \cdot \mathbf{P})^2 + (\mathbf{S}_3 \cdot \mathbf{P})^2 + p_w^2 = 1. \qquad (15.54)$$

If we notice that $1 - p_w^2 = \mathbf{P} \cdot \mathbf{P}$, this expression can be written as

$$\mathscr{R}': \quad (\mathbf{S}_1 \cdot \mathbf{P})\mathbf{S}_1 + (\mathbf{S}_2 \cdot \mathbf{P})\mathbf{S}_2 + (\mathbf{S}_3 \cdot \mathbf{P})\mathbf{S}_3 = \mathbf{P}, \qquad (15.55)$$

which states that the vector sum of the projections of $\mathbf{P}$ on the three joint directions is equal to $\mathbf{P}$. This equation holds for any three perpendicular directions.

We now substitute the expressions of the joint variables in the dual part of (15.51). Notice that due to the fact that the last component of the dual quaternion in (15.49) is equal to zero, a spherical joint cannot perform the most general relative displacement.

If we express the dual part of each joint axis as $\mathbf{S}_i^0 = \mathbf{c} \times \mathbf{S}_i$, where $\mathbf{c}$ is the common intersection point, the dual part of the equations becomes

$$\mathscr{M}: \quad (\mathbf{S}_1 \cdot \mathbf{P})\mathbf{c} \times \mathbf{S}_1 + (\mathbf{S}_2 \cdot \mathbf{P})\mathbf{c} \times \mathbf{S}_2 + (\mathbf{S}_3 \cdot \mathbf{P})\mathbf{c} \times \mathbf{S}_3 = \mathbf{P}^0, \qquad (15.56)$$

and this is equal to

$$\mathscr{M}': \quad \mathbf{c} \times ((\mathbf{S}_1 \cdot \mathbf{P})\mathbf{S}_1 + (\mathbf{S}_2 \cdot \mathbf{P})\mathbf{S}_2 + (\mathbf{S}_3 \cdot \mathbf{P})\mathbf{S}_3) = \mathbf{P}^0. \qquad (15.57)$$

Observe that the expression in parentheses is the left-hand side of (15.55). Use this to obtain

$$\mathscr{M}'': \quad \mathbf{c} \times \mathbf{P} = \mathbf{P}^0. \qquad (15.58)$$

This set of three equations specifies two out of the three coordinates of the point. At least two relative positions need to be defined to fully specify this point. However, as noted previously, they cannot be general positions if we want an exact solution.

Summarizing, for the spherical joint, only the coordinates of the intersection point $\mathbf{c}$ and the three joint angles are design variables. This gives a different counting from the one obtained by solving for three perpendicular and intersecting revolute joints.

15.3.4.3 The PP Joint

When a serial robot is to be designed with two consecutive prismatic joints, these can be made to be coplanar, since the location of a prismatic joint is not a design parameter. The set of displacements produced by the two prismatic joints forms the subgroup T_P of planar translations on a plane P. The subgroup has dimension 2, and two more parameters are needed to define the direction normal to the plane; for synthesis purposes, the location of the plane is again arbitrary.

We could create the PP joint by using two prismatic joints, which is a total of four joint parameters plus two joint slides. We are going to show that both directions do not appear independently in the equations. Let $\mathbf{S}_1$ and $\mathbf{S}_2$ be the directions of the two prismatic joints. We create the displacements of the PP joint as the dual quaternion product

$$\hat{S}(d_1, d_2) = \hat{S}_1(d_1)\hat{S}_2(d_2). \tag{15.59}$$

When expanded, it yields

$$\hat{S}^0(d_1, d_2) = 1 + \varepsilon\left(\frac{d_1}{2}\mathbf{S}_1 + \frac{d_2}{2}\mathbf{S}_2\right). \tag{15.60}$$

We solve linearly for the joint variables d_1 and d_2 in the dual part of the design equations,

$$\left[\tfrac{1}{2}\mathbf{S}_1 \ \tfrac{1}{2}\mathbf{S}_2\right]\begin{Bmatrix} d_1 \\ d_2 \end{Bmatrix} = \mathbf{P}^0. \tag{15.61}$$

For the system to have a solution, the determinant of the augmented matrix must be zero. This yields the simplified design equation

$$\mathcal{M}: \quad (\mathbf{S}_1 \times \mathbf{S}_2)\cdot\mathbf{P}^0 = 0. \tag{15.62}$$

The parameters of the prismatic joints always appear as a cross product, and we can substitute it by the common normal, $\mathbf{S}_1 \times \mathbf{S}_2 = \mathbf{N}$. Within the plane defined by $\mathbf{N}$, any two independent directions can be used to define the joint axes.

Summarizing, for synthesis purposes, the design parameters for two consecutive prismatic joints are the two slides and the vector $\mathbf{N}$ defining the normal direction to $\mathbf{S}_1$ and $\mathbf{S}_2$. It coincides with the parameters needed to define the subgroup T_P.

Other cases in which the number of design parameters is less than those obtained by imposing extra constraints on the joint axes can be found in a similar way.

15.4 Assembling the Design Equations

The structure of the Clifford algebra design equations provides a systematic approach to assembling the design equations for a broad range of serial chains. The strategy is to formulate the design equations for the nC serial chain, and then (i) restrict the joint variables to form prismatic or sliding joints, and (ii) impose geometric conditions on the axes to account for specific geometry and to form Hooke's or spherical joints. The result is a systematic way of defining the design equations for a broad range of chains. Here we present the procedure for the 3C serial chain, but it has been implemented in our numerical solver for the 2C, 4C, and 5C cases as well.

15.4.1 The 3C Chain

The Clifford algebra form of the relative kinematics equations for the 3C chain can be written as

$$\hat{D}(\Delta\hat{\theta}) = \left(c\frac{\Delta\hat{\theta}_1}{2} + s\frac{\Delta\hat{\theta}_1}{2}S_1\right)\left(c\frac{\Delta\hat{\theta}_2}{2} + s\frac{\Delta\hat{\theta}_2}{2}S_2\right)\left(c\frac{\Delta\hat{\theta}_3}{2} + s\frac{\Delta\hat{\theta}_3}{2}S_3\right),$$

(15.63)

where $S_i = \mathbf{S}_i + \mathbf{S}_i^\circ \varepsilon$ define the joint axes in the reference position, and $\Delta\hat{\theta}_i = \Delta\theta_i + \Delta d_i$ define the rotation and slide of the cylindric joint around the ith axis.

Expand the right side of (15.63) using the Clifford product to obtain

$$\hat{D}(\Delta\hat{\theta}) = (\hat{c}_1\hat{c}_2 - \hat{s}_1\hat{s}_2S_1 \cdot S_2 + \hat{s}_1\hat{c}_2S_1 + \hat{c}_1\hat{s}_2S_2 + \hat{s}_1\hat{s}_2S_1 \times S_2)(\hat{c}_3 + \hat{s}_3S_3),$$

$$= \hat{c}_1\hat{c}_2\hat{c}_3 - \hat{s}_1\hat{s}_2\hat{c}_3S_1 \cdot S_2 - \hat{s}_1\hat{c}_2\hat{s}_3S_1 \cdot S_3 - \hat{c}_1\hat{s}_2\hat{s}_3S_2 \cdot S_3 - \hat{s}_1\hat{s}_2\hat{s}_3S_1 \times S_2 \cdot S_3$$

$$+ \hat{s}_1\hat{c}_2\hat{c}_3S_1 + \hat{c}_1\hat{s}_2\hat{c}_3S_2 + \hat{c}_1\hat{c}_2\hat{s}_3S_3 + \hat{s}_1\hat{s}_2\hat{c}_3S_1 \times S_2 + \hat{s}_1\hat{c}_2\hat{s}_3S_1 \times S_3$$

$$+ \hat{c}_1\hat{s}_2\hat{s}_3S_2 \times S_3 + \hat{s}_1\hat{s}_2\hat{s}_3((S_1 \times S_2) \times S_3 - (S_1 \cdot S_2)S_3).$$

(15.64)

For convenience, we have introduced the notation $\hat{c}_i = \cos\frac{\Delta\hat{\theta}_i}{2}$ and $\hat{s}_i = \sin\frac{\Delta\hat{\theta}_i}{2}$.

Equation (15.64) can be written in matrix form to emphasize that it is the linear combination of the eight monomials formed as products of the joint angles, which we assemble into an array in reverse lexicographic order obtained by reading right to left,

$$\hat{V} = (\hat{c}_1\hat{c}_2\hat{c}_3, \hat{s}_1\hat{c}_2\hat{c}_3, \hat{c}_1\hat{s}_2\hat{c}_3, \hat{c}_1\hat{c}_2\hat{s}_3, \hat{s}_1\hat{s}_2\hat{c}_3, \hat{s}_1\hat{c}_2\hat{s}_3, \hat{c}_1\hat{s}_2\hat{s}_3, \hat{s}_1\hat{s}_2\hat{s}_3)^T.$$

(15.65)

To do this, we must introduce the vector form of the dual unit quaternion $\hat{Q} = \cos\frac{\Delta\hat{\theta}}{2} + \sin\frac{\Delta\hat{\theta}}{2}S$ given by

$$\hat{Q} = \begin{Bmatrix} \sin\frac{\Delta\hat{\theta}}{2}(S_x + S_x^\circ\varepsilon) \\ \sin\frac{\Delta\hat{\theta}}{2}(S_y + S_y^\circ\varepsilon) \\ \sin\frac{\Delta\hat{\theta}}{2}(S_z + S_z^\circ\varepsilon) \\ \cos\frac{\Delta\hat{\theta}}{2} \end{Bmatrix} = \begin{Bmatrix} \sin\frac{\Delta\hat{\theta}}{2}S \\ \cos\frac{\Delta\hat{\theta}}{2} \end{Bmatrix}.$$

(15.66)

Collecting terms in (15.64), we obtain the matrix equation

$$\hat{D}(\Delta\hat{\theta})$$

$$= \begin{bmatrix} 0 & S_1 & S_2 & S_3 & S_1 \times S_1 & S_1 \times S_3 & S_2 \times S_3 & -(S_1 \cdot S_2)S_3 + (S_1 \times S_2) \times S_3 \\ 1 & 0 & 0 & 0 & -S_1 \cdot S_2 & -S_1 \cdot S_3 & -S_2 \cdot S_3 & -S_1 \times S_2 \cdot S_3 \end{bmatrix} \hat{V}.$$

(15.67)

The Clifford algebra notation is compact in that each column of this matrix actually forms a column of four dual coefficients, or eight real coefficients if we write the dual components of the dual quaternion after the real components, forming an eight-dimensional vector. Similarly, each of the monomials in $\hat{\mathbf{V}}$ expands into four real terms, which we can list as

$$\mathbf{M} = \left(\mathbf{V}, \frac{\Delta d_1}{2}\mathbf{V}, \frac{\Delta d_2}{2}\mathbf{V}, \frac{\Delta d_3}{2}\mathbf{V}\right),$$
(15.68)

where $\mathbf{V}$ is the array of real parts of $\hat{\mathbf{V}}$. Thus, (15.67) expands to an 8×32 matrix equation. The number k of joint variable monomials in an nC serial chain is given by

$$k = (n+1)2^n.$$
(15.69)

Thus, these equations become 8×12 for 2C, 8×80 for 4C, and 8×192 for 5C chains.

The kinematics equations (15.67) can be used directly for the design of a 3C chain. In what follows, we specialize these equations to obtain design equations for a variety of special serial chains.

15.4.2 RCC, RRC, and RRR Chains

The ith cylindric joint in the 3C chain is converted to a revolute joint simply by setting $\Delta d_i = 0$. This can be done in seven different ways to define three permutations of the RRC chain, three permutations of the RRC chain, and the RRR chain.

For example, the monomials in (15.67) that define the RCC, CRC, and CCR chains are given by

$$\text{RCC}: \quad \mathbf{M} = \left(\mathbf{V}, \frac{\Delta d_2}{2}\mathbf{V}, \frac{\Delta d_3}{2}\mathbf{V}\right),$$

$$\text{CRC}: \quad \mathbf{M} = \left(\mathbf{V}, \frac{\Delta d_1}{2}\mathbf{V}, \frac{\Delta d_3}{2}\mathbf{V}\right),$$

$$\text{CCR}: \quad \mathbf{M} = \left(\mathbf{V}, \frac{\Delta d_1}{2}\mathbf{V}, \frac{\Delta d_2}{2}\mathbf{V}\right).$$
(15.70)

Similarly, the RRC, RCR and CRR chains have the monomials

$$\text{RRC}: \quad \mathbf{M} = \left(\mathbf{V}, \frac{\Delta d_3}{2}\mathbf{V}\right),$$

$$\text{RCR}: \quad \mathbf{M} = \left(\mathbf{V}, \frac{\Delta d_2}{2}\mathbf{V}\right),$$

$$\text{CRR}: \quad \mathbf{M} = \left(\mathbf{V}, \frac{\Delta d_1}{2}\mathbf{V}\right).$$
(15.71)

Finally, the RRR chain is defined by the monomial list

$$\text{RRR}: \quad \mathbf{M} = \mathbf{V}. \tag{15.72}$$

Notice that if an nC chain is specialized to have r revolute joints, then the number of monomials is given by

$$k = (n - r + 1)2^n. \tag{15.73}$$

15.4.3 PCC, PPC, and PPP Chains

A two-step process is required to convert the ith cylindric joint to a prismatic joint. The first step is to set $\Delta \theta_i = 0$. The second step consists in specializing the joint axis $\mathbf{S}_i = \mathsf{S}_i$, so that its dual part is zero. This latter constraint arises because the pure translation defined by a prismatic joint depends only on the direction, not the location in space, of its axis.

In order to define the monomials for the three permutations of the PCC chain, we introduce $\mathbf{W}_1 = (c_1 c_2 c_3, c_1 s_2 c_3, c_1 c_2 s_3, c_1 s_2 s_3)$, and similarly define $\mathbf{W}_2$ and $\mathbf{W}_3$, where the subscript i indicates that we make $s_i = 0$. This allows us to define the arrays of monomials

$$\text{PCC}: \quad \mathbf{M} = \left(\mathbf{W}_1, \frac{\Delta d_1}{2} \mathbf{W}_1, \frac{\Delta d_2}{2} \mathbf{W}_1, \frac{\Delta d_3}{2} \mathbf{W}_1 \right),$$

$$\text{CPC}: \quad \mathbf{M} = \left(\mathbf{W}_2, \frac{\Delta d_1}{2} \mathbf{W}_2, \frac{\Delta d_2}{2} \mathbf{W}_2, \frac{\Delta d_3}{2} \mathbf{W}_2 \right),$$

$$\text{CCP}: \quad \mathbf{M} = \left(\mathbf{W}_3, \frac{\Delta d_1}{2} \mathbf{W}_3, \frac{\Delta d_2}{2} \mathbf{W}_3, \frac{\Delta d_3}{2} \mathbf{W}_3 \right). \tag{15.74}$$

The monomials for the three permuations of the the PPC chain are easily determined by introducing the set of monomials $\mathbf{W}_{12} = (c_1 c_2 c_3, c_1 c_2 s_3)$, and similarly $\mathbf{W}_{13}$ and $\mathbf{W}_{23}$,

$$\text{PPC}: \quad \mathbf{M} = \left(\mathbf{W}_{12}, \frac{\Delta d_1}{2} \mathbf{W}_{12}, \frac{\Delta d_2}{2} \mathbf{W}_{12}, \frac{\Delta d_3}{2} \mathbf{W}_{12} \right),$$

$$\text{PCP}: \quad \mathbf{M} = \left(\mathbf{W}_{13}, \frac{\Delta d_1}{2} \mathbf{W}_{13}, \frac{\Delta d_2}{2} \mathbf{W}_{13}, \frac{\Delta d_3}{2} \mathbf{W}_{13} \right),$$

$$\text{CPP}: \quad \mathbf{M} = \left(\mathbf{W}_{23}, \frac{\Delta d_1}{2} \mathbf{W}_{23}, \frac{\Delta d_2}{2} \mathbf{W}_{23}, \frac{\Delta d_3}{2} \mathbf{W}_{23} \right). \tag{15.75}$$

Finally, the PPP chain is defined by the monomial list

$$\text{PPP}: \quad \mathbf{M} = \left((c_1 c_2 c_3), \frac{\Delta d_1}{2} (c_1 c_2 c_3), \frac{\Delta d_2}{2} (c_1 c_2 c_3), \frac{\Delta d_3}{2} (c_1 c_2 c_3) \right). \tag{15.76}$$

The number of monomials in an nC chain with p of the joints restricted to be prismatic is seen to be

$$k = (n+1)2^{n-p}. \tag{15.77}$$

Table 15.2 summarizes the constraints needed to transform the C joint into the most common types of joints. Notice that for the spherical joint and other special cases, we use the approach of adding constraints between consecutive joint axes. This will not yield the minimum set of joint parameters, but it gives satisfactory results with the numerical solver.

Table 15.2 Constraints that specialize C joints to R, P, T and S joints.

Joint	Axes	Constraints
R	S_i	$\Delta d_i = 0$
P	S_i	$\Delta \theta_i = 0$
C	S_i	none
T	S_i, S_{i+1}	$\Delta d_i = 0$, $\Delta d_{i+1} = 0$, $S_i \cdot S_{i+1} = 0$
S	S_i, S_{i+1}, S_{i+2}	$\Delta d_i = 0$, $\Delta d_{i+1} = 0$, $\Delta d_{i+2} = 0$,
		$S_i \cdot S_{i+1} = 0$, $S_{i+1} \cdot S_{i+2} = 0$, $S_i \cdot S_{i+2} = 0$

This approach to the formulation of the design equations for special cases of the CCC chain can be extended to any nC chain.

15.5 The Synthesis of 5C and Related Chains

In this section, we present a numerical synthesis algorithm that uses the Clifford algebra exponential design equations for the 5C serial chain, see Figure 15.2. The special cases of this chain include robots with up to five joints and up to ten degrees of freedom.

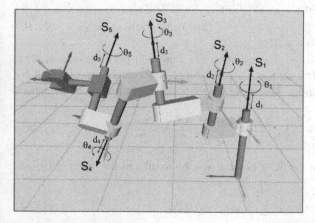

Fig. 15.2 The 5C serial robot.

The design equations for a specific serial robot are be obtained from the 5C robot equations by imposing conditions on some of the axes or joint variables. The kinematics equations for the 5C robot are given by

$$\hat{Q}_{5C} = e^{\frac{\Delta \hat{\theta}_1}{2} \mathsf{S}_1} e^{\frac{\Delta \hat{\theta}_2}{2} \mathsf{S}_2} e^{\frac{\Delta \hat{\theta}_3}{2} \mathsf{S}_3} e^{\frac{\Delta \hat{\theta}_4}{2} \mathsf{S}_4} e^{\frac{\Delta \hat{\theta}_5}{2} \mathsf{S}_5}, \qquad (15.78)$$

or

$$\hat{Q}_{5C} = \left(\cos \frac{\Delta \hat{\theta}_1}{2} + \sin \frac{\Delta \hat{\theta}_1}{2} \mathsf{S}_1 \right) \left(\cos \frac{\Delta \hat{\theta}_2}{2} + \sin \frac{\Delta \hat{\theta}_2}{2} \mathsf{S}_2 \right) \cdots$$
$$\left(\cos \frac{\Delta \hat{\theta}_5}{2} + \sin \frac{\Delta \hat{\theta}_5}{2} \mathsf{S}_5 \right). \quad (15.79)$$

The kinematics equations for a serial chain consisting of revolute R, prismatic P, universal T, cylindrical C, or spherical S joints can be obtained from the 5C robot using the approach presented in the previous section. For example, the kinematics equations of the TPR serial chain are obtained by requiring the axes S_1 and S_2 to be perpendicular and coincident, which is obtained by setting the joint variables d_1, d_2, θ_3, and d_4 to zero. The extra joint is eliminated by setting θ_5 and d_5 to zero. Other joints, like the helical H and planar E joints can also be modeled by imposing constraints on the axes and joint parameters.

In order to facilitate the specialization of the general 5C robot to a specific serial chain topology,the kinematics equations are organized as a linear combination of the products of joint angles and slides, which form the *monomials* of these equations with coefficients that are given by the structural parameters of the chain. In this way, the kinematics equations of the 5C serial chain are a linear combination of 192 monomials, which can be organized into six sets of 32 products of sines and cosines of the $\Delta \theta_i$ joint angles, given by

$$\mathbf{V} = (s_1 s_2 s_3 s_4 s_5, (s_1 s_2 s_3 s_4 c_5)_5, (s_1 s_2 s_3 c_4 c_5)_{10}, (s_1 s_2 c_3 c_4 c_5)_{10},$$
$$(s_1 c_2 c_3 c_4 c_5)_5, c_1 c_2 c_3 c_4 c_5), \quad (15.80)$$

where $c_i = \cos \frac{\Delta \theta_i}{2}$, $s_i = \sin \frac{\Delta \theta_i}{2}$. The notation $()_j$ denotes j permutations of each set of sines and cosines. The remaining five sets of monomials are obtained by multiplying $\mathbf{V}$ by the joint slides $\frac{\Delta d_i}{2}$, so we have a total set of monomials $\mathbf{M}$, where

$$\mathbf{M} = \left(\mathbf{V}, \frac{\Delta d_1}{2} \mathbf{V}, \frac{\Delta d_2}{2} \mathbf{V}, \frac{\Delta d_3}{2} \mathbf{V}, \frac{\Delta d_4}{2} \mathbf{V}, \frac{\Delta d_5}{2} \mathbf{V} \right). \qquad (15.81)$$

The kinematics equations of the 5C robot can now be written as the linear combination

$$\hat{Q}_{5C} = \sum_{i=1}^{192} \mathbf{K}_i m_i, \quad m_i \in \mathbf{M}. \qquad (15.82)$$

The coefficients $\mathbf{K}_i$ are 8-dimensional vectors containing the structural variables defining the joint axes.

This equation is adjusted to accommodate a revolute or prismatic joint inserted as the jth joint axis by selecting the nonzero components of the vector $\mathbf{M}$. Notice that if the jth C joint is restricted to be a revolute joint, then the slide Δd_j is zero, which eliminates 32 components in $\mathbf{M}$. Similarly, if this joint is replaced by a prismatic joint the angle becomes $\Delta \theta_j = 0$, which eliminates 16 terms from the vector $\mathbf{V}$.

In order to construct these equations, start with the array $L_{5C} = \{1, 2, \ldots, 192\}$ of indices that denote the components of $\mathbf{M}$ for the general 5C chain, sorted as shown above. Next define the arrays L_{R_j}, L_{P_j} and L_{C_j} that denote the nonzero components of $\mathbf{M}$ for the cases when joint j is either a revolute, prismatic, or cylindric joint, given by

$$L_{R_j} = \left\{ i : \left(\cos \frac{\Delta \theta_j}{2} \wedge \sin \frac{\Delta \theta_j}{2} \right) \in m_i \vee \frac{\Delta d_j}{2} \notin m_i \right\},$$

$$L_{P_j} = \left\{ i : \left(\frac{\Delta d_j}{2} \wedge \cos \frac{\Delta \theta_j}{2} \right) \in m_i \vee \sin \frac{\Delta \theta_j}{2} \notin m_i \right\},$$

$$L_{C_j} = \left\{ i : \left(\frac{\Delta d_j}{2} \wedge \cos \frac{\Delta \theta_j}{2} \wedge \sin \frac{\Delta \theta_j}{2} \right) \in m_i \right\}, \tag{15.83}$$

where $\wedge$ and $\vee$ are the logical or and logical and operations, respectively. Finally, compute the array of indices L for a specific serial chain topology by intersecting the arrays obtained for all of the joints, that is,

$$L = \bigcap_{j=1}^{5} (L_{R_j} \cup L_{P_j} \cup L_{C_j}), \tag{15.84}$$

where $L_{P_j} = \emptyset$ and $L_{C_j} = \emptyset$ if j is a revolute joint, for example.

The kinematics equations for the specific serial chain is now given by

$$\hat{Q} = \sum_{i \in L} \mathbf{K}_i m_i. \tag{15.85}$$

The synthesis equations for the chain are obtained by equating the kinematics equations in (15.85) to the set of task positions $\hat{P}_{1i}$, that is,

$$\hat{Q} = \hat{P}_{1i}, \quad i = 2, \ldots m, \tag{15.86}$$

where the maximum number of task positions, m, is obtained for the chosen topology using equations (15.43) and (15.44). Additional constraint equations may be added to account for the specialized geometry of T and S joints or for any other geometric constraint present in the robot.

These synthesis equations are solved to determine the joint axes S_i in the reference configuration, as well as for values for the joint variables that ensure that the serial chain reaches each of the task positions.

15.5.1 The Synthesis Process

It is possible to automate the generation of the synthesis equations as cases of the four classes of 2C, 3C, 4C, and 5C related serial chains. The synthesis equations can then be solved numerically given a random start value. The input data consists of a set of task positions and topology of the serial chain. The topology of the chains is used to construct its kinematics equations $\hat{Q}$. These equations are set equal to the task positions $\hat{P}_{1i}$ to yield the synthesis equations as the difference $\hat{Q} - \hat{P}_{1i}$, $i = 2, \dots, m$. The numerical solver finds values for the components of the joint axes and joint variables that minimize this difference.

It is not necessary that the numerical solver use the minimum set of design equations as defined by (15.43). In fact, it is convenient to use all $8(m-1)+c$ synthesis equations. For the cases of 3R, 4R, and 5R serial chains this approach introduces 2, 8, and 30 redundant equations, respectively. Experience shows that the additional equations enhance the convergence of the numerical algorithm.

15.6 Example: Design of a CCS Chain

In order to demonstrate the kinematic synthesis of spatial serial chains we begin by identifying a goal trajectory that was determined using the Bézier interpolation of a set of spatial key frames. From this trajectory we select twelve positions shown in Figure 15.3 to define the task. Next we specify the topology of the chain to be the seven-degree-of-freedom CCS serial chain. Notice that this is the special case of a 5C serial chain, obtained by requiring the last three C joints to be restricted to revolute joints with axes that intersect in a single point.

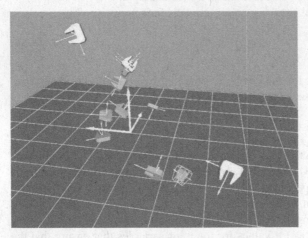

Fig. 15.3 The 12 positions defining the task.

The CCS chain consists of a shoulder and elbow that allow both rotation and translation about and along the axes, combined with a spherical wrist Figure 15.4. In addition, we constrain the joint angles of the shoulder C joint to have specific values for the rotation and translation in each of the task positions. See Table 15.3 for the task positions and the angle specification.

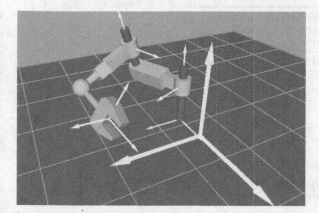

Fig. 15.4 The CCS serial chain.

Table 15.3 Task positions and values for the first joint angles

Task	Dual quaternion coordinates
1	$(0.02, -0.15, 0.58, 0.80, 0.66, -0.30, 0.09, -0.14)$
2	$(0.06, -0.37, 0.38, 0.85, 0.77, -0.38, 0.20, -0.31)$
3	$(-0.03, -0.44, -0.90, 0.03, -0.38, 0.77, -0.34, 0.67)$
4	$(0.41, -0.29, -0.41, 0.76, 0.32, 0.37, 0.14, 0.04)$
5	$(-0.04, -0.43, -0.90, 0.05, -0.39, 0.74, -0.30, 0.64)$
6	$(0.28, -0.01, -0.36, 0.89, -0.07, -0.06, 0.11, 0.06)$
7	$(0.36, 0.35, 0.48, 0.72, -0.43, -0.54, -0.12, 0.56)$
8	$(-0.18, 0.79, 0.03, 0.58, -0.73, -0.94, -0.51, 1.07)$
9	$(-0.29, -0.54, 0.55, 0.57, 0.87, -0.26, 1.106, -0.87)$
10	$(-0.93, -0.19, 0.26, 0.19, 0.27, -0.07, 0.10, 1.09)$
11	$(-0.37, 0.09, 0.73, 0.56, 0.37, -0.78, -0.51, 1.04)$

Joint Parameters	Values
θ_1	$j(\pi/11), \quad j = 1, \ldots, 11$
d_1	$j(0.02), \quad j = 1, \ldots, 11$

For this example, we run the Java software twice to obtain two different solutions. The first solution took two iterations of the solver with a total time of 91 seconds. The second solution took 61 seconds and one iteration. See Table 15.4 for the coordinates of the joint axes. Figure 15.5 shows one of the resulting robots moving along the desired task.

Table 15.4 CCS chains designed to perform the specified task.

Solution	Joint Axis	Value
1	1	$(0.14, 0.49, 0.86, -3.59, -0.45, 0.85)$
	2	$(-0.24, 0.92, 0.29, -2.44, -0.73, 0.26)$
	wrist	$(0.23, 1.21, -0.12)$
2	1	$(0.72, -0.50, 0.48, 0.40, 0.03, -0.57)$
	2	$(-0.10, 0.52, -0.85, -2.53, 0.17, 0.41)$
	wrist	$(0.05, 0.72, -0.94)$

15.7 Planar Serial Chains

We now specialize the kinematics equations defined above to the case of planar serial chains. It is convenient for our purposes to focus on chains consisting only of revolute joints, the nR chain.

The Plücker coordinates of the axis of a typical revolute joint in a planar chain are given by $\mathsf{J} = (\vec{k}, \mathbf{C} \times \vec{k})$, where $\vec{k} = (0,0,1)$ is directed along the z-axis of the base frame, and $\mathbf{C} = (c_x, c_y, 0)$ is the point of intersection of this axis with the x-y plane. The associated twist matrix $\hat{J}$ is

$$\hat{J} = \begin{bmatrix} 0 & -1 & 0 & -c_y \\ 1 & 0 & 0 & c_x \\ 0 & 0 & 0 & 0 \\ 0 & 0 & 0 & 0 \end{bmatrix}. \tag{15.87}$$

Let the transformation to the base of the chain be a translation by the vector $\mathbf{G} = (g_x, g_y, 0)$. Then he zero configuration of the nR planar chain has the points $\mathbf{C}_i$, $i = 1, \ldots, n$, on the joint axes J_i distributed along a line parallel to the x-axis Figure 15.6, such that

$$\mathbf{C}_1 = \begin{Bmatrix} g_x \\ g_y \\ 0 \end{Bmatrix}, \ \mathbf{C}_2 = \begin{Bmatrix} g_z + a_{12} \\ g_y \\ 0 \end{Bmatrix}, \ \ldots, \ \mathbf{C}_n = \begin{Bmatrix} g_x + a_{12} + a_{23} + \cdots + a_{n-1,n} \\ g_y \\ 0 \end{Bmatrix}. \tag{15.88}$$

Substituting these points into (15.87), we obtain a twist matrix $\hat{J}_i$ for each revolute joint and the product of exponentials kinematics equations

$$[D(\vec{\theta})] = e^{\Delta\theta_1 \hat{J}_1} e^{\Delta\theta_2 \hat{J}_2} \cdots e^{\Delta\theta_n \hat{J}_n} [D_0]. \tag{15.89}$$

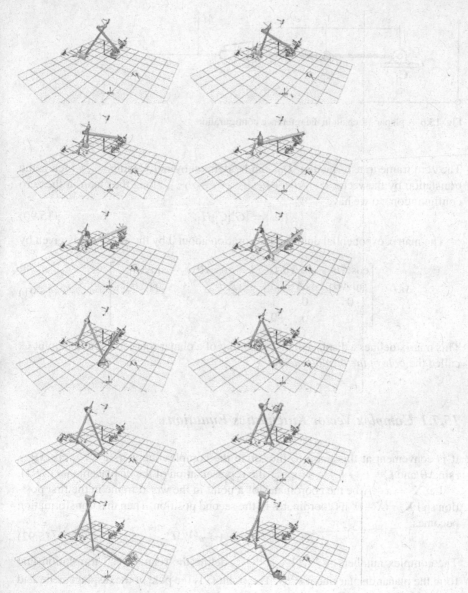

Fig. 15.5 A solution CCS serial chain.

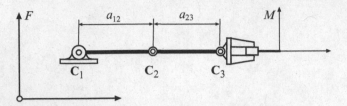

Fig. 15.6 A planar 3R chain in the reference configuration.

The zero frame transformation $[D_0]$ can be defined by introducing $[C]$, which is the translation by the vector $\mathbf{c} = (a_{12} + a_{23} + \cdots + a_{n-1,n})\vec{\imath}$ along the chain in the zero configuration, so we have

$$[D_0] = [G][C][H]. \tag{15.90}$$

The matrix exponential defining the rotation about J by the angle $\Delta\theta$ is given by

$$e^{\Delta\theta\hat{\jmath}} = \begin{bmatrix} \cos\Delta\theta & -\sin\Delta\theta & 0 & (1-\cos\Delta\theta)c_x + \sin\Delta\theta c_y \\ \sin\Delta\theta & \cos\Delta\theta & 0 & -\sin\Delta\theta c_x + (1-\cos\Delta\theta)c_y \\ 0 & 0 & 1 & 0 \\ 0 & 0 & 0 & 1 \end{bmatrix}. \tag{15.91}$$

This matrix defines a displacement consisting of a planar rotation about the point $\mathbf{C}$, called the *pole of the displacement*.

15.7.1 Complex Vector Kinematics Equations

It is convenient at this point to introduce the complex numbers $e^{i\Delta\theta} = \cos\Delta\theta + i\sin\Delta\theta$ and $\mathbf{C} = c_x + ic_y$ to simplify the representation of the displacement (15.91).

Let $\mathbf{X}_1 = x + iy$ be the coordinates of a point in the world frame in the first position and $\mathbf{X}_2 = X + iY$ its coordinates in the second position. Then this transformation becomes

$$\mathbf{X}_2 = e^{i\Delta\theta}\mathbf{X}_1 + (1 - e^{i\Delta\theta})\mathbf{C}. \tag{15.92}$$

The complex numbers $[e^{i\Delta\theta}, (1 - e^{i\Delta\theta})\mathbf{C}]$ define the rotation and translation that form the planar displacement $e^{\Delta\theta\hat{\jmath}}$. The point $\mathbf{C}$ is the pole of the displacement, and the translation vector $\mathbf{D}$ associated with this displacement is given by

$$\mathbf{D} = (1 - e^{i\Delta\theta})\mathbf{C}. \tag{15.93}$$

The composition of the exponentials $e^{\theta_1\hat{C}_1}$ and $e^{\theta_2\hat{C}_2}$ that define rotations about the points $\mathbf{C}_1$ and $\mathbf{C}_2$, respectively, yields

$$e^{\phi\hat{P}} = e^{\theta_1\hat{C}_1}e^{\theta_2\hat{C}_2}, \tag{15.94}$$

or

$$[e^{i\phi}, (1 - e^{i\phi})\mathbf{P}] = [e^{i\theta_1}, (1 - e^{i\theta_1})\mathbf{C}_1][e^{i\theta_2}, (1 - e^{i\theta_2})\mathbf{C}_2]$$
$$= [e^{i(\theta_1 + \theta_2)}, (1 - e^{i\theta_1})\mathbf{C}_1 + e^{i\theta_1}((1 - e^{i\theta_2})\mathbf{C}_2]. \qquad (15.95)$$

Here $\mathbf{P}$ denotes the pole of the composite displacement.

The complex form of the relative kinematics equations (15.9) is seen to be

$$[D(\Delta\vec{\theta})] = [e^{i\Delta\theta_1}, (1 - e^{i\Delta\theta_1})\mathbf{C}_1][e^{i\Delta\theta_2}, (1 - e^{i\Delta\theta_2})\mathbf{C}_2] \cdots [e^{i\Delta\theta_n}, (1 - e^{i\Delta\theta_n})\mathbf{C}_n]. \qquad (15.96)$$

If we define the relative displacement of the end-effector to be $[D] = [e^{i\Delta\phi}, (1 - e^{i\Delta\phi})\mathbf{P}]$, then we can expand this equation and equate the rotation and translation components to obtain

$$e^{i\Delta\phi} = e^{i\Delta\theta_1}e^{i\Delta\theta_2}\cdots e^{i\Delta\theta_n} = e^{i(\Delta\theta_1 + \Delta\theta_2 + \cdots + \Delta\theta_n)},$$
$$(1 - e^{i\Delta\phi})\mathbf{P} = (1 - e^{i\Delta\theta_1})\mathbf{C}_1 + e^{i\Delta\theta_1}(1 - e^{i\Delta\theta_2})\mathbf{C}_2 + \cdots$$
$$+ e^{i(\Delta\theta_1 + \Delta\theta_2 + \cdots + \Delta\theta_{n-1})}(1 - e^{i\Delta\theta_n})\mathbf{C}_n. \qquad (15.97)$$

These complex vector equations can be used to design planar nR serial chains. We will see shortly that they are exactly Sandor and Erdman's standard form equations. However, in the next section we introduce an equivalent set of design equations using the Clifford algebra form of the kinematics equations.

15.7.2 The Even Clifford Algebra $C^+(P^2)$

The even Clifford algebra of the projective plane P^2 is a generalization of complex numbers. It is a vector space with a product operation that is linked to a scalar product. The elements of this Clifford algebra can be identified with the complex vectors that define points in the plane, and with rotations and translations of these coordinates.

Using homogeneous coordinates of points in the projective plane as the vectors and a degenerate scalar product, we obtain an eight-dimensional Clifford algebra, $C(P^2)$. This Clifford algebra has an even subalgebra, $C^+(P^2)$, which is a set of four-dimensional elements of the form

$$A = a_1 i\varepsilon + a_2 j\varepsilon + a_3 k + a_4. \qquad (15.98)$$

The basis elements $i\varepsilon$, $j\varepsilon$, k and 1 satisfy the following multiplication table:

	$i\varepsilon$	$j\varepsilon$	k	1
$i\varepsilon$	0	0	$-j\varepsilon$	$i\varepsilon$
$j\varepsilon$	0	0	$i\varepsilon$	$j\varepsilon$
k	$j\varepsilon$	$-i\varepsilon$	-1	k
1	$i\varepsilon$	$j\varepsilon$	k	1

(15.99)

Notice that the set of Clifford algebra elements $\mathbf{z} = x + ky$ formed using the basis element k ($k^2 = -1$) is isomorphic to the usual set of complex numbers. This means that we have $e^{k\theta} = \cos\theta + k\sin\theta$.

Translation by the vector $\mathbf{d} = d_x + kd_y$ and rotation by the angle ϕ are represented by the Clifford algebra elements

$$T(\mathbf{d}) = 1 + \frac{1}{2}\mathbf{d}i\varepsilon \quad \text{and} \quad R(\phi) = e^{k\phi/2}, \tag{15.100}$$

and a general planar displacement $D = T(\mathbf{d})R(\phi)$ is given by

$$D = \left(1 + \frac{1}{2}\mathbf{d}i\varepsilon\right)e^{k\phi/2}. \tag{15.101}$$

A displacement defined to be a rotation by $\Delta\theta$ about a point $\mathbf{C}$ has the associated Clifford algebra element

$$D = \left(1 + \frac{1}{2}(1 - e^{k\Delta\theta})\mathbf{C}i\varepsilon\right)e^{k\Delta\theta/2}, \tag{15.102}$$

which is the Clifford algebra version of the matrix exponential (15.91). Expand this equation to obtain the four-dimensional vector

$$
\begin{aligned}
D &= \frac{1}{2}(e^{-k\Delta\theta/2} - e^{k\Delta\theta/2})\mathbf{C}i\varepsilon + e^{k\Delta\theta/2} \\
&= -\sin\frac{\Delta\theta}{2}\mathbf{C}j\varepsilon + e^{k\Delta\theta/2} \\
&= c_y\sin\frac{\Delta\theta}{2}i\varepsilon - c_x\sin\frac{\Delta\theta}{2}j\varepsilon + \sin\frac{\Delta\theta}{2}k + \cos\frac{\Delta\theta}{2}.
\end{aligned}
\tag{15.103}
$$

The components of this vector form the kinematic mapping used by Bottema and Roth [5] to study planar displacements. Also see Ravani and Roth [100].

15.7.3 Clifford Algebra Kinematics Equations

The relative kinematics equations of an nR planar chain (15.96) can be written in terms of the Clifford algebra elements (15.103) to define

$$-\sin\frac{\Delta\phi}{2}\mathbf{P}j\varepsilon + e^{k\Delta\phi/2} = \left(-\sin\frac{\Delta\theta_{1j}}{2}\mathbf{C}_1 j\varepsilon + e^{k\Delta\theta_1/2}\right)\left(-\sin\frac{\Delta\theta_2}{2}\mathbf{C}_2 j\varepsilon + e^{k\Delta\theta_2/2}\right)$$

$$\cdots \left(-\sin\frac{\Delta\theta_n}{2}\mathbf{C}_n j\varepsilon + e^{k\Delta\theta_n/2}\right). \tag{15.104}$$

Expand this equation and equate coefficients of the basis elements to obtain

$$e^{k\Delta\phi/2} = e^{k(\Delta\theta_1+\Delta\theta_2+\cdots+\Delta\theta_n)/2},$$

$$\sin\frac{\Delta\phi}{2}\mathbf{P} = \sin\frac{\Delta\theta_1}{2}\mathbf{C}_1 e^{-k(\Delta\theta_2+\cdots+\Delta\theta_n)/2} + e^{k\Delta\theta_1/2}\sin\frac{\Delta\theta_2}{2}\mathbf{C}_2 e^{-k(\Delta\theta_3+\cdots+\Delta\theta_n)/2}$$

$$+\cdots+ e^{k(\Delta\theta_1+\Delta\theta_2+\cdots+\Delta\theta_{n-1})/2}\sin\frac{\Delta\theta_n}{2}\mathbf{C}_n. \tag{15.105}$$

These equations are equivalent to the complex vector equations presented above. In fact, multiplication of (15.105) by $e^{k\Delta\phi/2}$ yields the equations (15.97). Note that we must replace k by the usual complex number i.

15.8 Design Equations for the Planar nR Chain

The goal of our design problem is to determine the dimensions of the planar nR chain that can position a tool held by its end-effector in a given set of task positions. The location of the base of the robot, the position of the tool frame, as well as the link dimensions and joint angles are considered to be design variables.

15.8.1 Relative Kinematics Equations for Specified Task Positions

Identify a set of planar task positions $[P_j]$, $j = 1,\ldots,m$. Then, the physical dimensions of the chain are defined by the requirement that for each position $[P_j]$ there be a joint parameter vector $\vec{\theta}_j$ such that the kinematics equations of the chain yield

$$[P_j] = [D(\vec{\theta}_j)], \quad i = 1,\ldots,m. \tag{15.106}$$

Now choose $[P_1]$ as the reference position and compute the relative displacements $[P_j][P_1]^{-1} = [P_{1j}]$, $j = 2,\ldots,m$. This formulation of the linkage design equations can be found in Suh and Radcliffe [134]. The result is the relative kinematics equations

$$[P_{1j}] = e^{\Delta\theta_{1j}\hat{J}_1} e^{\Delta\theta_{2j}\hat{J}_2}\cdots e^{\Delta\theta_{nj}\hat{J}_n}, \quad j = 2,\ldots,m, \tag{15.107}$$

where

$$\Delta\vec{\theta}_j = \vec{\theta}_j - \vec{\theta}_1 = (\Delta\theta_{1j},\ldots,\Delta\theta_{nj}).$$

The complex number form of (15.107) yields the equations

$$e^{i\Delta\phi_j} = e^{i(\Delta\theta_{1j}+\Delta\theta_{2j}+\cdots+\Delta\theta_{nj})},$$

$$(1 - e^{i\Delta\phi_j})\mathbf{P}_{1j} = (1 - e^{i\Delta\theta_{1j}})\mathbf{C}_1 + e^{i\Delta\theta_{1j}}(1 - e^{i\Delta\theta_{2j}})\mathbf{C}_2 + \cdots$$
$$+ e^{i(\Delta\theta_{1j}+\Delta\theta_{2j}+\cdots+\Delta\theta_{n-1,j})}(1 - e^{i\Delta\theta_{nj}})\mathbf{C}_n, \quad j = 2,\dots,m,$$

$$(15.108)$$

where $\Delta\phi_j = \phi_j - \phi_1$ and $\mathbf{P}_{1j}$ is the pole of the relative displacement $[P_{1j}]$. These are the equations we use to design the planar nR chain.

In terms of elements of the Clifford algebra we obtain the equivalent set of design equations,

$$e^{k\Delta\phi_j/2} = e^{k(\Delta\theta_{1j}+\Delta\theta_{2j}+\cdots+\Delta\theta_{nj})/2},$$

$$\sin\frac{\Delta\phi_j}{2}\mathbf{P}_{1j} = \sin\frac{\Delta\theta_{1j}}{2}\mathbf{C}_1 e^{-k(\Delta\theta_{2j}+\cdots+\Delta\theta_{nj})/2} + e^{k\Delta\theta_{1j}/2}\sin\frac{\Delta\theta_{2j}}{2}\mathbf{C}_2 e^{-k(\Delta\theta_{3j}+\cdots+\Delta\theta_{nj})/2}$$

$$+\cdots+ e^{k(\Delta\theta_{1j}+\Delta\theta_{2j}+\cdots+\Delta\theta_{n-1,j})/2}\sin\frac{\Delta\theta_{nj}}{2}\mathbf{C}_n, \quad j = 2,\dots,m.$$

$$(15.109)$$

The equations (15.109) allow the introduction of $\sin\frac{\Delta\theta_{ij}}{2}$ and $\cos\frac{\Delta\theta_{ij}}{2}$ as algebraic unknowns, so these equations can be solved for the various joint angles as well as the coordinates of the joints. This is demonstrated below in our algebraic solution of the five-position synthesis of a planar RR chain.

15.8.2 The Number of Design Positions and Free Parameters

If we specify m task positions, then equations (15.108) provide $m - 1$ rotation and $2(m-1)$ translation equations. The unknowns consist of the $n(m-1)$ relative joint angles and the $2n$ coordinates $\mathbf{C}_i, i = 1,\dots,n$.

It is useful to notice that the rotation equations are solved independently, which means that they determine $m - 1$ of the relative joint angles. Thus, we have $2(m-1)$ translation equations to solve for $(n-1)(m-1)$ joint variables and $2n$ coordinates $\mathbf{C}_i$, that is,

$$E = 2n + (n-1)(m-1) - 2(m-1) = m(n-3) + n + 3, \qquad (15.110)$$

where E excess of unknowns over equations.

Notice that except for $n = 1$ and $n = 2$ the excess of variables over equations is greater than zero. For $n = 1$, we see that $m = 2$ yields an exact formula for what is equivalent to the pole of a relative displacement. For $n = 2$, we find that an exact solution is possible for $m = 5$, which is Burmester's result that a RR chain can be designed to reach five specified positions (Burmester [7], Hartenberg and Denavit [46]).

Now consider the case $n = 3$, which has six unknown coordinates $\mathbf{C}_i$, $i = 1,2,3$, and $2(m-1)$ joint variables that are determined by $2(m-1)$ equations. The excess is $E = 6$ no matter how many positions are specified. In order to formulate this design problem, we specify the $m-1$ relative joint angles around $\mathbf{C}_1$. This is equivalent to adding $m-1$ design equations, which means that (15.110) takes the form $E = 6 - (m-1)$. The result is that given seven positions $m = 7$, we obtain a set of equations that determine the six coordinates $\mathbf{C}_i$, $i = 1,2,3$.

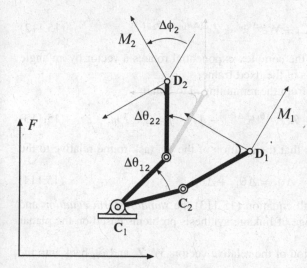

Fig. 15.7 Two positions of a planar RR chain.

15.8.3 The Standard-Form Equations

The synthesis of planar RR chains is the primary step in the design of four-bar linkages, which are constructed by joining the end links of two RR chains to form the floating link, or coupler. Specializing the relative kinematics equations (15.108) to this case, we obtain

$$e^{i\Delta\phi_j} = e^{i(\Delta\theta_{1j}+\Delta\theta_{2j})},$$
$$(1-e^{i\Delta\phi_j})\mathbf{P}_{1j} = (1-e^{i\Delta\theta_{1j}})\mathbf{C}_1 + e^{i\Delta\theta_{1j}}(1-e^{i\Delta\theta_{2j}})\mathbf{C}_2, \quad j = 2,\dots,m. \quad (15.111)$$

We now show that this is the standard-form equation used by Sandor and Erdman for planar mechanism synthesis.

The standard-form equation is obtained by equating the relative displacement vector between two positions to the difference of vectors along the chain in the two positions. See Figure 15.7. Let $\mathbf{C}_1$ be the fixed pivot and $\mathbf{C}_2$ the moving pivot when the tool frame of the RR chain is aligned with the first position.

Introduce the relative vectors $\mathbf{W} = \mathbf{C}_2 - \mathbf{C}_1$ and $\mathbf{Z} = \mathbf{D}_1 - \mathbf{C}_2$, where $\mathbf{D}_1$ is the translation vector to the first task position. We can now form the vector equations

$$\mathbf{D}_1 = \mathbf{C}_1 + \mathbf{W} + \mathbf{Z},$$
$$\mathbf{D}_2 = \mathbf{C}_1 + \mathbf{W}e^{i\Delta\theta_{12}} + \mathbf{Z}e^{i(\Delta\theta_{12}+\Delta\theta_{22})},$$
$$\ldots$$
$$\mathbf{D}_m = \mathbf{C}_1 + \mathbf{W}e^{i\Delta\theta_{1m}} + \mathbf{Z}e^{i(\Delta\theta_{1m}+\Delta\theta_{2m})}. \tag{15.112}$$

Recall that multiplication by the complex exponential rotates a vector by an angle measured relative to the x-axis of the fixed frame.

Subtract the first equation from the remaining m to obtain

$$\delta_{1j} = \mathbf{W}(e^{i\Delta\theta_{1j}} - 1) + \mathbf{Z}(e^{i(\Delta\theta_{1j}+\Delta\theta_{2j})} - 1), \quad j = 2,\ldots,m, \tag{15.113}$$

where $\delta_{1j} = \mathbf{D}_j - \mathbf{D}_1$. Notice that the rotation of the jth task frame relative to the first position is

$$\Delta\phi_j = \Delta\theta_{1j} + \Delta\theta_{2j}. \tag{15.114}$$

Sandor and Erdman [112] call equation (15.113) the *standard-form equation* and they use it to formulate a range of linkage synthesis problems based on the planar RR chain.

Now substitute the definition of the relative vectors $\mathbf{W}$, $\mathbf{Z}$, and δ_{ij} back into the standard-form equation to obtain

$$\mathbf{D}_j - \mathbf{D}_1 = (\mathbf{C}_2 - \mathbf{C}_1)(e^{i\Delta\theta_{1j}} - 1) + (\mathbf{D}_1 - \mathbf{C}_2)(e^{i(\Delta\theta_{1j}+\Delta\theta_{2j})} - 1),$$

and simplify to obtain

$$\mathbf{D}_j - \mathbf{D}_1 e^{i\Delta\phi_j} = (1 - e^{i\Delta\theta_{1j}})\mathbf{C}_1 + e^{i\Delta\theta_{1j}}(1 - e^{i\Delta\theta_{2j}})\mathbf{C}_2, \quad j = 1,\ldots,m. \tag{15.115}$$

In order to show that this equation is identical to (15.111), we compute the pole $\mathbf{P}_{1j}$ in terms of the translation vectors $\mathbf{D}_j$ and $\mathbf{D}_1$.

Let $[D_j] = [e^{i\phi_j}, \mathbf{D}_j]$, $j = 1,\ldots,m$, and compute

$$[D_{1j}] = [D_j][D_1]^{-1} = [e^{i(\phi_j-\phi_1)}, \mathbf{D}_j - \mathbf{D}_1 e^{i(\phi_j-\phi_1)}]. \tag{15.116}$$

Now the pole $\mathbf{P}_{1j}$ of this relative displacement is defined as the point that has the same coordinates before and after the displacement, which means that it satisfies the condition

$$\mathbf{P}_{1j} = e^{i(\phi_j-\phi_1)}\mathbf{P}_{1j} + \mathbf{D}_j - \mathbf{D}_1 e^{i(\phi_j-\phi_1)}. \tag{15.117}$$

Thus, we obtain

$$(1 - e^{i\Delta\phi_j})\mathbf{P}_{1j} = \mathbf{D}_j - \mathbf{D}_1 e^{\Delta\phi_j}, \tag{15.118}$$

and substituting this into (15.115), we find that the relative kinematics equations (15.111) are exactly Sandor and Erdman's standard-form equations.

15.8.4 Synthesis of 3R Serial Chains

The planar 3R robot has three degrees of freedom and can reach any set of positions within its workspace boundary. The design equations for m task positions take the form

$$e^{i\Delta\phi_j} = e^{i(\Delta\theta_{1j}+\Delta\theta_{2j}+\Delta\theta_{3j})},$$

$$(1 - e^{i\Delta\phi_j})\mathbf{P}_{1j} = (1 - e^{i\Delta\theta_{1j}})\mathbf{C}_1 + e^{i\Delta\theta_{1j}}(1 - e^{i\Delta\theta_{2j}})\mathbf{C}_2 + e^{i(\Delta\theta_{1j}+\Delta\theta_{2j})}(1 - e^{i\Delta\theta_{3j}})\mathbf{C}_3,$$

$$j = 2,\ldots,m. \qquad (15.119)$$

We consider the design of this chain for three, five, and seven task positions with the condition that the relative joint angles around $\mathbf{C}_1$ be specified by the designer.

15.8.4.1 Three Task Positions

If we specify three task positions, the result is four translation design equations, or two complex equations, which determine the six coordinates of $\mathbf{C}_i$ and the $2(3-1) = 4$ relative joint angles around $\mathbf{C}_1$ and $\mathbf{C}_2$. The joint angles around $\mathbf{C}_3$ are determined by the rotation design equations.

If we specify the four unknown relative joint angles and $\mathbf{C}_1$, then these four design equations are linear in the coordinates of $\mathbf{C}_2$ and $\mathbf{C}_3$. The result is two complex linear equations in two complex unknowns,

$$\kappa_{12} = e^{i\Delta\theta_{12}}(1 - e^{i\Delta\theta_{22}})\mathbf{C}_2 + e^{i(\Delta\theta_{12}+\Delta\theta_{22})}(1 - e^{i\Delta\theta_{32}})\mathbf{C}_3,$$

$$\kappa_{13} = e^{i\Delta\theta_{13}}(1 - e^{i\Delta\theta_{23}})\mathbf{C}_2 + e^{i(\Delta\theta_{13}+\Delta\theta_{23})}(1 - e^{i\Delta\theta_{33}})\mathbf{C}_3, \qquad (15.120)$$

where κ_{1j} are the known complex numbers

$$\kappa_{1j} = (1 - e^{i\Delta\phi_j})\mathbf{P}_{1j} - (1 - e^{i\Delta\theta_{1j}})\mathbf{C}_1. \qquad (15.121)$$

15.8.4.2 Five Task Positions

If five task positions are specified, then we have eight translation design equations in fourteen unknowns, the six coordinates $\mathbf{C}_i$ and eight relative joint angles. Now specify the coordinates of $\mathbf{C}_1$ and the four relative angles around it to define six parameters. The result is the four complex equations

$$\kappa_{12} = e^{i\Delta\theta_{12}}(1 - e^{i\Delta\theta_{22}})\mathbf{C}_2 + e^{i(\Delta\theta_{12}+\Delta\theta_{22})}(1 - e^{i\Delta\theta_{32}})\mathbf{C}_3,$$
$$\dots$$
$$\kappa_{15} = e^{i\Delta\theta_{15}}(1 - e^{i\Delta\theta_{25}})\mathbf{C}_2 + e^{i(\Delta\theta_{15}+\Delta\theta_{25})}(1 - e^{i\Delta\theta_{35}})\mathbf{C}_3, \qquad (15.122)$$

where κ_{1j} are known complex numbers defined by (15.121). These equations have exactly the same structure as Sandor and Erdman's standard-form equations (15.115) for five-position synthesis and are solved in the same way.

15.8.4.3 Seven task positions

If seven task positions are specified as well as the six relative joint angles around $\mathbf{C}_1$, then we obtain 12 translation design equations in the twelve unknowns consisting of the six joint coordinates $\mathbf{C}_i$, and six relative joint angles around $\mathbf{C}_2$. The result is six complex equations

$$(1 - e^{i\Delta\phi_2})\mathbf{P}_{12} = (1 - e^{i\Delta\theta_{12}})\mathbf{C}_1 + e^{i\Delta\theta_{12}}(1 - e^{i\Delta\theta_{22}})\mathbf{C}_2 + e^{i(\Delta\theta_{12}+\Delta\theta_{22})}(1 - e^{i\Delta\theta_{32}})\mathbf{C}_3,$$
$$\dots$$
$$(1 - e^{i\Delta\phi_7})\mathbf{P}_{17} = (1 - e^{i\Delta\theta_{17}})\mathbf{C}_1 + e^{i\Delta\theta_{17}}(1 - e^{i\Delta\theta_{27}})\mathbf{C}_2 + e^{i(\Delta\theta_{17}+\Delta\theta_{27})}(1 - e^{i\Delta\theta_{37}})\mathbf{C}_3.$$
$$(15.123)$$

This problem has been solved using homotopy continuation by Subbian and Flugrad [131].

15.8.5 Single DOF Coupled Serial Chains

Krovi et al. [59] expand the standard-form equations to nR chains in which the joints are coupled by cable transmissions so the system has one degree of freedom. They call the chain a *single-degree-of-freedom coupled serial chain*, or SDCSC. We formulate an equivalent form of their design equations using the relative kinematics equations (15.108).

Consider a planar nR serial chain in which each joint is connected to ground through a series of cables and pulleys located at each joint. Let each pulley have the same diameter and the cables routed through the links so they form parallelogram linkages. The result is n drive pulleys at the base of the chain that control the angle α_i of the ith link relative to the x-axis of the world frame, which means that each joint angle is given by

$$\theta_i = \alpha_i - \alpha_{i-1}. \qquad (15.124)$$

We now introduce a single drive angle β such that each joint angle is given by the relation $\theta_i = R_i\beta$, where R_i denotes a constant speed ratio. The relations (15.124) yield that the transmission matrix $[C]$ to the base drive angles is given by

$$\begin{Bmatrix} \alpha_1 \\ \alpha_2 \\ \vdots \\ \alpha_n \end{Bmatrix} = \begin{bmatrix} 1 & 0 & \cdots & 0 \\ 1 & 1 & \cdots & 0 \\ \vdots & & \cdots & \vdots \\ 1 & 1 & \cdots & 1 \end{bmatrix} \begin{bmatrix} R_1 \\ R_2, \\ \vdots \\ R_n \end{bmatrix} \beta, \qquad (15.125)$$

or

$$\vec{\alpha} = [C][R]\beta, \qquad (15.126)$$

where $[R]$ is the column matrix formed by the speed ratios. Our formulation differs slightly from Krovi et al. [59] in that we have added the drive variable β and therefore an additional speed ratio R_1.

Consider the design of an nR chain in which the speed ratios R_i, $i = 1, \ldots, n$, are specfied. Substitute these speed ratios into the rotation term of the design equations (15.108) to obtain

$$e^{i\Delta\phi_j} = e^{i(R_1+R_2+\cdots+R_m)\Delta\beta_j}, \quad j = 2, \ldots, m, \qquad (15.127)$$

where $\Delta\beta_j = \beta_j - \beta_1$ is the relative rotation of the drive angle. We find for each relative task position that

$$\Delta\beta_j = \frac{\Delta\phi_j}{R_1 + R_2 + \cdots + R_n}. \qquad (15.128)$$

Substitute this into the translation terms of (15.108) to define a linear equation in the coordinates $\mathbf{C}_i$, $i = 1, \ldots, n$, for each relative task position,

$$(1 - e^{i\Delta\phi_j})\mathbf{P}_{1j} = (1 - e^{iR_1\Delta\beta j})\mathbf{C}_1 + e^{iR_1\Delta\beta j}(1 - e^{iR_2\Delta\beta j})\mathbf{C}_2 + \cdots$$
$$+ e^{i(R_1+R_2+\cdots+R_{n-1})\Delta\beta j}(1 - e^{iR_n\Delta\beta j})\mathbf{C}_n, \quad j = 2, \ldots, m. \quad (15.129)$$

Given $m = n + 1$ task positions, we can solve these equations for the n complex unknowns $\mathbf{C}_i$. The result is a coupled serial nR chain designed to reach $n + 1$ arbitrarily specified task positions.

15.9 Algebraic Solution of the RR Design Equations

In this section we solve the standard-form equations for five-position synthesis using the Clifford algebra design equations given by

$$-\sin\frac{\Delta\phi_j}{2}\mathbf{P}_{1j}j\varepsilon + e^{k\Delta\phi_j/2}$$
$$= (-\sin\frac{\Delta\theta_{1j}}{2}\mathbf{C}_1 j\varepsilon + e^{k\Delta\theta_{1j}/2})(-\sin\frac{\Delta\theta_{2j}}{2}\mathbf{C}_2 j\varepsilon + e^{k\Delta\theta_{2j}/2}),$$
$$j = 2, 3, 4, 5. \quad (15.130)$$

For convenience we relabel the coordinates of the fixed and moving pivots so that $\mathbf{C}_1 = \mathbf{G} = (g_x, g_y)$ and $\mathbf{C}_2 = \mathbf{W} = (w_x, w_y)$. Similarly, label the joint angles $\theta_1 = \beta$ and $\theta_2 = \alpha$, then we have

$$-\sin\frac{\Delta\phi_j}{2}\mathbf{P}_{1j}j\varepsilon + e^{k\Delta\phi_j/2} = (-\sin\frac{\Delta\beta_j}{2}\mathbf{G}j\varepsilon + e^{k\Delta\beta_j/2})(-\sin\frac{\Delta\alpha_j}{2}\mathbf{W}j\varepsilon + e^{k\Delta\alpha_j/2}),$$

$$j = 2,3,4,5. \quad (15.131)$$

We now formulate these equations in component form.

15.9.1 Matrix Form of the RR Design Equations

Expand and collect the components of the Clifford algebra elements in the design equations (15.131) to define the arrays

$$\left\{\begin{array}{c} s\frac{\Delta\phi_j}{2}p_{xj} \\ s\frac{\Delta\phi_j}{2}p_{yj} \\ s\frac{\Delta\phi_j}{2} \\ c\frac{\Delta\phi_j}{2} \end{array}\right\} = \left\{\begin{array}{c} s\frac{\Delta\beta_j}{2}c\frac{\Delta\alpha_j}{2}g_y + c\frac{\Delta\beta_j}{2}s\frac{\Delta\alpha_j}{2}w_y - s\frac{\Delta\beta_j}{2}s\frac{\Delta\alpha_j}{2}(g_x - w_x) \\ -s\frac{\Delta\beta_j}{2}c\frac{\Delta\alpha_j}{2}g_x - c\frac{\Delta\beta_j}{2}s\frac{\Delta\alpha_j}{2}w_x - s\frac{\Delta\beta_j}{2}s\frac{\Delta\alpha_j}{2}(g_y - w_y) \\ s\frac{\Delta\beta_j}{2}c\frac{\Delta\alpha_j}{2} + c\frac{\Delta\beta_j}{2}s\frac{\Delta\alpha_j}{2} \\ c\frac{\Delta\beta_j}{2}c\frac{\Delta\alpha_j}{2} - s\frac{\Delta\beta_j}{2}s\frac{\Delta\alpha_j}{2} \end{array}\right\},$$

$$j = 2,3,4,5, \quad (15.132)$$

where s and c denote the sine and cosine functions. This equation can be written in the matrix form

$$\left\{\begin{array}{c} s\frac{\Delta\phi_j}{2}p_{xj} \\ s\frac{\Delta\phi_j}{2}p_{yj} \\ s\frac{\Delta\phi_j}{2} \\ c\frac{\Delta\phi_j}{2} \end{array}\right\} = \begin{bmatrix} g_y & w_y & w_x - g_x & 0 \\ -g_x & -w_x & w_y - g_y & 0 \\ 1 & 1 & 0 & 0 \\ 0 & 0 & -1 & 1 \end{bmatrix} \left\{\begin{array}{c} s\frac{\Delta\beta_j}{2}c\frac{\Delta\alpha_j}{2} \\ c\frac{\Delta\beta_j}{2}s\frac{\Delta\alpha_j}{2} \\ s\frac{\Delta\beta_j}{2}s\frac{\Delta\alpha_j}{2} \\ c\frac{\Delta\beta_j}{2}c\frac{\Delta\alpha_j}{2} \end{array}\right\},$$

$$j = 2,3,4,5. \quad (15.133)$$

This matrix equation can be inverted to define the joint variables in terms of the joint coordinates $\mathbf{G}$ and $\mathbf{W}$, that is,

$$\frac{1}{R^2}\begin{bmatrix} -(w_y - g_y) & w_x - g_x & \mathbf{W}\cdot(\mathbf{W} - \mathbf{G}) & 0 \\ w_y - g_y & -(w_x - g_x) & -\mathbf{G}\cdot(\mathbf{W} - \mathbf{G}) & 0 \\ w_x - g_x & w_y - g_y & g_x w_y - g_y w_x & 0 \\ w_x - g_x & \cdot\ w_y - g_y & g_x w_y - g_y w_x & R^2 \end{bmatrix} \left\{\begin{array}{c} s\frac{\Delta\phi_j}{2}p_{xj} \\ s\frac{\Delta\phi_j}{2}p_{yj} \\ s\frac{\Delta\phi_j}{2} \\ c\frac{\Delta\phi_j}{2} \end{array}\right\} = \left\{\begin{array}{c} s\frac{\Delta\beta_j}{2}c\frac{\Delta\alpha_j}{2} \\ c\frac{\Delta\beta_j}{2}s\frac{\Delta\alpha_j}{2} \\ s\frac{\Delta\beta_j}{2}s\frac{\Delta\alpha_j}{2} \\ c\frac{\Delta\beta_j}{2}c\frac{\Delta\alpha_j}{2} \end{array}\right\},$$

$$(15.134)$$

where R is the distance between the two joints, that is, $R^2 = (\mathbf{W} - \mathbf{G})^2$. This solves the inverse kinematics problem for the RR chain.

The joint variables can be eliminated from this equation using the identity

$$\frac{\sin \frac{\Delta \beta_j}{2} \sin \frac{\Delta \alpha_j}{2}}{\cos \frac{\Delta \beta_j}{2} \sin \frac{\Delta \alpha_j}{2}} = \frac{\sin \frac{\Delta \beta_j}{2} \cos \frac{\Delta \alpha_j}{2}}{\cos \frac{\Delta \beta_j}{2} \cos \frac{\Delta \alpha_j}{2}}. \tag{15.135}$$

The result is a set of four quadratic design equations

$$\begin{aligned} \mathscr{D}_j : \ & (g_x w_y - g_y w_x) c \frac{\Delta \phi_j}{2} + (g_x w_x + g_y w_y) s \frac{\Delta \phi_j}{2} \\ & + g_x \left(-p_{xj} c \frac{\Delta \phi_j}{2} + p_{yj} s \frac{\Delta \phi_j}{2} \right) + g_y \left(-p_{yj} c \frac{\Delta \phi_j}{2} - p_{xj} s \frac{\Delta \phi_j}{2} \right) \\ & + w_x \left(p_{xj} c \frac{\Delta \phi_j}{2} + p_{yj} s \frac{\Delta \phi_j}{2} \right) + w_y \left(p_{yj} c \frac{\Delta \phi_j}{2} - p_{xj} s \frac{\Delta \phi_j}{2} \right) \\ & + (p_{xj}^2 + p_{yj}^2) s \frac{\Delta \phi_j}{2} = 0, \quad j = 2, 3, 4, 5. \end{aligned} \tag{15.136}$$

These equations are bilinear in the unknowns g_x, g_y, w_x, and w_y, so we can collect their coefficients to define the linear equations

$$\begin{bmatrix} A_2 & B_2 & C_2 \\ \vdots & \vdots & \vdots \\ A_5 & B_5 & C_5 \end{bmatrix} \begin{Bmatrix} w_x \\ w_y \\ 1 \end{Bmatrix} = \begin{Bmatrix} 0 \\ \vdots \\ 0 \end{Bmatrix}, \tag{15.137}$$

where

$$\begin{aligned} A_j &= (p_{xj} - g_y) c \frac{\Delta \phi_j}{2} + (p_{yj} + g_x) s \frac{\Delta \phi_j}{2}, \\ B_j &= (p_{yj} + g_x) c \frac{\Delta \phi_j}{2} + (-p_{xj} + g_y) s \frac{\Delta \phi_j}{2}, \\ C_j &= (p_{xj}^2 + p_{yj}^2) s \frac{\Delta \phi_j}{2} - (g_x p_{xj} + g_y p_{yj}) c \frac{\Delta \phi_j}{2} + (g_x p_{yj} - g_y p_{xj}) s \frac{\Delta \phi_j}{2}. \end{aligned} \tag{15.138}$$

These equations are solved by computing the four 3×3 minors M_j obtained by removing row j. This yields the four polynomials $\mathscr{R}_j(g_x, g_y)$ obtained previously for the synthesis of planar RR chains. They are solved in the same way to determine as many as four design candidates for the RR chain $\mathbf{GW}$.

15.10 Summary

The exponential form of the kinematics equations of the chain are reformulated using Clifford algebra exponentials to obtain an efficient and systematic set of design equations. The design equations for a spatial serial chain are shown to be obtained as a specialization of the design equations for an nC serial chain. This process yields 130 equations in 130 unknowns for the design of a 5R chain that reaches 21 task positions. As an example, we formulate and solve the design equations for a CCS serial chain so that it reaches a 12-position task. The complete solution to these synthesis equations have not been achieved as yet.

15.11 References

This chapter is based on the work of Perez and McCarthy [97]. McCarthy [76] describes the use of Clifford algebras in kinematics that is a generalization of the complex number formulation of Sandor and Erdman [112]—also see Krovi et al. [59] and Perez and McCarthy [95]. Herve [48] used subgroups of the Lie group of rigid body displacements to formulate robotic systems with desired workspace properties. Wenger [150] describes the benefits of new serial chain topologies that allow reconfiguration within the workspace. Lee and Mavroidis [62, 63] formulate and solve the design equations for a 3R spatial chain that reaches four arbitrarily specified positions. Perez and McCarthy [96] formulate and solve the Clifford algebra synthesis equations for spatial serial chains.

Chapter 16
Platform Manipulators

In this chapter we consider six-degree-of-freedom systems consisting of a platform supported by multiple serial chains, called parallel, or platform, manipulators. Walking machines, mechanical fingers manipulating an object, and vehicle simulator platforms are all examples of platform manipulators. The Jacobian of these systems defines the contribution of each actuator to the resultant force and torque applied to the platform.

Our focus is on platform manipulators that have the property that each actuator generates a pure force acting on the platform. For these systems the columns of the Jacobian are the Plücker vectors of lines in space. And singular configurations are associated with linearly dependent sets of lines, that we called *line-based singularities*. Our goal is a geometric classification of these line-based singularities.

16.1 Introduction

Merlet [81] introduced a classification of line-based singularities, drawn from the work Dandurand [21] and others, in order to study the singular configurations of a triangular simplified symmetric manipulator, which has a structure similar to Figure 16.1. He was able to identify these configurations by inspection of the various ways in which the axes of the linear actuators could form linearly dependent sets of lines. A similar approach was used by Collins and Long [14] to analyze a pantograph-based hand-controller; also see the study by Notash [91] of uncertainty configurations in parallel manipulators.

This geometric method of analysis provides insight to the structure of the Jacobian matrix for these systems, which may often be complex, and can by-pass the computation of the determinant to identify singular configurations. It draws on classical results of line geometry, which can be found in Jessop [53], Salmon [111], and Woods [153], tailored to the features of these robotic systems.

J.M. McCarthy and G.S. Soh, *Geometric Design of Linkages*, Interdisciplinary Applied Mathematics 11, DOI 10.1007/978-1-4419-7892-9_16,
© Springer Science+Business Media, LLC 2011

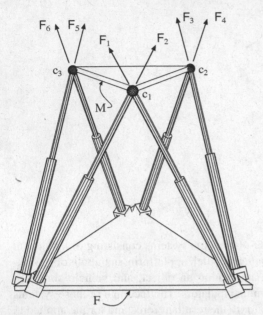

Fig. 16.1 An example of a Stewart platform. A singular configuration occurs when the lines F_i become linearly dependent.

16.1.1 Twists and Wrenches

The angular velocity **w** and linear velocity **d** of a moving body are three-dimensional vectors that can be assembled into a six-vector, called a *twist*. Similarly, the resultant force **f** and torque **t** acting on a body can be assembled into another six-vector, called a *wrench*, Figure 16.2.

The screw Jacobian of a serial chain defines the twist of the end-effector in terms of the partial twists at each joint of the chain. The principle of virtual work shows that this Jacobian also relates the wrench at the end-effector to the torque applied at each actuator.

16.1.1.1 Twists

The velocity of the end-link M of an open chain is obtained from its kinematics equations $[D(\bar{\theta})] = [A(\bar{\theta}), \mathbf{d}(\bar{\theta})]$ by computing

$$[S] = [\dot{D}(t)][D(t)^{-1}] = \begin{bmatrix} \dot{A}A^T & -\dot{A}A^T\mathbf{d} + \dot{\mathbf{d}} \\ 0\,0\,0 & 0 \end{bmatrix}. \tag{16.1}$$

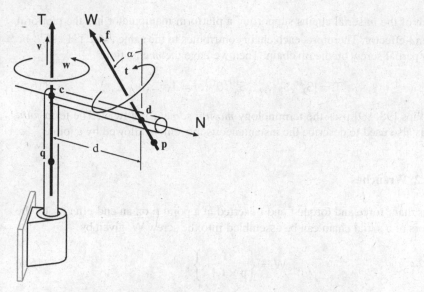

Fig. 16.2 A wrench $W = (\mathbf{f}, \mathbf{p} \times \mathbf{f} + \mathbf{t})^T$ acting on twist $T = (\mathbf{w}, \mathbf{q} \times \mathbf{w} + \mathbf{v})^T$.

This matrix is termed an element of the Lie algebra so(3) by Murray et al. [88]. From the components of $[S]$ we construct the angular velocity vector $\mathbf{w}$ and linear velocity $\mathbf{v} = \dot{\mathbf{d}}(t)$. The six-vector $T = (\mathbf{w}, \mathbf{q} \times \mathbf{w} + \mathbf{v})^T$ is called the *twist* of the motion.

This matrix takes a particularly simply form for the link transformation matrix $[T_i] = [Z(\theta_i, d_i)][X(\alpha_{ij}, a_{ij})]$ used in the kinematics equations for a serial chain. Assume for the moment that both θ_i and d_i are variable, then we have

$$[S_i] = ([\dot{Z}_i][X_{ij}])([Z_i][X_{ij}])^{-1} = \begin{bmatrix} 0 & -\dot{\theta}_i & 0 & 0 \\ \dot{\theta}_i & 0 & 0 & 0 \\ 0 & 0 & 0 & \dot{d}_i \\ 0 & 0 & 0 & 0 \end{bmatrix}. \tag{16.2}$$

From this equation we see that the twist associated with a revolute joint is $S_i = (\dot{\theta}_i \vec{k}, 0)^T$ and for a prismatic joint it is $S_i = (0, \dot{d}_i \vec{k})^T$.

The kinematics equations of a robot arm allow the expansion of (16.1) in terms of the *partial twists* S_j associated with each joint parameter θ_j, given by

$$T = \left\{ \begin{matrix} \mathbf{w} \\ \mathbf{q} \times \mathbf{w} + \mathbf{v} \end{matrix} \right\} = [S_1, S_2, \ldots, S_6]\dot{\vec{\theta}}, \tag{16.3}$$

where

$$S_j = [\hat{T}_1] \cdots [\hat{T}_j] K. \tag{16.4}$$

Note that here $K = (\vec{k}, 0)^T$ for a revolute joint and $K = (0, \vec{k})^T$ for a prismatic joint.

Each of the m serial chains supporting a platform manipulator has the platform as its end-effector. Therefore, each chain contributes to the same twist T. Let $S_j^{(i)}$ be the jth partial screw on the ith chain. Then we have the m equations

$$T = [S_1^{(i)}, S_2^{(i)}, \dots, S_6^{(i)}] \dot{\theta}^{(i)}, \, i = 1, \dots, m. \tag{16.5}$$

Phillips [98, 99] uses the terminology *motion screw* for a twist. The term *joint screw* is also used to describe the instantaneous movement allowed by a joint.

16.1.1.2 Wrenches

The resultant force and torque $\mathbf{f}$ and $\mathbf{t}$ exerted at a point $\mathbf{p}$ on an end-effector by the actuators of a serial chain can be assembled into the screw W, given by

$$W = \left\{ \begin{matrix} \mathbf{f} \\ \mathbf{p} \times \mathbf{f} + \mathbf{t} \end{matrix} \right\}, \tag{16.6}$$

called a *wrench*. A wrench for which $\mathbf{t} = 0$ is a pure force. In this case $F = (\mathbf{f}, \mathbf{p} \times \mathbf{f})^T$ is the Plücker vector of the line of action of the force. The screw $M = (0, \mathbf{t})^T$ is a pure torque.

The total wrench W applied to a platform supported by m serial chains is the sum of the individual wrenches

$$W = \sum_{i=1}^{m} W^{(i)}. \tag{16.7}$$

We use the term *actuator screw* for wrenches that represent the force-torque contribution of an actuator. Philips refers to the normalized version of this wrench as an *action screw*.

16.1.2 Virtual Work

The work done by a wrench $W = (\mathbf{f}, \mathbf{p} \times \mathbf{f} + \mathbf{t})^T$ as it moves through a twist $T = (\mathbf{w}, \mathbf{q} \times \mathbf{w} + \mathbf{v})^T$ over a virtual time period δt is given by

$$\delta W = \big(\mathbf{f} \cdot (\mathbf{q} \times \mathbf{w} + \mathbf{v}) + (\mathbf{p} \times \mathbf{f} + \mathbf{t}) \cdot \mathbf{w} \big) \delta t$$
$$= \big(\mathbf{f} \cdot \mathbf{v} + \mathbf{t} \cdot \mathbf{w} - (\mathbf{p} - \mathbf{q}) \cdot (\mathbf{w} \times \mathbf{f}) \big) \delta t = \mathscr{W} \delta t. \tag{16.8}$$

The instantaneous quantity $\mathscr{W}$ is sometimes called the infinitesimal work or the rate of work done. We call it the *virtual work* of W acting on T with the understanding that it is associated with a virtual time period.

To simplify the computation of virtual work, we follow Kumar [60] and use the 6×6 matrix $[\Pi]$ (12.34) to interchange the vector components of a screw, that is,

$$[\Pi]\mathsf{S} = \begin{bmatrix} 0 & I \\ I & 0 \end{bmatrix} \begin{Bmatrix} \mathbf{s} \\ \mathbf{u} \end{Bmatrix} = \begin{Bmatrix} \mathbf{u} \\ \mathbf{s} \end{Bmatrix}. \tag{16.9}$$

Lipkin and Duffy [68] describe this as the transformation of a screw from ray coordinates to axial coordinates.

For convenience, we introduce the notation $\check{\mathsf{T}} = [\Pi]\mathsf{T}$ to simplify the equations that follow. This notation allows us to write the virtual work in the form

$$\mathsf{W}^T \check{\mathsf{T}} = \mathbf{f} \cdot \mathbf{v} + \mathbf{t} \cdot \mathbf{w} - (\mathbf{p} - \mathbf{q}) \cdot (\mathbf{w} \times \mathbf{f}). \tag{16.10}$$

If the virtual work of a wrench acting on a twist is zero, then the two screws are said to be *reciprocal*.

Consider for example a pure force applied to a body connected to ground by a revolute joint. In this case we have $\mathsf{F} = (\mathbf{f}, \mathbf{p} \times \mathbf{f})^T$ and the twist $\mathsf{S} = (\dot{\theta}\mathbf{S}, \dot{\theta}\mathbf{q} \times \mathbf{S})^T$. The virtual work is

$$\mathsf{F}^T \check{\mathsf{S}} = -\dot{\theta}(\mathbf{p} - \mathbf{q}) \cdot (\mathbf{S} \times \mathbf{f}). \tag{16.11}$$

Notice that this virtual work is zero if the lines F and S intersect, or are parallel.

16.2 The Jacobian of a Platform Manipulator

The Jacobian for a single serial chain manipulator is defined by Craig [15] and Murray et al. [88]. See (11.57). For our purposes we use the closely related matrix (11.55) that defines the twist $\check{\mathsf{T}}$ as

$$\check{\mathsf{T}} = [\check{\mathsf{S}}_1, \check{\mathsf{S}}_2, \ldots, \check{\mathsf{S}}_6] = [J]\dot{\theta}. \tag{16.12}$$

This is termed the *screw Jacobian* by Tsai [136]. The difference between this matrix and the usual Jacobian lies solely in the choice of reference point for the partial twists. The screw Jacobian uses the origin of the fixed frame F rather than the origin of the moving frame M.

The principle of virtual work yields the important result that the Jacobian also relates joint torques $\bar{\tau} = (\tau_1, \ldots, \tau_6)^T$ of a serial chain to the resultant wrench W on the end-effector. To see this we compute the virtual work done by each of the actuators $\dot{\theta}_i \tau_i \delta t$ and equate their sum to the virtual work done by W on the end-effector twist T. The result is

$$\bar{\tau}^T \dot{\theta} \delta t = \mathsf{W}^T \check{\mathsf{T}} \delta t. \tag{16.13}$$

Substitute (16.12) into this equation and equate the coefficients of $\dot{\theta}\delta t$ to obtain

$$\bar{\tau} = [J^T]\mathsf{W}. \tag{16.14}$$

For a platform manipulator supported by m serial chains, we have the set of equations

$$\breve{\mathsf{T}} = \left[\breve{\mathsf{S}}_1^{(i)}, \breve{\mathsf{S}}_2^{(i)}, \ldots, \breve{\mathsf{S}}_6^{(i)} \right] \dot{\theta}^{(i)} = [J^{(i)}] \dot{\theta}^{(i)}, \, i = 1, \ldots, m, \tag{16.15}$$

where T, the twist of the platform, is the same for each serial chain. Let the resultant wrench applied by the ith supporting chain of a platform manipulator be $\mathsf{W}^{(i)}$. Then we have the joint torques $\bar{\tau}^{(i)} = (\tau_1^{(i)}, \ldots, \tau_6^{(i)})^T$, $i = 1, \ldots, m$, given by

$$\bar{\tau}^{(1)} = [J^{(1)T}]\mathsf{W}^{(1)}, \, \bar{\tau}^{(2)} = [J^{(2)T}]\mathsf{W}^{(2)}, \ldots, \bar{\tau}^{(m)} = [J^{(m)T}]\mathsf{W}^{(m)}. \tag{16.16}$$

Invert each of these equations to determine the applied wrench $\mathsf{W}^{(i)}$ in terms of the applied joint torques $\bar{\tau}^{(i)}$, that is,

$$\mathsf{W}^{(i)} = [J^{(i)T}]^{-1}\bar{\tau}^{(i)} = [\mathsf{F}_1^{(i)}, \mathsf{F}_2^{(i)}, \ldots, \mathsf{F}_6^{(i)}]\bar{\tau}^{(i)}. \tag{16.17}$$

The wrench $\mathsf{F}_j^{(i)}$ represents the contribution of the jth actuator of the ith chain.

The resultant force and torque applied to the platform is the sum

$$\mathsf{W} = \sum_{i=1}^{m} \mathsf{W}^{(i)} = \left[[J^{(1)T}]^{-1}, [J^{(2)T}]^{-1}, \ldots, [J^{(m)T}]^{-1} \right] \left\{ \begin{matrix} \bar{\tau}^{(1)} \\ \vdots \\ \bar{\tau}^{(m)} \end{matrix} \right\}. \tag{16.18}$$

Substitute $[J^{(i)T}]^{-1}$ from (16.17) into this to obtain

$$\mathsf{W} = \left[\mathsf{F}_1^{(1)}, \mathsf{F}_2^{(1)}, \ldots, \mathsf{F}_6^{(1)}; \ldots; \mathsf{F}_1^{(m)}, \mathsf{F}_2^{(m)}, \ldots, \mathsf{F}_6^{(m)} \right] \left\{ \begin{matrix} \bar{\tau}^{(1)} \\ \vdots \\ \bar{\tau}^{(6)} \end{matrix} \right\}, \tag{16.19}$$

or

$$\mathsf{W} = [\Gamma]\bar{\tau}. \tag{16.20}$$

The matrix $[\Gamma]$ is called the Jacobian of the platform manipulator. The transpose of $[\Gamma]$ defines the joint rates of the manipulator in terms of the desired twist.

The wrench W depends on the configuration of each of the serial chains supporting the platform manipulator. If the rank of $[\Gamma]$ is less than six then the manipulator is said to be in a *singular* configuration.

16.3 Conditions for Line-Based Singularities

We now focus our attention on platform manipulators that have a total of six actuators, and introduce conditions that ensure that the actuator screws are pure forces. For these systems it is possible to characterize singular configurations in terms of the geometry of linearly dependent sets of lines. We call these *line-based singularities*. For serial chains consisting of hinges and sliders, we find a convenient way to determine the lines of action of these forces.

Each arm supporting the platform of the manipulator system is assumed to have the structure of a serial robotic arm that can allow six-degree-of-freedom movement of the platform. We further assume that each arm has at least one joint at which a nonzero torque is applied. Otherwise, it does not contribute to defining singular configurations. Thus, our platforms are supported by at most six serial chains.

Finally, we require that each of the m the serial chains has an unactuated spherical wrist. The center c_i of this wrist is considered to be the point of attachment of the ith chain to the platform. This ensures that each chain applies only a pure force to the platform, which means that no column of the Jacobian $[\Gamma]$ has a torque term. Thus, each column is the Plücker vector of a line. Systems that satisfy these conditions must have at least three supporting serial chains to resist an external force and torque, and $3 \le m \le 6$.

These kinematic conditions are typical of many platform manipulator systems, particularly those based on the 6TPS Stewart platform, for example Figure 16.1. Figure 16.3 shows the structure and actuation schemes for platform manipulators that have purely line-based singularities. The point of attachment on the platform and base can be generally located, and it is possible to apply the actuator forces with one or more serial chains. Figure 16.4 shows a 3RRRS platform manipulator that has line-based singularities.

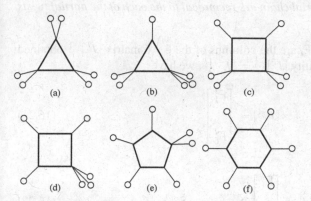

Fig. 16.3 The six basic designs and actuation schemes for platform manipulators with line-based singularities. The circles denote the actuator forces of supporting chains.

16.3.1 Locating the Lines of Action

If the first three joints of the supporting serial chain are constructed using revolute or prismatic joints, then their axes combine with the attachment point c to locate the axes of the forces F_i, $i = 1, 2, 3$, of the chain. This makes it possible to determine the line of action of these forces by inspection.

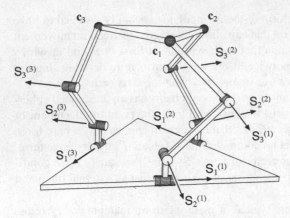

Fig. 16.4 The 3-RRRS platform manipulator has line-based singularities.

To show this, we first derive a fundamental relationship between the actuator screws F_i and joint screws S_j for a general six-degree-of-freedom serial chain.

Theorem 19 (The Reciprocal Screw Theorem). *The actuator screws F_i of a six-degree-of-freedom spatial serial chain are reciprocal to the each of the partial twists S_j for $i \neq j$.*

Proof. The actuator screws F_i are the columns of the 6×6 matrix $[J^T]^{-1}$ obtained from (16.14). Using the identity $[J^T]^{-1} = [J^{-1}]^T$, we have

$$[J^{-1}] = \begin{bmatrix} F_1^T \\ \vdots \\ F_6^T \end{bmatrix}. \tag{16.21}$$

Now compute

$$[J^{-1}][J] = \begin{bmatrix} F_1^T \\ \vdots \\ F_6^T \end{bmatrix} [\check{S}_1, \check{S}_2, \dots, \check{S}_6] = [I]. \tag{16.22}$$

Thus, $F_i^T \check{S}_j = 0$ for $i \neq j$. Each actuator screw F_i generates zero virtual work when it acts on the joint screw S_j for $i \neq j$, that is, these screws are reciprocal. $\square$

In our case the forces $F_i = (\mathbf{f}, \mathbf{p} \times \mathbf{f})^T$ and axes $S_j = (\mathbf{s}, \mathbf{r} \times \mathbf{s})^T$ are the Plücker vectors of two lines. This means that the condition that these screws are reciprocal $F_i^T \check{S}_j = 0$ is also the condition (16.11) that the lines intersect, or are parallel.

Consider the line of action of F_1, which is generated by an actuator at the first joint. It must intersect the five axes $S_2, \dots, S_6$. Notice that for our systems the last three axes intersect in the attachment point $\mathbf{c}$. Each line F_i must pass through this point. Furthermore, consider the plane defined by $\mathbf{c}$ and S_3. The axis S_2 intersects this plane in a point $\mathbf{p}$. The line joining $\mathbf{c}$ and $\mathbf{p}$ is uniquely determined, and must be the line of action of F_1. It may happen that S_2 is parallel to the plane defined by S_3 and **c**. In this case F_1 is the line in the plane through **c** parallel to S_2, which is said to intersect S_2 at infinity. A similar construction yields the line of action for the force generated by each actuator of the chain.

16.4 Classification of Line-Based Singularities

A platform manipulator is in a singular configuration when any one of the lines of action of the forces on the platform becomes linearly dependent on the others. We say that the singularity is of *type n* if this line F is dependent on no fewer than n other lines. In what follows, we describe the geometry of linearly dependent sets of lines that define the singular configurations for platform manipulators. Merlet's notation (Merlet [81]) is used for the general classes, though we include additional subcases. Figures 16.5 and 16.6 provide illustrations of these various distributions of lines.

A platform manipulator with six actuated joints has a Jacobian of the form

$$W = [F_1, F_2, \ldots, F_6] \begin{Bmatrix} \tau_1 \\ \vdots \\ \tau_6 \end{Bmatrix} = [\Gamma] \bar{\tau}, \qquad (16.23)$$

where the actuated joints are now numbered $i = 1, \ldots, 6$. Notice that if the platform meets the conditions above, then the columns of this Jacobian have the form $F_i = (\mathbf{f}_i, \mathbf{p}_i \times \mathbf{f}_i)^T$ which are the Plücker vectors of the forces applied to the platform. These actuator screws define the line of action of the forces and have the property that each is reciprocal to itself, that is, $F_i^T \check{F}_i = 0$.

16.4.1 Type-1 Singularities

If any one of the six actuator screws F in a platform manipulator is linearly dependent on one of the remaining five, denoted by F_1, then the system is in a type-1 singularity. This is equivalent to the condition that

$$\mathscr{T}_1: \quad F = k_1 F_1 \qquad (16.24)$$

for some scalar k_1. It is easy to see that for this to occur F and F_1 must define the same line.

16.4.2 Type-2 Singularities

A type-2 singularity occurs when one of the actuator screws F is linearly dependent
on two others, denoted by F_1 and F_2, and not on either one independently. This
means that nonzero scalars k_1 and k_2 exist such that

$$\mathscr{T}_2: \quad F = k_1 F_1 + k_2 F_2. \tag{16.25}$$

In general, this equation defines a set of screws known as a *two-system*. However,
we are concerned only with actuator screws F that are Plücker vectors of a line.
Thus, we have the additional requirement

$$F^T \check{F} = k_1^2 (F_1^T \check{F}_1) + 2k_1 k_2 (F_1^T \check{F}_2) + k_2^2 (F_2^T \check{F}_2) = 0. \tag{16.26}$$

Since F_1 and F_2 are lines, the terms $F_i^T \check{F}_i$ are zero, and we find that F cannot be a line
unless $F_1^T \check{F}_2 = 0$. This means that the two lines F_1 and F_2 must intersect, (16.11).
When this is true, the lines of $\mathscr{T}_2$ lie in the plane defined by F_1 and F_2 and pass
through their point of intersection. This is known as a *pencil* of lines.

16.4.2.1 Type 2a

Two skew lines are often identified as a linearly dependent set of lines, however,
because two skew lines cannot generate a third line that is linearly dependent on
both of them, they are not, strictly speaking, a type-2 singularity. A third line must
actually coincide with one of the two lines, as in the type-1 singularity.

16.4.2.2 Type 2b

A type-2 singularity requires that F_1 and F_2 intersect in a point **p**. In this case the
two lines define a plane, and $\mathscr{T}_2$ is the pencil of lines in this plane through **p**. There
are two cases.

1. If F_1 and F_2 are not parallel, then **p** is a finite point and all the lines of $\mathscr{T}_2$ pass
   through **p**.
2. If F_1 and F_2 are parallel, then they are said to intersect at infinity and $\mathscr{T}_2$ consists
   of the lines parallel to F_1 and F_2.

16.4.3 Type-3 Singularities

When one actuator screw F is dependent on three other actuator screws F_1, F_2,
and F_3, the platform manipulator is in a type-3 singularity. This occurs when k_i,
$i = 1, 2, 3$, exist, so F is given by

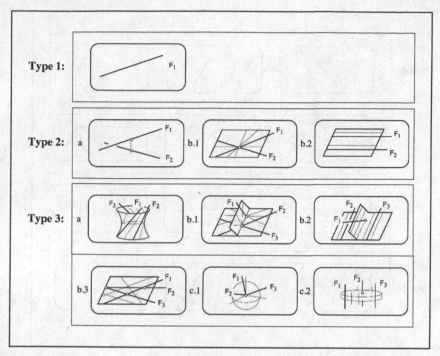

Fig. 16.5 A classification of sets of lines linearly dependent on one, two, and three given lines.

$$\mathscr{T}_3 : \quad \mathsf{F} = k_1 \mathsf{F}_1 + k_2 \mathsf{F}_2 + k_3 \mathsf{F}_3. \tag{16.27}$$

The requirement that F also be a line yields the relation

$$\mathsf{F}^T \check{\mathsf{F}} = 2k_1 k_2 (\mathsf{F}_1^T \check{\mathsf{F}}_2) + 2k_1 k_3 (\mathsf{F}_1^T \check{\mathsf{F}}_3) + 2k_2 k_3 (\mathsf{F}_2^T \check{\mathsf{F}}_3) = 0. \tag{16.28}$$

Notice that we have dropped the terms $\mathsf{F}_i^T \check{\mathsf{F}}_i = 0$. Equation (16.28) is easily solved to obtain

$$k_3 = \frac{-k_1 k_2 (\mathsf{F}_1^T \check{\mathsf{F}}_2)}{k_1 (\mathsf{F}_1^T \check{\mathsf{F}}_3) + k_2 (\mathsf{F}_2^T \check{\mathsf{F}}_3)}. \tag{16.29}$$

This result combines with (16.27) to define $\mathscr{T}_3$ as a one-dimensional set of lines that is known to be a quadric surface $\mathscr{Q}$.

Another view of this quadric $\mathscr{Q}$ is obtained by considering the lines $\mathsf{L} = (\mathbf{s}, \mathbf{r} \times \mathbf{s})^T$ that intersect F_1, F_2, and F_3, given by the matrix equation

$$[\mathsf{F}_1, \mathsf{F}_2, \mathsf{F}_3]^T \check{\mathsf{L}} = 0. \tag{16.30}$$

Any line L satisfying this equation intersects all the lines of $\mathscr{T}_3$ that form $\mathscr{Q}$, and therefore must lie on $\mathscr{Q}$. This provides a convenient way to derive its algebraic expression in terms of the point coordinates $\mathbf{r} = (x, y, z)^T$. Given the lines $\mathsf{F}_i = (\mathbf{f}_i, \mathbf{p}_i \times \mathbf{f}_i)$, we write (16.30) in the form

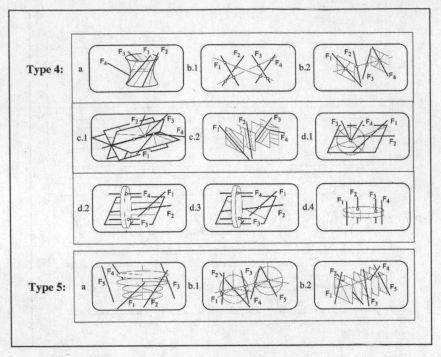

Fig. 16.6 A classification of sets of lines linearly dependent on four and five given lines.

$$\begin{bmatrix} \left((\mathbf{p}_1 - \mathbf{r}) \times \mathbf{f}_1\right)^T \\ \left((\mathbf{p}_2 - \mathbf{r}) \times \mathbf{f}_2\right)^T \\ \left((\mathbf{p}_3 - \mathbf{r}) \times \mathbf{f}_3\right)^T \end{bmatrix} \begin{Bmatrix} s_x \\ s_y \\ s_z \end{Bmatrix} = 0. \qquad (16.31)$$

This equation has a solution for $\mathbf{s} = (s_x, s_y, s_z)^T$ only if the determinant of the coefficient matrix is zero. This determinant can be expressed as the triple product

$$\mathscr{Q}: \ \left((\mathbf{p}_1 - \mathbf{r}) \times \mathbf{f}_1\right) \cdot \left((\mathbf{p}_2 - \mathbf{r}) \times \mathbf{f}_2\right) \times \left((\mathbf{p}_3 - \mathbf{r}) \times \mathbf{f}_3\right)) = 0. \qquad (16.32)$$

The cubic terms cancel because $(\mathbf{r} \times \mathbf{f}_1) \cdot (\mathbf{r} \times \mathbf{f}_2) \times (\mathbf{r} \times \mathbf{f}_3) = 0$, and (16.32) is the equation of the quadric surface $\mathscr{Q}$. The lines L and F define the two separate sets of rulings on this quadric, known as *reguli*.

16.4.3.1 Type 3a

In general, the set of lines $\mathscr{T}_3$ is the regulus of lines F defined by (16.29) lying on the quadric $\mathscr{Q}$ defined by (16.32).

16.4.3.2 Type 3b

If any two of the three lines F_1, F_2, and F_3 intersect, then the quadric $\mathscr{Q}$ degenerates to a pair of planes. To see this, let $\mathbf{p} = \mathbf{p}_2 = \mathbf{p}_3$. Then (16.32) takes the form

$$\big((\mathbf{r}-\mathbf{p}) \cdot (\mathbf{p}-\mathbf{p}_1) \times \mathbf{f}_1\big)\big((\mathbf{r}-\mathbf{p}) \cdot \mathbf{f}_2 \times \mathbf{f}_3\big) = 0, \qquad (16.33)$$

which is the product of two linear equations in the coordinates of $\mathbf{r} = (x, y, z)^T$. These equations define the two planes

$$\mathscr{P}_1 : (\mathbf{r}-\mathbf{p}) \cdot (\mathbf{p}-\mathbf{p}_1) \times \mathbf{f}_1 = 0,$$
$$\mathscr{P}_2 : (\mathbf{r}-\mathbf{p}) \cdot \mathbf{f}_2 \times \mathbf{f}_3 = 0. \qquad (16.34)$$

We can distinguish three cases:

1. If the planes are distinct and not parallel, then their intersection is a finite line, denoted by L. This line must pass through the point of intersection $\mathbf{p}$ of F_2 and F_3 and intersect F_1 in a point $\mathbf{a}$. The set $\mathscr{T}_3$ consists of all the lines through $\mathbf{a}$ in $\mathscr{P}_1$ and all those through $\mathbf{p}$ in $\mathscr{P}_2$. This case includes the situation when two of the lines lines, say F_1 and F_2, both intersect the third line F_3. The two planar pencils are defined by the pairs of lines F_1 and F_3, and F_2 and F_3.
2. If the planes are distinct and parallel, then their intersection L is at infinity. This occurs when two of the three lines are parallel. In this case $\mathscr{T}_3$ is two planes of parallel lines.
3. If the three lines lie in one plane, then the quadric degenerates to two coincident planes. In this case, $\mathscr{T}_3$ consists of all lines in that plane.

16.4.3.3 Type 3c

If the three lines F_1, F_2, and F_3 intersect in the same point $\mathbf{p} = \mathbf{p}_1 = \mathbf{p}_2 = \mathbf{p}_3$, then the quadric degenerates to a bundle of lines through $\mathbf{p}$. There are two cases depending on whether this point is finite or at infinity.

1. If the point of intersection $\mathbf{p}$ is not at infinity, then equation $\mathscr{T}_3$ is the set of all lines through $\mathbf{p}$.
2. If the three lines are parallel, then the point of intersection $\mathbf{p}$ is at infinity. Then all lines in space parallel to these three lines form $\mathscr{T}_3$.

16.4.4 Type-4 Singularities

The type-4 singularities occur when one actuator screw F is linearly dependent on no fewer than four other actuator screws, that is,

$$\mathscr{T}_4 : \quad F = k_1 F_1 + k_2 F_2 + k_3 F_3 + k_4 F_4. \qquad (16.35)$$

This is a two-dimensional set of lines known as a *linear congruence*.

We analyze this set of lines by considering the screws $S = (s, u)^T$ that are reciprocal to all four lines F_i, $i = 1, 2, 3, 4$, and therefore are reciprocal to the entire set $\mathcal{T}_4$. The screws S must satisfy the four linear conditions $S^T \check{F}_i = 0$, $i = 1, 2, 3, 4$, which we write in matrix form as

$$[\check{F}_1, \check{F}_2, \check{F}_3, \check{F}_4]^T S = 0. \tag{16.36}$$

Let $[A]$ be the 4×4 submatrix formed from the first four columns of the 4×6 coefficient matrix in (16.36), and let b_1 and b_2 be the fifth and sixth columns, so we have $[\check{F}_1, \check{F}_2, \check{F}_3, \check{F}_4]^T = [A, b_1, b_2]$. Equation (16.36) can now be solved to determine

$$S_1 = \left\{ \begin{matrix} -[A^{-1}]b_1 \\ 1 \\ 0 \end{matrix} \right\} \quad \text{and} \quad S_2 = \left\{ \begin{matrix} -[A^{-1}]b_2 \\ 0 \\ 1 \end{matrix} \right\}. \tag{16.37}$$

Actually, any linear combination $L = sS_1 + tS_2$ of these two screws will satisfy (16.36). This is known as *two-system* of screws.

We now consider whether or not the two-system spanned by S_1 and S_2 contains any lines. This is determined by the roots of the quadratic equation

$$L^T \check{L} = s^2 (S_1^T \check{S}_1) + 2st (S_1^T \check{S}_2) + t^2 (S_2^T \check{S}_2) = 0. \tag{16.38}$$

If the roots are imaginary, then there are no real lines in the two-system, in which case $\mathcal{T}_4$ is called an *elliptic* linear congruence. If there are two real roots, then two lines exist that intersect every line of $\mathcal{T}_4$, which is termed a *hyperbolic* linear congruence. Finally, if (16.38) yields a double root, then $\mathcal{T}_4$ is a *parabolic* linear congruence.

16.4.4.1 Type 4a

If the two roots of (16.38) are imaginary, then the two-system generated by S_1 and S_2 contains no lines. In this case, the set of lines $\mathcal{T}_4$ is an *elliptic linear congruence*. Hunt [50] shows that these lines form concentric hyperboloids about the common normal to the axes of the two screws S_1 and S_2. The relationship between these screws and the distribution of the lines of $\mathcal{T}_4$ deserves further study.

16.4.4.2 Type 4b

If the two roots of (16.38) are real and distinct, then two lines L_a and L_b exist that intersect the entire set of lines $\mathcal{T}_4$, termed a *hyperbolic congruence*. Assume that these two lines are skew. Then we have two cases depending on whether or not one of these lines lies at infinity. A line at infinity has the Plücker vector of the form $L = (0, v)^T$. This is also described as a screw with infinite pitch.

1. If the two lines L_a and L_b are finite and skew, then (16.35) is the set of lines that intersect these two lines.
2. If one of the lines, say L_b, is at infinity, then it has the form $L_b = (0, \mathbf{v})^T$. In this case, L_a intersects the set of parallel planes orthogonal to $\mathbf{v}$, through L_b, in a series of points. The congruence consists of the pencil of lines through this point on each parallel plane.

The cases in which the two lines L_a and L_b intersect each other are usually identified separately. We consider these cases below.

16.4.4.3 Type 4c

If (16.38) has a double root, then we have a double line $L = L_a = L_b$. In this case, the set $\mathscr{T}_4$ is known as a *parabolic linear congruence*. We can distinguish two cases, depending on whether or not this double line is at infinity.

1. Each point on the finite line L is the vertex of a planar pencil of lines. Each pencil lies in a different plane passing through the line L.
2. If the two lines coincide at infinity, that is, $L_a = L_b = (0, \mathbf{v})^T$, then all lines in planes orthogonal to $\mathbf{v}$ form the congruence.

16.4.4.4 Type 4d

The situations in which the two lines of the hyperbolic congruence intersect are termed *degenerate*. We identify four subcases depending on whether one or both of the lines are finite or at infinity.

1. If the two lines defined by (16.38) intersect, then $\mathscr{T}_4$ consists of all the lines in the plane defined by the two lines L_a and L_b, and all lines in space that pass through their point of intersection.
2. If the lines L_a and L_b are parallel, which means that they intersect at a point at infinity, then $\mathscr{T}_4$ consists of all lines in the plane containing these two lines, and all lines in space parallel to them.
3. If one line L_a is finite and the other line $L_b = (0, \mathbf{v})^T$ is at infinity, then L_a is contained entirely in one of the parallel planes orthogonal to $\mathbf{v}$. In this case the congruence consists of the lines parallel to L_a, together with all lines in the plane that contains L_a.
4. If both lines lie at infinity, then we have $L_a = (0, \mathbf{v}_a)^T$ and $L_b = (0, \mathbf{v}_b)^T$, in which case the congruence is formed by all lines in the direction $\mathbf{s} = \mathbf{v}_a \times \mathbf{v}_b$. For this to occur the four lines F_i, $i = 1, 2, 3, 4$, must be parallel, which means that they form a type-3c.2 singular configuration. Thus, we conclude that this case cannot occur independently.

16.4.5 Type-5 Singularities

The final singularity that we consider occurs when an actuator screw is a linear combination of the remaining five actuator screws. In this case, we have

$$\mathscr{T}_5: \quad \mathsf{F} = k_1 \mathsf{F}_1 + k_2 \mathsf{F}_2 + k_3 \mathsf{F}_3 + k_4 \mathsf{F}_4 + k_5 \mathsf{F}_5. \tag{16.39}$$

Let the six minors of the 6×5 matrix $[F] = [\mathsf{F}_1, \mathsf{F}_2, \mathsf{F}_3, \mathsf{F}_4, \mathsf{F}_5]$ be denoted by M_i—the ith minor is the determinant of the 5×5 matrix obtained by removing row i from $[F]$. Thus, the set of lines $\mathscr{T}_5$ satisfy the linear equation

$$M_1 f_1 + M_2 f_2 + M_3 f_3 + M_4 f_4 + M_5 f_5 + M_6 f_6 = 0. \tag{16.40}$$

This is the equation of a *linear complex*. Assemble the six coefficients M_i into the screw $\mathsf{M} = (M_4, M_5, M_6, M_1, M_2, M_3)^T$ with axis $\mathsf{L} = (\mathbf{s}, \mathbf{r} \times \mathbf{s})^T$ and pitch μ; that is, $\mathsf{M} = (\mathbf{s}, \mathbf{r} \times \mathbf{s} + \mu \mathbf{s})^T$. Then (16.40) can be written in terms of M and $\mathsf{F} = (\mathbf{f}, \mathbf{p} \times \mathbf{f})$ to obtain

$$\mathsf{M}^T \check{\mathsf{F}} = \mathsf{L}^T \check{\mathsf{F}} + \mu \mathbf{s} \cdot \mathbf{f} = 0. \tag{16.41}$$

Let the distance from L to F along their common normal be d and the angle about the common normal be α. Then (16.41) reduces to

$$\mathsf{M}^T \check{\mathsf{F}} = -d \sin \alpha + \mu \cos \alpha = 0, \quad \text{or} \quad d \tan \alpha = \mu. \tag{16.42}$$

This equation shows that lines of this complex that are a distance d from the axis L lie at the angle $\alpha = \arctan(\mu/d)$ about the common normal.

16.4.5.1 Type 5a

In general, the complex $\mathscr{T}_5$ is the set of lines tangent to helices with the line L, described above, as its axis. If the radius of the helix is d, then its lead is $2\pi d^2/\mu$, Woods [153].

16.4.5.2 Type 5b

If the components of the screw M are such that they form the Plücker coordinates of a line, that is, $\mathsf{M}^T \check{\mathsf{M}} = 0$, then $\mathscr{T}_5$ is a *special linear complex*. There are two cases depending on whether or not M is at infinity.

1. If M is a real line, then $\mathscr{T}_5$ consists of all lines in space that intersect M.
2. If M has the form $\mathsf{M} = (0, \mathbf{v})^T$, which means that it lies at infinity, then $\mathscr{T}_5$ consists of all lines in the set of parallel planes orthogonal to $\mathbf{v}$. This means that the common normal to every pair of lines in the complex is parallel to the direction $\mathbf{v}$.

16.5 Examples Singularities

Singularities occur in platform manipulators either as a result of a singular configuration within a supporting chain or due to interdependence of actuator screws on separate chains. For the systems we are considering the serial chain singularities can be only of type 1 or 2, that is, either two actuator screws fall on the same line, or three lie on the same plane.

If we assume that no supporting chain is in a singular configuration, then we can determine the following examples of singular configurations in the basic platform designs shown in Figure 16.3:

1. A type-1 singularity occurs when any two forces F_i and F_j applied at the attachment points c_i and c_j align with the segment $c_i c_j$.
2. A type-2b.1 has the lines of action of three forces in the same plane and passing through the same point. This can occur in designs a, b, c, and e as follows. Let c_i be the attachment point of a supporting chain that that has two actuators. The lines of action of the two forces at c_i define a plane. If any other attachment point c_j lies in this plane, and if the line of action of a force F_j at this point lies along the segment $c_i c_j$, then the configuration is singular.
3. The type-3c.1 singularity has the lines of action of four forces passing though one point. This is easily seen in designs b and d. Let c_i be the point supported by three actuator screws. This point has the lines of action of three actuator screws passing through it. The singularity occurs when the line of action of a force at another attachment point c_j lies along the segment $c_i c_j$.
4. The degenerate hyperbolic congruence of type-4d.1 can also be seen in designs b and d. In this case the lines of action of five forces intersect two lines. Let c_i be the point supported by the chain with three actuators. Then the singularity occurs when the lines of action of the forces at two other attachment points c_j and c_k lie in the plane defined by the triangle $\triangle c_i c_j c_k$. The five actuator screws intersect the two edges of the platform.
5. The type-5b.1 singularity occurs when all six lines of action of the forces on the platform pass through the same line. This is easily seen in designs a and b. For design a the plane of the platform must coincide with both planes formed by the actuator screws at two attachment points. For design b the actuator screws of the one and two actuator supporting chains must lie in the plane of the platform. In both cases, all six actuator screws intersect the line formed by an edge of the platform.

16.6 Summary

This chapter has examined platform manipulators that have Jacobians in which each column is a Plücker vector of a line. Each of these lines represents the force of an actuator on the platform and singular configurations for these systems occur when these lines are linearly dependent. This is called a line-based singularity. A type-n singularity occurs when the line of action of one actuator force is a linear combination of at least n other actuator forces on the platform. The geometric description these linear combinations of lines provides insight to the geometry of singular configurations.

16.7 References

Song and Waldron [124] present the design of walking machines and Mason and Salisbury [75] consider the design and control of mechanical hands. The platform used for vehicle simulators is described by Stewart [126], also see Fichter [33]. Chablat et al. [8] and Li et al. [64] demonstrate different approaches to the design of specialized parallel platforms that optimize position and velocity performance throughout the workspace. Our formulation for the Jacobian of these systems follows Kumar [60].

The results in this chapter are a special case of a broader study of screws introduced by Ball [2]. Hunt [50] provides a survey of linearly dependent sets of screws important to mechanism theory. Gibson and Hunt [39, 40] and Martinez and Duffy [73, 74] provide a more detailed look at these screw systems. Phillips [98, 99] provides a machine-based perspective of screw theory.

Exercises

1. Determine the Jacobian for the 3-RRS platform robot shown in Figure 16.4.
2. Consider the TRS robot that is connected to a platform such that the spherical joint is unactuated. Determine the line of action of each of the actuator screws from the condition that it must intersect the remaining five joint axes.
3. Formulate the linear combination of the Plücker coordinates of two lines F_1 and F_2 and show that their linear combination is, in general, a screw not a line. Determine the condition that ensures that it is a line.
4. Show that the set of lines reciprocal to three arbitrary lines is a quadric surface that also contains these lines.
5. Consider a cube held by a four fingered grip such that it applies only normal forces to each of the faces. Determine the set of twists allowed by arbitrarily located fingers. Under what conditions do the applied forces become linearly dependent.
6. Show that the lines of the complex $\mathcal{T}_5$ are orthogonal to the velocities of points in a body moving so its twist has the axis L and pitch μ.

Appendix A
Solving Constraint Equations

A.1 The 4R Linkage Constraint

The analysis of planar and spherical 4R linkages, the spatial RSSR, and Bennett's linkage all yield a constraint equation between the output angle ψ and the input angle θ that takes the form

$$A(\theta)\cos\psi + B(\theta)\sin\psi = C(\theta). \tag{A.1}$$

There are two ways to solve this equation the *trigonometric solution* and the *tan-half-angle* technique.

The trigonometric solution begins by dividing both sides of (A.1) by $\sqrt{A^2 + B^2}$. This allows us to introduce the angle δ such that

$$\cos\delta = \frac{A}{\sqrt{A^2 + B^2}} \quad \text{and} \quad \sin\delta = \frac{B}{\sqrt{A^2 + B^2}}. \tag{A.2}$$

Notice that $\delta = \arctan(B/A)$. The left side of (A.1) takes the form

$$\cos\delta\cos\psi + \sin\delta\sin\psi = \cos(\psi - \delta). \tag{A.3}$$

Thus, the right side must be the cosine of an angle κ,

$$\cos\kappa = \frac{C}{\sqrt{A^2 + B^2}}. \tag{A.4}$$

Because κ and its negative have the same cosine, we have that $\delta + \kappa$ and $\delta - \kappa$ are both solutions to (A.1). This combines with the definition of δ to yield

$$\psi = \arctan\left(\frac{B}{A}\right) \pm \arccos\left(\frac{C}{\sqrt{A^2 + B^2}}\right). \tag{A.5}$$

J.M. McCarthy and G.S. Soh, *Geometric Design of Linkages*, Interdisciplinary Applied Mathematics 11, DOI 10.1007/978-1-4419-7892-9,
© Springer Science+Business Media, LLC 2011

Notice that the angle κ exists only if $-1 \leq \cos \kappa \leq 1$. Therefore, $C^2 \leq A^2 + B^2$, or equivalently, $A^2 + B^2 - C^2 \geq 0$ must be satisfied for a solution to exist.

In this formulation δ is the angle to a diagonal of the quadrilateral formed by the linkages. The angle κ is measured on either side of this diagonal to define the output angle ψ.

The tan-half-angle technique uses a transformation of variables to convert $\sin \psi$ and $\cos \psi$ into algebraic functions. Introduce the parameter $y = \tan(\psi/2)$, which allows us to define

$$\cos \psi = \frac{1 - y^2}{1 + y^2} \quad \text{and} \quad \sin \psi = \frac{2y}{1 + y^2}. \tag{A.6}$$

Substitute this into (A.1) to obtain

$$(A + C)y^2 - 2By - (A - C) = 0. \tag{A.7}$$

This equation is solved using the quadratic formula to obtain

$$\tan \frac{\psi}{2} = \frac{B \pm \sqrt{A^2 + B^2 - C^2}}{A + C}. \tag{A.8}$$

In order to have a real solution we must have $A^2 + B^2 - C^2 \geq 0$. This is the same condition obtained above for the trigonometric solution.

The constraint equations for these linkages are each special cases of a general equation that we can write as

$$(a_1 \cos \theta + a_2 \sin \theta - a_3) \cos \psi + (b_1 \cos \theta + b_2 \sin \theta - b_3) \sin \psi$$
$$- (c_1 \cos \theta + c_2 \sin \theta - c_3) = 0, \tag{A.9}$$

where a_i, b_i, and c_i are constants. Introduce the tan-half-angle parameters $x = \tan(\theta/2)$ and $y = \tan(\psi/2)$ so we have

$$((a_1 + c_1 + a_3 + c_3)x^2 - 2(a_2 + c_2)x - (a_1 + c_1 - a_3 - c_3))y^2$$
$$- 2((b_1 + b_3)x^2 - 2b_2 x - (b_1 - b_3))y$$
$$- ((a_1 - c_1 + a_3 - c_3)x^2 - 2(a_2 - c_2)x - (a_1 - c_1 - a_3 + c_3)) = 0. \tag{A.10}$$

This is a biquadratic equation in the unknowns x and y.

A.2 The Platform Constraint Equations

The 3RR planar and 3RR spherical platforms can be viewed formed from two 4R linkages $\mathbf{OAB_1C_1}$ and $\mathbf{OAB_2C_2}$ driven by the same crank $\mathbf{OA}$. Let θ be the angle of the input crank $\mathbf{OA}$ and let ϕ be the angle of the coupler at $\mathbf{A}$. The constraint equations for the two 4R linkages yield

$$A_1(\theta)\cos\phi + B_1(\theta)\sin\phi = C_1(\theta),$$
$$A_2(\theta)\cos\phi + B_2(\theta)\sin\phi = C_2(\theta). \tag{A.11}$$

We now present two ways to solve these equations. The first eliminates $\sin\phi$ and $\cos\phi$ linearly, and the second uses the resultant to solve simultaneous biquadratic equations.

For the first solution let $x = \cos\phi$ and $y = \sin\phi$ and solve the resulting linear equations using Cramer's rule to obtain

$$x = \frac{C_1 B_2 - C_2 B_1}{A_1 B_2 - A_2 B_1} \quad \text{and} \quad y = \frac{A_2 C_1 - A_1 C_2}{A_1 B_2 - A_2 B_1}. \tag{A.12}$$

In order for these equations to define a solution ϕ they must satisfy the identity $x^2 + y^2 = 1$. This yields an equation in $\cos\theta$ and $\sin\theta$, given by

$$(C_1 B_2 - C_2 B_1)^2 + (A_2 C_1 - A_1 C_2)^2 - (A_1 B_2 - A_2 B_1)^2 = 0. \tag{A.13}$$

Introduce the tan-half-angle parameter $x = \tan(\theta/2)$ so this equation becomes the polynomial $P(x) = 0$. For each root x_j of $P(x)$, compute θ_j and determine the coefficients A_{ij}, B_{ij}, and C_{ij} of the platform equations (A.11). Solve either one equations determine ϕ_j.

An alternative approach transforms (A.11) into a pair of biquadratic equations by introducing the tan-half-angle parameters $x = \tan(\theta/2)$ and $y = \tan(\psi/2)$. This yields the equations

$$D_1 y^2 + E_1 y + F_1 = 0,$$
$$D_2 y^2 + E_2 y + F_2 = 0, \tag{A.14}$$

where

$$D_i = d_{1i} x^2 + d_{2i} x + d_{3i},$$
$$E_i = e_{1i} x^2 + e_{2i} x + e_{3i},$$
$$F_i = f_{1i} x^2 + f_{2i} x + f_{3i}. \tag{A.15}$$

To solve these equations, we introduce a second pair of equations obtained by multiplying both by y and assemble the four equations into the matrix equation

$$\begin{bmatrix} 0 & D_1 & E_1 & F_1 \\ 0 & D_2 & E_2 & F_2 \\ D_1 & E_1 & F_1 & 0 \\ D_2 & E_2 & F_2 & 0 \end{bmatrix} \begin{Bmatrix} y^3 \\ y^2 \\ y \\ 1 \end{Bmatrix} = \begin{Bmatrix} 0 \\ 0 \\ 0 \\ 0 \end{Bmatrix}. \tag{A.16}$$

This equation can be solved for the vector $(y^3, y^2, y, 1)^T$ only if the coefficient matrix $[M]$ has determinant zero. Expand this determinant to obtain an eighth-degree polynomial $P(x)$ in the parameter x. The roots of this polynomial define x_j for which (A.16) can be solved to determine y_j.

Appendix B
Graphical Constructions

The following constructions use a straightedge to draw lines and a compass to construct circles and measure distances. They are useful in the graphical synthesis of planar RR chains.

B.1 Perpendicular Bisector

Given two points P^1 and P^2, we construct the perpendicular bisector L of the segment P^1P^2, Figure B.1, as follows:

1. Construct circles C_1 and C_2 centered on P^1 and P^2 with radii equal to or greater than one-half the length of P^1P^2.
2. C_1 and C_2 intersect in two points. Join these points to form the perpendicular bisector L.

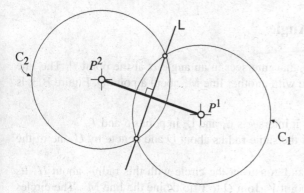

Fig. B.1 Construction of the perpendicular bisector L of the segment P_1P_2.

B.2 Circle Through Three Points

Given the triangle $\triangle P^1 P^2 P^3$, the center C of the circle that circumscribes this triangle, Figure B.2, is given by the construction:

1. Construct the perpendicular bisectors L_1 and L_2 to the segments $P^1 P^2$ and $P^2 P^3$.
2. The intersection of the lines L_1 and L_2 defines the center C of the circle through the three points.

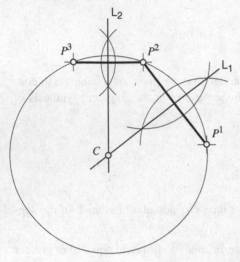

Fig. B.2 Construction of the circle through three points P^1, P^2, and P^3.

B.3 Duplication of an Angle

Consider two lines L_1 and L_2 that intersect in an angle α at the point P. The line M_2 that makes the same angle with another line M_1 about a point Q, Figure B.3, is constructed as follows:

1. Draw a circle C_P such that it intersects L_1 and L_2 in points S and T.
2. Construct a circle C_Q with the same radius about Q and denote by U one of the intersections with M_1.
3. Measure the distance ST and construct the circle with this radius about U. Its intersection with C_Q is a point V. Join Q to V to define the line M_2. The circles intersect in two points, so choose the one that provides the same orientation for M_2 as L_2 relative to L_1.

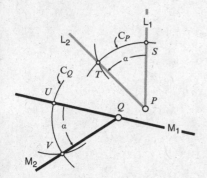

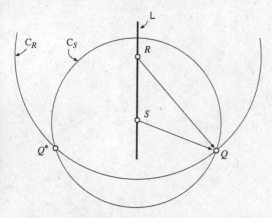

Fig. B.3 Duplication of the angle α about the point P from the line M_1 to determine the line M_2.

Fig. B.4 Reflection of the point Q through the line L.

B.4 Reflection of a Point Through a Line

In order to find the reflection of the point Q through the line L, Figure B.4, we use the construction:

1. Select two points R and S on L and construct the circles C_R and C_S with radii RQ and SQ, respectively.
2. The circles C_R and C_S intersect in two points. One is Q and the other is its reflection Q^*.

Appendix C
Spherical Trigonometry

Consider the spherical triangle $\triangle S_1 S_2 S_3$, where the axes are labeled in a counter-clockwise sense around the triangle, see Figure C.1. Associated with the side $S_i S_j$ we can define the normal vector $N_{ij} = S_i \times S_j / |S_i \times S_j|$. The angular dimension α_{ij} of this side is defined by the equations

$$\cos \alpha_{ij} = S_i \cdot S_j. \tag{C.1}$$

Thus, we can compute the three angles α_{12}, α_{23}, and α_{31}, which we consider to have a positive magnitude between 0 and π, thus $\alpha_{ij} = \alpha_{ji}$. The sense of the angles α_{ij} will be determined as needed relative to the normal vector N_{ij}.

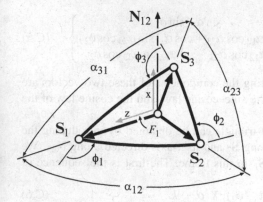

Fig. C.1 The frame F_1 has its x-axis along N_{12} and its z-axis along S_1.

At each vertex S_i we denote the exterior dihedral angle by ϕ_i, which is defined by the formula

$$\tan \phi_i = \frac{N_{ki} \times N_{ij} \cdot S_i}{N_{ki} \cdot N_{ij}} = \frac{S_k \times S_i \cdot S_j}{(S_k \times S_i) \cdot (S_i \times S_j)}. \tag{C.2}$$

The indices (i,j,k) in this equation are any one of the three cyclic permutations $(1,2,3)$, $(2,3,1)$, or $(3,1,2)$.

In the following derivations, we distinguish the F-frame equations from the B-frame. These are simply two different ways to formulate the same equations. We will label as F_i the frame that has its z-axis along the vertex $\mathbf{S}_i$ and the normal vector $\mathbf{N}_{ij}$ as its x-axis. The frame B_i will also have $\mathbf{S}_i$ as its z-axis, but its x-axis will now be the normal vector $\mathbf{N}_{ki}$.

C.1 The F-Frame Formulas

Our goal is to obtain trigonometric identities for the triangle $\triangle\mathbf{S}_1\mathbf{S}_2\mathbf{S}_3$. We begin with the frame F_1, aligned with the side $\mathbf{S}_1\mathbf{S}_2$, so the z-axis is along $\mathbf{S}_1$ and $\mathbf{N}_{12}$ is the x-axis. In this frame, we can determine two equations for the coordinates of $\mathbf{S}_3$ in terms of the dimensions of the triangle. The first is defined by the sequence of rotations

$$^1\mathbf{S}_3 = [Z(\pi - \phi_1)][X(\alpha_{31})]\vec{k}, \tag{C.3}$$

where $\pi - \phi_1$ is the interior angle at $\mathbf{S}_1$, and $\alpha_{31} = \alpha_{13}$ is the angular length of the side $\mathbf{S}_1\mathbf{S}_3$. Recall that $\vec{k} = (0,0,1)^T$. The superscript preceding $\mathbf{S}_3$ denotes the coordinate frame F_1 in which we are computing these coordinates. The second equation is given by

$$^1\mathbf{S}_3 = [X(\alpha_{12})][Z(\phi_2)][X(\alpha_{23})]\vec{k}. \tag{C.4}$$

Expand these equations to obtain

$$^1\mathbf{S}_3 = \left\{ \begin{matrix} \sin\alpha_{31}\sin\phi_1 \\ \sin\alpha_{31}\cos\phi_1 \\ \cos\alpha_{31} \end{matrix} \right\} = \left\{ \begin{matrix} \sin\alpha_{23}\sin\phi_2 \\ -(\sin\alpha_{12}\cos\alpha_{23} + \cos\alpha_{12}\sin\alpha_{23}\cos\phi_2) \\ \cos\alpha_{12}\cos\alpha_{23} - \sin\alpha_{12}\sin\alpha_{23}\cos\phi_2 \end{matrix} \right\}. \tag{C.5}$$

The three identities obtained by equating the components of these two vectors are known, respectively, as the sine law, the sine–cosine law, and the cosine law of the spherical triangle.

A different set of relations for this triangle can be obtained by introducing the coordinate frame F_2 with its z-axis along $\mathbf{S}_2$ and its x-axis directed along $\mathbf{N}_{23}$. We now consider the two definitions of $\mathbf{S}_1$ in this frame. The first is the sequence of rotations

$$^2\mathbf{S}_1 = [Z(\pi - \phi_2)][X(\alpha_{12})]\vec{k}. \tag{C.6}$$

The second way to determine $\mathbf{S}_1$ in F_2 is given by

$$^2\mathbf{S}_1 = [X(\alpha_{23})][Z(\phi_3)][X(\alpha_{31}]\vec{k}. \tag{C.7}$$

Expanding these equations, we obtain the identity

$$
{}^2\mathbf{S}_1 = \left\{ \begin{array}{c} \sin\alpha_{12}\sin\phi_2 \\ \sin\alpha_{12}\cos\phi_2 \\ \cos\alpha_{12} \end{array} \right\} = \left\{ \begin{array}{c} \sin\alpha_{31}\sin\phi_3 \\ -(\sin\alpha_{23}\cos\alpha_{31} + \cos\alpha_{23}\sin\alpha_{31}\cos\phi_3) \\ \cos\alpha_{23}\cos\alpha_{31} - \sin\alpha_{23}\sin\alpha_{31}\cos\phi_3 \end{array} \right\}. \quad (C.8)
$$

Notice that these equations can be obtained from (C.5) by permuting the indices $(1,2,3)$ to form $(2,3,1)$.

Finally, we can obtain a third set of identities by computing the components of $\mathbf{S}_2$ in the frame F_3 located with its z-axis along $\mathbf{S}_3$ and its x-axis along $\mathbf{N}_{31}$. The same derivation as above yields the identities

$$
{}^3\mathbf{S}_2 = \left\{ \begin{array}{c} \sin\alpha_{23}\sin\phi_3 \\ \sin\alpha_{23}\cos\phi_3 \\ \cos\alpha_{31} \end{array} \right\} = \left\{ \begin{array}{c} \sin\alpha_{12}\sin\phi_1 \\ -(\sin\alpha_{31}\cos\alpha_{12} + \cos\alpha_{31}\sin\alpha_{12}\cos\phi_1) \\ \cos\alpha_{31}\cos\alpha_{12} - \sin\alpha_{31}\sin\alpha_{12}\cos\phi_1 \end{array} \right\}. \quad (C.9)
$$

Notice that these equations can be obtained from (C.8) by again applying the cyclic permutation $(1,2,3) \mapsto (2,3,1)$.

Crane and Duffy [16] introduce the symbols $(\bar{X}_j, \bar{Y}_j, \bar{Z}_j)$ defined by

$$
\left\{ \begin{array}{c} \bar{X}_j \\ \bar{Y}_j \\ \bar{Z}_j \end{array} \right\} = \left\{ \begin{array}{c} \sin\alpha_{jk}\sin\phi_j \\ -(\sin\alpha_{ij}\cos\alpha_{jk} + \cos\alpha_{ij}\sin\alpha_{jk}\cos\phi_j) \\ \cos\alpha_{ij}\cos\alpha_{jk} - \sin\alpha_{ij}\sin\alpha_{jk}\cos\phi_j \end{array} \right\}. \quad (C.10)
$$

Comparing this to our equations above, we have

$$
{}^1\mathbf{S}_3 = \left\{ \begin{array}{c} \bar{X}_2 \\ \bar{Y}_2 \\ \bar{Z}_2 \end{array} \right\}, \quad {}^2\mathbf{S}_1 = \left\{ \begin{array}{c} \bar{X}_3 \\ \bar{Y}_3 \\ \bar{Z}_3 \end{array} \right\}, \quad \text{and} \quad {}^3\mathbf{S}_2 = \left\{ \begin{array}{c} \bar{X}_1 \\ \bar{Y}_1 \\ \bar{Z}_1 \end{array} \right\}. \quad (C.11)
$$

Also from our calculations above, we have

$$
{}^j\mathbf{S}_i = \left\{ \begin{array}{c} \sin\alpha_{ij}\sin\phi_j \\ \sin\alpha_{ij}\cos\phi_j \\ \cos\alpha_{ij} \end{array} \right\}. \quad (C.12)
$$

Thus, we obtain Crane and Duffy's compact form for these identities:

$$
\left\{ \begin{array}{c} \bar{X}_k \\ \bar{Y}_k \\ \bar{Z}_k \end{array} \right\} = \left\{ \begin{array}{c} \sin\alpha_{ij}\sin\phi_j \\ \sin\alpha_{ij}\cos\phi_j \\ \cos\alpha_{ij} \end{array} \right\}, \quad (C.13)
$$

where the indices (i,j,k) are any one of the cyclic permutations $(1,2,3)$, $(2,3,1)$, or $(3,1,2)$.

C.2 The B-Frame Formulas

We now perform the same analysis but with a different set of reference frames. Let B_1 be the reference frame aligned with side S_3S_1, so its z-axis is along S_1 and its x-axis is the normal vector N_{31}, Figure C.2. In this case, we determine two equations for the coordinates of S_2, rather than S_3 as we did above. The first equation is defined by the sequence of rotations

$$^1S_2 = [Z(\phi_1)][X(\alpha_{12})]\vec{k}. \tag{C.14}$$

The superscript preceding S_2 denotes the coordinate frame B_1. The second equation is given by

$$^1S_2 = [X(-\alpha_{31})][Z(\pi - \phi_3)][X(\alpha_{23})]\vec{k}. \tag{C.15}$$

Expand these equations to obtain

$$^1S_2 = \left\{ \begin{array}{c} \sin\alpha_{12}\sin\phi_1 \\ -\sin\alpha_{12}\cos\phi_1 \\ \cos\alpha_{12} \end{array} \right\} = \left\{ \begin{array}{c} \sin\alpha_{23}\sin\phi_3 \\ \sin\alpha_{31}\cos\alpha_{23} + \cos\alpha_{31}\sin\alpha_{23}\cos\phi_3 \\ \cos\alpha_{31}\cos\alpha_{23} - \sin\alpha_{31}\sin\alpha_{23}\cos\phi_3 \end{array} \right\}. \tag{C.16}$$

The three identities obtained by equating the components of these two vectors are alternative forms for the sine law, the sine–cosine law, and the cosine law of the spherical triangle.

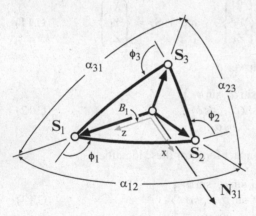

Fig. C.2 The frame B_1 has its x-axis along N_{31} and its z-axis along S_1.

Following the same procedure, we obtain the B-frame versions of $^2S_3$ and $^3S_1$. We can also get these results using the permutations $(2,3,1)$ and $(3,1,2)$ of the indices $(1,2,3)$. This results in the formulas

$$^2S_3 = \left\{ \begin{array}{c} \sin\alpha_{23}\sin\phi_2 \\ -\sin\alpha_{23}\cos\phi_2 \\ \cos\alpha_{23} \end{array} \right\} = \left\{ \begin{array}{c} \sin\alpha_{31}\sin\phi_1 \\ \sin\alpha_{12}\cos\alpha_{31} + \cos\alpha_{12}\sin\alpha_{31}\cos\phi_1 \\ \cos\alpha_{12}\cos\alpha_{31} - \sin\alpha_{12}\sin\alpha_{31}\cos\phi_1 \end{array} \right\} \tag{C.17}$$

and

$$^3S_1 = \left\{ \begin{array}{c} \sin\alpha_{31}\sin\phi_3 \\ -\sin\alpha_{31}\cos\phi_3 \\ \cos\alpha_{31} \end{array} \right\} = \left\{ \begin{array}{c} \sin\alpha_{12}\sin\phi_2 \\ \sin\alpha_{23}\cos\alpha_{12} + \cos\alpha_{23}\sin\alpha_{12}\cos\phi_2 \\ \cos\alpha_{23}\cos\alpha_{12} - \sin\alpha_{23}\sin\alpha_{12}\cos\phi_2 \end{array} \right\}. \quad \text{(C.18)}$$

Crane and Duffy introduce symbols (X_j, Y_j, Z_j) defined by the equations

$$\left\{ \begin{array}{c} X_j \\ Y_j \\ Z_j \end{array} \right\} = \left\{ \begin{array}{c} \sin\alpha_{ij}\sin\phi_j \\ -(\sin\alpha_{jk}\cos\alpha_{ij} + \cos\alpha_{jk}\sin\alpha_{ij}\cos\phi_j) \\ \cos\alpha_{jk}\cos\alpha_{ij} - \sin\alpha_{jk}\sin\alpha_{ij}\cos\phi_j \end{array} \right\}. \quad \text{(C.19)}$$

Comparing these equations to our results above, we see that

$$^1S_2 = \left\{ \begin{array}{c} X_3 \\ -Y_3 \\ Z_3 \end{array} \right\}, \quad ^2S_3 = \left\{ \begin{array}{c} X_1 \\ -Y_1 \\ Z_1 \end{array} \right\}, \quad \text{and} \quad ^3S_1 = \left\{ \begin{array}{c} X_2 \\ -Y_2 \\ Z_2 \end{array} \right\}. \quad \text{(C.20)}$$

Also from our calculations above, we have

$$^iS_j = \left\{ \begin{array}{c} \sin\alpha_{ij}\sin\phi_i \\ -\sin\alpha_{ij}\cos\phi_i \\ \cos\alpha_{ij} \end{array} \right\}. \quad \text{(C.21)}$$

In Crane and Duffy's notation these identities become

$$\left\{ \begin{array}{c} X_k \\ Y_k \\ Z_k \end{array} \right\} = \left\{ \begin{array}{c} \sin\alpha_{ij}\sin\phi_i \\ \sin\alpha_{ij}\cos\phi_i \\ \cos\alpha_{ij} \end{array} \right\}, \quad \text{(C.22)}$$

where the indices (i, j, k) are any one of the cyclic permutations $(1,2,3)$, $(2,3,1)$, or $(3,1,2)$. Notice that we have canceled the negative signs in the y-components of these equations.

C.3 Summary

The result of this analysis is two sets of three vector identities relating the vertex angles and sides of a spherical triangle. The first set of equations is (C.5), (C.8), and (C.9). The second set is (C.16), (C.17), and (C.18). The notation of Crane and Duffy allows these sets of equations to be written compactly as (C.13) and (C.22), respectively.

It is important to notice that associated with this triangle $\triangle S_1S_2S_3$ is its polar triangle $\triangle N_{12}N_{23}N_{31}$. We may analyze this triangle in exactly the same way as above to obtain two more sets of three vector identities. See Crane and Duffy for a complete listing of these identities, and for similar identities for the spherical quadrilateral, pentagon, hexagon, and heptagon.

Appendix D
Operations with Dual Numbers

The standard form of a screw $W = (k\mathbf{s}, k\mathbf{c} \times \mathbf{s} + kp\mathbf{s})^T$ is simplified by defining the multiplication between the ordered pair (k, kp) and the line $L = (\mathbf{s}, \mathbf{c} \times \mathbf{s})^T$ so that

$$W = (k\mathbf{s}, k\mathbf{c} \times \mathbf{s} + kp\mathbf{s})^T = (k, kp)(\mathbf{s}, \mathbf{c} \times \mathbf{s})^T. \tag{D.1}$$

We view this as a product of the dual scalar $\hat{k} = (k, kp)$ and the dual vector $L = (\mathbf{s}, \mathbf{c} \times \mathbf{s})^T$. The dual vector is equivalent to a vector of dual scalars, so (D.1) implies that the product of two dual scalars $\hat{a} = (a, a^\circ)$ and $\hat{b} = (b, b^\circ)$ is given by

$$\hat{a}\hat{b} = (a, a^\circ)(b, b^\circ) = (ab, a^\circ b + ab^\circ). \tag{D.2}$$

This may be taken as the definition of multiplication operation for dual numbers. However, in what follows we will define it in a way that extends easily to functions $F(\hat{a})$ of a dual number.

Consider the function $a(\varepsilon)$ of a real parameter ε. Define the dual number $\hat{a}$ associated with $a(\varepsilon)$ to be the pair constructed from this function and its derivative $a'(\varepsilon)$, both evaluated at $\varepsilon = 0$, that is, $a(0) = a$ and $a'(0) = a^\circ$. Therefore,

$$\hat{a} = \left(a(\varepsilon), \frac{da}{d\varepsilon} \right)\bigg|_{\varepsilon=0} = (a, a^\circ). \tag{D.3}$$

While there are many choices for $a(\varepsilon)$, the simplest is $a(\varepsilon) = a + \varepsilon a^\circ$.

Using this approach, we define the addition and subtraction of two dual numbers as the dual numbers associated with the functions $a(\varepsilon) + b(\varepsilon)$ and $a(\varepsilon) - b(\varepsilon)$, that is,

$$\hat{a} + \hat{b} = (a+b, a^\circ + b^\circ) \quad \text{and} \quad \hat{a} - \hat{b} = (a-b, a^\circ - b^\circ). \tag{D.4}$$

We can now see that (D.2) is the dual number obtained from the function

$$a(\varepsilon)b(\varepsilon) = (a + \varepsilon a^\circ)(b + \varepsilon b^\circ) = ab + \varepsilon(ab^\circ + a^\circ b) + \varepsilon^2 a^\circ b^\circ \tag{D.5}$$

by evaluating it and its derivative at $\varepsilon = 0$. A similar computation yields the division of two dual numbers as

$$\frac{\hat{a}}{\hat{b}} = \left(\frac{a}{b}, \frac{a^\circ b - ab^\circ}{b^2}\right). \tag{D.6}$$

A differentiable function of a dual number $F(\hat{a})$ can be evaluated using the chain rule to obtain

$$F(\hat{a}) = \left(F(a(\varepsilon)), F'(a(\varepsilon))\frac{da(\varepsilon)}{d\varepsilon}\right)\Bigg|_{\varepsilon=0} = \left(F(a), F'(a)a^\circ\right). \tag{D.7}$$

For example, in order to evaluate $\hat{a}^{1/2}$ we consider the function $F(\hat{a}) = (a + \varepsilon a^\circ)^{1/2}$ and compute

$$F(a) = a^{1/2}, \quad F'(a)a^\circ = \frac{1}{2}a^{-1/2}a^\circ. \tag{D.8}$$

Thus, we have

$$\hat{a}^{1/2} = \left(a^{1/2}, \frac{a^\circ}{2a^{1/2}}\right). \tag{D.9}$$

Other examples are the trigonometric functions of a dual angle $\hat{\theta} = (\theta, d)$, given by

$$\cos\hat{\theta} = (\cos\theta, -d\sin\theta),$$
$$\sin\hat{\theta} = (\sin\theta, d\cos\theta),$$
$$\tan\hat{\theta} = \left(\tan\theta, \frac{d}{\cos^2\theta}\right). \tag{D.10}$$

These computations show that the set of dual numbers forms a commutative ring that we can use to define dual vectors and dual matrices. The result is *dual vector algebra*, which is used to manipulate the coordinates of lines and screws.

Appendix E
Kinematics Equations

E.1 The Planar RR Chain

We now show how to use the kinematics equation of a planar RR chain to obtain the relative displacement $[T(\phi_{ij}, \mathbf{P}_{ij})]$ of the end-link as the composition of rotations about the fixed and moving pivots. The kinematics equations of an RR chain define the displacement $[D]$ of M relative to F as the composition of local transformations

$$[D(\alpha, \beta)] = [G][Z(\beta)][X(a)][Z(\alpha)][H]. \tag{E.1}$$

The 3×3 matrices $[Z(\beta)]$ and $[Z(\alpha)]$ define rotations about the z-axis and $[X(a)]$ is a pure translation along the x-axis. The transformations $[G]$ and $[H]$ are displacements from F to the fixed pivot and from the moving pivot to M.

Compute the relative displacement of the moving body M from position i to position j, given by

$$
\begin{aligned}
[D_{ij}] &= [D_j][D_i^{-1}] \\
&= [G][Z(\beta_j)][X(a)][Z(\alpha_j)][H][H^{-1}][Z(\alpha_i)^{-1}][X(a)^{-1}][Z(\beta_i)^{-1}][G^{-1}].
\end{aligned}
\tag{E.2}
$$

Define the relative angles $\alpha_{ij} = \alpha_j - \alpha_i$ and $\beta_{ij} = \beta_j - \beta_i$ and introduce the identity displacement $[I] = [Z(\beta_i)^{-1}][G^{-1}][G][Z(\beta_i)]$ in this equation to obtain

$$
\begin{aligned}
[D_{ij}] &= \left([G][Z(\beta_{ij})][G^{-1}]\right)\left([G][Z(\beta_i)][X(a)][Z(\alpha_{ij})][X(a)^{-1}][Z(\beta_i)^{-1}][G^{-1}]\right) \\
&= [T(\beta_{ij}, \mathbf{G})][T(\alpha_{ij}, \mathbf{W}^i)],
\end{aligned}
\tag{E.3}
$$

where

$$
\begin{aligned}
[T(\beta_{ij}, \mathbf{G})] &= [G][Z(\beta_{ij})][G^{-1}], \\
[T(\alpha_{ij}, \mathbf{W}_i)] &= [G][Z(\beta_i)][X(a)][Z(\alpha_{ij})][X(a)^{-1}][Z(\beta_i)^{-1}][G^{-1}].
\end{aligned}
\tag{E.4}
$$

We now show that the transformations $[T(\beta_{ij}, \mathbf{G})]$ and $[T(\alpha_{ij}, \mathbf{W}^i)]$ are rotations about the respective poles $\mathbf{G}$ and $\mathbf{W}^i$. First notice that

$$\mathbf{G} = [G]\vec{k} \quad \text{and} \quad \mathbf{W}^i = [G][Z(\beta_i)][X(a)]\vec{k}, \tag{E.5}$$

where $\vec{k} = (0,0,1)^T$ is the homogeneous coordinates of the origin of F. We now show that the pole of the displacement $[D][A][D^{-1}]$, where $[A]$ is a rotation matrix and $[D] = [B, \mathbf{b}]$ is a general planar displacement, is $[D]\vec{k} = \mathbf{b}$. The composition of displacements $[D][A][D^{-1}]$ becomes

$$[D][A][D^{-1}] = [B, \mathbf{b}][A, \mathbf{0}][B^T, -B^T\mathbf{b}]$$
$$= \left[BAB^T, [I - BAB^T]\mathbf{b}\right] = \left[A, [I - A]\mathbf{b}\right]. \tag{E.6}$$

The equality of the rotation matrices $[B][A][B^T] = [A]$ results from the addition of planar rotations. This is the definition of a planar displacement in terms of its pole.

Thus, equation (E.3), which equates the relative displacement of the floating link of an RR dyad to the composition of rotations about its fixed and moving axes, is the transformation equation associated with the planar dyad triangle.

A similar derivation yields the equation

$$[D_{ij}] = [T(\alpha_{ij}, \mathbf{W}^j)][T(\beta_{ij}, \mathbf{G})], \tag{E.7}$$

where

$$[T(\beta_{ij}, \mathbf{G})] = [G][T(\beta_{ij})][G^{-1}], \tag{E.8}$$
$$[T(\alpha_{ij}, \mathbf{W}^j)] = [G][Z(\beta_j)][X(a)][Z(\alpha_{ij})][X(a)^{-1}][Z(\beta_j)^{-1}][G^{-1}]. \tag{E.9}$$

Invert equation (E.1) of an RR dyad to obtain the position of the fixed frame relative to the moving frame

$$[D(\alpha, \beta)^{-1}] = [H^{-1}][Z(\alpha)^{-1}][X(a)^{-1}][Z(\beta)^{-1}][G^{-1}]. \tag{E.10}$$

For the pair of positions M_i and M_k of the moving body, we have the inverse relative displacements $[T_{ik}^\dagger] = [T_k^{-1}][T_i]$, where

$$[D_{ik}^\dagger] = [H^{-1}][Z(\alpha_k)^{-1}][X(a)^{-1}][Z(\beta_k)^{-1}][G^{-1}][G][Z(\beta_i)][X(a)][Z(\alpha_i)][H]. \tag{E.11}$$

An analysis identical to that discussed above for RR dyads yields the equation

$$[D_{ik}^\dagger] = [T(-\alpha_{ik}, \mathbf{w})][T(-\beta_{ik}, \mathbf{g}^i)] = [T(-\beta_{ik}, \mathbf{g}^k)][T(-\alpha_{ik}, \mathbf{w})]. \tag{E.12}$$

The transformation $[T(-\alpha_{ik}, \mathbf{w})]$ is a displacement with the moving pivot $\mathbf{w}$ in M as its pole. Similarly, $[T(-\beta_{ik}, \mathbf{g}^i)]$ is a displacement with pole $\mathbf{g}^i$, that is, the point in M corresponding to the fixed pivot in the ith position.

We now compute $[D_{ik}^j]$ using the first equation in (E.12),

$$[D_{ik}^j] = [T_j][T(-\alpha_{ik}, \mathbf{w})][T(-\beta_{jk}, \mathbf{g}^i)][T_j^{-1}]. \tag{E.13}$$

This transformation $[T_j][T(-\alpha_{ik}, \mathbf{w})][T_j^{-1}]$ changes the coordinates of the moving pivot $\mathbf{w}$ to $\mathbf{W}^j$. Similarly, $[T_j][T(-\beta_{jk}, \mathbf{g}^i)][T_j^{-1}]$ transforms the point $\mathbf{g}^i$ in M to $\mathbf{G}^i$ in F, so we have

$$[D_{ik}^j] = [T(-\alpha_{ik}, \mathbf{W}^j)][T(-\beta_{ik}, \mathbf{G}^i)]. \tag{E.14}$$

E.2 The Spherical RR Chain

The kinematics equation of a spherical RR chain can be used to derive the relative rotation $[R(\phi_{ij}, \mathbf{S}_{ij})]$ of the end-link M as the composition of rotations about the fixed and moving axes of the chain. The kinematics equations define the orientation $[R]$ of M relative to F as the product of local transformations

$$[R(\alpha, \beta)] = [G][Z(\beta)][X(\rho)][Z(\alpha)][H], \tag{E.15}$$

where $[Z(\beta)]$ and $[Z(\alpha)]$ are coordinate rotations about z-axis, and $[X(\rho)]$ is the coordinate rotation around the x-axis. The rotations $[G]$ and $[H]$ are transformations from F to the fixed axis $\mathbf{G}$ and from the moving axis $\mathbf{W}$ to M, respectively.

The relative rotation $[R_{ij}]$ of M from orientation M_i to M_j is given by

$$\begin{aligned}
[R_{ij}] &= [R_j][R_i^T] \\
&= [G][Z(\beta_j)][X(\rho)][Z(\alpha_j)][H][H^T][Z(\alpha_i)^T][X(\rho)^T][Z(\beta_i)^T][G^T].
\end{aligned} \tag{E.16}$$

Define the relative angles $\alpha_{ij} = \alpha_j - \alpha_i$ and $\beta_{ij} = \beta_j - \beta_i$, and introduce the identity $[I] = [Z(\beta_i)^T][G^T][G][Z(\beta_i)]$ to obtain

$$\begin{aligned}
[R_{ij}] &= \left([G][Z(\beta_{ij})][G^T]\right)\left([G][Z(\beta_i)][X(\rho)][Z(\alpha_{ij})][X(\rho)^T][Z(\beta_i)^T][G^T]\right) \\
&= [A(\beta_{ij}, \mathbf{G})][A(\alpha_{ij}, \mathbf{W}^i)],
\end{aligned} \tag{E.17}$$

where

$$\begin{aligned}
[A(\beta_{ij}, \mathbf{G})] &= [G][Z(\beta_{ij})][G^T] \\
[A(\alpha_{ij}, \mathbf{W}^i)] &= [G][Z(\beta_i)][X(\rho)][Z(\alpha_{ij})][X(\rho)^T][Z(\beta_i)^T][G^T].
\end{aligned} \tag{E.18}$$

We now show that $[A(\beta_{ij}, \mathbf{G})]$ and $[A(\alpha_{ij}, \mathbf{W}^i)]$ are rotations about poles $\mathbf{G}$ and $\mathbf{W}^i$, respectively. First notice that

$$\mathbf{G} = [G]\vec{k} \quad \text{and} \quad \mathbf{W}^i = [G][Z(\beta_i)][X(\rho)]\vec{k}, \tag{E.19}$$

where $\vec{k} = (0, 0, 1)^T$. In light of these relations, all we have to show is that the rotation matrix $[B][A][B^T]$ has $[B]\mathbf{S}$ as its rotation axis if the rotation $[A]$ has the rotation axis $\mathbf{S}$. To see this, simply check the definition

$$[BAB^T - I][B]\mathbf{S} = [BA - B]\mathbf{S} = [B][A - I]\mathbf{S} = 0. \tag{E.20}$$

The last equality results because $\mathbf{S}$ is the rotation axis of $[A]$.

Thus, (E.18) equates the relative rotation to the composition of rotations about the fixed and moving axes and defines the spherical dyad triangle.

A similar derivation yields the equivalent relation

$$[R_{ij}] = [A(\alpha_{ij}, \mathbf{W}^j)][A(\beta_{ij}, \mathbf{G})], \tag{E.21}$$

where

$$[A(\beta_{ij}, \mathbf{G})] = [G][Z(\beta_{ij})][G^T],$$
$$[A(\alpha_{ij}, \mathbf{W}^j)] = [G][Z(\beta_j)][X(\rho)][Z(\alpha_{ij})][X(\rho)^T][Z(\beta_j)^T][G^T]. \tag{E.22}$$

Now consider the inverse relative rotation $[R_{ik}^\dagger] = [R_k^T][R_i]$ for the spherical RR chain, given by

$$[R_{ik}^\dagger] = [H^T][Z(\alpha_k)^T][X(\rho)^T][Z(\beta_k)^T][G^T][G][Z(\beta_i)][X(\rho)][Z(\alpha_i)][H]. \tag{E.23}$$

An analysis identical to that discussed above yields the identities

$$[R_{ik}^\dagger] = [A(-\alpha_{ik}, \mathbf{w})][A(-\beta_{ik}, \mathbf{g}^i)] = [A(-\beta_{ik}, \mathbf{g}^k)][A(-\alpha_{ik}, \mathbf{w})]. \tag{E.24}$$

The rotation $[A(-\alpha_{ik}, \mathbf{w})]$ has the moving axis $\mathbf{w}$ in M as its rotation axis. Similarly, $[A(-\beta_{ik}, \mathbf{g}^i)]$ has as its rotation axis $\mathbf{g}^i$, which is the fixed axis for the dyad in the ith position of M.

Transform these equations to F with M in position M_j to define

$$[R_{ik}^j] = [R_j][A(-\alpha_{ik}, \mathbf{w})][A(-\beta_{ik}, \mathbf{g}^i)][R_j^T]. \tag{E.25}$$

The transformation $[R_j][A(-\alpha_{ik}, \mathbf{w})][R_j^T]$ changes the coordinates of the moving pivot $\mathbf{w}$ to $\mathbf{W}^j$. Similarly, $[R_j][A(-\beta_{ik}, \mathbf{g}^i)][R_j^T]$ transforms the point $\mathbf{g}^i$ in M to $\mathbf{G}^i$ in F, so we have

$$[R_{ik}^j] = [A(-\alpha_{ik}, \mathbf{W}^j)][A(-\beta_{ik}, \mathbf{G}^i)]. \tag{E.26}$$

E.3 The CC Chain

Here we show that the kinematics equations of the CC chain can be used to define the relative transformation $[D_{ij}]$ as the composition of screw displacements about the fixed and moving axes of the chain. The kinematics equation equates the spatial displacement $[D]$ of the moving body M to the sequence of relative displacements along the chain,

$$[D(\hat{\alpha}, \hat{\beta})] = [G][Z(\hat{\beta})][X(\hat{\rho})][Z(\hat{\alpha})][H], \tag{E.27}$$

where $[Z(\hat{\beta})]$ and $[Z(\hat{\alpha})]$ are the coordinate screw displacements about the fixed and moving axes, and $[X(\hat{\rho})]$ is a screw displacement along the crank.

The relative transformation $[D_{ij}]$ of M as it moves from position M_i to M_j is given by

$$[D_{ij}] = [D_j][D_i^{-1}]$$
$$= [G][Z(\hat{\beta}_j)][X(\hat{\rho})][Z(\hat{\alpha}_j)][H][H^{-1}][Z(\hat{\alpha}_i)^{-1}][X(\hat{\rho})^{-1}][Z(\hat{\beta}_i)^{-1}][G^{-1}].$$
(E.28)

We simplify this expression by defining $\hat{\alpha}_{ij} = \hat{\alpha}_j - \hat{\alpha}_i$, and introduce the identity $[I] = [Z(\hat{\beta}_i)^{-1}][G^{-1}][G][Z(\hat{\beta}_i)] = I$ to obtain

$$[D_{ij}] = ([G][Z(\hat{\beta}_{ij})][G^{-1}]) \, ([Z(\hat{\beta}_i)][X(\hat{\rho})][Z(\hat{\alpha}_{ij})][X(\hat{\rho})^{-1}][Z(\hat{\beta}_i)^{-1}][G^{-1}])$$
$$= [T(\hat{\beta}_{ij}, \mathsf{G})][T(\hat{\alpha}_{ij}, \mathsf{W}^i)],$$
(E.29)

where

$$[T(\hat{\beta}_{ij}, \mathsf{G})] = [G][Z(\hat{\beta}_{ij})][G^{-1}],$$
$$[T(\hat{\alpha}_{ij}, \mathsf{W}^i)] = [G][Z(\hat{\beta}_i)][X(\hat{\rho})][Z(\hat{\alpha}_{ij})][X(\hat{\rho})]^{-1}[Z(\hat{\beta}_i)^{-1}][G^{-1}].$$
(E.30)

We now show that the transformations $[T(\hat{\beta}_{ij}, \mathsf{G})]$ and $[T(\hat{\alpha}_{ij}, \mathsf{W}^i)]$ are screw displacements about the axes G and W^i measured in F. To see this, consider k to be the screw along the z-axis of the fixed frame. Then G and W^i are obtained from the screw transformations

$$\mathsf{G} = [\hat{G}]\mathsf{k} \quad \text{and} \quad \mathsf{W}^i = [\hat{G}][\hat{Z}(\hat{\beta}_i)][\hat{X}(\hat{\rho})]\mathsf{k}.$$
(E.31)

Notice that the screw transformation $[\hat{B}][\hat{A}][\hat{B}^{-1}]$ has the screw axis $[\hat{B}]\mathsf{S}$, because

$$[\hat{B}\hat{A}\hat{B}^{-1} - I][\hat{B}]\mathsf{S} = [\hat{B}\hat{A} - \hat{B}]\mathsf{S} = [\hat{B}][\hat{A} - I]\mathsf{S} = 0.$$
(E.32)

The last equality arises because S is the screw axis of $[\hat{A}]$.

Therefore, we can conclude that (E.30) is the matrix transformation associated with the spatial dyad triangle.

A similar derivation yields the equation

$$[D_{ij}] = [T(\hat{\alpha}_{ij}, \mathsf{W}^j)][T(\hat{\beta}_{ij}, \mathsf{G})],$$
(E.33)

where

$$[T(\hat{\beta}_{ij}, \mathsf{G})] = [G][Z(\hat{\beta}_{ij})][G^{-1}],$$
$$[T(\hat{\alpha}_{ij}, \mathsf{W}^j)] = [G][Z(\beta_j)][X(\hat{\rho})][Z(\hat{\alpha}_{ij})][X(\hat{\rho})^{-1}][Z(\hat{\beta}_j)^{-1}][G^{-1}].$$
(E.34)

References

1. Altmann, S. L., "Hamilton, Rodrigues, and the quaternion scandal," *Math. Mag.* 62(5):291–308, 1989.
2. Ball, R. S., *A Treatise on the Theory of Screws,* The University Press, Cambridge, England, 1900.
3. Bernshtein, D. N., "The number of roots of a system of equations," *Functional Analysis and Its Applications,* 9(3):183–185, 1975.
4. Beyer, R., *Kinematische Getriebesynthese.* Springer-Verlag, Berlin, 1953, (Trans. Kuenzel, H. (1963) as *The kinematic synthesis of mechanisms,* Chapman and Hall, London).
5. Bottema, O., and Roth, B., *Theoretical Kinematics,* North Holland Press, NY, 1979, (reprinted by Dover Publications).
6. Bouma, W., Fudos, I., Hoffmann, C. M., Cai, J., Paige, R., "A geometric constraint solver," *CAD* 27:487–501, 1995.
7. Burmester, L., *Lehrbuch der Kinematik,* Verlag Von Arthur Felix, Leipzig, Germany, 1886.
8. Chablat, D., Wenger, P., Majou, F., and Merlet, J. P., "An Interval Analysis Based Study for the Design and the Comparison of Three-Degrees-of-Freedom Parallel Kinematic Machines," *The International Journal of Robotics Research,* 23(6): 615-624, 2004.
9. Chase, T. R., "Burmester Theory for Four Precision Positions: An Extended Discourse with Application to the Dimensional Synthesis of Arbitrary Planar Linkages," *PhD thesis, Department of Mechanical Engineering, University of Minnesota, Minneapolis,* 1984.
10. Chen, P., and Roth, B., "Design Equations for the Finitely and Infinitesimally Separated Position Synthesis of Binary Links and Combined Link Chains," *ASME Journal of Engineering for Industry,* 91:209–219, 1969.
11. Cheng, H., and Gupta, K. C., "An Historical Note on Finite Rotations," *ASME Journal of Applied Mechanics,* 56: 139–145, 1989.
12. Chiang, C. H., *Kinematics of Spherical Mechanisms,* Cambridge Univ. Press, Cambridge, UK, 1988.
13. Clifford, W. K. *Mathematical Papers,* (ed. R. Tucker), London: Macmillan, 1882.
14. Collins, C. L., and Long, G. L., "The singularity analysis of an in-parallel hand controller for force-reflected teleoperation," *IEEE Trans. on Robotics and Automation,* 11(5):661–669, 1995.
15. Craig, J. J., *Introduction to robotics, mechanics, and control* Addison-Wesley Publ. Co., Reading MA, 1986.
16. Crane, C. D., and Duffy, J., *Kinematic Analysis of Robot Manipulators.* Cambridge University Press, Cambridge, UK, 1998.
17. Crossley, F. R. E., "The Permutations of Kinematic Chains of Eight Members or Less from the Graph-Theoretic Viewpoint," *Developments in Theoretical and Applied Mechanics.* Pergamon Press, Oxford, 2:467–486, 1965.
18. Crossley, F. R. E., "3-D Mechanisms," *Machine Design.* pp. 175–179, 1955.

433

19. Dai, J. S., and Rees-Jones, J., "Mobility in metamorphic mechanisms of foldable/erectable kinds," *ASME Journal of Mechanical Design*, 121(3): 375–382, 1999.
20. Dai, J. S., and Rees-Jones, J., "Matrix representation of topological changes in metamorphic mechanisms," *ASME Journal of Mechanical Design*, 127(4):610–619, 2005.
21. Dandurand, A., "The Rigidity of Compound Spatial Grids," *Structural Topology*, 10:43–55, 1984.
22. Denavit, J. and Hartenberg, R. S., "A Kinematic Notation for Lower-Pair Mechanisms Based on Matrices," *Journal of Applied Mechanics*, 22:215–221, 1955.
23. DiGregorio, R., "On the Polynomial Solution of the Synthesis of Five Plane-Sphere Contacts or PPS Chains That Guide a Rigid Body Through Six Assigned Poses," *Proceedings of the ASME 2005 International Design Engineering Technical Conferences*, September 2428, 2005 , Long Beach, California, USA. (doi:10.1115/DETC2005-84788)
24. Dimarogonas, A., "The Origins of the Theory of Machines and Mechanisms," *Modern Kinematics: Developments in the Last Forty Years*, (A. G. Erdman, ed.) John Wiley and Sons, New York, 1993.
25. Dimentberg, F. M., *The screw calculus and its applications in mechanics*, (in Russian) Moscow, 1965, (English trans: AD680993, Clearing house for Federal Technical and Scientific Information, Virginia).
26. Dobrovolskii, V. V., "Synthesis of Spherical Mechanisms," (in Russian) Akadamiia Nauk, SSSR Trudi Seminara po Teorii Ashin, Mechanizmov, pp. 5–20, 1943.
27. Dowler, H. J., Duffy, J., and Tesar, D., "A generalized study of four and five multiply separated positions in spherical kinematics," *Mechanism and Machine Theory*, 13(4): 409–436, 1978.
28. Erdman, A. G., and Gustafson, J. E., "LINCAGES: Linkage Interactive Computer Analysis and Graphically Enhanced Synthesis Package." ASME Paper No. 77-DTC-5, 1977.
29. Erdman A. G., "Three and Four Precision Point Kinematic Synthesis of Planar Linkage," *Mechanism and Machine Theory*, 16(3):227–245, 1981.
30. Erdman, A. G., and Sandor, G. N., *Mechanism Design: Analysis and Synthesis, Vol. 1*. Prentice-Hall, New Jersey, 1997.
31. Faugere, J. C., and Lazard, D., "Combinatorial Classes of Parallel Manipulators," *Mechanism and Machine Theory*, 30(6):756–776, 1995.
32. Filemon, E., "Useful Ranges of Center-point Curves for Design of Crank-and-Rocker Linkages," *Mechanism and Machine Theory*, 7:47–53, 1972.
33. Fichter, E. F., "A Stewart Platform-based Manipulator: General Theory and Practical Construction," *Int. Journal of Robotics Research*, 5(2):157–182, 1986.
34. Fischer, I. S., *Dual-Number Methods in Kinematics, Statics, and Dynamics*, CRC Press, Boca Raton, 1999.
35. Foster, D. E. and Cipra, R. J., "An Automatic Method for Finding the Assembly Configurations of Planar Non-Single-Input-Dyadic Mechanisms," *Journal of Mechanical Design*, 124(1):58–67, 2002.
36. Freudenstein, F. and Dobryankyj, L., "On a Theory for the Type Synthesis of Mechanisms," *Proc. 11th Conf. on Applied Mechanics*, pp. 420–428, 1964.
37. Freudenstein, F., and Sandor, G. N., On the Burmester Points of a Plane. *ASME Journal of Applied Mechanics*. 28(1):41–49, 1961.
38. Furlong, T., Vance, J., and Larochelle, P., "Spherical Mechanism Synthesis in Virtual Reality", *Proc. ASME Design Engineering Technical Conference*, Atlanta, GA, September 13-16, 1998.
39. Gibson, C. G., and Hunt, K. H., "Geometry of screw systems—1, screws: Genesis and geometry," *Mechanism and Machine Theory*, 25(1):1–10, 1990.
40. Gibson, C. G., and Hunt, K. H., "Geometry of screw systems—2, classification of screw systems," *Mechanism and Machine Theory*, 25(1):11–27, 1990.
41. Gupta, K. C., 1986, "Kinematic analysis of manipulators using the zero reference position description," *International Journal of Robotics Research*, 5(2):5-13.
42. Hall, A. S., Jr., *Kinematics and Linkage Design*. Prentice-Hall, Inc., Englewoods Cliffs, New Jersey, 1961.
43. Hamilton, W. R., *Elements of Quaternions*, reprinted by Chelsea Press, 1969.

44. Harrisberger, L., "A Number Synthesis Survey of Three-dimensional Mechanisms," *ASME Journal of Engingineering for Industry*, 87B:213–220, 1965.
45. Harrisberger, L., and Soni, A. H. "A Survey of Three-Dimensional Mechanisms with One General Constraint," *ASME Paper No. 66-Mech-44*, 1966.
46. Hartenberg, R. S., and Denavit, J., *Kinematic Synthesis of Linkages*. McGraw-Hill, New York, 1964.
47. Hernandez, S., Bai, S., and Angeles, J., "The Design of a Chain of Spherical Stephenson Mechanisms for a Gearless Robotic Pitch-Roll Wrist," *Journal of Mechanical Design*, 128(2):422-429, 2006.
48. Herve, J. M., "The Lie group of rigid body displacements, a fundamental tool for mechanism design," *Mechanism and Machine Theory*, 34:717-730, 1999.
49. Huang, C., "The cylindroid associated with finite motions of a Bennett mechanisms," *Proc. ASME Design Engineering Technical Conference,* Irvine, CA, September 1996.
50. Hunt, K. H., *Kinematic Geometry of Mechanisms,* Clarendon Press, Oxford, 1978.
51. Husty, M. L., "An Algorithm for Solving the Direct Kinematics of General Stewart-Gough Platforms," *Mechanism and Machine Theory*, 31(4):365–379, 1996.
52. Innocenti, C., "Polynomial Solution of the Spatial Burmester Problem," Mechanism Synthesis and Analysis, ASME DE-Vol. 70, 1994.
53. Jessop, C. M., *A Treatise on the Line Complex,* Cambridge, England, 1903, (reprinted by Chelsea Publishing Company, New York, New York, 1969).
54. Kaufman, R. E., "Mechanism Design by Computer," *Machine Design.* pp. 94–100, October 1978.
55. Kim, H. S. and Tsai, L. W., "Kinematic Synthesis of a Spatial 3-RPS Parallel Manipulator," *ASME Journal of Mechanical Design*, 125(1):92–97, 2003.
56. Kimbrell, J. T., *Kinematics Analysis and Synthesis,* McGraw-Hill, Inc. New York, 1991.
57. Kota, S. (ed.), "Type Synthesis and Creative Design" *Modern Kinematics: Developments in the Last Forty Years* (A. G. Erdman, ed.), John Wiley and Sons, New York, 1993.
58. Krishnaprasad, P. S., and Yang, R., "On the geometry and dynamics of floating four-bar linkages," *Dynamics and Stability of Systems*, 9:19–45, 1994.
59. Krovi, V., Ananthasuresh, G. K., and Kumar, V., "Kinematic and Kinetostatic Synthesis of Planar Coupled Serial Chain Mechanisms," *ASME Journal of Mechanical Design*, 124(2):301-312, 2002.
60. Kumar, V., "Instantaneous Kinematics of Parallel-Chain Robotic Mechanisms," *Journal of Mechanical Design*, 114(3):349–358, 1992.
61. Larochelle, P., Dooley, J., Murray, A., McCarthy, J. M., "*SPHINX*–Software for Synthesizing Spherical 4R Mechanisms," *Proc. of the 1993 NSF Design and Manufacturing Systems Conference,* Univ. of North Carolina at Charlotte, 1:607–611, 1993.
62. Lee, E., and Mavroidis, C., "Solving the Geometric Design Problem of Spatial 3R Robot Manipulators Using Polynomial Homotopy Continuation," *ASME J. Mechanical Design*, 124(4):652-661, 2002.
63. Lee, E., and Mavroidis, C., "Geometric Design of 3R Manipulators for Reaching Four End-Effector Spatial Poses," *The International Journal of Robotics Research*, 23(3): 247-254, 2004.
64. Li, Z., Li, M., Chetwynd, D. G., and Gosselin, C. M., "Conceptual Design and Dimensional Synthesis of novel 2-DOF Translational Parallel Robot for Pick and Place Operations," *ASME Journal of Mechanical Design*, 126(3):449-455, 2004.
65. Li, D., Zhang, Z., and McCarthy, J. M., "A Constraint Graph Representation of Metamorphic Linkages," *Mechanism and Machine Theory*, 2010.
66. Liao, Q. and McCarthy, J. M., "On the Seven Position Synthesis of a 5-SS Platform Linkage," *ASME Journal of Mechanical Design* 123(1):74-79, 2001.
67. Lin C. S., and Erdman A. G., "Dimensional synthesis of planar triads: Motion generation with prescribed timing for six precision positions," *Mechanism and Machine Theory*, 22(5):411–419, 1987.
68. Lipkin, H., and Duffy, J., "The Elliptic Polarity of Screws," *ASME J. of Mechanisms, Transmissions, and Automation in Design* 107(3):377–387, 1985.

69. Luck, K., and Modler, K. H., *Konstruktionslehre der Getrebe* (in German), Akademie-Verlag, Berlin, 1990.

70. Mabie, H. H., and Reinholtz, C. F., *Mechanisms and Dynamics of Machinery,* John Wiley, New York, 1987.

71. Mallik, A. K., Ghosh, A., and Dittrich, G., *Kinematic Analysis and Synthesis of Mechanisms,* CRC Press, Boca Raton, 1994.

72. Manocha, D. and Krishnan, S., "Solving Algebraic Systems using Matrix Computations," *ACM Sigsam Bulletin,* 30(4):4–21, 1996.

73. Martinez, J. M. R., and Duffy, J., "Classification of screw systems – i. one- and two-systems," *Mechanism and Machine Theory,* 27(9):459–470, 1992.

74. Martinez, J. M. R., and Duffy, J., "Classification of screw systems – ii. three-systems," *Mechanism and Machine Theory,* 27(9):471–490, 1992.

75. Mason, M. T., and Salisbury, J. K., 1985, *Robot Hands and the Mechanics of Manipulation,* Cambridge, MA: MIT Press.

76. McCarthy, J. M., *An Introduction to Theoretical Kinematics,* MIT Press, Cambridge, Mass., 1990.

77. McCarthy, J. M., "A Parameterization of the Central Axis Congruence Associated with Four Positions of a Rigid Body in Space," *ASME Journal of Mechanical Design,* 115(3):547–551, 1993.

78. McCarthy, J. M., "The Synthesis of Planar RR and Spatial CC Chains and the Equation of a Triangle," *ASME J. of Mechanical Design and J. of Vibration and Acoustics,* 117(B):101–106, June 1995.

79. McCarthy, J. M., and Ahlers, S. G., "Dimensional Synthesis of Robots Using a Double Quaternion Formulation of the Workspace," *9th. International Symposium of Robotics Research ISRR'99,* pp. 1-6, Snowbird, Utah, October, 1999.

80. McCarthy, J. M., and Bodduluri, R. M. C, "Avoiding Singular Configurations in Finite Position Synthesis of Spherical 4R Linkages," *Mechanism and Machine Theory,* 35:451–462, 2000.

81. Merlet, J. P., "Singular Configurations of Parallel Manipulators and Grassmann Geometry," *The International Journal of Robotics Research,* 8(5):45–56, 1989.

82. Meyer zur Capellen, W., Dittrich, G., and Janssen, B., "Systematik und Kinematik ebener und sphärische Vierlenkgetribe," Forshungsbericht des Landes Nordrhein-Westfalen, No. 1611, 1965.

83. Mirth, J. A. and Chase, T. R., "Circuits and branches of single-degree-of freedom planar linkages," *Journal of Mechanical Design,* 115(2):223–230, 1993.

84. Morgan, A. P, Sommese, A. J, and Wampler, C. W., 1995, "A Product-Decomposition Bound for Bezout Numbers," *SIAM J. of Numerical Analysis,* 32(4):1308-1325.

85. Murray, A. P., and McCarthy, J. M., "Five Position Synthesis of Spatial CC Dyads," *Proc. ASME Mechanisms Conference* Minneapolis, MN, Sept. 1994.

86. Murray, A. P., and McCarthy, J. M., "A Linkage Type Map for Spherical Four Position Synthesis," *Proc. 1995 ASME Design Engineering Technical Conferences,* Boston, MA, Sept. 1995.

87. Murray, A. P., and Larochelle, P. M., "A Classification Scheme for Planar 4R, Spherical 4R and Spatial RCCC Linkages to Facilitate Computer Animation," *Proc. 1998 ASME Design Engineering Technical Conferences: Mechanisms Conference,* Atlanta, Georgia, 1998.

88. Murray, R. M., Li, X., and Sastry, S., *A Mathematical Introduction to Robotic Manipulation,* CRC Press, Boca Raton, LA, 1995.

89. Nielsen, J. and Roth, B., "Elimination Methods for Spatial Synthesis," *Computational Kinematics,* (eds. J. P. Merlet and B. Ravani), Vol. 40 of Solid Mechanics and Its Applications, pp. 51-62, Kluwer Academic Publishers, 1995.

90. Nielsen, J., and Roth, B., "Solving the Input/Output Problem for Planar Mechanisms," *Journal of Mechanical Design,* 121(2):206–211, 1999.

91. Notash, L., "Uncertainty Configurations of Parallel Manipulators," *Mechanisms and Machine Theory,* 33(1):123–138, 1998.

92. Paul, R. P., *Robot manipulators: mathematics programming and control,* MIT Press, Cambridge, MA, 1981.

93. Pennock, G. R., and Yang, A. T. "Application of Dual-Number Matrices to the Inverse Kinematics Problem of Robot Manipulators," *ASME J. of Mechanisms, Transmissions, and Automation in Design* 107(2):201–208, 1985.

94. Perez, A., and McCarthy, J. M., "Dual Quaternion Synthesis of Constrained Robotic Systems," *ASME Journal of Mechanical Design*, 126(3):425-435, 2004.

95. Perez, A., and McCarthy, J. M., "Clifford Algebra Exponentials and Planar Linkage Synthesis Equations," *ASME Journal of Mechanical Design*, 127(5):931-940, 2005. (doi:10.1115/1.1904047)

96. Perez, A., and McCarthy, J. M., "Geometric Design of RRP, RPR and PRR Serial Chains," *Mechanism and Machine Theory*, 40(11):1294-1311, 2005. (doi:10.1016/j.mechmachtheory.2004.12.023)

97. Perez-Gracia, A., and McCarthy, J. M. , "The Kinematic Synthesis of Spatial Serial Chains Using Clifford Algebra Exponentials," *Proc. IMechE Part C: Journal of Mechanical Engineering Science*, 220(7):953-968 (16), July 2006. (doi: 10.1243/09544062JMES166)

98. Phillips, J., *Freedom in Machinery: Volume 1, Introducing Screw Theory,* Cambridge University Press, New York, NY, 1984.

99. Phillips, J., *Freedom in Machinery: Volume 2, Screw Theory Exemplified.* Cambridge University Press, New York, NY, 1990.

100. Ravani, B., and Roth, B., "Motion Synthesis Using Kinematic Mapping," *ASME J. of Mechanisms, Transmissions, and Automation in Design*, 105(3):460–467, 1983.

101. Reuleaux, F., *The Kinematics of Machinery—Outline of a Theory of Machines,* Translated by A. B. W. Kennedy, MacMillan and Co., London, 1876.

102. Raghavan, M., and Roth, B., "Solving Polynomial Systems for the Kinematic Analysis and Synthesis of Mechanisms and Robot Manipulators," *ASME Journal of Mechanical Design*, 117B:71-79, June 1995.

103. Raghavan, M., "Suspension Mechanism Synthesis for Linear Toe Curves," *Proceedings fo the Design Engineering Technical Conferences*, paper no. DETC2002/MECH-34305, Sept. 29-Oct. 2, Montreal, Canada, 2002.

104. Roth, B., "On the Screw Axes and Other Special Lines Associated With Spatial Displacements of a Rigid Body," *ASME Journal of Engineering for Industry,* February, 102–110, 1967.

105. Roth, B., "Kinematics of Motion through Finitely Separated Positions," *ASME Journal of Applied Mechanics*, 34E:591–598, 1967.

106. Roth, B., "Finite Position Theory Applied to Mechanism Synthesis," *ASME Journal of Applied Mechanics*, 34E:599–605, 1967.

107. Roth, B., "The Design of Binary Cranks with Revolute, Cylindric and Prismatic Joints," *Journal of Mechanisms,* 3:61–72, 1968.

108. Roth, B., and Freudenstein, F., "Synthesis of Path-Generating Mechanisms by Numerical Methods," *ASME Journal of Engineering for Industry* 85B(3):298–306, 1963.

109. Ruth, D. A., and McCarthy, J. M., "SphinxPC: An Implementation of Four Position Synthesis for Planar and Spherical Linkages," *Proc. of the ASME Design Engineering Technical Conferences,* Sacramento, CA, Sept. 14-17, 1997.

110. Ruth, D. A., and McCarthy, J. M., "The Design of Spherical 4R Linkages for Four Specified Orientations," *Computational Methods in Mechanisms,* (ed. J. Angeles), Springer-Verlag, 1998.

111. Salmon, G., *Analytic Geometry of Three Dimensions,* Vols. 1 and 2, revised by R. A. P. Rogers, Seventh ed. by C. H. Rowe, 1927, (reprinted by Chelsea Publ. Co. NY 1965.)

112. Sandor, G. N., and Erdman, A. G., *Advanced Mechanism Design: Analysis and Synthesis,* Vol. 2, Prentice-Hall, Inc., Englewoods Cliffs, New Jersey, 1984.

113. Schoenflies, A., *Geometrie der Bewegung in Synthetischer Darstellung,* Leipzig, 1886. (See also the French translation: *La Géométrie du Mouvement,* Paris, 1983.)

114. Shigley, J. E., and Uicker, J. J. Jr., *Theory of Machines and Mechanisms,* McGraw-Hill, Inc., New York, 1995.

115. Sommese, A. J. and Wampler, C. W., *The Numerical Solution of Systems of Polynomials Arising in Engineering and Science*, World Scientific Publishing, 2005.

116. Soh, G. S., Perez, A., and McCarthy, J. M., "The Kinematic Synthesis of Mechancially Constrained Planar 3R Chains," *Proceedings of the first European Conference on Mechanism Science (EUCOMES)*, 2006.

117. Soh, G. S., and McCarthy, J. M., "Synthesis of Mechanically Constrained Planar N-R Serial Chain Robots," *Advances in Robot Kinematics*, 2006.

118. Soh, G. S., and McCarthy, J. M., "Synthesis of Mechanically Constrained Planar 2-RRR Planar Parallel Robots," *Proceedings of the 12th IFToMM World Congress*, 2007.

119. Soh, G. S., and McCarthy, J. M., "Synthesis of Constrained nR Planar Robots to Reach Five Task Positions," *Proceedings of Robotics: Science and System*, 2007.

120. Soh, G. S., and McCarthy, J. M., "The Synthesis of Six-Bar Linkages as Constrained Planar 3R Chains," *Mechanism Machine Theory*, 43(2):160–170, 2008.

121. Soh, G. S., and McCarthy, J. M., "Five Position Synthesis of Spherical Six-Bar Linkages," *Proceedings of the ASME International Design Engineering Technical Conference*, 2008.

122. Soh, G. S., and McCarthy, J. M., "Synthesis and Analysis of a Constrained Spherical Parallel Manipulator," *Advances in Robot Kinematics*, 2008.

123. Soh, G. S., and McCarthy, J. M., "Parametric Design of a Spherical Eight-Bar Linkage based on a Spherical Parallel Manipulator," *ASME Journal of Mechanism and Robotics*, 1(1):011004, 2008.

124. Song, S. M., and Waldron, K. J., *Machines that Walk: The Adaptive Suspension Vehicle*, Cambridge, MA: MIT Press, 1988.

125. Sreenath, N., Oh, Y. G., Krishnaprasad, P. S., and Marsden, J. E., "The dynamics of coupled planar rigid bodies, Part 1: Reduction, equilibria and stability," *Dynamics and Stability of Systems*, 3:25–49, 1988.

126. Stewart, D., "A Platform with Six Degree-of-Freedom," *Proc. of the Institution of Mechanical Engineers*, 180(15):371–386, 1965.

127. Su, H. J., Liao, Q. Z., and Liang, C. G., Direct Positional Analysis for a Kind of 5-5 Platform In-Parallel Robotic Mechanism, *Mechanism and Machine Theory*, 34:285–301, 1999.

128. Su, H. J., McCarthy, J. M., and Watson, L. T. , "Generalized Linear Product Homotopy Algorithms and the Computation of Reachable Surfaces," *ASME Journal of Computers and Information Science and Engineering*, 4(3):226-234, September 2004.

129. Su, H. J., Wampler, C. W., McCarthy, J. M., "Geometric Design of Cylindric PRS Serial Chains," *ASME Journal of Mechanical Design*, 126(2):269–278, 2004. (doi:10.1115/1.1667965)

130. Su, H. S., McCarthy, J. M., Sosonkina, M., and Watson, L. T., "Algorithm 857: POLSYS-GLPa parallel general linear product homotopy code for solving polynomial systems of equations," *ACM Transactions on Mathematical Software*, 32(4):561–579, 2006.

131. Subbian, T., and Flugrad, D. R., 1994, "6 and 7 Position Triad Synthesis using Continuation Methods," *Journal of Mechanical Design*, 116(2):660-665.

132. Suh, C. H., "Design of Space Mechanisms for Rigid Body Guidance," *ASME Journal of Engineering for Industry*, 90(3):499–506, 1967.

133. Suh, C. H., and Radcliffe, C. W., "Synthesis of spherical mechanisms with the use of the displacement matrix," *ASME Journal of Engineering for Industry*, 89(2):215–222, 1967.

134. Suh, C. H., and Radcliffe, C. W., *Kinematics and Mechanism Design*, John Wiley and Sons, New York, 1978.

135. Tsai, L. W., and Roth, B., "A note on the design of revolute-revolute cranks," *Mechanism and Machine Theory*, 8:23–31, 1973.

136. Tsai, L. W., *Robot Analysis: The Mechanics of Serial and Parallel Manipulators*, John Wiley and Sons, New York, 1999.

137. Tsai, L. W., and Morgan, A. P., "Solving the kinematics of the most general six- and five-degree-of-freedom manipulators by continuation methods," *ASME Journal of Mechanisms, Transmissions, and Automation in Design*, 107:189–200, 1985.

138. Tsai, L. W., *Mechanism Design: Enumeration of Kinematic Structures According to Function*. Florida: CRC Press LLC, 2001.

139. Verschelde, J, and Haegemans, A., "The GBQ-Algorithm for Constructing Start Systems of Homotopies for Polynomial Systems," *SIAM J. Numerical Analysis*, 30(2):583-594, 1993.

140. Verschelde, J., 1999, "Algorithm 795: PHCpack: A general purpose solver for polynomial systems by homotopy continuation," *ACM Transactions on Mathematical Software*, 25(2): 251276, 1999.

141. Waldron, K. J., "Symmetric Overconstrained Linkages," *ASME Journal of Engineering for Industry,* 91B:158–164, 1969.

142. Waldron, K. J., "Elimination of the Branch Problem in Graphical Burmester Mechanisms Synthesis for Four Finitely Separated Positions," *ASME Journal of Engineering for Industry,* 98B:176–182, 1976.

143. Waldron, K. J., "Graphical Solution of the Branch and Order Problems of Linkage Synthesis for Multiply Separated Positions," *ASME Journal of Engineering for Industry*, 99B(3):591–597, 1977.

144. Waldron, K. J., and Kinzel, G. L., *Kinematics and Dynamics, and Design of Machinery,* John Wiley and Sons, New York, 1999.

145. Waldron, K. J., and Song, S. M., "Theoretical and Numerical Improvements to an Interactive Linkage Design Program, RECSYN." *Proc. of the Seventh Applied Mechanisms Conference,* Kansas City, MO, Dec: 8.1–8.8, 1981.

146. Wampler, C. W., Morgan, A. P., and Sommese, A. J., "Complete Solution of the Nine-Point Path Synthesis Problem for Four-Bar Linkages," *ASME Journal of Mechanical Design,* 114(1):153–159, 1992.

147. Wampler, C. W., "Solving the kinematics of planar mechanisms," *Journal of Mechanical Design*, 121(3):387–391, 1999.

148. Wampler, C. W., "Solving the Kinematics of Planar Mechanisms by Dixon Determinant and a Complex Plane Formulation," *Journal of Mechanical Design*, 123(3):382–387, 2001.

149. Wampler, C. W. "Displacement Analysis of Spherical Mechanisms Having Three or Fewer Loops," *Journal of Mechanical Design*, 126(1):93–100, 2004.

150. Wenger, P., "Some guidelines for the kinematic design of new manipulators," *Mechanism and Machine Theory*, 35:437-449, 2000.

151. Wise, S. M., Sommese, A. J., and Watson, L. T., "Algorithm 801: POL- SYS PLP: A Partitioned Linear Product Homotopy Code for Solving Polynomial Systems of Equations," *ACM Transactions on Mathathematical Software* 26(1):176-200, 2000.

152. Woo, L., and Freudenstein, F., "Application of line geometry to theoretical kinematics and the kinematic analysis of mechanical systems," *Journal of Mechanisms,* 5:417–460, 1970.

153. Woods, F. S., *Higher Geometry: An Introduction to Advanced Methods in Analytic Geometry,* Ginn and Company, 1922, (reprinted by Dover Publ. 1961).

154. Yang, A. T., and Freudenstein, F., "Application of Dual-Number Quaternion Algebra to the Analysis of Spatial Mechanisms," *ASME Journal of Applied Mechanics*, 86:300–308, 1964.

155. Yang, A. T., "Displacement Analysis of Spatial Five-Link Mechanisms using 3×3 Matrices with Dual-number Elements," *ASME Journal of Engineering for Industry*, 91:152–157, 1969.

156. Yu, H. C., "The Bennett Linkage, its Associated Tetrahedron and the Hyperboloid of its Axes," *Mechanism and Machine Theory,* 16:105–114, 1981.

Index